SPRINGER PROCEEDINGS IN PHYSICS 130

SPRINGER PROCEEDINGS IN PHYSICS

113 **Theoretical and Numerical Unsaturated Soil Mechanics**
Editor: T. Schanz

114 **Advances in Medical Engineering**
Editor: T.M. Buzug

115 **X-Ray Lasers 2006**
Proceedings of the 10th International Conference, August 20–25, 2006, Berlin, Germany
Editors: P.V. Nickles, K.A. Janulewicz

116 **Lasers in the Conservation of Artworks**
LACONA VI Proceedings, Vienna, Austria, Sept. 21–25, 2005
Editors: J. Nimmrichter, W. Kautek, M. Schreiner

117 **Advances in Turbulence XI**
Proceedings of the 11th EUROMECH European Turbulence Conference, June 25–28, 2007, Porto, Portugal
Editors: J.M.L.M. Palma and A. Silva Lopes

118 **The Standard Model and Beyond**
Proceedings of the 2nd International Summer School in High Energy Physics, Mğla, 25–30 September 2006
Editors: M. Serin, T. Aliev, N.K. Pak

119 **Narrow Gap Semiconductors 2007**
Proceedings of the 13th International Conference, 8–12 July, 2007, Guildford, UK
Editors: B. Murdin, S. Clowes

120 **Microscopy of Semiconducting Materials 2007**
Proceedings of the 15th Conference, 2–5 April 2007, Cambridge, UK
Editors: A.G. Cullis, P.A. Midgley

121 **Time Domain Methods in Electrodynamics**
A Tribute to Wolfgang J. R. Hoefer
Editors: P. Russer, U. Siart

122 **Advances in Nanoscale Magnetism**
Proceedings of the International Conference on Nanoscale Magnetism ICNM-2007, June 25–29, Istanbul, Turkey
Editors: B. Aktas, F. Mikailov

123 **Computer Simulation Studies in Condensed-Matter Physics XIX**
Editors: D.P. Landau, S.P. Lewis, and H.-B. Schüttler

124 **EKC2008 Proceedings of the EU-Korea Conference on Science and Technology**
Editor: S.-D. Yoo

125 **Computer Simulation Studies in Condensed-Matter Physics XX**
Editors: D.P. Landau, S.P. Lewis, and H.-B. Schüttler

126 **Vibration Problems ICOVP 2007**
Editors: E. Inan, D. Sengupta, M.M. Banerjee, B. Mukhopadhyay, and H. Demiray

127 **Physics and Engineering of New Materials**
Editors: D.T. Cat, A. Pucci, and K.R. Wandelt

128 **Ultrasonic Wave Propagation in Non Homogeneous Media**
Editors: A. Leger, M. Deschamps

129 **Interface Controlled Organic Thin Films**
Editors: H.-G. Rubahn.; H. Sitter; G. Horowitz; K. Al-Shamery

130 **X-Ray Lasers 2008**
Proceedings of the 11th International Conference on X-Ray Lasers, 17-22 August 2008, Belfast, UK
Editors: C.L.S. Lewis; D. Riley

For other titles published in this series, go to
http://www.springer.com/series/361

Ciaran L.S. Lewis Dave Riley
(Eds.)

X-Ray Lasers 2008

Proceedings of the
11th International Conference,
August 17–22, 2008, Belfast, UK

Springer

Professor Ciaran L.S. Lewis
Department of Physics and Astronomy
Queen's University Belfast
University Road
Belfast BT7 1NN
UK

Dr. Dave Riley
Department of Physics and Astronomy
Queen's University Belfast
University Road
Belfast BT7 1NN
UK

A C.I.P. Catalogue record for this book is available from the Library of Congress.

ISSN 0930-8989
ISBN-13 978-1-4020-9923-6 (HB)
ISBN-13 978-1-4020-9924-3 (e-book)

Published by Springer,
P.O. Box 17, 3300 AA Dordrecht, The Netherlands
In association with
Canopus Academic Publishing Limited,
15 Nelson Parade, Bedminster, Bristol, BS3 4HY, UK

www.springer.com

Preface

The 11th International Conference on X-Ray Lasers was held at the Queen's University of Belfast between 17th and 22nd of August 2008 with talks presented in the Emeleus Lecture Theatre, just off the relaxing central campus quad area. This is the second time the conference has been hosted in the UK and the first time in Ireland. In a series of oral and poster sessions, delegates reported on a wide range of developments in the area of generation and application of soft X-ray radiation sources.

The meeting was listed by the University in the programme of major events for its 2008 Centenary Year celebrations, in recognition of the importance of hosting this international conference. XRL2008 was attended by ~100 delegates from ~15 countries with many spouses, friends and children also attending and enjoying the *craic*. While for some it was an unbroken record of eleven conferences, for others it was a first time – indicating the lasting vitality of the subject area.

As reported in these proceedings, the conference covered the generation of plasma-based X-ray lasers, the traditional focus of the series, but also many evolving applications and techniques for improving beam qualities which demonstrate the maturity of the field. In addition, several contributions were included to illustrate the competing alternative approaches that are also evolving in parallel. This led to interesting and valuable discussions highlighting the special rôle each approach can play. Perhaps the most notable development is the large percentage of contributions reporting applications of X-ray lasers in a wide range of areas including plasma physics, chemistry, biology and material science.

In addition to the main science schedule, delegates survived some damp Irish weather to enjoy a wide-ranging social events programme which included a trip to the Giant's Causeway, a stroll over the Carrick-a-Rede rope bridge and an evening at the Ulster Folk and Transport Museum. There were many opportunities for renewal and forging of lasting friendships within this relatively small but always enthusiastic community. Perhaps the highlight was the Conference Dinner held in the magnificent Great Hall of the University where both young and old enjoyed demonstrations of traditional Irish music and dance as part of the after-dinner entertainment.

Many colleagues and students contributed to the successful organisation of the conference but, in particular, we wish to thank Donna Convery (Eventus-QUB), Liam McGuire (CMS-QUB) and Jenny McCabe (Physics-QUB). In addition we acknowledge generous sponsorship from Andor Technology, Coherent, Office Depot, the Institute of Physics and the Belfast Visitor and Convention Bureau.

Finally, we would like to thank the delegates and contributors, who ensure the success of this series, for their willingness to participate and comply with deadlines (for the most part!). We hope these proceedings provide a useful record of their efforts.

Ciaran L.S. Lewis
Dave Riley

Contents

Part 1 – Progress in X-Ray Laser Facilities and Infrastructures

Part 1 – Progress in X-Ray Laser Facilities and Infrastructures

Recent Progress in X-Ray Laser Research in JAEA

T. Kawachi[1], M. Kishimoto[1], M. Kado[1], N. Hasegawa[1], M. Tanaka[1], Y. Ochi[1], M. Nishikino[1], M. Ishino[1], T. Imazono[1], T. Ohba[1], Y. Kunieda[1], A. Faenov[1], T. Pikuz[1], K. Namikawa[2], S. Namba[3], Y. Kato[4], H. Nishimura[5], N. Sarukura[5], M. Kando[1], Y. Fukuda[1], H. Kotaki[1], A. Pirozhkov[1], J. Ma[1], A. Sagisaka[1], M. Mori[1], J. Koga[1], S. Bulanov[1], H. Daido[1], and T. Tajima[1]

[1] Quantum Science Directorate, Japan Atomic Energy Agency (JAEA)
[2] Department of Physics, Tokyo Gakugei University
[3] Department of Engineering, Hiroshima University
[4] The Graduate School for the Creation of New Photonics Industries
[5] Institute of Laser Engineering (ILE), Osaka University

Abstract. Recent progress in x-ray laser (XRL) research in Japan Atomic Energy Agency (JAEA) is reviewed. The repetition-rate of the x-ray laser has been improved from each 20 minutes to 10 seconds (0.1 Hz) by installing new driver laser, TOPAZ, which allows us to promote the applications of fully spatial coherent 13.9 nm laser in the wide variety of research fields such as material science, single-shot x-ray holography and atomic physics. In order to improve the present performance of the x-ray lasers, we have investigated the possibilities of the enhancement of the peak brilliance using v-groove target and the generation of circularly polarized x-ray laser under a strong magnetic field. Towards shorter wavelength x-ray lasers, we have investigated several schemes. One is the use of reflection of the light by relativistic plasma mirror driven by laser-wake-field, and the other is photo-pumping scheme using Kα emission from a solid target.

1 Introduction

Advent of transient collisional excitation (TCE) laser makes it possible for us to realize compact coherent soft x-ray lasers [1-3]. The repetition-rate of these lasers in the gain-saturation regime has been improved up to 5~10 Hz [1, 2], and now we are on the stage to use this novel soft x-ray sources as powerful scientific tools in the wide variety of research fields.

In Japan Atomic Energy Agency (JAEA), we have firstly demonstrated fully spatial coherent x-ray laser beam at the wavelength of 13.9 nm by the method of double target geometry, in which the first gain medium works as the soft x-ray oscillator, and the second gain medium works as soft x-ray amplifier [4]. Succeeding optimization of the pumping condition such as the pumping intensity, traveling wave, temporal separation and shapes of the pre- and main-pulses allows us to obtain high quality, intense x-ray laser beam.

The typical parameters of the 13.9 nm laser are the beam divergence of better than 1 mrad, 1 μJ output energy and more than 10^9 photons in the coherent volume [5].

Using this high quality laser, the application experiments have been intensively promoted in the research fields of material science and atomic and molecular physics. However the extension of the applications to single-shot x-ray diffraction imaging, x-ray laser ablation and nano-fabrication, requires further improvement in the performance of the XRLs, *e.g.*, the repetition rate, output energy, controllable polarization and the lasing in the shorter wavelength region. In the following, the recent progress of XRL research program of JAEA is described in terms of these topics.

2 New Driver Laser System: TOPAZ

New driver laser system, TOPAZ stands for Two OPtical Amplifiers using Zigzag slab. TOPAZ laser consists of the oscillator, pulse stretcher, OPCPA preamplifier, prepulse generator, zigzag slab Nd:glass power amplifiers, pulse compressor, and optics for producing the line focus on the target. The two beam lines are indispensable for generating fully spatial coherent x-ray laser beam using double target geometry, and high contrast pulse is required for photo-pumping experiment using Kα line emission described later.

The oscillator is a mode-locked Ti:sapphire laser (Spectra-Physics; TSUNAMI) pumped by 10-W diode-pumped solid state laser (Spectra-Physics; Millenia). The central wavelength is 1053 nm and the spectral bandwidth is 20 nm in the full width at the half maximum (FWHM). The oscillation frequency is 80 MHz with typical power of 300 mW (~ 4 nJ/pulse). The pulse stretcher consists of a diffraction grating with 1740 grooves/mm, a spherical mirror with focal length of 1500 mm, which generate the frequency chirp of 250 ps/nm. The spectral bandwidth after the pulse stretcher is 8nm, which is limited by the optics size.

The stretched pulse is amplified by OPCPA. The pump source is 532-nm, 10Hz repetition-rate, Q-switched YAG laser with seeder (Continuum; Powerlite Precision II) with 8-ns duration and 700-mJ energy. Total amplification gain reaches around 10^6 by using four BBO crystals at the pump intensity of 100 MW/cm^2. More than 10 mJ output energy with the energy fluctuation of 7.6 % rms is obtained. The contrast ratio of the amplified laser pulse to the background is better then 10^4.

The output of OPCPA, which has circular beam profile, is cut to square shape with 10 mm x 10 mm by a serrated aperture and is amplified by the first zigzag slab 6-pass amplifier.

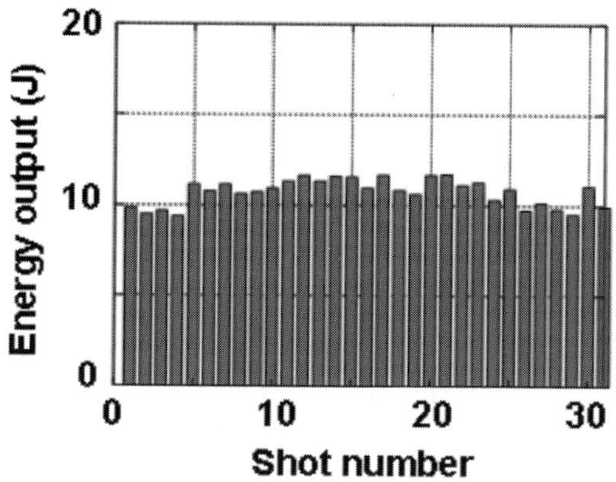

Fig. 1 Energy output and stability of TOPAZ

After the 6-pass amplifier, the height of the laser beam is expanded to 90 mm, and the laser pulse is amplified by double-pass zigzag slab amplifier. The output energy is 10 J after the final amplifier for each beam line. Figure 1 shows the output energy from TOPAZ laser under the operation of 0.1 Hz repetition-rate. From the far-field image of the beam pattern, we could not find any distortion originated from the thermal effect.

We have already demonstrated nickel-like silver laser at a wavelength of 13.9 nm using this new driver laser [6], and the preparation of application experiments using fully spatial coherent XRL driven by TOPAZ is carried on [7].

3 Application of the 13.9nm laser

In this section, we describe the application experiments using the fully spatial coherent 13.9 nm laser under the collaborations with universities, research institutes, and other research sections of JAEA.

A couple of years ago, we have taken the pico-second snap-shot of domain-structure of ferro-electric substances, $BaTiO_3$ [8, 9], by use of x-ray laser speckle technique. The following pump & probe speckle measurement revealed the temporal-correlation of "fluctuation" in the phase-transition of this materials [10]. The series of these measurements shows us that XRLs are the powerful tool to study *non-periodic* ultra-fast phenomena such as the domain fluctuation under the phase transition. The fluctuation plays a decisive role in the phase transition of the materials, and it includes the fluctuation in atomic structure, charge distribution, and spin distribution, and so on. Our next goal is to extend the observation to the fluctuation in the charge distribution of high-temperature superconductors.

The study of optical property of materials is another interesting application. In the research fields relevant to next generation lithography, the development of efficient and fast imaging scintillator devices with sufficient size is one of the key issues for lithographic applications. Zinc oxide (ZnO), a wide gap semiconductor, is one of the promising materials for the scintillators, and recent progress in the fabrication technique enables us to obtain large-size homogeneous crystal at a low cost.

Since the property of the fluorescence spectrum of ZnO for the EUV light pumping, *e.g.*, the wavelength and the lifetime, has not been well-known, therefore we tried to characterize the fluorescence by the method of soft x-ray laser induced fluorescence (X-LIF) spectroscopy. Obtained fluorescence had a peak at around 380 nm. It was sufficiently intense and the lifetime was short enough ($t\sim$ 10 ns), furthermore these optical properties were virtually the same with the case pumped by 351 nm UV laser. This implied that ZnO crystal was suitable for the fast-scintillator device for the UV-EUV region [11].

In atom and molecules physics, interaction between Xe cluster and intense soft x-ray pulse has been studies by the collaboration with Hiroshima university. The 13.9 nm laser pulse with sub micro joules and 7 ps-duration irradiated the Xe cluster target, and the production rate of several ionic stages of Xe ions were measured by the method of time of flight. Our result showed the production rate of Xe^{3+} ions dominated that of Xe^{2+}, which contradicted to the result obtained in synchrotron radiation source (SR). This was due to that XRL photon flux is larger by 6 orders of magnitude than that of the SR. Under such the condition, more than 10% of atoms in the cluster were inner-shell-ionized, and this together with the following auto-ionization process formed virtually solid state density plasma before the Coulomb explosion. Our quantitative estimation indicated that substantial ionization level lowering of Xe ions in high density plasma enhanced the production channel of Xe^{3+} ions. This study is closely connected to the physics of strongly coupled plasma or warm dense matter. [12, 13].

In the research field of x-ray laser imaging, we demonstrated single-shot Fourier transform holography. The 13.9 nm XRL beam was focused by a Fresnel zone plate (FZP) with a 50 nm-thickness Au zone fabricated on a 0.75×0.75 mm^2 silicon nitride (Si_3N_4) membrane with a thickness of 100 nm. The diameter of the FZP was 0.434 mm, and the total zone number and outermost zone width were 1700 and 64 nm, respectively. The focal length was 2 mm for the 13.9 nm laser. The focal spot size was 66 nm, and the focused beam was used as the reference beam. A test grid pattern with 2 microns period was put in the focal plane with a certain displacement (several tens of microns) from the focal position to avoid the illumination by the reference beam. The 0th order light, which was passing through the zone-plate, illuminated the test patterns, and the wave-front of the 0th-order light was distorted.

This distorted wave and the reference beam were interfered each other. The interference pattern was recorded by the x-ray CCD at the distance of 0.23 m from the sample. Figure 2 shows the image of the test patterns taken by optical microscope; (a), the raw-data of single-shot hologram; (b), and the reconstructed intensity image; (c). As shown in Figure 2 (c), the vertical and horizontal 1 μm line-and-space pattern could be virtually resolved [14]. In this experiment, the energy of the 13.9 nm laser on the FZP was 0.1 μJ due to the poor throughput (10%) of the system. Our estimation showed that in order to obtain clear single-shot hologram, at least 10^{11} photons on the sample were needed. Therefore the improvement of the output energy of the XRL was strongly desired for this purpose.

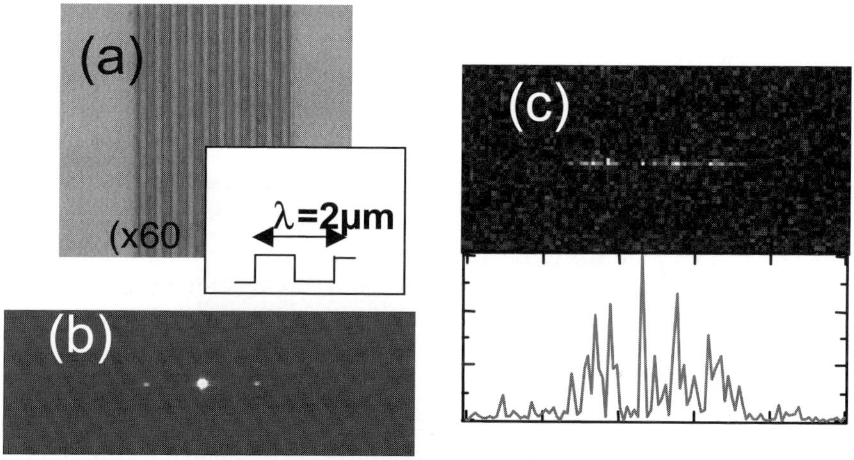

Fig. 2 Images of the test pattern taken by optical microscope; (a), single-shot Fourier transform hologram; (b), and the reconstruction image; (c).

4 Improvement of Performance of X-ray laser

4.1 Experiment towards circularly polarized x-ray laser beam

Circularly polarized soft x-ray source is promising tool to study the chilarity of stereoisomer of molecules in pharmacology and circular dichronism of magnetic substances in material science. In the collisional excitation laser, dominant lasing line is $J = 0$ to 1 transition, where J is total angular momentum quantum number of the lasing levels. Here we assume classical dipole oscillation model. If we take the quantization axis parallel to the x-ray laser propagation direction, z-direction, π-component corresponds to the electric dipole oscillation parallel to the quantization axis, and σ-components are the dipole oscillation perpendicular to z. This means that observed x-ray laser beam is the mixture of the σ-components, *i.e.*, the right-hand circular polarization component and left hand component

Consider that an external magnetic field is applied along the quantization axis. If the magnetic field is strong enough, Zeeman shift of the m_J=+1 and -1 sublevels of the lower lasing level becomes larger than the linewidth of the lasing line. Since the linewidth is typically $\Delta\lambda/\lambda \sim 10^{-4}$, the required strength of the magnetic field is ~ 40 T. This value can be achieved without any difficulties by using pulse power magnet system.

We conducted an experiment to extract the circularly polarized components of the Ni-like Mo XRL. The thin rod Mo slab target was set at the center of the magnet solenoid coil with the magnetic field of 20 T. The grazing incidence pumping (GRIP) scheme was employed to generate the XRL gain medium. The spectral profile of the XRL was measured by a high-resolution spectrometer (HIREFS) with the spectral resolution of 12.7 mA. We put an entrance slit just after the gain medium plasma, and the image of the slit was relayed to the position of x-ray CCD as the detector.

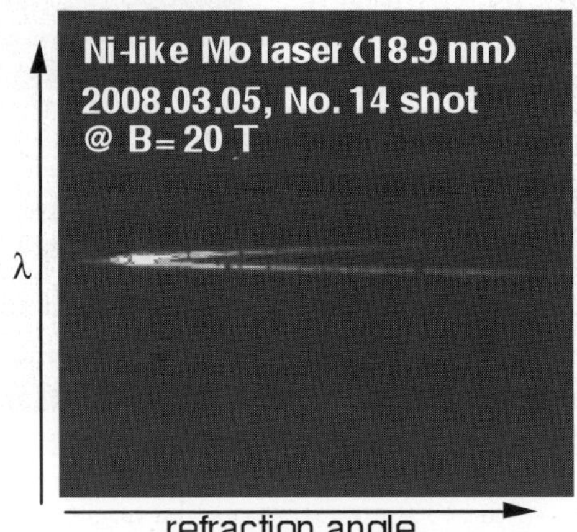

Fig. 3 Ni-like Mo XRL spectrum under the external magnetic field. Expected magnetic field strength derived from the applied voltage is 20 T.

Figure 3 shows the obtained typical spectrum of the Ni-like Mo XRL under the external magnetic field. The separation of the line is clearly shown and is a function of the refraction angle of the XRL beam. Dctailed explanation of the experiment is described in another paper of this proceedings [15].

4.2 Improvement of beam divergence and pumping efficiency using v-groove target

In a view point of single-shot exposure experiments, more intense x-ray lasers are desired. For an example, in order to obtain clear image of the nano-structure by the method of single-shot soft x-ray hologram or diffraction imaging, more than 10^{11} spatially coherent photons may be required. In the TCE lasers, the output intensity under the gain-saturation regime is order of 10^{10} Wcm^{-2}, which may give the limitation of the output photon number under the typical size of the gain region. Therefore the generation of large-size gain region with a calm density gradient is key issue to increase the coherent photon number of the XRLs .

In order to increase the size of the gain region, confinement of the pumping energy into the plasma is indispensable. V-groove target may reduce the free expansion of the plasma or radiation cooling, as the result the pumping energy is confined effectively in the inside of the groove.

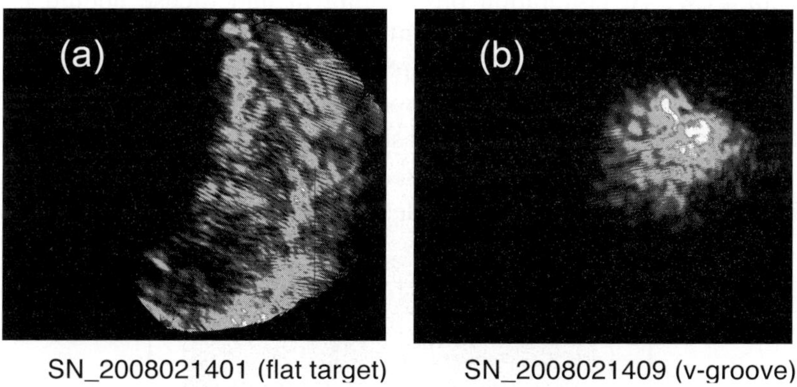

Fig. 4 Far-field patterns of the 13.9 nm laser with flat target (a) and v-groove target (b).

V-groove target with 150 μm-depth and 200 μm-width was irradiated by double pulses of CPA Nd:glass laser with the pumping energy of 10 J. Spatial distribution of the x-ray emission in several KeV range observed by an x-ray pinhole camera showed that the area of x-ray emission became 7 times larger, and the energy conversion efficiency was enhanced by more than one order of magnitude compared with the case of the flat target. It should be noted that in the case of v-groove target, the best result was obtained under the defocus condition, *i.e.*, the best focusing position was 800-900 μm before the groove.

With the information obtained by this preliminary test, the v-groove silver target with length of 6 mm was irradiated by our CPA Nd:glass laser. Figure 4 shows comparison of the far-field pattern of the Ni-like Ag laser with the flat target; (a) and the v-groove target; (b). The beam divergence was improved from 10 x 20 mrad^2 to 5 x 5 mrad^2, although the output energy decreases by only a factor of 2. The narrower beam divergence in Figure 4(b) may be due to that the beam propagation direction is limited by the shape of the groove and that large gain region with a calm density gradient is obtained. Present result indicates the potential of this target to improve the output energy in the double targets geometry.

4.3 New scheme for shorter wavelength x-ray lasers

In this subsection, we describe attempts towards shorter wavelength x-ray lasers. One is the frequency up-shift using the reflection from a relativistic plasma mirror, and the second is photo-pumping scheme using K-α line.

Relativistic plasma mirror is called as "Flying mirror", and the use of this mirror for generating ultra-short coherent x-ray pulse is proposed in [16]. Flying mirror is formed by a breaking wake field created by an intense laser pulse with 2TW, 76 fs-duration propagating in underdense helium plasma, and the source pulse (IR laser pulse) with the duration of 76 fs collide with the mirror in the direction of 45 deg with respect to the direction of the propagation of the mirror. Reflection of the source pulse by the moving mirror induces the frequency up-shift and pulse shortening by a factor of ~ $4\gamma^2\cos(\theta/2)$ due to the double Doppler effect, where γ and θ is the relativistic gamma factor of the flying mirror and the incident angle, respectively.

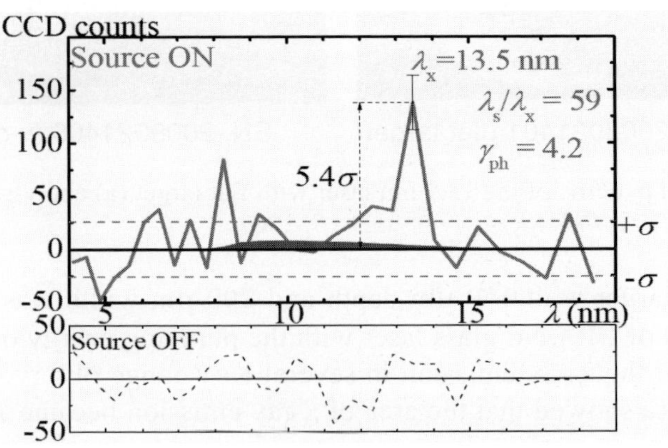

Fig. 5 Obtained spectrum of the reflected light by flying mirror.

Typical reflected spectrum of the source pulse is shown in Figure 5. Peak was obtained at around 13.5 nm, which corresponded to the upshift factor of 59 or γ= 4.2 [17]. This scheme has potential to generate atto-second coherent soft x-ray pulse by use of larger γ factor or more intense driver laser such as 100 TW class laser.

In the photo-pumping x-ray laser scheme, spectral line emission from particular ions is absorbed by different element ions to create the population inversion in the latter. The success of this scheme as an x-ray laser depends upon exact spectral matching between the emission line and the absorption line [18,19]. However, the use of "emitter" ions and "absorber" ions involves a technical difficulty: the emitter and the absorber ions should be located as close as possible so that the pumping emission reaches to the absorber ions efficiently. At the same time, the electron temperature should be high for the emitter to increase the emissivity of the ions, whereas the lower temperature is favorable for the absorber ions to avoid the "thermal" population in the lower lasing level, which reduces the amplification gain. This implies that the use of "emitter" ions and "absorber" ions is not practical under usual laser irradiation geometry [20].

We propose the use of Kα line from a solid target as the emitter coupled with the laser-produced plasma as the absorber. We focus the precise wavelength matching of aluminum Kα line (0.833816 nm) and resonance line $2p^6$-$2p^5 4d$ ($J = 1$) of neon-like zinc ions (0.83400 nm) and calculate the temporal evolution of the excited level population of the neon-like zinc ions. The calculated result shows that substantial amplification gain in the transition of $2p^5 3p$ – $2p^5 4d$ line at a wavelength of 3.5 nm can be generated in this scheme [21].

References

1. Luther B M *et al.*, *Opt. Lett.* 30, 165, 2005
2. Sebban S *et al.*, *Phys. Rev. Lett.* 89, 2002
3. Kawachi T *et al.*, *Phys. Rev. A* 66, 033815, 2002
4. Nishikino M *et al.*, *Phys. Rev. A* 68, R061802, 2003
5. Nishikino M *et al.,* 47, 1129, 2008
6. Ochi Y *et al.*, *Appl. Opt.* 46, 1500, 2007.
7. Ochi Y *et al.*, *proceedings of this conference*
8. Tai R, *et al.*, *Phys. Rev. Lett.* 89, 257602, 2002
9. Tai R *et al.*, *Phys. Rev. Lett.* 93, 087601, 2004
10. Namikawa *et al.*, *proceedings of this conference*
11. Tanaka M *et al.*, *Appl. Phys. Lett.* 91, 231117, 2008, and this conference proceedings
12. Namba S *et al.*, Phys. Rev. Lett. 99, 043004, 2007.
13. Namba S *et al.*, this conference proceedings
14. Nishikino M *et al.*, this conference proceedings
15. Hasegawa N *et al.*, this conference proceedings
16. Bulanov S V, *et al.*, *Phys. Rev. Lett.* 91, 085001, 2003

17. Kando M, *et al.*, *Phys. Rev. Lett.*, 99, 135001, 2007
18. Nilsen J, Biersdorfer P, *et al.*, *Phys. Rev. A* 50 2143, 1994
19. Elliott S R, Biersdorfer P, MacGowan B J, and Nilsen J, *Phys. Rev. A* 52: 2689, 1995
20. Kato Y and Kawachi T, "Prospect of Laser-Driven X-Ray Lasers for Extension to Shorter Wavelengths", *Progress in Ultrafast Intense Laser Science IV*, 231-247, (Berlin: Springer) 2008
21. Kawachi T and Kato Y, *this conference proceedings*

Recent Advances on LASERIX Facility: Development of XUV Sources System and Applications. Perspectives from 2008 to 2010.

D. Ros [1,2], S. Kazamias [1,2], O. Guilbaud [1,2], J. Habib [1,2], B. Zielbauer [1,2], M. Pittman [1,2], G. Jamelot [1], A. Klisnick [1], J.-C. Lagron [1,2], D. Joyeux [3], S. de Rossi [3], F. Delmotte [3], S. Lacombe [4], E. Porcel [4], C. Lesech [4], A. M. Penhoat [5], A. Touati [5].

[1] LIXAM, Bât 350, Université Paris-Sud / CNRS, Orsay, France
[2] LASERIX, Centre Laser Université Paris-Sud, France
[3] LCFIO, Institut d'Optique, UMR 8501, bât 503, 91405 Orsay, France
[4] LCAM, Bât 351, Université Paris-Sud / CNRS, Orsay, France
[5] Institut de Minéralogie et de Physique des Milieux Condensés, Université Paris 6/CNRS, France.

Abstract. LASERIX is a high-power laser facility leading to *High-repetition-rate XUV laser pumped by Titanium:Sapphire laser.* The aim of this laser facility is to offer Soft XRLs in the 30-7 nm range and auxiliary IR beam that could also be used to produce synchronized XUV sources. This experimental configuration highly enhances the scientific opportunities of the facility, giving thus the opportunity to realize both X-ray laser experiments and more generally pump/probe experiments, mixing IR and XUV sources. In this contribution, the main results concerning both the development of XUV sources (in the seeded or ASE mode) and their use for applications (irradiation of DNA samples) are presented.

1 Introduction and context

Early X-ray laser actions were obtained in high-power laser facilities intended to inertial fusion studies. Since the first demonstration of the laboratory X-ray lasers 20 years ago [1], there has been significant progress in demonstration of X-ray amplification based on various pumping schemes, characterizing and improving their performances and developing XRL applications. Nevertheless, the low access and low repetition rate of the large laser facilities are not well adapted to improve the development of short wavelength lasers and those of their applications. Considering this context and the international experiment of the LIXAM team, we obtained a financial support (4.2 M€) to build a laser facility devoted to the development of XRL mainly emitting in the 30-10 nm range and of their applications, particularly investigations on XRL interaction with matter.

The main technology of the LASERIX driver is based on Ti:Sa crystals [2, 3]. Indeed, due to their large line width, Ti: Sa lasers may emit much shorter pulses (in the range of few tens of fs) than Nd-glass ones (up than 300 fs). The general architecture of the Ti:Sa laser is schematically represented in Figure 1.

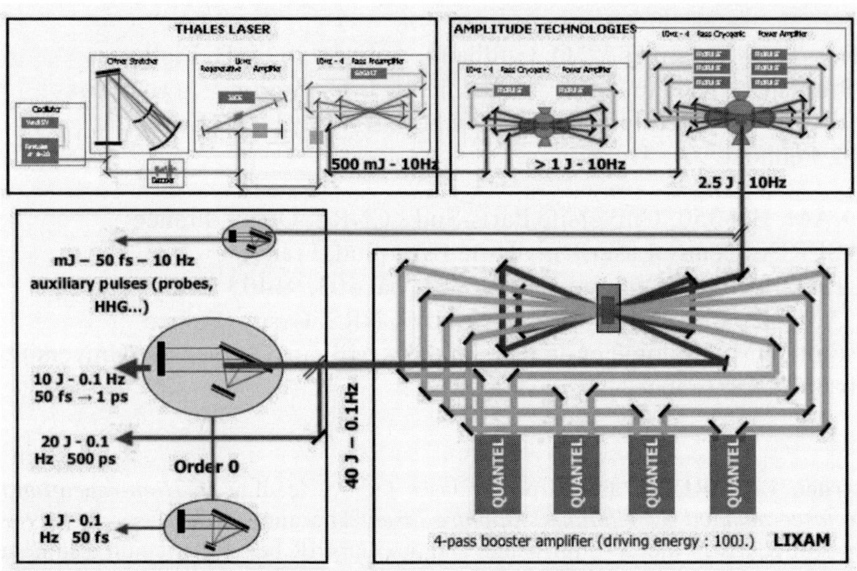

Fig. 1 Schematic view of the LASERIX driver architecture

The front-end is designed as a customized laser based on standard modules, developed by several French companies (THALES LASER and AMPLITUDE TECHNOLOGIES). It is composed by two main parts : one for the shaping and pre-amplification of the oscillator pulse, the other for two cryogenic amplifiers. The output energy at the front-end is more than 2J at the 10Hz repetition rate. The front-end beam (2.5 J) is then injected in the main amplifier, which is composed by the large Ti:Sa crystal (diameter 100 mm), shown in Figure 2. The crystal is pumped by a 4-module Nd:glass laser delivering 100 Joules of 2ω green light, developed by the French laser company QUANTEL. The energy deposition on each side of the crystal is homogenized using lens arrays. The crystal is held in a mount in which a special liquid is circulating all around to cool the crystal and limit the transverse lasing. After 4 successive passes through the crystal, the expected output before compression is ≅ 40 Joules at the repetition rate of 0.1 Hz. Basically, as shown in Figure 1, the 40-joule beam is divided in two parts, respectively 20 Joules of 500 ps and 10 Joules of 50 fs-1 ps (after compression). Besides, two more beams are offered at the final stage. Thus, the zero-order rejected by the compressor may be itself compressed to give a beam of ≅ 1 J in 50 fs.

Besides, a weak part of the energy at the exit of the front-end, ≅50 mJ in 50 fs at the repetition rate of 10Hz, can be offered to the users

Fig. 2 The large titanium-doped sapphire crystal (diameter : 100 mm) of the LASERIX driver amplifier is shown on the left. On the right image, we can see the crystal and its mount, including a liquid all around the mount.

The first step of the development of our laser facility consisted in the production of 40 joules at 0.1 Hz repetition rate. To achieve this goal, we hardly worked both on the limitation of the transverse lasing effect and on the homogenisation of the pumping of the Ti:Sa crystals. This work has been successful and then we made in 2006 the demonstration of the production of 32 pumping Joules using 3 over 4-module Nd:glass laser [4].

Fig. 3 Picture of the "transient" experimental area.

The second step was ordered considering the status of the project in 2006. Indeed, we were supposed to move in a new building at the LOA (ENSTA, Palaiseau) especially dedicated to several laser facility programs. But due to a delay in the building's construction, we had to stay two more years in a transient building. As shown in Figure 3, this building was just large enough for the pumping laser. Thus, we decided after the validation of the pumping system to remove the final stage of amplification to get free an area for the development of XUV sources and their uses for applications.

2 Development of XUV sources

The LASERIX configuration that was used for the development of the XUV sources is the low energy/high repetition rate part of the full system, as described on part one of this paper (see also Figure 1). Typically, we used 2 Joules of uncompressed infrared energy per pulse coming out from the last Ti:Sa amplifier stage of the front-end. The final amplified beam is equally split into two new beams. The first one remains uncompressed (700 ps) and the second one enters an in-vacuum compressor providing durations varying from 40 fs to several 10 ps.

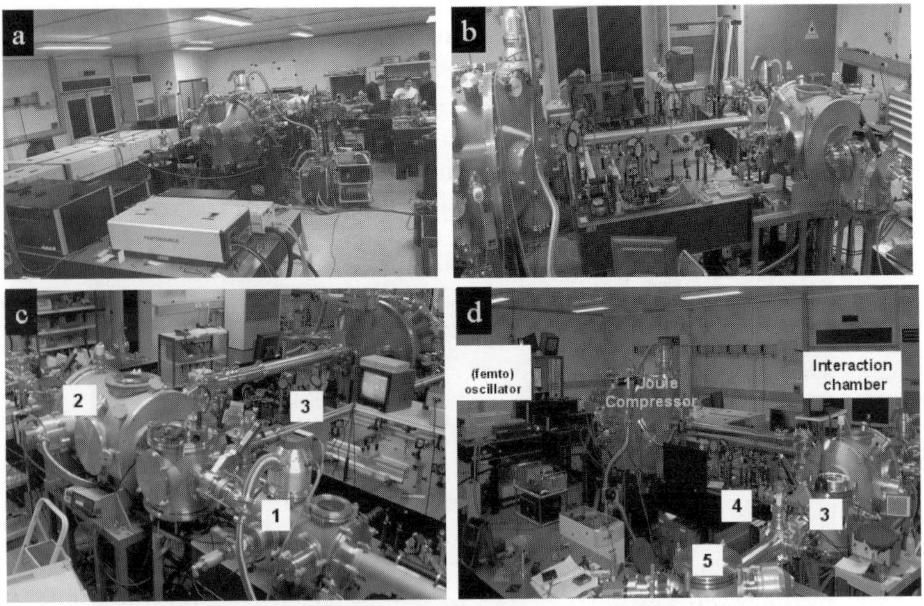

Fig. 4 Experimental set-up for the XRL generation

The Figure 4-a gives a general view of the arrangement of the experimental area, including the pumping laser until the front-end part, the compressor chamber (at the centre of the image) and the XUV sources investigations zone (at the back of the image). Different view of the experimental area for development of XUV sources are presented on Figures 4-b, 4-c and 4d.

During our first investigations at 10 Hz repetition rate, we planed to produce both XUV sources based on generation of high Harmonics and X-Ray lasers using Transient Collisionnal Excitation scheme. The materials indicated by the numbers 1 and 2 in the Figure 4-c are respectively the chamber for the production of HHG sources and the target chamber for the X-ray lasers. Several diagnostics were used to monitor the energy, as indicated in Figure 4-d, the source size and the beam uniformity of the x-ray laser, or the HHG source: the near-field imaging system (identified by the number 3), far-field system (identified by the number 4), grating spectrometer. Near field images were obtained using an XUV aspherical mirror (f=500mm) forming the image of the exit aperture of the source on an XUV CCD camera with a magnification of 13 (following the tube, identified by the number 3 in the Figure 4-c).

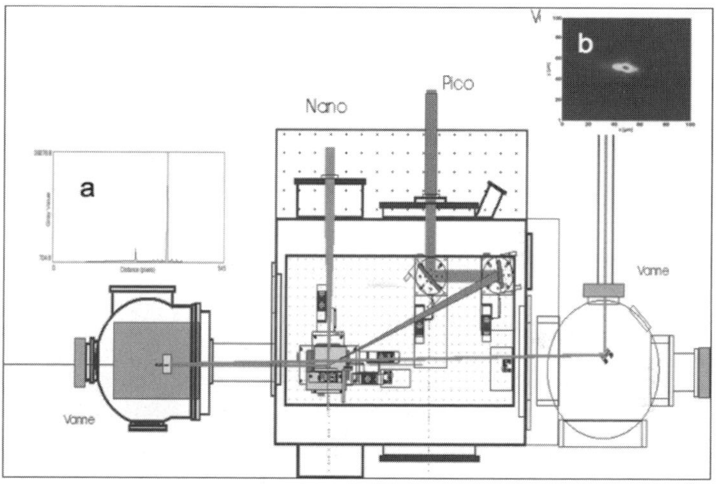

Fig. 5 Experimental set-up for the XRL generation. Typical results about the spectrum (a) and near-field image (b) are shown on picture.

We focused our investigations on the development of Mo Ni-like X-ray laser emitting at 18.9 nm, following our recent investigations at the Lund laser Center, in collaboration with Pf. C.-G. Wahlstrom, concerning the Grazing Incidence Pumping [5-7]. According to our previous work, the two pumping beams are focussed into lines with 3 to 4 mm length and 100 μm width. As indicated on Figure 5, the first one (uncompressed beam, called in figure "nano") arrives on the 4 mm long Molybdenum target at normal incidence while the second one (called in figure "pico") is incident after 400 ps at a grazing angle of 19° (the best GRIP configuration that we considered). The line foci optimization is done in IR with a high resolution imaging device (called RILF system). The XRL optimization in terms of pulse energy and stability at 10 Hz is done using an XUV near field imaging system.

A wavefront division interferometer, developed by the Laboratoire Charles Fabry (Institut d'Optique) was also placed in permanent position in the experience to perform interferometric measurement connected to temporal coherence studies [8]. This interferometer is achromatic and has a good transmission on a large range of wavelength. As we can see in the Figure 6 (top of the figure), the incident beam is reflected into two beamlets that are interfering in far-field. The path difference between them can be changed by moving vertically one of the dihedron of the interferometer. The interferometer is 4 m away from the source to ensure a high level of spatial coherence and a good intensity uniformity in the selected part of the beam. A multilayer mirror placed before the interferometer generates a spectral selection around the soft x-ray laser wavelength.

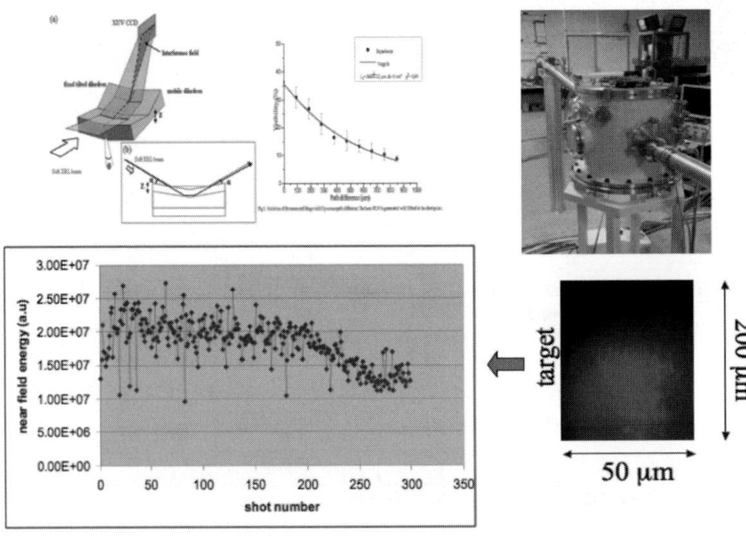

Fig. 6 Experimental set-up for the temporal coherence studies (top of the figure). Typical results about neard field intensity during 300 shots on the same target position.

All these diagnostics are very useful because they both assume a precise characterization of the source and control it from shot to shot during application experiments. Thus, we can observe in the Figure 6 (bottom of the image) the variation of the intensity of near-field image during 300 shots on the same target position. We did 10 Hz films of the soft x-ray near-field source. Good energy stability is the result of a specific optimization that includes a study of the effect of pre-pulses in the uncompressed beam. In the best configuration, the same energy level can be maintained over approximately 200 shots, after which a slow decrease occurs. We can then conclude that for 200 shots, the source is enough stable for applications.

Finally, before investigate applications, we developed an air compressor and used a leak of the front-end laser for generate High Harmonics sources. As shown in the Figure 7, we success to produce XUV emissions using Ar and Ne gas.

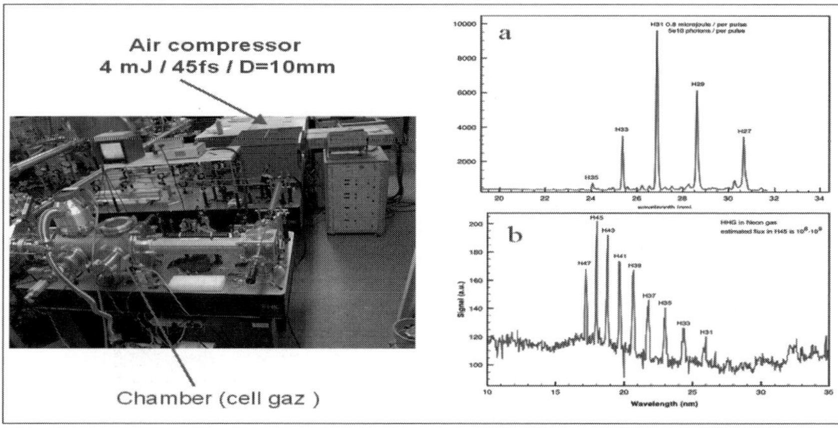

Fig. 7 Picture of the air compressor and typical results of HHG spectra using Ar gas (a) and Ne gas (b).

The next step will be to seed these HHG source in an X-ray laser plasma and then obtain an amplified XUV source with high coherence quality and high intensity. This configuration both leads to reduce the pumping energy and enhances the optical properties of the XUV sources for applications.

3 Use of XUV sources for applications

In the goal to illustrate the interest of the 10Hz X-ray laser at 18.9 nm, we investigate the irradiation of DNA samples. This experiment enrolls as addition to a previous scientific collaboration investigated at LULI (École Polytechnique, France) [9] and PALS (ASTERIX, Pragha) [10], with the purpose of identifying the physical mechanisms underlying the formation of DNA lesions induced by ionising radiation. Soft x-ray photons can be used as dosimetry tool for specific types of radiation damage and biological response. The strand breaks damage in the DNA plasmids in function of the dose. The model K predicts the break event in DNA molecule, to the K or L innershell ionisation of component DNA atoms [11, 12].

The previous experiments were realised with a repetition rate about 3 shots per hour. The main interest of this new investigation consists on the use of 10 Hz repetition-rate source, particularly useful for more detailed investigations. In order to quantify properly the number of photons on sample, we used a new system, developed by B. Zielbauer and located in the application chamber (see Figure 8).

Fig. 8 Picture of the general environment of the application chamber (a) and the inside of the application chamber (b).

Indeed, an on-line measurement of the photon number delivered on target during the irradiation has been used to monitor the irradiation process. This device, shown in the Figure 8-b, is composed of a copper grid with more than 80 % XUV transmission connected to a picoamperemeter. The photocurrent caused by the photoelectric effect of the soft x-ray photons on the copper surface has been cross-calibrated with a photodiode and we performed a statistic analysis to check the linearity of the process. The shot to shot energy fluctuations are of the order of 20 %, as shown in the figure 6, but we checked that from one 200 shot dose accumulation to another one, the integrated dose fluctuations are limited to less than 5 %. The dose that had to be delivered on the DNA samples to observe significant amount of DSB is so high that a large number of shots are to be sent: This is of the order of a few ten thousands for 200 nJ energy per shot in the soft x-ray beam. We thus decided to irradiate DNA samples by shooting 200 times on the same Molybdenum target groove, then moving it by 200 microns in height and redo the same operation over the entire available target height of 50 mm. The DNA samples stay under vacuum continuously. Plasmid DNA (pBr322 Euromedex) was deposited and dried on Mylar foils under controlled atmosphere to form a thin homogeneous sample of 4 mm^2 surface. Several samples were placed in the irradiation chamber on a multi-target holder wheel and aligned under vacuum in the XRL beam using a combination of visible and XUV cameras. The soft x-ray laser beam was filtered by a 200 nm Al filter and focussed by an f = 300 mm spherical multilayer mirror aligned slightly off-axis to match the sample surface. Surface doses of up to 500 kGray were delivered during a time of 90 minutes operation at a repetition rate of 10 Hz and average pulse energy of 200 nJ with Mo target surface changes every 20 s. After irradiation, the DNA was re-dissolved from the Mylar surface and analyzed via electrophoresis.

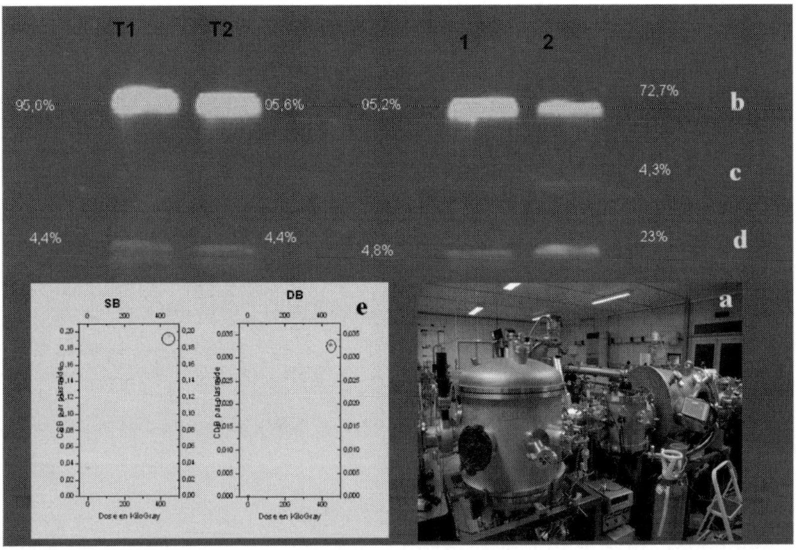

Fig. 9 Picture of the application chamber (a). Picture typical results of the 1% agarose gel, indicating the percentage of non-damaged DNA (b), single break (d) and double break (c).

Significant single and double strand break amounts were obtained. Figure 9 shows one example of a 90 000 shot irradiation. T1 and T2 denominate DNA samples not placed inside the interaction chamber, 1 is the control (witness sample) that remained under vacuum during all the irradiation time and 2 is the irradiated sample.

As compared to the control (witness), sample 2 shows 23% SSB with 5% noise (SNR>4) and 4.3% DSB with undetectable noise for the control under vacuum. This type of irradiation was repeated three times and produced very similar results. We thus have the proof of principle for this kind of experiment: the occurrence of double strand breaks not present in control samples having undergone the same treatment except for the XRL irradiation can be seen as a clear effect of the XRL irradiation. This confirms the presence of specific effects in DNA damages as previously seen with low energy ions [13, 14] and electrons.

4 Conclusion and perspectives

The LASERIX facility assumes that experimental stabilization and characterization of soft x-ray irradiation is now sufficient to enter in a phase of systematic and quantitative study applications as irradiation. We now plane to move (first semester of 2009) in a new building to install the full energy facility. Thus, we will purchase to develop shorter and higher X-ray laser sources, to investigate pump/probe experiments.

Acknowledgment

LASERIX is an X-ray laser facility of the Université Paris-Sud. The financial support of the Conseil Général de l'Essonne and the Ministère de la Recherche under the Contrat de Plan Etat-Régions 2000-2006 is gratefully acknowledged.

References

1. D.L. Matthews et al, Phys. Rev. Lett. **54**, 110, 1985.
2. G. Jamelot et al., in X-Ray Lasers 2004, IOP Conf Series N° 186, 677 (2005).
3. D. Ros at al., in X-Ray lasers 2006, Springer
4. M. Pittman et al., Opt. Lett.
5. K. Cassou and al., Opt. Lett. (2007)
6. P. Lindau and al., Opt. Express (2007).
7. S. Kazamias and al., Phys. Rev. A (2007).
8. J. Habib et al., these proceeding (2008).
9. D. Ros et al. proceedinf of 8th ICXL (2004)
10. K. Cassou et al., *J. Phys. IV (2004)*
11. Chétioui A. et. al, Int. J. Radiat. Biol., 65, 511-522, (1994).
12. Hervé du Penhoat M.A et. al, Rad. Res.,151, 649-658,(1999).
13. S. Lacombe et al., Physics in Medecine and Biology 49 N65 (2004)
14. B. Boudaïffa, et al. Science 287, 1658-1660 (2000).

Recent Progress in Grazing-Incidence-Pumped X-Ray Lasers at Uni-BE

J.E. Balmer, M. Grünig, C. Imesch, and F. Staub

Institute of Applied Physics, University of Bern, Switzerland

Abstract. We report on recent progress achieved in x-ray laser research at the Institute of Applied Physics of the University of Bern. Using the existing Nd:glass CPA (chirped-pulse amplification) laser system and the grazing-incidence pumping (GRIP) scheme, saturated x-ray lasing has been obtained on the 4d → 4p, J = 0-1 lines of Sn and Sb at wavelengths of 11.9 and 11.4 nm, respectively, using pumping energies of only 2 J (Sn) and 2.5 J (Sb). The experiments on Sn were done with 2-ps duration pump pulses, while for the Sb experiments the pulse duration was 7 ps. In both cases, multi-microjoule output was measured for a grazing angle of 45° using a grazing-incidence prepulse of the same duration as the main pulse, having 2.8% of its energy, and preceding it by 4.4 ns.

1 Introduction

Important progress towards higher efficiency, reduced size, and higher repetition rate of soft-x-ray lasers has been achieved in recent years by the introduction of the grazing-incidence pumping (GRIP) technique [1,2], which has resulted in a substantial reduction of the pump pulse energy required for saturation to ~1 J at wavelengths down to 13.2 nm [3,4] and substantial gain down to 10.9 nm [3]. At the same time, x-ray lasing has been demonstrated at repetition rates as high as 10 Hz using Ti:sapphire laser systems as the pump laser sources. This is a major step forward as it enables x-ray lasers to become a very useful tool for numerous diagnostic applications.

In the GRIP scheme, a preformed plasma is generated in a first step, much as in conventional, normal-incidence pumping, by irradiating the target with a long (~200-ps) laser pulse at normal incidence. The plasma is allowed to expand for a given time to allow the electron density gradients to relax. The short (~ps) pump pulse then irradiates the expanding plasma plume at a grazing angle, θ, chosen such that the turning point (or apex) density of the pump radiation, n_e – given by refraction as $n_e = n_{ec}\sin^2\theta$, where n_{ec} is the critical density for the pump laser wavelength – coincides with the density for which maximum gain is predicted for a given x-ray laser transition. Absorption of the pump energy thus occurs directly and very efficiently into this region which is rapidly heated up to the temperatures required to produce strong collisional excitation of the upper laser level. In this sense, the GRIP scheme is an extension of the transient-collisional-excitation (TCE) scheme to non-normal incidence. Moreover, the short pulse beam is inherently a travelling wave with a travelling wave (TW) speed slightly (at least for small θ) above c, i.e., $v = c/\cos\theta$ ($v \approx 1.06c$ for $\theta = 20°$).

When moving towards shorter wavelengths, the optimum density increases, and so will the optimum grazing angle in the GRIP geometry. In the case of Ne-like ions, the scaling of the optimum density, n_{opt}, with the nuclear charge number, Z, can be approximated by $n_{opt} = 4{\cdot}10^{15}\,(Z-9)^{3.75}$ cm^{-3} [5]. For the Ni-like ions, however, no simple scaling law is known to date. Experimentally, optimum grazing angles $\theta_{opt} \sim 20°$ and $\theta_{opt} \sim 23°$ have been reported for Pd and Cd, respectively, corresponding to densities of $\sim 2{\cdot}10^{20}$ and $\sim 3{\cdot}10^{20}$ cm^{-3}, respectively, for the pump wavelength of 800 nm [3,4]. For our case of a Nd:glass laser driver ($\lambda = 1054$ nm), this translates into $\theta_{opt} \sim 27°$ and $\theta_{opt} \sim 31°$ for Pd and Cd, respectively.

It is the purpose of this contribution to investigate the characteristics of the GRIP x-ray laser scheme at the pump wavelength of 1054 nm and to find the minimum pump energy required for saturated lasing on the 4d – 4p transition in Ni-like Sn and Sb at 11.9 and 11.4 nm, respectively, using the GRIP technique. It is shown that gain saturation is achieved with slightly more than 2 J (Sn) and 2.5 J (Sb) for pump pulses ~2 ps (7 ps) in duration and irradiation at a grazing angle of 45°. At higher pump energies (up to 5 J), x-ray pulse energies of up to ~12 μJ were measured. The results were obtained after a systematic optimization of both the grazing angle of incidence and the prepulse configuration (time delay, amplitude).

2 Experimental Setup

The experiments were conducted using the CPA upgrade to the existing Nd:glass laser system operated at our laboratory for a number of years. At the front end, the upgrade includes a commercial 100-mW, 200-fs Nd:glass oscillator followed by a double-pass, single-grating pulse stretcher and a diode-pumped Nd:glass regenerative amplifier producing a pulse energy of ~1 mJ. The 200-fs pulses from the oscillator are stretched to ~1 ns. Due to clipping in the stretcher and gain narrowing during the amplification process, the bandwidth and the duration of the pulses are reduced to ~2 nm and ~500 ps, respectively, at this point. The pulses are further amplified in the 5-stage main amplifier chain consisting of flashlamp-pumped Nd:glass rods having diameters of 10, 16, 25, 45, and 90 mm.

The maximum energy output of the system is limited by the damage threshold of the 190 × 350 mm^2 size, 1740-line/mm compressor gratings, given as

250 mJ/cm^2 by the manufacturer. For a beam diameter of 80 mm, a maximum of ~8 J was available for x-ray laser experiments, taking into account the 4-pass geometry of the compressor, the diffraction efficiency of the gratings of ~90%, and the 2:1 peak-to-average ratio of the beam fluence distribution. The pulse duration of the compressed pulse was set to ~2 ps FWHM for the experiments on Sn. For the Sb experiments (and scaling towards sub-10 nm wavelengths), the beam diameter was increased to 120 mm, and the pulse duration was set to ~7 ps FWHM to reduce the problem of increasing TW velocity mismatch at larger grazing angles.

The experimental setup used for the x-ray laser experiments is shown schematically in Fig. 1. The main pulse beam is focused at variable grazing angles using a multilayer-coated f = 60.9 cm (24") parabolic mirror. The width of the line focus was measured to be 50 μm (FWHM) at full power. The measured length of the line focus was 8.5 mm (FWHM) at a grazing angle of 30° and 12.5 mm (FWHM) at 45°. Prepulses of variable amplitude and timing are generated by inserting beam-splitters into the double-pass path of the 90-mm amplifier. The prepulses propagate along the main pulse beam line, thus producing compressed 2-ps pulses. Optionally, a long-duration (500-ps) prepulse can be introduced along a separate beam line.

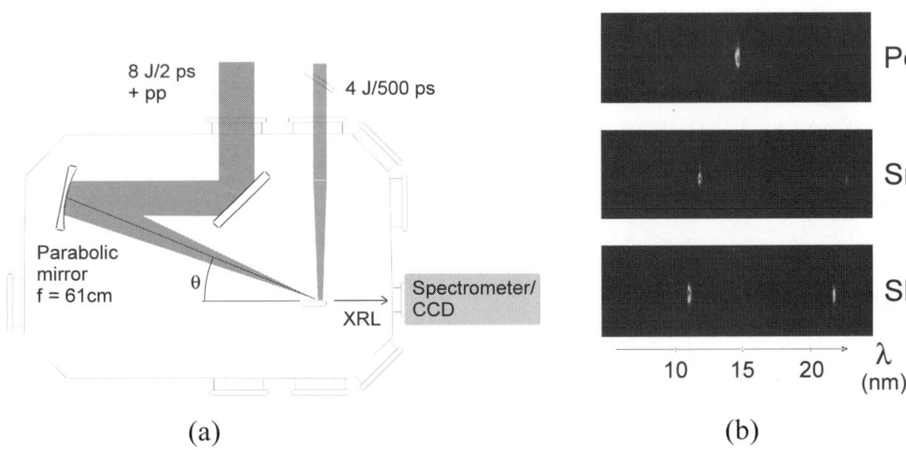

Fig. 1 (a) Experimental setup for the grazing-incidence pumped x-ray laser experiments and (b) sample spectra showing the Pd, Sn, and Sb lasing lines

The main diagnostics of the plasma emission was an on-axis, time-integrating spectrometer that consists of a 1200-lines/mm, aberration-corrected, concave Hitachi grating (radius of curvature: 5649 mm), working at a grazing-incidence angle of 3°. The grating disperses the incident radiation onto a 40-mm diameter P20 phosphor screen, which was imaged to a cooled CCD camera having a pixel size of 23 × 23 μm^2. The wavelength coverage of the spectrometer was between ~4 and 26 nm with a spectral resolution of ~0.2 nm. The relatively poor resolution is a consequence of the slitless operation of the spectrometer.

3 Results for Sn (and Pd) lasing

Based on previous results on the optimisation of the prepulse configuration of the Pd and Sn x-ray lasers [6], a configuration was chosen in which a 4.5% prepulse irradiated the target 1.6 ns before the main pulse, both having a pulse duration of 2 ps. With these parameters fixed, the x-ray laser output was measured for grazing angles between 22.5 and 45°, the latter being limited by mechanical constraints in the present pumping configuration (s. Fig. 1).

The results of the angular variation of the 14.7-nm Pd line intensity and the 11.9-nm Sn line intensity are shown in Fig. 2 for a main pulse energy of nominally 5 J. In both cases, starting at a grazing angle of 22.5°, the output intensity is seen to increase, at a gradually slower rate, up to the limiting GRIP angle of 45°, which corresponds to an apex density of $5{\cdot}10^{20}$ cm^{-3}. The second order polynomial fit (solid line) implies that the optimum GRIP angle is very close to 45° in the case of Sn (Fig. 2b). This is a surprising result, as one would rather expect a maximum at an angle around ~35° and a decreasing intensity above this.

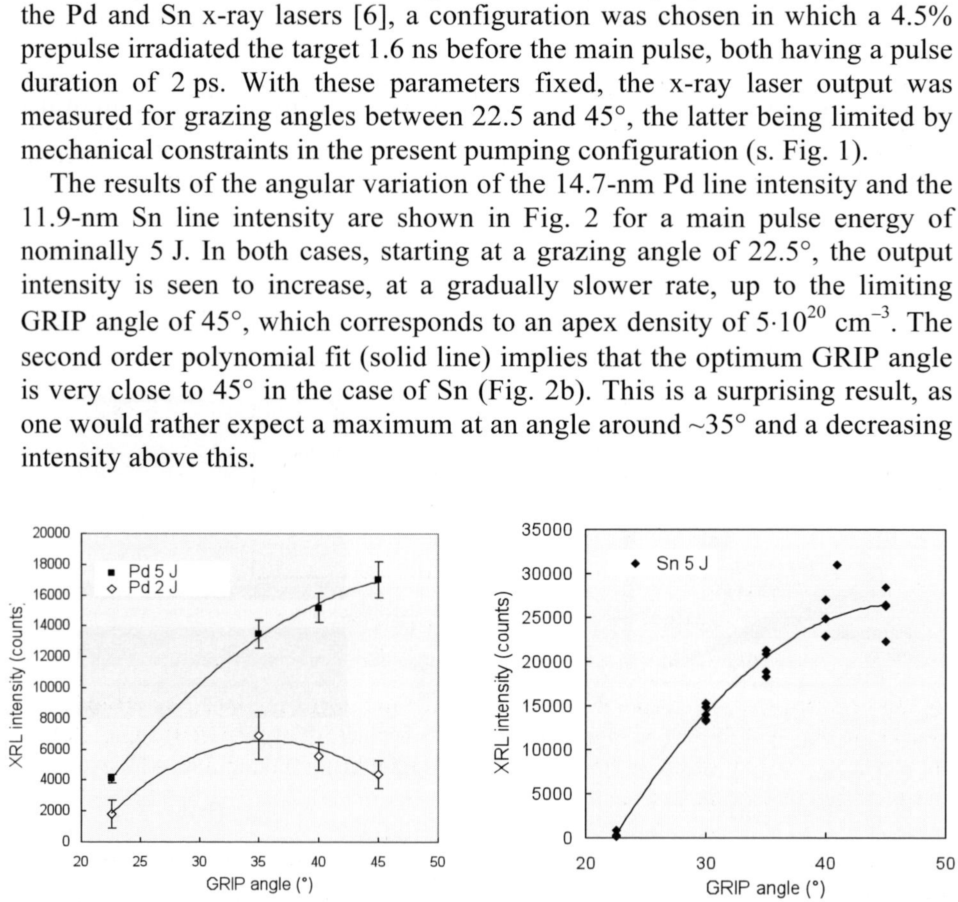

(a) (b)

Fig. 2 (a) Measured 14.7-nm line intensity of Ni-like Pd versus pump grazing angle. Solid line shows a polynomial fit of second order, (b) same for the 11.9-nm Sn line.

This is the case for the low-energy shots (2 J) on Pd (Fig. 2a) and implies that pumping far above threshold may increase the angular range of high gain. Numerical simulations of the Sn laser under very similar conditions have recently been reported which indeed show a larger range of high gain at higher pump energy [7]. The main reason, however, may be found in the fact that for off-axis focusing, using either spherical or parabolic mirrors, the length of the line focus increases with increasing GRIP angle.

For a beam of uniform irradiance, the length of the line focus produced by a spherical mirror, L, can be written as [8]

$$L = \frac{f}{\sqrt{\left|1-\{\sin(\theta/2)+d/4f\}^2\right|}} - \frac{f}{\sqrt{\left|1-\{\sin(\theta/2)-d/4f\}^2\right|}}$$

At high f-numbers, this formula also holds to very good approximation for parabolic-mirror focusing. Fig. 3 shows the measured and calculated length of the line focus as a function of the GRIP angle for a uniform irradiance beam (solid line) and for a top-hat beam having 86% of the energy within its $1/e^2$ diameter. It is seen that the length of the line focus increases from 8.5 mm (FWHM) at a grazing angle of 30° to 12.5 mm (FWHM) at 45°.

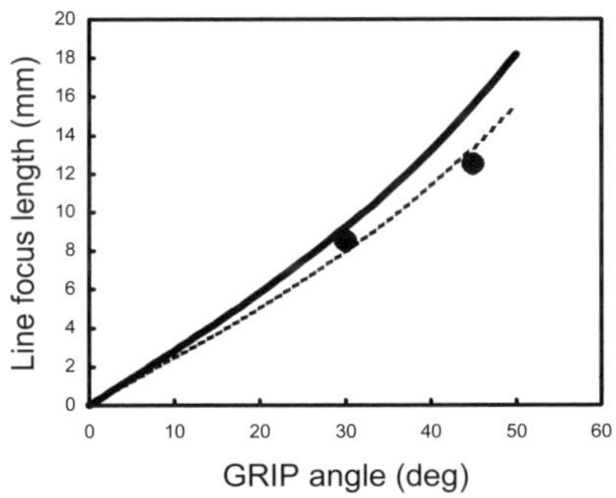

Fig. 3 Measured (dots) and calculated length of the line focus produced by off-axis focusing using a parabolic (or spherical) mirror as a function of the GRIP angle. Solid curve is for a uniform beam, dotted curve for a top-hat beam having 86% of the energy within its $1/e^2$ diameter.

The effect of increased length of the gain medium can be estimated by considering the scaling of the x-ray laser output with gain length, using an approximation of the Linford formula [9]

$$I_{XRL} \propto \frac{(e^{g_0 L}-1)^{3/2}}{(g_0 L e^{g_0 L})^{1/2}} \approx \frac{e^{g_0 L}}{(g_0 L)} \approx e^{g_0 L}$$

for $g_0 L > 2$, but below saturation.

In this approximation, the problem reduces to estimating the scaling of g_0L with pump irradiance I_p. For a given pump energy and line focus geometry, I_p is easily seen to scale as L^{-1}. The small-signal gain coefficient g_0 at line center, on the other hand, is well approximated by [10]

$$g_0 = f(0)\frac{\lambda^2}{8\pi}A_{ul}N_u \approx f(0)\frac{\lambda^2}{8\pi}A_{ul}\tau_u R_{gu},$$

where $f(0)$ denotes the peak value of the line profile, A_{ul} the transition probability for spontaneous emission from the upper to the lower laser level, τ_u the lifetime of the level, and $R_{gu} = N_e N_i K_{gu}$ the monopole excitation rate from the ground state to the upper laser level, with K_{gu} being the corresponding rate coefficient. Daido et al. [11] showed that the latter can be written as

$$K_{gu} = 1.58\times10^{-5}\frac{\gamma_{gu}(E_{gu}/T_e)\exp(-E_{gu}/T_e)}{E_{gu}\sqrt{T_e}} \quad \mathrm{cm^3s^{-1}},$$

where the excitation energy E_{gu} and the electron temperature T_e are measured in eV, and $\gamma_{gu}(E_{gu}/T_e)$ varies only slowly with electron temperature. The monopole excitation rate coefficient for Ni-like ions of Sn, calculated by using the tabulated values for γ_{gu} [11], is plotted in Fig. 4 as a function of the electron temperature T_e. It is seen that the curve is well approximated by a linear function $K_{gu} \propto T_e$ in the range of temperatures between ~200-400 eV. Above this range, the scaling with T_e is weaker than linear.

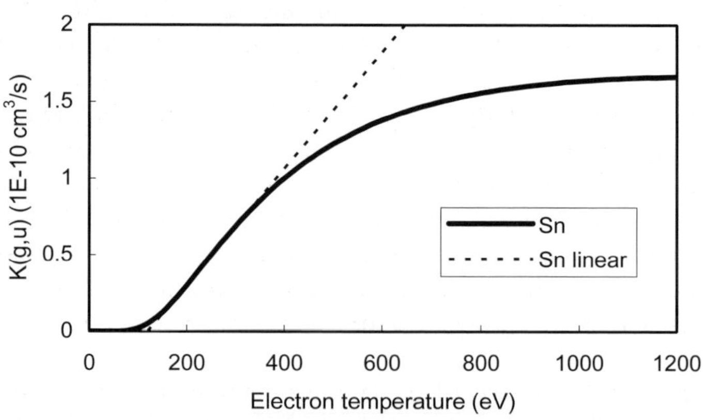

Fig. 4 Monopole excitation rate coefficient from the ground state to the upper laser level for Ni-like ions of Sn as a function of electron temperature.

Having established the scaling of g_0 with temperature, we need to know the scaling of temperature with pump irradiance. Since absorption of the pump irradiance in GRIP geometry occurs along the entire beam path of the pump laser well below critical density, the self-regulating model seems appropriate [12], according to which $T_e \propto \sqrt{I_p}$. Using this and assuming the fraction of Ni-like ions to be constant, we finally get

$$g_0 \propto n_e K_{gu} \propto n_e T_e \propto n_e \sqrt{I_p} \propto n_e / \sqrt{L}$$

and therefore

$$I_{XRL} \propto e^{g_0 L} \propto e^{\propto n_e \sqrt{L}} .$$

This shows that the (unsaturated) x-ray laser output intensity scales exponentially with an exponent that is proportional to n_e and $\sqrt{L}$. This qualitatively explains the increased x-ray laser output at large GRIP angles as being caused by both the increased length of the line focus and the increased apex electron density. In addition, it implies that it is advantageous to increase the line focus length as much as possible as long as the pump irradiance is kept above some "threshold" value.

4 Results for Sb lasing

On the route towards lasing at shorter x-ray wavelengths we subsequently chose nickel-like antimony (Z = 51) as the lasing medium. Sb is the neighbour element to Sn (Z = 51), and so its 4d-4p lasing wavelength is slightly shorter than that of Sn (11.4 vs. 11.9 nm). A first series of experiments was conducted using a commercially available, 25.5-mm wide thin foil of Sb as the target. Although weak x-ray laser action at 11.4 nm was observed in this way, the main pulse energy had to be increased to ~10 J. It was obvious that the flatness of the target was a problem; in fact this was already observable by visual inspection. Better targets were subsequently obtained by coating a ~1-μm thick layer of Sb onto diamond-machined, 25-mm wide brass slabs. With these targets, intense x-ray laser output was observed with a main pulse energy of ~2.5 J.

After careful optimization of the pumping parameters, a configuration was chosen for the gain measurements, in which a 2.8% prepulse irradiated the target 4.4 ns before the 2.5-J main pulse, both incident at the GRIP angle of 45° [13]. As the pump laser beam diameter had been increased to 120 mm in view of extracting energies up to 20 J for the sub-10-nm experiments, the length of the line focus increased accordingly and was measured as 19.6 mm.

As a consequence, the traveling-wave velocity mismatch becomes more serious and so it was decided to increase the pumping pulse duration to ~7 ps. The results of the gain measurement are shown in Fig. 5. The 11.4-nm line intensity is seen to increase exponentially as a function of target length until it rolls off into saturation for target lengths above ~10 mm. At short lengths (4 and 5 mm) the intensity was below the threshold of the detector for some shots. Therefore the averaged points in this range tend to be overestimated. Nevertheless, the gain measurement shows a clear roll-off effect, which is a sign of saturation. A fit to the data gives a small-signal gain coefficient of 40 cm^{-1} and a gain-length product of ~16.5 at saturation. More details on how to confirm the achievement of true gain saturation can be found in Ref. [13].

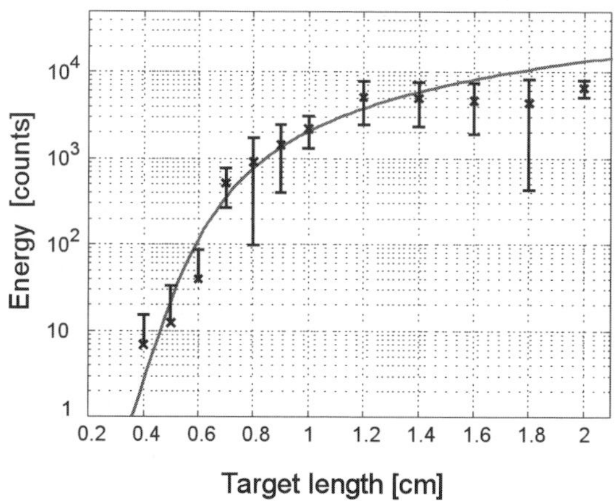

Fig. 5 Measurements of the output energy of the 11.4-nm lasing line in Ni-like Sb as a function of the target length.

Acknowledgements

The authors would like to acknowledge the technical assistance of B. Locher and W. Lüscher for target preparation. This work was supported in part by the Swiss National Science Foundation.

References

1. Shlyaptsev, V.N. et al.: 'Numerical study of transient and capillary x-ray lasers and their applications', *Proc. SPIE*, Vol. **5197**, 221-228, 2003
2. Keenan, R. et al.: 'Efficient pumping scheme for high-average brightness collisonal x-ray lasers', *ibid.*, 213-220
3. Wang, Y. et al.: 'Demonstration of high-repetition-rate tabletop soft-x-ray lasers

with saturated output at wavelengths down to 13.9 nm and gain down to 10.9 nm', *Phys. Rev. A* **72**, 053807-1-7, 2005
4. Rocca J.J. et al.: 'Saturated 13.2 nm high-repetition-rate laser in nickel-like cadmium', *Opt. Lett.* **30**, 2581-2583, 2005
5. Elton, R.C.: *X-Ray Lasers* (Academic Press, San Diego, 1990), p. 106
6. Grünig, M., Imesch, C., Staub, F., and Balmer, J.E.: 'Saturated x-ray lasing in Ni-like Sn at 11.9 nm using the GRIP scheme', *Opt. Commun.* (accepted for publication)
7. Kacar, E. and Demir, P.: 'Simulations of grazing-incidence-pumped Ni-like Sn X-ray laser at 11.9 nm', *Opt. Commun.* 281, 718. 2008
8. Ursescu, D. et al.: 'Focusing System for a Non-Normal Incidence Pumped, Sub-10 nm TCE XRL', *Proc. 10th Int. Conf. on X-Ray Lasers*, p. 69, Springer, 2007
9. Elton, R.C.: *X-ray lasers*, Academic Press, 1990
10. Tallents, G.J.: 'The physics of soft x-ray lasers pumped by electron collisions in laser plasmas', *J. Phys. D: Appl. Phys.* 36, R259-R276, 2003
11. Daido, H. et al.: 'Atomic number scaling of the nickel-like soft x-ray lasers', *Int. J. Mod. Phys.* 11, 945-990, 1997
12. Pert, G.J.: 'Optimizing the performance of the nickel-like collisionally pumped x-ray lasers. II. Lasers for the wavelength range 50-100 ', Phys. Rev. A 75, 023808-1-15, 2007
13. Imesch, C. Grünig, M., Staub, F., and Balmer, J.E.: 'Saturated x-ray lasing in Ni-like Sb at 11.4 nm using the GRIP scheme', in preparation

Review on Recent High Intensity Physics Experiments Relevant to X-Ray and Quantum Beam Generation at JAEA

H. Daido[1], A. Pirozhkov[1], M. Nishiuchi[1], A. Yogo[1], S. Orimo[1], K. Ogura[1], A. Sagisaka[1], I. Daito[1], M. Mori[1], M. Ikegami[1], H.Kiriyama[1], H. Okada[1], S. Bulanov[1], T. Esirkepov[1], S. Kanazawa[1], S. Kondo[1], T. Shimomura[1], M. Tanoue[1], Y. Nakai[1], H. Sasao[1], D. Wakai[1], P. Bolton[1], Y. Fukuda[1], A. Faenov[1,5], T. Pikuz[1,5], M. Suzuki[1], M. Tampo[1], H. Sakaki[1], T. Tajima[1], S. Kawanishi[1], T. Kawachi[1], M. Nishikino[1]

I. W. Choi[2], C. M. Kim[2], T. M. Jeong[2], N. Hafz[2], T. J. Yu[2], J. H. Sung[2], Y.-C. Noh[2], D.-K. Ko[2] and J. Lee[2]

Y. Oishi[3], K. Nemoto[3], T. Nayuki[3], T. Fujii[3]

H. Nagatomo[4], K. Nagai[4], H. Nishimura[4]

[1] Advanced Photon Research Center and Photo-Medical Research Center, Japan Atomic Energy Agency (JAEA)
[2] Advanced Phonic Research Institute, Gwangju Institute of Science and Technology (GIST)
[3] Central Research Institute of Electric Power Industry (CRIEPI)
[4] Institute of Laser Engineering (ILE), Osaka University
[5] Joint Institute for High Temperature of the Russian Academy of Sciences, Moscow, Russia

Abstract. The authors describe firstly the lasers for high intensity physics experiments at JAEA including J-KAREN and JLITE-X lasers which can deliver 100TW and 10TW laser power, respectively. Secondly the authors describe demonstration of flying mirror technique which will become a new technique to make a coherent tunable x-ray source. Thirdly a femto-second laser driven incoherent soft x-ray source and its imaging applications for nano-structures are described. The 4th topic includes femto-second laser driven multiple quantum beam generation and its applications using a laser driven thin foil target.

1 Introduction

Recently ultra-high intensity laser driven x-ray as well as particle beams and electromagnetic waves become popular. At the Advanced Photon Research Center (APRC), we have systematically performed high intensity physics experiments using J-KAREN, an Optical Parametric Chirped Pulse Amplification system coupled with a Titanium Sapphire laser amplifiers with laser power of up to ~100TW [1, 2]. Further amplification to obtain higher energy and intensity using a glass laser pumping system will be scheduled for target shooting. At the APRC, smaller laser system called JLITE-X [3] which usually delivers several TW in a few tens of femto-seconds is also used for high intensity physics experiments.

Using these laser systems, we have performed laser driven x-ray generation and its applications. Firstly we describe the result of relativistic flying mirror [4-7] which may grow up to be a tunable coherent x-ray source. From this point of view, this technique is alternative one to high order harmonic generation from an ultra-high peak power laser irradiated ultra-thin foil target [8, 9]. Another topic is an intense incoherent x-ray source from a cluster target driven by such a high intensity laser [10]. We have performed production of x-ray source and x-ray imaging experiments for useful applications.

We have also performed multiple quantum beam generation by an intense laser irradiated thin foil target such as x-ray, energetic particles as well as electro-magnetic waves such as THz waves. Simultaneous imaging of a test sample using x-ray and protons [11], THz wave and protons [12] and protons and electrons [13] have been demonstrated. In this talk we describe future prospect of such ultra-high intensity laser applications.

2 Relativistic Flying Mirror for a Tunable Coherent X-ray Source

As we know from the Special Theory of Relativity, the frequency of the light reflected from a moving mirror is multiplied by the factor four-gamma-squared (this approximation is valid for large gamma). The pulse duration is divided by the same factor. For example, for a moderate gamma value of 5, we obtain approximately 100 times pulse shortening. What is interesting in the case of perfect reflection? The pulse shape remains same. The pulse is just squeezed in time, but it preserves such characteristics as coherence, number of cycles, presence of the chirp, polarization and so on. If we have a perfect relativistic mirror, it would be an ideal source of ultra-short pulses. The question is: How can we implement such a relativistic mirror? The technique is the Relativistic Flying Mirror, which was firstly proposed by Bulanov and co-authors [4].

When a short relativistic-irradiance pulse propagates in an underdense plasma (Fig. 1), it can generate a wake wave. This wake wave, in particular, can accelerate charged particles, as was proposed by Tajima and Dawson [14]. Under the appropriate condition, namely, near the wave-breaking threshold, the electron density has sharp spikes, with the peak density much larger than the original plasma density. These spikes move at the phase velocity which is equal to the group velocity of the driver pulse, which is close to the velocity of light in vacuum. Now, if we send another laser pulse, namely, the source pulse in the counter-propagating direction, part of its energy will be reflected back from the Relativistic Flying Mirror. The important thing is that the reflected pulse frequency is much higher and duration is much shorter than the original one.

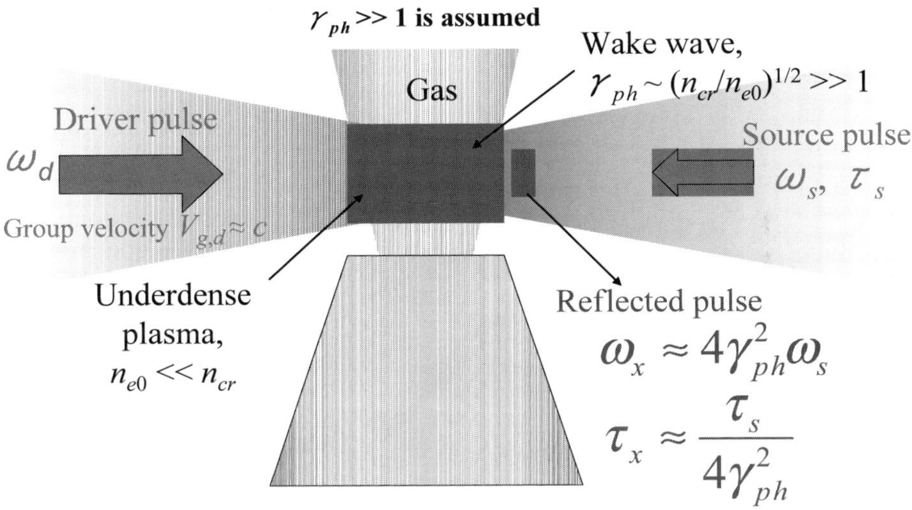

Fig. 1 Schematic diagram of the Flying Mirror Technique.

Figure 2 shows the simplified experimental setup [6, 7]. The driver pulse has the energy around 200 mJ and the duration of 76 fs. It is focused to the gas jet which has a rectangular shape 10 by 1.2 mm. The maximum plasma density is around 5×10^{19} cm^{-3}. The driver pulse produces the wake wave, which can be monitored by the electron generation. The source pulse with the energy of approximately 12 mJ is focused to the region of the wake wave. The incidence angle is 45 degrees. A signal reflected by the Flying Mirror is detected by a grazing-incidence spectrometer comprising a toroidal mirror, a slit, a grating, and a back-illuminated CCD. Due to the relativistic effects, the reflected signal and driver pulse propagates almost in the same direction. Therefore, we have to shield the spectrometer against the strong laser light, which can contain hundred mJ. We used two Mo/C multilayer filters for this purpose. In order to align two laser pulses, we used a lot of diagnostics including a shadowgraph, an interferometry, a source focal spot monitor, a transmitted spectrum measurement, an optical plasma imaging, measurement of the scattered driver pulse spectrum, and so on. The reason to have so many instruments is the alignment difficulty: we have to align two laser pulses in such a way, that the driver pulse generates a good wake wave, and the source pulse is focused to the region of the wave breaking at an appropriate time. The situation becomes more complicated due to the presence of large volume of gas and plasma with not negligible density, which actually significantly affects the laser propagation. Figure 3 shows the snap-shot of the two colliding pulses after we made a lot of efforts to align the laser pulses in plasma. Figure 4 shows the reflected signal dependence on the alignment accuracy in time and space. The result confirms the first experimental demonstration of the reflection from the flying mirror.

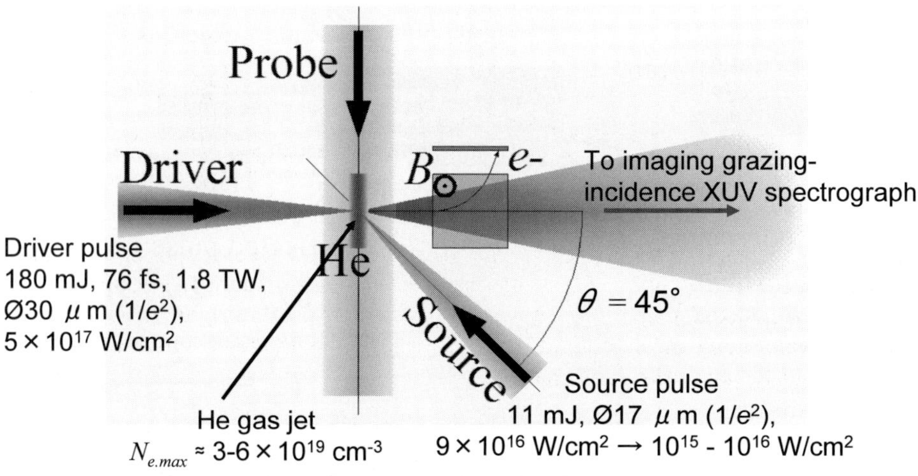

Fig. 2 Simplified view of the experimental setup.

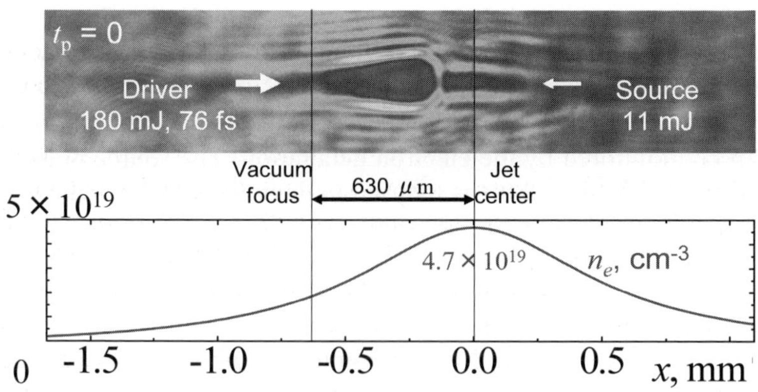

Fig. 3 Shadowgram at which a driver pulse collides with a source pulse in a plasma. Bottom panel shows the plasma density profile.

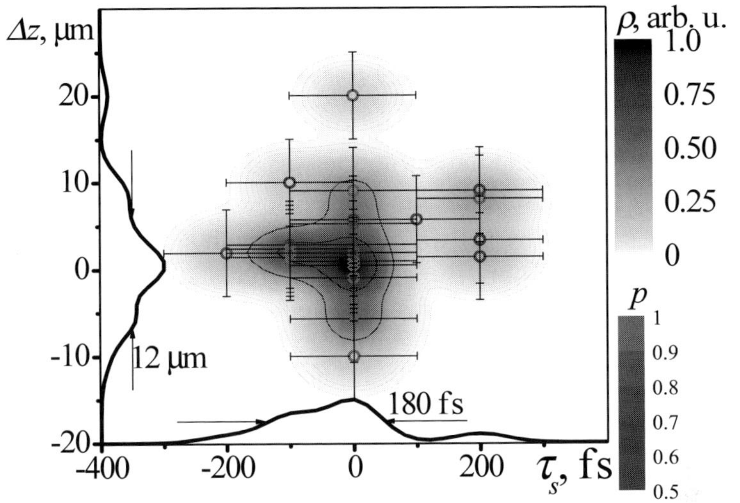

Fig. 4 Dependence of the reflected signal on the alignment accuracy in time (τ_s) and space (Δz, vertical misalignment). Dots show individual shots in which reflected signals were detected, color scale shows the probability that the signal is not noise. The gray scale shows the smoothed signal distribution, with the cross-section widths of 180 fs and 12 μm in time and space.

3 Experimental Study on femto-second laser driven soft X-ray source

For soft x-ray imaging, Fukuda, Faenov and Pikuz et al. have systematically performed experiments on soft X-ray generation using femto-second-laser-irradiated clusters [10]. The experiment has been performed using the JLITE-X Ti:sapphire laser facility at JAEA, which generate a 36 fs pulse with energy of 160 mJ on target [3]. Figure 5 shows specially designed supersonic gas jet nozzle [15] which is used to produce clusters with a diameter of 1 μm for pure CO2 and 0.5 μm for the mixed gas of He and CO2. The laser beam is focused about 1.5 mm above the nozzle orifice by an off-axis f /13 parabola with the spot size of around 50 μm giving a laser intensity of 4×1017 W/cm2 in vacuum. The spatially resolved X-ray spectra have been measured using two focusing spectrometers with spatial resolution [16]. To check the isotropy of X-ray production one of the spectrometers (FSSR-1 in Figure 5) equipped by Andor 440 CCD as an X-ray detector was placed at angle about 20 degrees above laser beam axis, and the second identical spectrometer (FSSR-2 in Figure 5) equipped by Andor 420 CCD was placed nearly perpendicularly to the laser beam propagation. Both spherically bent mica crystal spectrometers were placed at a distance of 220 mm from the laser plasma source and was centered at λ=18.85 Å, which corresponds to a Bragg angle of 18.8° for the first reflection order of the mica crystal. For such alignment spectra of Lyα (energy 653.7 eV) and Heβ (energy 665.7 eV) lines of Oxygen with simultaneously high spectral ($\delta\lambda/\lambda \sim 3000$) and spatial (in the case of using as an X-ray detector Andor 420 CCD the resolution was 40 μm and for Andor 440 X-ray CCD was 20 μm) resolution have been measured. Two layers of polypropylene filters covered by 0.1 μm of Al from both sides have been used to protect X-Ray detectors from visible and VUV radiation. Soft x-ray emission from the clusters produced from two different gas targets, i.e., 2–6 MPa of pure CO2 and 6 MPa of the mixed gas of He and CO2, have been measured. The number of photons for the emission lines of Lyα and Heβ of the oxygen ions produced with the mixed gas of He and CO2 is as high as 2.7×1010 and 2.8×1010 photons/sr pulse, respectively, which are four to seven times larger than that produced with the pure CO2 gas. The required number of laser shots is dramatically reduced to only 1000–1500 for obtaining high quality soft x-ray contact images. The experiment has been performed to reveal the large field of view and the large dynamic range in the nanostructure imaging using our soft x-ray source with the wide spatial-scale LiF crystal detector [17] as shown in Figure 6. Spatial resolution of 700 nm has been achieved with a wide field of view of a few cm. To demonstrate the accurate sensitivity of obtained images to the 100-nm-thick Mo film, traces along the different directions of images are presented in Figure 6. These spatial profiles show up to 5 different gray levels, which correspond to the open LiF crystal area and 4 layers of Mo film. Useful applications which are suitable for using this type of source are demonstrated.

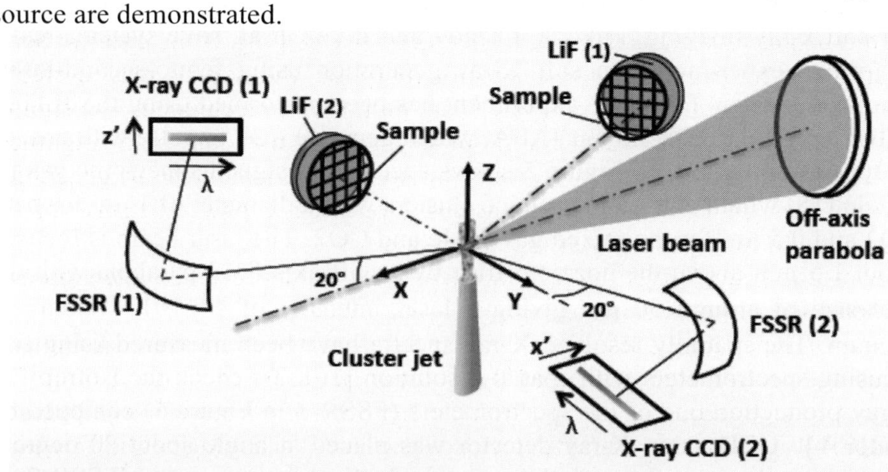

Fig. 5 Schematic view of the experimental setup. Two focusing spectrometers with spatial resolution FSSR (1) and FSSR (2) are placed as shown in the diagram.

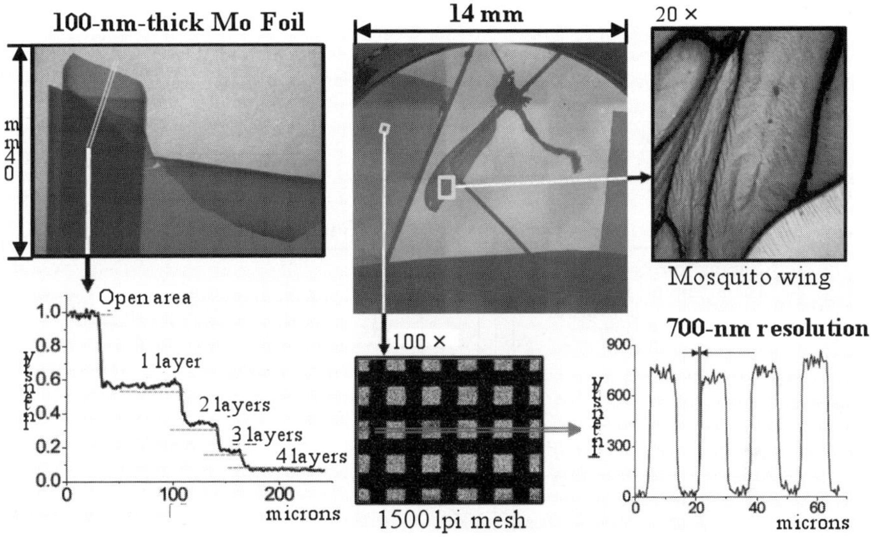

Fig. 6 High spatial resolution of soft x-ray contact imaging recorded on a few-cm scale LiF detector. At the left hand side, an image of 100-nm-thick Mo foil is visible. At the central picture, an image of a mosquito is visible and a higher magnification image can be seen at the right hand side.

4 Laser driven multiple quantum beam generation using a thin foil

Laser driven source delivers jitter-less multiple beams such as electron, x-ray, proton beams as well as the electro-magnetic waves [18]. If the specific beam intensity is high enough for a specific purpose, the source may perfectly match to the femto-second multiple beam pump probe technique. The significant advantage of the laser driven source includes not only high brightness with short pulse and small source size but also is the simultaneous generation of multiple beams. The disadvantage is low average power because of the restriction of high-power laser repetition rate.

A proton beam driven by a high-intensity laser has received attention as a compact ion source for medical applications [19, 20].

We have performed the high intensity laser-matter interaction experiments using a thin-foil target irradiated by Ti:sapphire lasers of Japan Atomic Energy Agency (JAEA), Central Research Institute of Electric Power Industry (CRIEPI) and Gwangju Institute of Science and Technology (GIST) with a peak irradiance of 10^{17} to 10^{20} W/cm^2. Figure 7 shows the various beams from an ultra-shot and ultra-high intensity laser driven thin foil. Details are found in the caption.

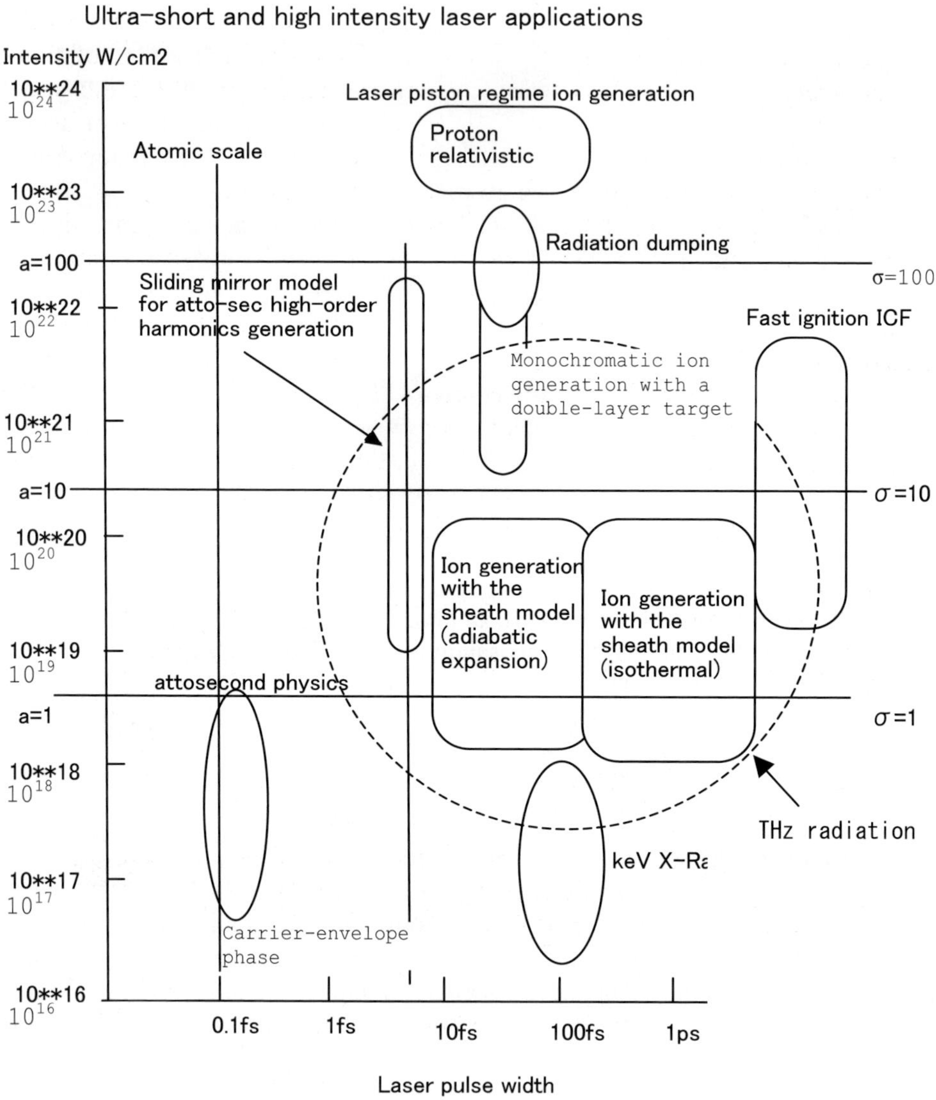

Fig. 7 Various symbolic phenomena in the ultra-high and ultra-short laser driven thin foil targets are shown. The vertical and the horizontal axes represent laser intensity on the target in the unit of W/cm^2 and laser pulse width in femto-second and pico-second, respectively. The notation a and σ are the normalized vector potential and normalized areal electron density equivalent to the product of the target thickness and density, respectively [9]; in the optimum conditions, a should be of the order of σ.

A tape target driver provides a fresh surface of a thin-foil at each laser shot. We have developed several on-line real time monitors such as a time-of-flight proton spectrometer which is placed behind the target and along the target normal as well as an interferometer for an electron density profile measurement of a preformed plasma, an x-ray pinhole camera, a THz radiation detector, and an UV spectrometer for harmonics measurement as shown in Figure 8. At GIST in Korea and at JAEA, we have tested simultaneous generation of x-ray and protons from a thin tape target resulting in a demonstration of projection images of the micro-meter-scale test pattern as a feasibility study of pump-probe experiment with intense multiple beams [11]. We have also tested intense THz radiation together with protons [12] which is originally developed as a projection proton shadowgraphy setup [13]. In summary, the authors have shown the preliminary experimental results on generation of laser-driven multiple beams for mainly simultaneous imaging applications. The authors expect that the technique opens up the ultra-fast imaging applications as well as ultra-fast pump-probe technique.

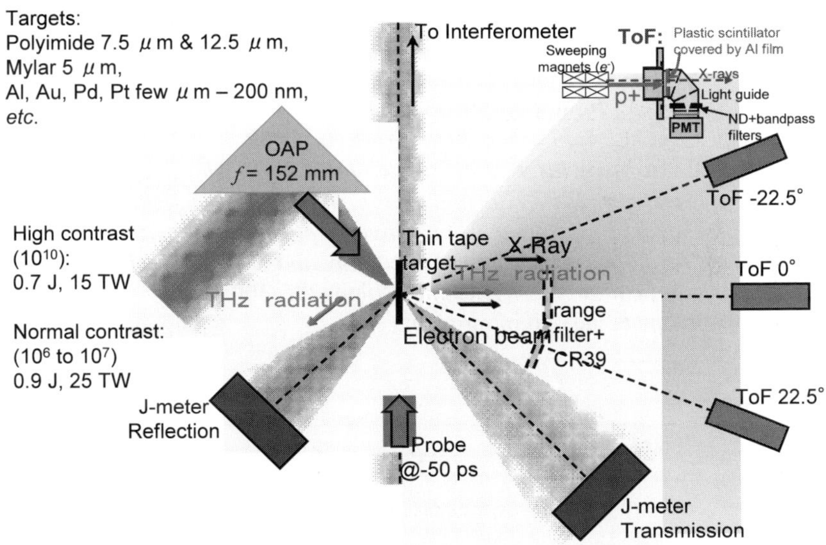

Fig. 8 Experimental setup for the multiple beam generation mainly for proton generation and related laser-plasma interaction diagnostics.

References

1. Kiriyama, H. et al.: *Opt. Lett,*. 32, 2315, 2007
2. Kiriyama, H. et al.: *Opt. Lett.*, 33, 645, 2008
3. Mori, M. et al.: *Laser Phys.*, 16, 1092, 2006
4. Bulanov, S. V. et al.: *Kratk. Soobshch. Fiz.* 6, 9, 1991
5. Bulanov, S. V., Esirkepov, T. Zh. and Tajima, T.: *Phys. Rev. Lett.* 91, 085001, 2003
6. Kando, M. et al.: *Phys. Rev. Lett.* 99, 135001, 2007
7. Pirozhkov, A. S. et al.: *Phys. Plasmas* 14, 123106, 2007
8. Dromey, B. et al.: *Nature Phys.* 2, 456, 2006; Teubner, U. et al.: *Phys. Rev. Lett.* 92, 185001, 2004
9. Pirozhkov, A. S. et al.: *Phys. Plasmas* 13, 013107, 2006; Pirozhkov, A. S. et al.: *Phys. Lett. A* 349, 256, 2006.
10. Fukuda, Y. et al. : *Appl. Phys. Lett.* 92, 121110, 2008
11. Orimo, S., Nishiuchi M. et al.: *Jpn. J. Appl. Phys.* 46, 5853, 2007
12. Sagisaka, A.et al. : *Appl.Phys.* B90, 373, 2008
13. Nishiuchi, M.et al., in preparation
14. Tajima, T and Dowson, J. M.: *Phys. Rev. Lett.* 43, 267, 1979
15. Boldarev, A. S. et al.: *Rev. Sci. Instrum.* 77, 083112, 2006
16. Faenov, A. Ya. et al.: *Phys. Scr.* 50, 333, 1994
17. Baldacchini, G. et al.: *Appl. Phys. Lett.* 80, 4810, 2002
18. Daido, H. et al.: *in X-ray lasers 2006, edited by P. V. Nickles and K. A. Janulewicz, (Springer, 2007), Springer Proc. in Physics* 115, 595-605 (2007)
19. Murakami, M. et al.: *in Laser-driven relativistic plasmas applied for science, industry, and medicine, edited by S. V. Bulanov and H. Daido, AIP conf. proc. No. 1024 (American Institute of Physics, 2008),* pp. 275-300.
20. http://wwwapr.kansai.jaea.go.jp/pmrc_en/

Towards an 100 Hz X-Ray Laser Station

J. Tümmler, H. Stiel, R. Jung, K.A. Janulewicz[1], P.V. Nickles and W. Sandner

Max Born Institut für Nichtlineare Optik und Kurzzeitspektroskopie (MBI), D-12489 Berlin, Germany
[1] Now at: Advanced Photonics Research Institute (APRI) at Gwangju Institute for Science and Technology (GIST), Republic of Korea

Abstract. During the last few years the optimization of pumping schemes of X-ray lasers (XRL) has reached a level where the required pump power could be provided by table-top or even by commercially available laser systems. But the stability of the XRL output signal is limited by that of the pumping lasers and also the repetition rate is at maximum about 10 Hz. Many envisioned applications would however benefit from an improvement of these crucial parameters. A way to overcome this situation could be the use of diode pumped solid state lasers (DPSSL) as drivers. Therefore we are developing a new 100 Hz DPSSL based on Yb:YAG thin disk and CPA technology. This system is based on newly developed efficient diode stacks for 100 Hz repetition rate. According to the common requirements of a transient collisional XRL (here in a grazing incidence pumping scheme -GRIP) the new laser driver has a double beam structure with one beam for plasma performing, delivering an energy at the target in the range of 200 mJ in 200 ps and a second one with > 500 mJ and < 5 ps to heat the plasma. The amplifier system consists of 4 amplifiers of different sizes. For the following XRL operation a water cooled Ag or Mo tape as target for 13.9 nm or 18.9 nm XRL emission was developed. The target speed can be adjusted to the driver laser repetition rate. Parallel to the commissioning the XRL station and first application experiments an upgrade of the driver laser is planned.

1 Introduction

Despite the fact that during the last few years the development of X-ray lasers has reached a level where commercially available laser systems can provide the required pump pulse parameters [1,2,3], there are so far no dedicated XRL stations like e.g. synchrotron radiation sources, to give access to this radiation with unique properties.

The usual approach on XRL development and experiments is to set up the X-ray laser, do some optimization concerning e.g. the required pump energy, the output stability, or the beam profile, and perform a pilot experiment to demonstrate the ability of XRLs. After the experiment is finished the XRL setup is dismantled and the pump laser is used for other experiments. A continuous use, maintenance and improvement of an X-ray laser can hardly be found. Our aim is to install an XRL station to provide the unique properties of XRL radiation in a reliable source not only for prove of principle experiments but also as a tool for standard investigations in science and technology.

In this paper we will describe the XRL station with special attention to the Yb:YAG pump laser. After an overview of the pump laser system and a short explanation of the thin disk technology we will give a detailed description of the single components.

2 Pump Laser Design

The pump laser as a driver laser for the XRL station should match a number of parameters. Beneath the usual requirements for every laser system like high stability and good beam profile the pump laser for an common transient collisional XRL needs to have two synchronized arms, one with lower energy and longer pulse duration (200 mJ @ 200 ps) to preform the plasma and one with higher energy for the short heating of the plasma (>500 mJ @ 5 ps). In order to make the XRL usable for applications requiring an averaging operation we aimed at a repetition rate up to 100 Hz.

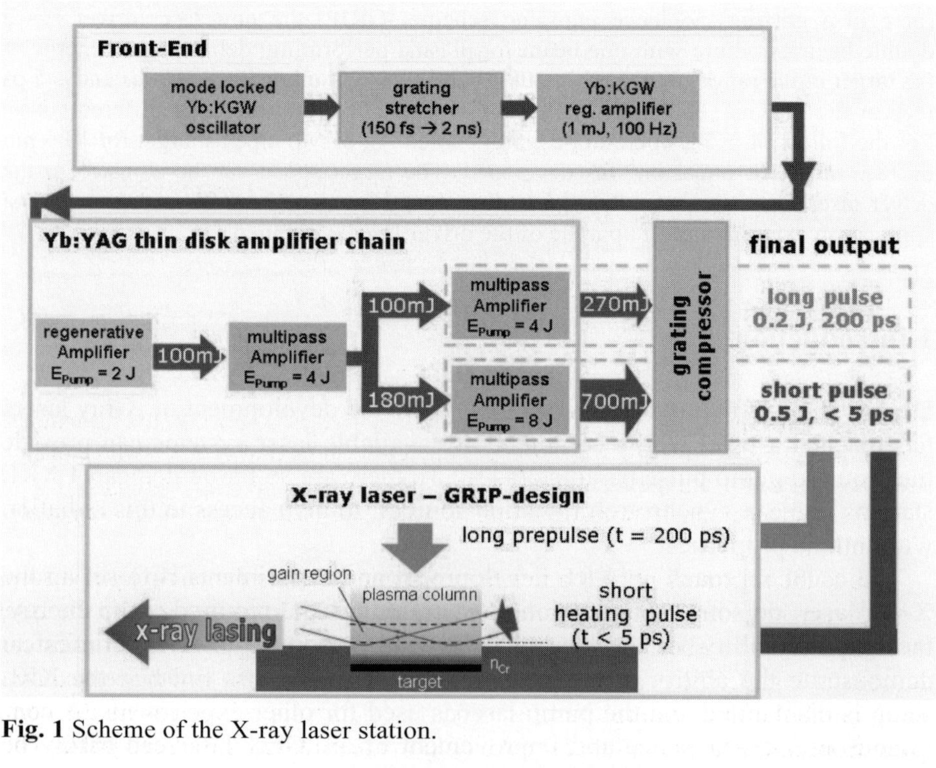

Fig. 1 Scheme of the X-ray laser station.

To realize these required parameters we designed the pump laser as a diode pumped CPA solid state laser in Yb:YAG thin disk technology with newly developed efficient diode stacks of ~1 kW peak power (1 J in 1 ms pulse duration) and up to 100 Hz repetition rate [3].

Fig. 1 shows the scheme of the XRL station. The laser consists of a front-end with oscillator and stretcher followed by an amplifier chain based on Yb:YAG thin disk technology. The last element is a grating compressor to compress the two laser pulses to the required 200 ps and 5 ps, respectively.

Even though the system is not fully completed we will in the following describe the status of the project and report about the different components of the XRL station more in detail. But first we give a short overview of the thin disk technology used in the pump laser system.

2.1 Thin Disk Technology

The key-components of the pump laser system are the amplifier stages based on Yb:YAG thin disk technology [4, 5]. Characteristic features of this technology are very low thermal lens effects due to efficient cooling with the temperature gradient parallel to the beam direction, and good power scalability. Fig. 2 shows the common thin disk pumping scheme according to A. Giesen et al.

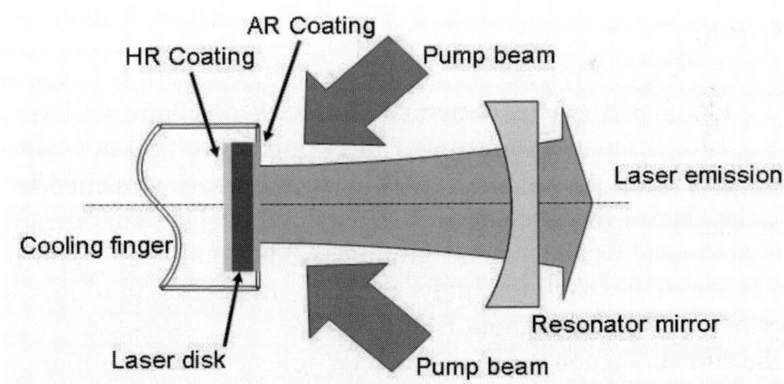

Fig. 2 Thin disk pumping scheme. The laser disk is used as a mirror in the resonator set up. It is pumped from the front side and cooled form behind over the whole disk area. For the pump and the seed wavelength it has a high reflectance coating on the back side and an antireflection coating on the front side (Courtesy of A. Giesen at al).

Due to the thickness of only a few hundred micrometers the pump light absorption is very low. Therefore the pump light is reflected up to 24 times to the laser disk until it is absorbed completely. A laser head including the pump optics can be seen in Fig. 3. The beam enters the laser head from behind and is focused by a large parabolic mirror onto the laser disk. The parabolic mirror covers the complete front of the laser module except a hole in the middle for the seed pulse. This parabola is removed in Fig. 3 to have a free look inside. From the disk the pump beam is reflected back to the parabola, displaced by the prisms and again reflected to the laser disk. In total the pump beam is reflected to the laser disk 24 times.

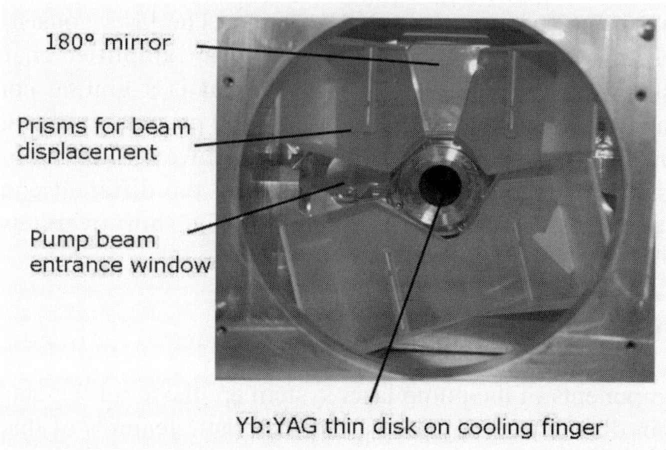

Fig. 3 View inside the thin disk laser head. Not visible in the picture is the parabolic mirror in front of the module that focuses the pump light to the laser disk. The prisms displace the pump light to reflect it 24 times to the laser disk.

2.2 Front-End

The front-end consists of a commercial Yb:KGW oscillator (PHAROS, Light Conversion Ltd. Lithuania) emitting 180 fs pulses at 1030 nm with 54 MHz repetition rate (mean power: 650 mW). The pulses are stretched to 2 ns. To minimize the size of the stretcher it has a folded design. That means the first grating is also used as the second grating. Additionally the beam passes the stretcher twice. Due to the highly efficient dielectric grating (diffraction efficiency > 95%) the throughput is still > 10%.

After the stretcher a commercial Yb:KGW regenerative amplifier (PHAROS, Light Conversion Ltd. Lithuania) is used to amplify the pulse to 1 mJ energy. The repetition rate of the regenerative amplifier can be set between 1 Hz and 1 kHz. Tab. 1 shows some beam parameters at the output of the front-end.

Tab. 1 Laser pulse parameters at the output of the front-end.

Pulse energy E	1 mJ
Repetition rate f	1 kHz
Pulse duration τ	2.2 ns
Bandwidth $\Delta\lambda$	2.8 nm

2.3 Thin Disk Amplifier Chain

The output of the front-end is further amplified by an amplifier chain based on Yb:YAG thin disk technology as described in Sec. 2.1. This amplifier chain consists of a regenerative amplifier (RA) and a number of multipass amplifiers (MP) pumped with increased power from stage to stage.

The complete system is pumped by specially developed fibre coupled laser diodes. These diodes deliver a maximum pulse energy of 1 J per fibre at 940 nm wavelength with a pulse duration of 1 ms and at a repetition rate of up to 100 Hz (10% duty cycle). They were developed at Ferdinand Braun Institute (FBH), Berlin, and match very well the requirements for pumping Yb:YAG.

2.3.1 Regenerative Amplifier

The front-end output pulse is adapted to the regenerative amplifier mode and coupled into the amplifier. The regenerative amplifier in a folded design has a length of 6.45 m measured between the two end-mirrors. The amplifier disk acts as one end-mirror with the in- and out-coupling Pockels cell in front of it. Instead of the easiest possible geometry with just a second end-mirror we use two additional spherical mirrors as a telescope inside the RA to increase the minimum beam diameter and reduce the power density on the optical components.

While the amplifier head is self-made the cooling finger with the laser disk is a component from TRUMPF Laser GmbH with much higher stability than the initially used Yb:YAG disk. With the Trumpf disk we reached an output pulse energy of 124 mJ @ 50 Hz in a stable beam. The required pump pulse energy was 1.6 J from 2 J available. Fig. 4 shows the measured output characteristic and stability. At maximum output pulse energy we measured an average pulse energy of 117 mJ +/- 2 mJ (rms) over a periode of 120 s.

The beam profile has an excellent shape with $M^2 = 1.2$ and a minimum focal spot diameter of 80 μm in x and 88 μm in y direction obtained with a f = 300 mm focussing lens. Fig. 5 shows the M^2 measurement according to the ISO standard with a SPIRICON M^2-200 meter (OPHIR Optronics Ltd). The distance between the CCD camera and an internal lens is varied and the beam diameter in x and y direction is measured.

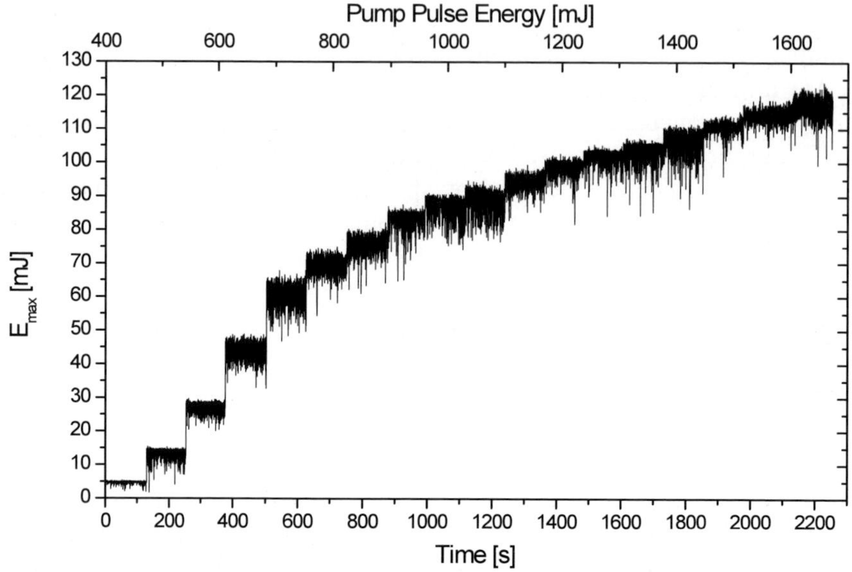

Fig. 4 Output characteristic of the regenerative amplifier: Output pulse energy vs. steps in pump pulse energy of 120 s duration. The mean output pulse energy at 1.6 J pump pulse energy is 117 mJ +/- 2 mJ (rms) with a maximum value of 124 mJ.

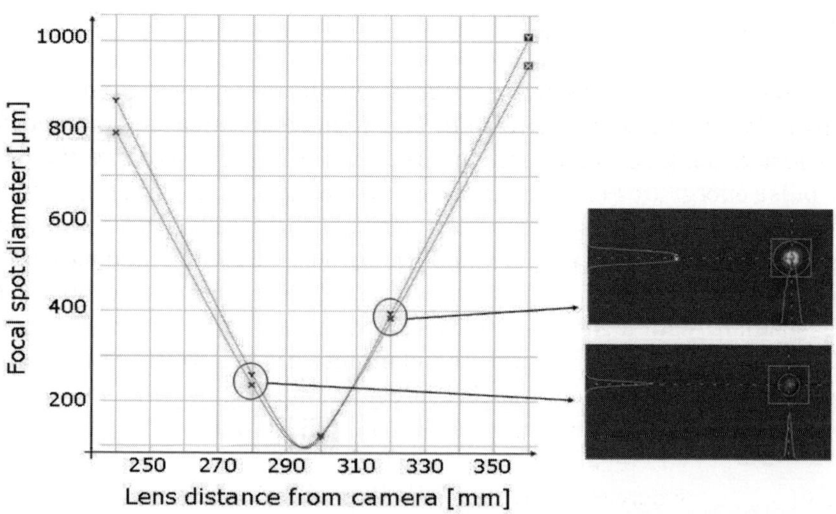

Fig. 5 M^2 measurement with ISO certified M^2-200 meter from SPIRICON, $M^2 = 1.2$, focal spot diameter x = 80 µm, y = 88 µm focussed with f = 300 mm lens.

2.3.2 Yb:YAG Thin Disk Multi Pass (MP) Amplifier

All MP amplifiers are based on the same technology/geometry that allows a modular upgrade of the system. The resonator is a multipass set-up with 12 passes across Yb:YAG thin disk (TRUMPF GmbH, Schramberg). The three MP amplifiers (cp. Fig. 1) vary in the applied diode pump energy (4 J and 8 J respectively) and accordingly in disk diameter and thickness.

Recently we have completed and tested the first MP amplifier pumped by 4 J pulses from the diode modules. At a seed energy from the RA of 75 mJ (65 % of max. RA output) an output pulse energy of more than 200 mJ with an pump pulse energy of 1.8 J (< 50% of max. pulse energy) has been reached reproducible. The maximum output pulse energy was about 320 mJ. The present investigations and preliminary results have shown that the first MP amplifier is at full pump energy and full seed energy capable to deliver output pulses with an energy of about 400 mJ However, for realising this output energy in the range of 400 mJ one has to redesign the resonator to get a larger beam diameter to prevent damage on the surface of the optical components.

The following MP amplifier (cp. Fig. 1) is a copy of the system described above except the Yb:YAG disk which has to be of larger diameter and the pump optics which allow to use up to 8 laser diode pump modules in order to reach the envisioned 1 J level. Both MP amplifier stages are currently under commissioning.

2.4 Compressor

The compressor is designed to accept two separate input beams (long and short pulse) to be compressed to different pulse durations as required for the transient XRL. Due to the large compression ratio from the ns to the ps range a compressor grating of 450 mm length is needed as second grating to accept the full spectral bandwidth. Like the stretcher the compressor is also equipped with dielectrically coated gratings (Lawrence Livermore National Labs) with more than 95% diffraction efficiency at 59° diffraction angle for 1030 nm. This gives a total transmission of more than 75%.

The two beams pass through the compressor at different heights and with separately adjustable path lengths to set the required pulse duration. Fig. 6 gives an overview of the beam pass through the compressor.

In the Yb:YAG regenerative amplifier the spectrum was narrowed from 2.8 nm to 1.5 nm whereas in the next MP amplifier stage(s) we did not find any further change in the spectrum. Our bandwidth limited pulse duration for the compressed pulses is therefore on the 1 ps scale. This assumption is supported by first measurements of the (non-optimized) compressed output pulse from the RA yielding a pulse duration of 1.2 ps.

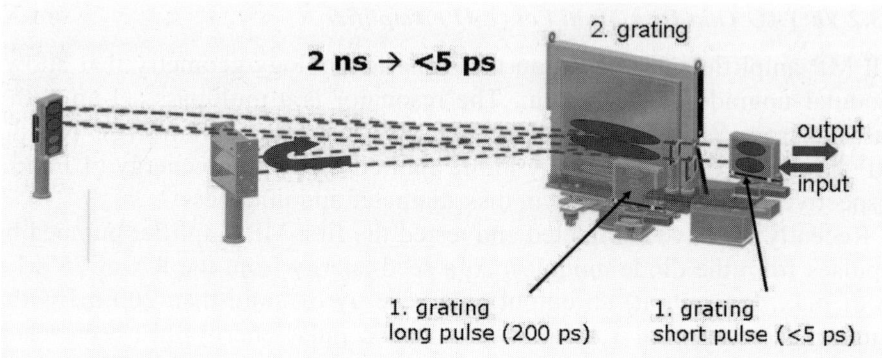

Fig. 6 Compressor set-up for simultaneous compression of two beams.

3 X-Ray Laser in GRIP geometry

With the above described Yb:YAG thin disk laser an X-ray laser in GRIP geometry will be available as user station.

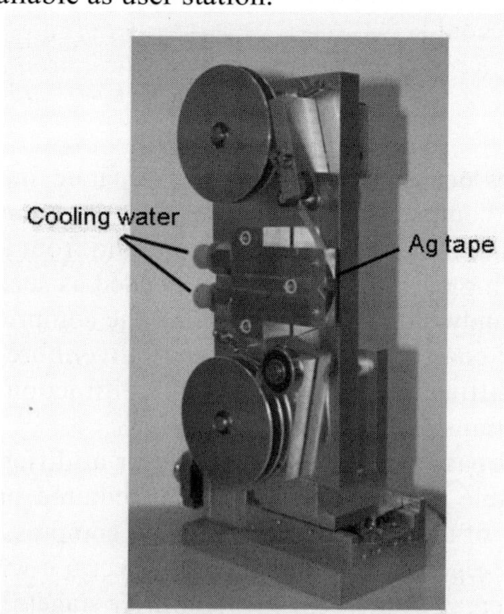

Fig. 7 Water cooled tape target for 100 Hz repetition rate.

As usual in the GRIP geometry the long pulse is line focused by a cylindrical and a spherical lens onto the target at normal incidence. The generated plasma column is then heated by the short pulse focused in grazing incidence by a spherical mirror into the plasma. As target for the planned 13.9 nm XRL station we use a continuously moving silver tape (Fig. 7). Its velocity can be adjusted to the pump laser repetition rate. At the line focus position the tape is guided above a water cooling unit capable to cool down a few hundred Watts.

All optical components will be protected against debris by thin glass plates or foil, where possible. The x-ray laser output will be guided through an aperture to reduce debris contamination on the following optical elements.

4 Summary and outlook

In the near future an x-ray laser station on the base of a saturated 13.9 nm Ag XRL with high average power in the range of 100 μW and high pulse stability will be available for scientific and industrial use. The fully diode pumped pump laser is especially designed as XRL driver laser operating in two beams with different pulse duration.

In the actual design the output pulse energy of the unique pump laser will be limited to about 700 mJ in the 5 ps pulse and to 200 mJ in the ~ 200 ps pulse. For further improvements of the output energy and/or shortening of the output wavelengths of the XRL a pump laser upgrade by an additional amplifier is planned to double the output energy in the long as well as the short pulse. These x-ray laser systems will be optimal for seeding experiments.

Acknowledgement

This work was supported by the ProFit program of the Land Berlin and EFRE of the European Union.

The collaboration with A. Giesen and J. Speiser, DLR Stuttgart and K. Wallmeroth and S. Schad, TRUMPF GmbH Schramberg is gratefully acknowledged.

References

1. Keenan, R. et al.: *'High-Repetition-Rate Grazing-Incidence Pumped X-Ray Laser Operating at 18.9 nm'*, Phys. Rev. Lett. 94, 103901, 2005
2. Tümmler, J. at al.: *'10 Hz grazing incidence pumped Ni-like Mo x-ray laser'*, Phys. Rev. E 72, 037401/1-4, 2005
3. Kim, H.T. et al.: *'Characteristics of a Ni-like silver x-ray laser pumped by a single profiled laser pulse'*, JOSA B, Vol. 25, Issue 7, pp. B76-B84, 2008

4. Fiebig, C., et al.: *'High-power, high-brightness 100 W QCW diode laser at 940 nm'*, Proc. SPIE 6456, 64560K, 2007
5. Giesen, A. et al.: *'Thin-Disk Lasers Come of Age'*, PHOTONICS SPECTRA, 52-58, May 2007
6. Killi, A. et al.: *'Current status and development trends of disk laser technology'*, Proc. SPIE 6871, 2008

Versatile High-Energy and Short-Pulse Operation of PHELIX

T. Kuehl [1,2] ,V. Bagnoud [1] , C. Bruske [1], S. Borneis [1], B. Ecker [1,2], U. Eisenbarth [1], J. Fils [1], S. Goette [1], T. Hahn [1], D. Hochhaus [1,2], D. Javorkova [1], F. Knobloch [1], M. Kreutz [1], S. Kunzer [1], T. Merz-Mantwill [1], E. Onkels [1], D. Reemts [1], A. Tauschwitz [1], K. Witte [1] , B. Zielbauer [1], D. Zimmer [1,2]

[1] GSI Helmholtz Centre for Heavy-ion Research, Darmstadt, Germany
[2] Johannes Gutenberg-University, Mainz, Germany

Abstract. PHELIX (Petawatt High Energy Laser for Heavy Ion Experiments) is a hybrid Ti:Sapphire / Nd:Glass laser system using large aperture amplifiers from the former Nova and Phebus laser systems at Livermore and Limeil, respectively, designed to offer pulse energies in access of 1 kJ and output power in the petawatt range. It is aiming mainly on combined experiments in plasma physics [1] and atomic physics [2] together with the GSI accelerator facility, and in preparation for the new FAIR facility for antiproton and ion research. Both nanosecond and sub-picosecond pulses can be supplied. Presently pulse energies up to 500 J are used, at pulse durations between 1 and 15 ns. Compressed pulses down to 500 fs are achieved after full amplification. The maximum output energy after the pulse compressor is limited by the damage threshold of the final grating. For 10 to 50 ps pulses, the maximal throughput energy is 200 J. For the pulses around 500-fs duration, it is reduced to 120 J. A special arrangement allows for the preparation of pulse pairs, where the duration of the pulses can be individually controlled between 1 and 100 ps. Recent experiments included the preparation of a plasma target for the interaction with energetic heavy ions [3], and the pumping of a plasma x-ray laser [4].

PHELIX is a member of Laserlab Europe, offering access as a European infrastructure

1 The PHELIX laser system

The general architecture of PHELIX is a MOPA design with dedicated front-ends for long-pulse (1 to20 ns) and short pulse operation. Both front-ends produce pulses in the range of 20 to 50 mJ, followed by a common Nd:Glass pre-amplifier. From here, pulses of 1 to 10 Joule energy are injected into the two-pass main amplifier with a clear aperture of 30 cm.

For many of the future experiments planned for PHELIX, nanosecond pulses with special temporal shaping are required. These are created in the ns-front-end consisting of a fiber-oscillator, a fiber-based double pass amplifier, an amplitude modulator to tailor the temporal pulse shape, and a regenerative ring-amplifier and a beam shaping section to modulate the spatial laser beam profile. The temporal resolution of the modulator is better than 500 picoseconds, mainly limited by the driving electrical circuit. Except for the ring regenerative amplifier the PHELIX ns-frontend is based on fiber technology. This provides stable and robust operation without much further alignment and maintenance.

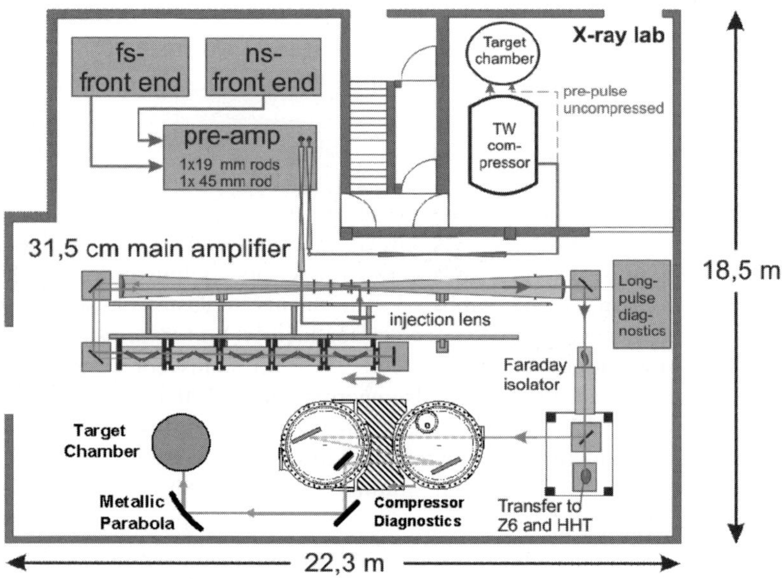

Fig. 1 Schematics of the PHELIX laser system.

For the short pulse option the chirped pulse amplification scheme is employed, with the goal to achieve pulse intensities in the petawatt range. This front-end is based on Ti:Sapphire as a gain medium, operating at the 1053 nm wavelength appropriate for the following Nd:Glass amplifiers. A 120 fs oscillator is followed by two regenerative amplifiers after stretching the pulses to > 1 ns. The pulse stretcher of PHELIX was rebuilt, recently. The essential new feature is adjustability of the stretching factor, thereby allowing to operate the compressor with fixed gratings. This way the pulse duration after recompression can be easily changed in a range from 0.5 to 10 ps.

The pre-amplifier consists of three Nd:Glass rod amplifiers of 19 mm and 45 mm. diameter. A flat-top beam profile is formed by a serrated aperture and the inherent spatial gain profile of the amplifiers. The shape can be selected to give a round profile, for optimal filling of the main amplifier components, or an elliptical shape optimised for the filling of the re-compressor gratings. Gain narrowing in the glass amplifiers is passively pre-compensated by a Lyot-filter. The 2-pass main amplifier uses Nd:Glass disks with a free aperture of 31.5 cm. Injection into this structure is done by geometric separation within the spatial filter. A limitation of the total output energy is given by the damage threshold of a pulsed Faraday isolator, which protects the system against reflected pulses from the target. After this the pulses can be directed either to the beam line towards the heavy-ion accelerator of GSI, or to the grating compressor to form short pulses.

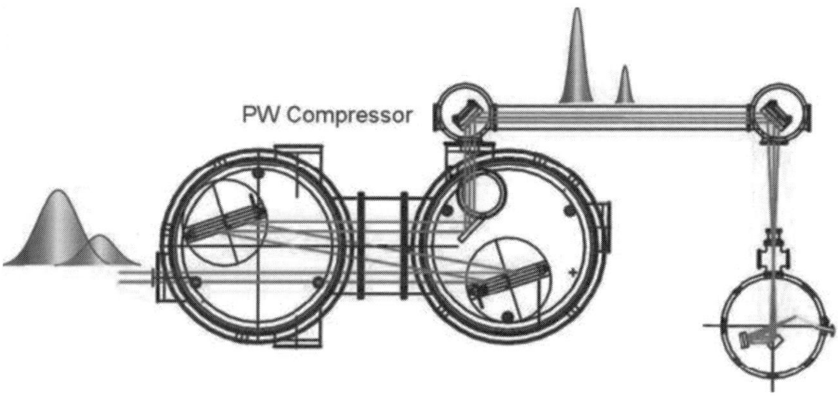

Fig. 2 Transformation of the double pulse in the pulse compressor. The stretched pulses are entering from the left, still partially overlapping in time. After compression two short pulses with the predetermined separation are formed.

2 Experimental set-up for pumping a short wavelength x-ray laser

For a first x-ray laser experiment [4] with Ni-like Samarium (6.8 nm and 7.3 nm) the capability to form double pulses already in the front-end of PHELIX was utilised. In this scheme, a Mach-Zehnder type beam splitting and recombination stage creates two pulses immediately after the pulse stretcher, where the beam diameter is below 5mm. These double pulses were amplified by the full PHELIX amplifier chain, and recompressed to produce two pulses with an adjustable delay of 0.1 to 10 ns. Figure 2 illustrates how the double pulses for pumping the x-ray laser were produced. The double pulse created in the front-end has a pulse separation of about 1 ns, which is much shorter than the duration of each of the stretched pulses. Only after these pulses passed the compressor optics and their duration is reduced, the real double pulse structure is revealed. With the help of a compact pre-compressor in one of the arms, completely different pulse durations for the two pulses can be chosen. The advantage lies in the fact, that these two pulses are focused absolutely identical in the beam delivering optics. A fast photodiode and alternatively a streak camera were used to determine the spacing of the pulses and the relative size of possible pre-pulses.

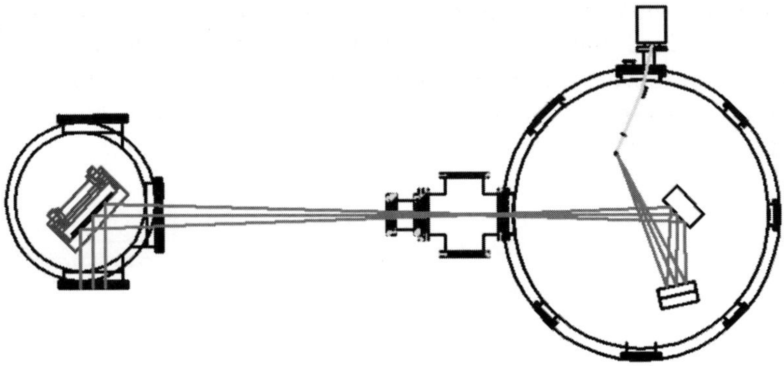

Fig. 3 The focusing scheme used for the generation of a line focus for the plasma x-ray laser. At first, the beam is reflected from a 90 degree off-axis parabolic mirror with 2 meter focal length. After the focus, a spherical mirror with 60 cm focal length is used under an incidence angle of 20 degrees. The insert shows the imprint on the target with a width of approximately 100 microns.

As shown in more detail in Fig. 3, the line focus on the x-ray laser target is produced using two focussing elements. At first, the beam is reflected from a 90 degree off-axis parabolic mirror with 2 meter focal length. After the point-focus, a spherical mirror with 60 cm focal length is used under an incidence angle of 12 degrees. In this way, relatively small optics can be used for the final focusing, despite the large beam diameter dictated by the low damage threshold of the compressor gratings.

At the given beam dimensions a line focus with a width of around 100 microns and a length of 8 mm is generated on the target at an incidence angle of 45 degrees. This line focus also has the property to exhibit a traveling wave characteristics with a speed of 1.4 c, well suited for the transient excitation scheme. The quality of the line focus is monitored with a microscope camera viewing the imprint on a glass screen which can be put at the exact location of the laser target. The targets themselves were fabricated at the GSI target laboratory by coating glass substrates with a layer of 2 – 3 micrometers of samarium. These targets were stored in a protective gas atmosphere, and mounted shortly before pump-down in the laser chamber. Positioning of the targets was controlled with a step-motor driven stage, which allowed the use of a single target for at least 5 laser shots choosing different locations on the target. The output of the x-ray laser was analyzed with a grating spectrometer using a 1200 lines/mm flat-field grating and a XUV CCD-camera. High energy x-ray and visible light background is rejected using zirconium coated carbon filters. Within the experimental campaign at the PHELIX laser, lasing in the Ni-like samarium system was achieved at pulse energies between 50 J and 100 J. A range of pulse configurations was tested with delays between 100 ps and 5 ns and pulse durations between 5 and 200 ps. Fig. 4 shows a spectrum with the Ni- like samarium lasing lines at 6.8 nm and 7.3 nm, registered at the second order of the grating, and a reference showing the carbon K-edge at 4.3 nm registered in third order at the same spectrometer settings.

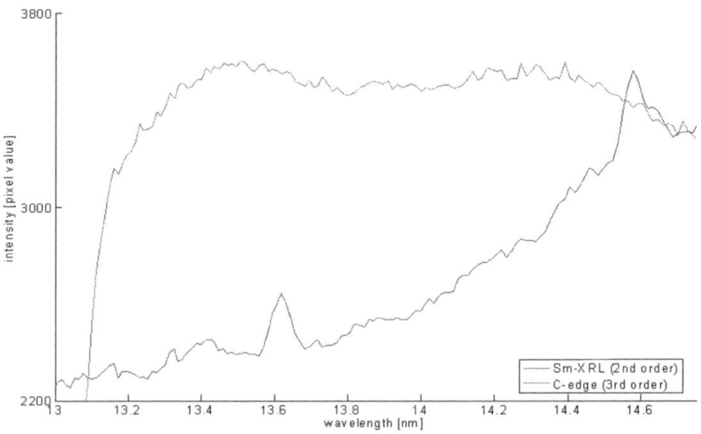

Fig. 4 Spectra showing the Ni-like Samarium x-ray laser lines at 6.8 nm and 7.3 nm (recorded in second order as 13.6 nm and 14.6 nm) together with a reference spectrum showing the carbon-edge at 4.36 nm (recorded in 3rd order as 13.1 nm) (Preliminary data by courtesy of [4]).

3 Conclusion

The PHELIX laser is able to operate with very different pulse durations and configurations. First results obtained with compressed double-pulses at PHELIX, unique for an high energy laser, demonstrate the feasibility of applying a simplified pumping scheme of higher energy x-ray lasers. This opens the way to a wider application of such x-ray sources for dedicated x-ray diagnostics of e.g. ion produced plasmas, as well as a future routine operation at facilities like LASERIX [5]. The experiment was supported through the Laserlab Europe Integrated Infrastructure Initiative.

References

1. Hoffmann, D. H. H.; Blazevic, A; Ni, P; Rosmej, O; Roth, M; Tahir, N; Tauschwitz, A; Udrea, S; Varentsov, D; Weyrich, K; Maron, Y.: 'Present and future perspectives for high energy density physics with intense heavy ion and laser beams', *Laser and Particle Beams 23 (3): 395-395, 2005*
2. T. Stoehlker et al.: 'Status and Perspectives of Atomic Physics Research at GSI: 'The new GSI Accelerator Project', *Nucl. Instr. and Meth. B, 205, 156, 2003*
3. A. Blazevich et al., *to be published*
4. D. Ros et al., *in preparation for Laser Partice Beams*
5. Kazamias, S; Cassou, K; Ros, D; et al.: 'High-cadence XUV laser pumped by a titanium: Sapphire laser, towards the LASERIX station', *Journal de Phys. IV, 138, 13, 2006*

Central Laser Facility High Power Laser Capabilities Applied to X-Ray Laser Science

M.M. Notley[1], N.B. Alexander[2], R. Heathcote[1], S. Blake[1], R.J. Clarke[1], J.L.Collier[1], P. Foster[1], S.J. Hawkes[1], C. Hernandez-Gomez[1], C.J. Hooker[1], D. Pepler[1], I.N. Ross[1], M. Streeter[1], G. Tallents[3], M. Tolley[1], T. Winstone[1], B. Wyborn[1] and D. Neely[1,3]

[1]Central Laser Facility, Rutherford Appleton Laboratory, Science and Technology facilities Council, Chilton, Didcot OX11 0QX UK
[2]General Atomics, PO Box 85608, San Diego, CA 92110
[3]Dept of Physics, University of Strathclyde, Glasgow, G4 0HG

Abstract. Soft X-ray lasers have been developed at the Central Laser Facility (CLF) of the Science and Technology Facilities Council over the last three decades by an active UK and international research community. The Vulcan Nd:glass Laser has been the primary drive system for these developments, providing pump pulses from nano- to pico- seconds, firing one shot every 20 minutes. An upgraded Vulcan facility is due to come on line in September 2008 with dual pico-second high energy beam lines (1 ps, 100 J and 10 ps, 500 J) combined with long pulse (80 ps, 40 J – 4 ns, 300 J) capability. Recent developments in laser technology have meant that the CLF has increased its ultra short pulse capabilities. The Ti:Sapphire Astra-Gemini laser facility, which came on-line in early 2008, gives 20 second shot turnaround times and ultra short pulses (30fs, 15J). The new opportunities which these drive systems enable for soft x-ray laser science, source development and applications will be presented.

1 Introduction

The Central Laser Facility (CLF) has been an important tool for researchers to develop x-ray laser science over the last three decades. With its versatile target area set-up and multiple high energy ns beam lines, it enabled research into the mechanisms of X-ray laser (XRL) production. Recombination[1] and collisional excitation x-ray lasers were the main areas of successful research with single[2] and double target irradiation[3] using 80ps @40J per pulse or 600ps @ 200 J Vulcan beam lines. A wide variety of saturated x-ray laser output[4] was achieved, paving the way to further techniques[5] along with initial application experiments using the XRL as a probe of plasma properties[6] and instabilities[7]. In the late 1990's, the travelling wave (transient) collisional excitation method[8] was introduced and further developed when after upgrading in 1997 the Chirped Pulse Amplification (CPA) arm of

Vulcan delivered energies of 100J in a ps pulse to target[9]. Since this time, the use of short optical pulses for pumping XRL output has been developed with the introduction of the Grazing Incidence Pumping (GRIP) scheme.

The high power lasers at the CLF have also grown in their capabilities, specifically in the recent year. A new mixture of long-, short-, shaped and pre-pulse modes are available with a variety of energies and higher repetition rate. These capabilities match the new methods being developed by the XRL community and provide potential for the future development of new applications.

2 High Power Laser Systems

2.1 Vulcan Developments

To accommodate soft X-ray laser experiments one of the first adaptations tested at the CLF was the generation of a near diffraction limited line focus using reflection of an expanding beam from a tilted spherical mirror[10]. For long pulses (60 ps – 1 ns) the beam was initially focussed using a transmissive doublet aspheric lens. With the introduction of ps drive pulses the lens was replaced by an off-axis parabolic mirror to avoid B-integral effects. Although the reflection from a tilted spherical mirror produced a near diffraction limited width line focus, it suffers from non uniform illumination uniformity along the length of the focus. To overcome this limitation, it was possible to illuminate a target from one side with up to five beams simultaneously (Fig. 1) and by arranging them in a suitable overlap[11], a high degree of uniformity could be achieved.

Over the last year the Vulcan laser and associated Target Area West have been upgraded to enable dual CPA delivery to our research user community. This dual output will enable either two beamlines at 1 ps, 100 J on target or one such beamline combined with another at 10 ps, 400 J on target. The beamlines will be 200mm square at 1 ps or 200 mm circular at 10 ps respectively. New laser pulse diagnostic systems[12] are being installed in the target area to enable reliable spectral, temporal and spatial beam qualities to be measured. Currently, one CPA beam can be frequency doubled using sub aperture crystals[13] to deliver ~ 30 J to target suitable for pumping short wavelength high Z soft X-ray lasers in a grazing incidence geometry. It is planned that with the acquisition of full aperture doubling crystal, up to 180 J could be provided in a ~6 ps pulse opening the possibility of saturated lasing in the water window.

The new beams are due to be commissioned in September 2008. Alongside this, is the capability to have six versatile 'long' pulse beamlines 80ps @ 40J – 4ns @300J per beam at 1053nm. There is also the option of pulse shaping using a new regenerative amplifier system able to provide pulse shaping at the longer pulse range >500ps up to 20ns, suitable for generating shock compressed matter where the XRL beam can be used as a probe source [14].

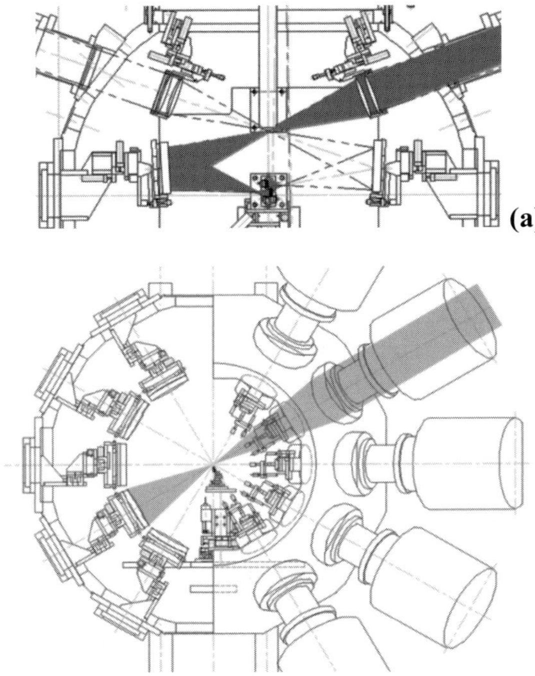

(a)

(b)

Fig. 1 Line focus multi-beam set-up 5 beam single sided target irradiation, (a) plan view, (b) side view.

2.2 Gemini Developments

Gemini is an upgrade [15] to the existing Astra high power laser which currently delivers 0.5J in 40fs at 10Hz, 800nm. A new laser bay with system pump lasers, amplification stages and two separate beamline compressors have been installed enabling the Gemini system to deliver up to 15J in 30fs per beam ultimately combining to 1PW, firing once every 20 seconds when fully commissioned.

A new target area was created underneath the laser bay in which a wide variety of set-ups are available. Currently available options include long (F20) and short (F2) focusing on target which can be configured in a variety of ways to suit experimental campaigns including near normal and grazing incidence line focus geometries. This system is capable of delivering up to 10^9 contrast, which can be further enhanced by the use of an optional plasma mirror[16] set-up on one of the beamlines to deliver 10^{12} contrast.

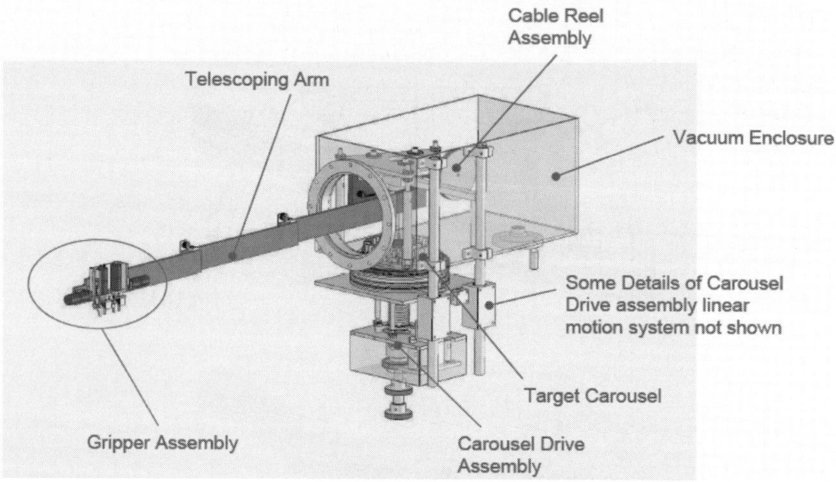

Fig. 2 Target inserter mechanism schematic showing arm for gripping carriers and the carousel to hold 50 carriers.

Another step change in the delivery of experiments to the Gemini Target Area is the installation of a remote target insertion device. This instrument is specified to be able to deliver one target carrier to the correct position to within +/- 2 microns. Each target carrier can hold up to 25 targets and a carousel can hold up to 50 carriers. This enables the delivery of up to 1250 targets without breaking vacuum. Fig. 2 shows a schematic of the inserter unit. By combining the target inserter technology with the relatively high shot throughput per day (~600 shots) and the multi-beam capability of Gemini it is hoped that a new era of x-ray laser development and applications will be opened up.

References

1. J. Zhang *et al.*, Physical Review Letters **74,** 1335 (1995).
2. D. Neely *et al.*, Institute of Physics Conference Series 37 (1992).
3. G. Cairns *et al.*, Applied Physics B-Lasers and Optics **58,** 51 (1994).
4. A. Carillon *et al.*, Physical Review Letters **68,** 2917 (1992).
5. R. E. Burge *et al.*, Journal of the Optical Society of America B-Optical Physics

15, 2515 (1998).
6. E. Wolfrum *et al.*, Physics of Plasmas **5,** 227 (1998).
7. D. H. Kalantar *et al.*, Physics of Plasmas **4,** 1985 (1997).
8. A. Klisnick *et al.*, Journal de Physique Iv **11,** 11 (2001).
9. C. L. S. Lewis *et al.*, Soft X-Ray Lasers and Applications Iii **3776,** 292 (1999).
10. I. N. Ross *et al.*, Applied Optics **26**, 1584 (1897)
11. D. Neely *et al.*, Optics Communications **87,** 231 (1992).
12. C. Stoeckl *et al.*, Review of Scientific Instruments **77,** (2006).
13. D. Neely *et al.*, Laser and Particle Beams **18,** 405 (2000).
14. M. H. Edwards *et al.*, Physical Review Letters **97,** (2006).
15. C. J. Hooker *et al.*, Journal de Physique Iv **133,** 673 (2006).
16. C. Ziener *et al.*, Journal of Applied Physics **93,** 768 (2003).

TARANIS: A Pump Source for X-Ray Lasers

G Nersisyan, T Dzelzainis, CLS Lewis, D Riley, R Ferrari, M Zepf, M Borghesi, L Romagnani. D Doria, D. Marlow, B Dromey

School of Mathematics and Physics, Queen's University Belfast, Belfast, BT7 1NN, UK

Abstract. A new laser system has recently been installed within the Centre for Plasma Physics (CCP) at the Queen's University Belfast (QUB) and is known as TARANIS (Terawatt Apparatus for Relativistic and Non-linear Interdisciplinary Science). It will support a wide-ranging science programme, including X-ray laser studies. The laser is a CPA Nd:Glass system capable of delivering 20J in ~600 fs in each of two beams. Alternatively, either beam can generate ~30J in 1 ns in uncompressed mode allowing the synchronous production of ps-ps, ps-ns or ns-ns pulse combinations for various X-ray laser pump scenarios. Although the system is limited to ~10 min shot cycles it offers scope for the systematic study of XRL schemes not always feasible at national laser facilities. Here, we present a brief report detailing the laser's parameters and its capabilities in relation to X-ray lasers. Also outlined is our plan for experimental X-ray laser work in QUB and some results from preliminary shots taken in preparation for a full X-ray laser experiment.

1 Introduction

A terawatt chirped-amplification (CPA) laser system known as TARANIS (Terawatt Apparatus for Relativistic and Nonlinear Interdisciplinary Science) is being operated in the CPP at the QUB installed by Coherent Ltd in 2006. This unique ultrafast Nd:Glass laser system can deliver intensities up to 10^{19} W/cm^2 in the focused spot at wavelength of 1053 nm. This apparatus along with a home-built synchronized femtosecond Ti:Sapphire laser system will be used in the research areas of laser-field accelerated ion generation, warm dense matter, pump-probe experiments and X-Ray lasers.

In order to achieve high energies at ultrashort pulse durations a CPA concept is employed [1]. Rapid developments in technology and laser science such as the design of stretcher-compressor systems and regenerative amplifiers, production of high quality optical components and coatings able to stand high energies and the discovery of new laser materials (particularly the Ti:Sapphire) crystal improved the progress of high power lasers moving them to petawatt power level.

2 Laser System

The front end of the Taranis terawatt laser is shown in Fig. 1 and starts with a commercial mode-locked Ti:Sapphire Mira oscillator which generates a train of transform-limited pulses of 120 fs at wavelength 1053 nm with a repetition frequency of 76 MHz. Although the gain of Ti:Sapphire crystal is not peaked at 1053 nm the laser delivers an average power of 400 mW.

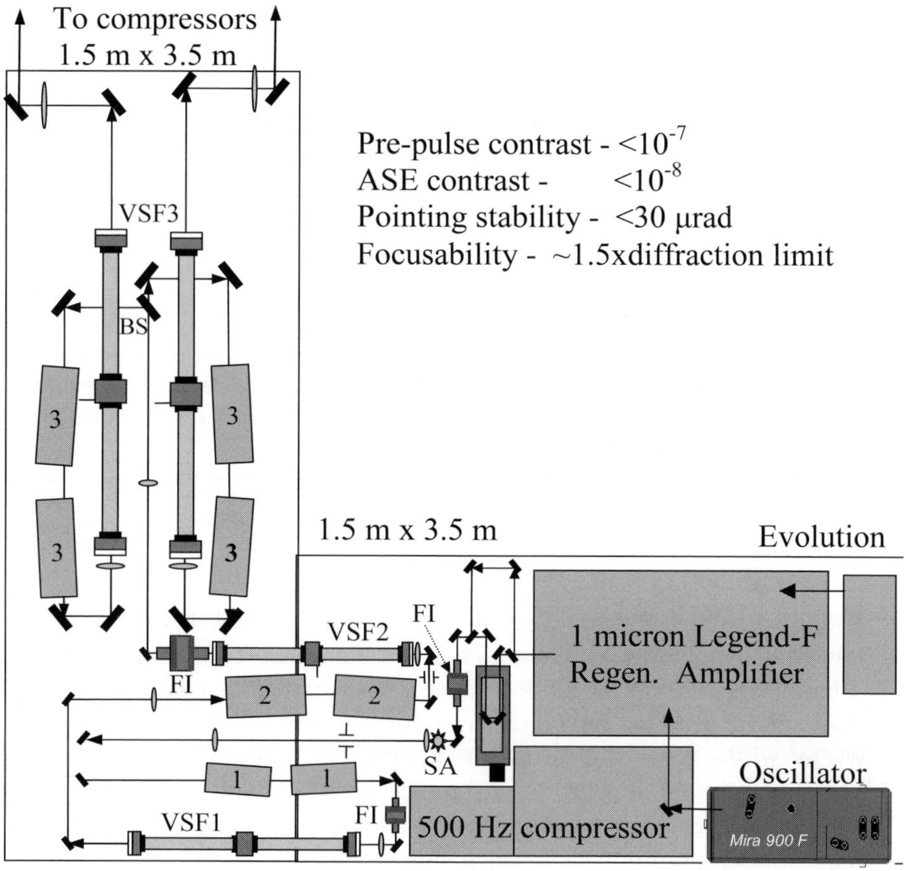

Fig. 1 Optical layout of the Taranis laser at QUB. 1, 2, 3 – laser heads with 9 mm, 25 mm and 50 mm diameter rods, respectively, VSF1, 2, 3 – vacuum spatial filters, FI – Faraday isolator, SA – serrated aperture, BS – beam splitter.

The train is injected into a folded all-reflective stretcher consisting of a spherical mirror (F=1524 mm), a diffraction grating (1740 lines/mm), a folding mirror located in the focal plane of the spherical mirror and a retro-mirror. To avoid disruption of the oscillator a Faraday isolator is placed after the oscillator. It is important to transfer the largest spectral bandwidth through the stretcher and avoid a hard clipping of the spectrum by optics and mounts. The cut-off in frequency results in a broadening of the pulse and a low contrast. However to reduce the cost, the bandpass of the stretcher was chosen equal to 4 times of the laser output bandwidth. The stretching factor is about 10^4.

The stretched pulse with a duration of 1.6 ns, enters the last stage of the front end; i.e. the Ti:Sapphire regenerative amplifier (RA) optimized for a high gain at 1053 nm. The RA is pumped by Evolution laser at 527 nm and with a repetition rate of 500 Hz from both sides of the Brewster-angle-cut Ti:Sapphire crystal. To improve the contrast of the RA a Pockels cell is used in the input of the RA as a pulse picker to select a single pulse from the oscillator train. The output of the RA is 0.7mJ per pulse at a stability of 3%. Although the gain of the laser crystal is peaked at 800 nm the special design of the oscillator and the RA allows high powers to be reached at 1053 nm. This is important for reduction of the ASE level from the RA and consequently in the output of the laser system. The high level of the output at 500 Hz rep rate of the front end makes it easy to align the main amplifier chain, the compressors, the targets and the diagnostics.

Further amplification of the front end output occurs in three stages consisting of phosphate-glass rod amplifiers pumped by flash lamps at a repetition rate of 1 shot per 10 min. Each amplification stage has two laser heads. The diameters of the rods are 9 mm, 25 mm and 50 mm with lengths 15 cm, 30 cm and 30 cm, respectively. Between the amplification stages there are vacuum spatial filters which magnify the beam diameter to fill the laser rods and image relay the beam profile at the output to the next stage of amplification. In the starting object plane of the image relaying sequence there is a serrated aperture with a diameter of 3.4 mm. Passing through such an imaging system the amplified pulse is spatially tailored by the serrated aperture, the hard truncation aperture and gain profiles in the laser rods. As a result of the "tailoring" and the image relaying, the final output beam has a near-uniform top-hat spatial profile with a diameter of 100 mm. In this case the energy extraction from the laser rods is the highest giving almost 30J per pulse after the last amplification stage. The gains in the 9 mm and 25 mm rods are 150 and 100, respectively. The small signal gain of the last amplification stage is 40 and the total gain of the Nd:Glass amplifies is $(4\text{-}6)\cdot 10^4$. Usually because of the gain narrowing in the amplification chain the pulse duration of the stretched pulse becomes shorter (~1 ns). The spectral width of the RA output reduces from 6.8 nm to 3 nm as well.

The Taranis laser is a flexible system consisting of two parallel final amplification stages allowing a combination of outputs with pulse durations starting from ns to ~600 fs and energies up to 30J. This suits the various research goals in the field of laser-matter interaction, particularly to study transient behaviours of plasmas and various X-Ray laser pump scenarios. The architecture of the laser enables access to three different target areas.

The intensity-dependent refractive index of various optical materials in the laser produces a phase retardation (B-integral) which can degrade the beam quality. It also results in a self-phase modulation limiting the compression of the pulse and producing wings in the temporal pulse. To decrease the B-integral the technique of CPA is usually applied and the estimated value of the B-integral for the Taranis laser system is 1.3.

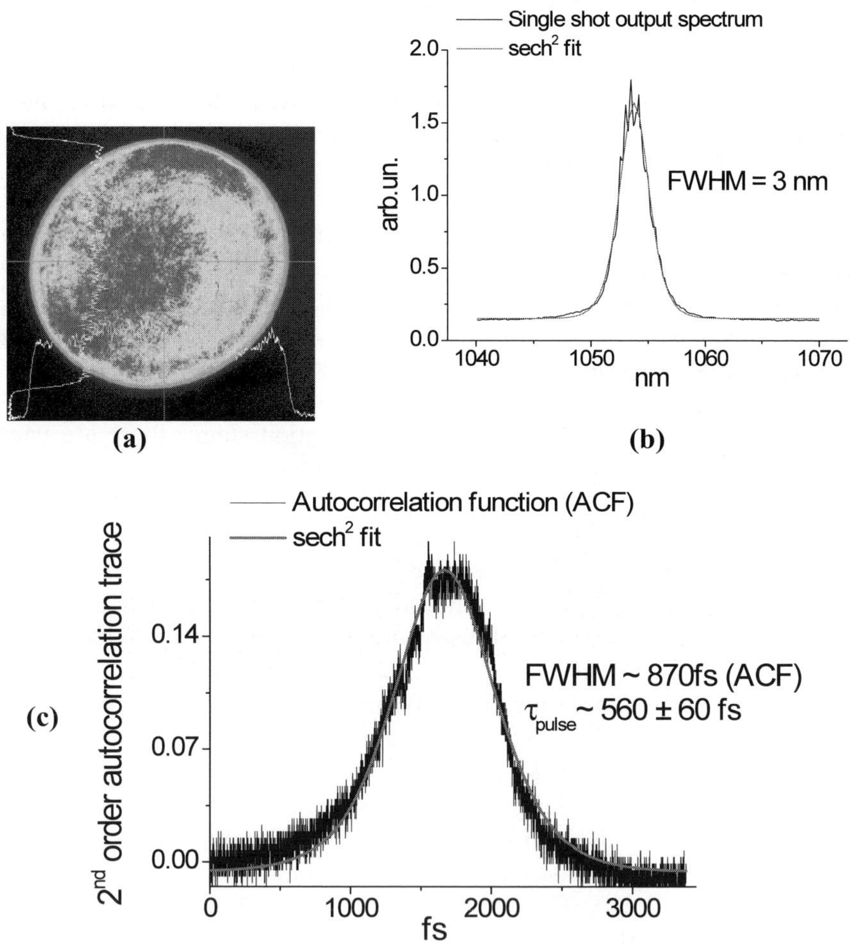

Fig. 2 A representative data of the Taranis laser, (a) near field profiles of the output beam, (b) single shot spectrum and (c) second order autocorrelation function (ACF).

The final stages of the laser are two large traditional double-pass grating vacuum compressors with a roof-mirror and a vacuum chamber from where the outputs of both compressors are directed to the target areas. The compressor gratings (210x400x50 mm) are positioned parallel to compensate the GDD of the stretcher and the GVD of the materials existing in the laser. The incidence angle of the input beam is 60° and the slant distance between the gratings is about 880 mm. The compressors are able to compress the amplified pulse down to 560 fs when an acousto-optic programmable dispersive filter (Dazzler) is used to correct high order phase distortions. The gratings provide an energy transmission of 60% through the compressors.

Different diagnostics are set to characterise and optimize the performance of the laser system and measure the energy, the near-field and far-field irradiances, the second order autocorrelation trace, the spectrum and the prepulse activity on each laser pulse shot. Some typical output characteristics are shown in Fig. 2. The near field irradiance taken after the compressor has almost a flat distribution. Assuming that the temporal profile of the pulse is sech^2 both output spectrum and the second order autocorrelation (ACF) are fitted by a sech^2 function. It is obvious that the fits well represent the central parts of the ACF and the spectrum except of the wings which are higher compare with the fitting function. The measured $\Delta\nu\Delta\tau$ product is 0.45 ± 0.06. Some important characteristics are presented in the Fig. 1.

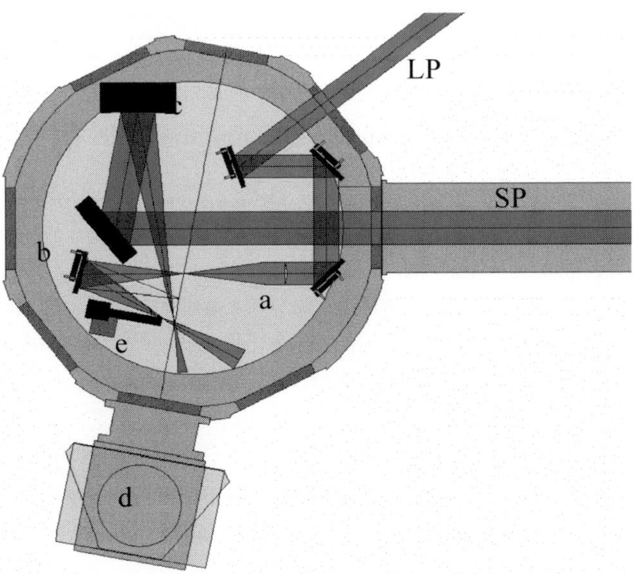

Fig. 3 Schematic of the GRIP laser set-up at QUB. **SP** is the short ps exciting pulse; **LP** is the long nsec plasma preparation pulse; **a** is a lens; **b**, **c** are tilted spherical mirrors; **d** is a flat-field spectrometer viewing the axial output; **e** is a crossed slit camera for X-ray emission.

3 The QUB X-Ray Laser

Initial X-ray Laser experiments to be conducted on TARANIS will investigate the GRIP scheme, first demonstrated by Keenan et al in 2005 [2] and characterise the output of various target materials as lasing media. The planned experimental set-up is shown in Fig. 3. The short pulse (SP) heats the plasma produced by the long pulse (LP) and creates a population inversion. The LP is focussed to 9 mm (absolute length) x 50μm by a combination of a F=300 mm lens and a F= 152 mm spherical mirror. The duration of the LP is ~1 ns and the energy can be up to 20 J. The SP, with a pulse duration of 1 ps and an energy available of up to 20 J is focussed to a line ~8 mm by an F=609 mm spherical mirror tilted at 10° to the incoming beam. The primary diagnostic of the X-ray laser will be a flat-field spectrometer employing a standard 1200 lines/mm Hitachi grating. Plasma diagnostics will consist of a spatially resolving crystal spectrometer (not pictured, above plane of diagram) and a crossed slit camera.

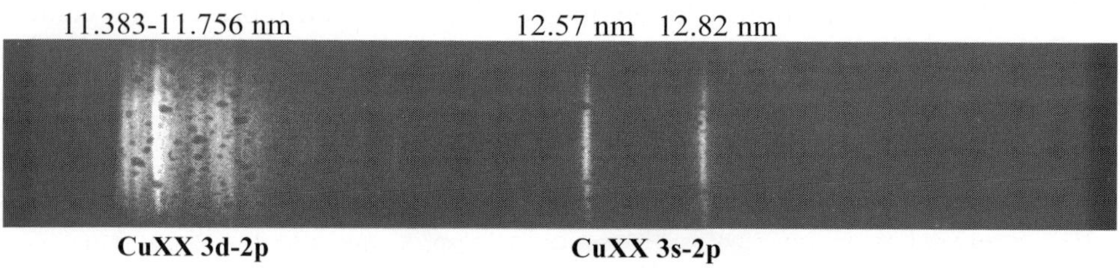

Fig. 4. Spatially resolved X-ray emission spectrum from laser produced Cu plasma.

Preparations for the XRL campaign have been made by generating and optimising the pre-plasma conditions. The plasma produced from a Cu target by the LP (1 ns, 5J) is used to observe the uniformity of the line focus. The spectrum shown in Fig. 4 is obtained with a spatially resolved crystal spectrometer. The intensity profile of the x-ray emission along the line with a Gaussian fit has a FWHM value corresponding to 6 mm.

References

1. Backus, S., et al.: "High power ultrafast lasers", *Rev. of Sci. Insrt.*, 96, 1207-1223, 1998
2. Keenan, et al.: "High-repetition-rate grazing-incidence pumped X-Ray laser operating at 18.9 nm", *Phys. Rev. Lett.*, 94, 103901-1 – 103901-4, 2005

Photon Frontier Network

Y. Kato[1], M. Gonokami[2], R. Kodama[3], Y. Sano[4], S. Yagi[5] and T. Yabuzaki[6]

[1] The Graduate School for the Creation of New Photonics Industries
[2] Department of Applied Physics, University of Tokyo
[3] Graduate School of Engineering, Osaka University
[4] Toshiba Corporation
[5] Mitsubishi Electric Corporation
[6] Osaka Electro-Communication University

Abstract. A new 10-year program "Photon Frontier Network" has been started in 2008 in Japan in order to establish strong bases in research and education in optical science and technology. This network is formed by many scientists belonging to various organizations in broad range of fields, and with industrial participation. The basic structure and objectives of this program are described.

1 Introduction

Optical science is called "enabling technology", since it has provided innovative approaches in broad range of fields from basic science to industry and medicine. In recent years, there have been significant advances in the development of new light sources and also in the understandings of the nature of light. Due to these advances, many new possibilities are emerging which may provide breakthroughs in various fields.

In order to extend these rapid advances to various fields, it is important to foster young scientists who have strong bases in the frontier of optical science and have broad scopes in the applications. However, there has not been coordinated approach to education in optical science, although optical science is regarded as an important basis in almost all fields of science and technology. We think that favourable environment exists at present, since optical science has been attracting many graduate students due to various interesting themes to work on.

In order to establish strong bases in the research and education in optical science, a new 10-year program "Photon Frontier Network" has been started in 2008 by the Ministry of Education, Culture, Sports, Science and Technology (MEXT). This network is formed by many laboratories in broad range of fields and organizations, with industrial participation. In this article, after a brief review of the historical background in optical science policy in Japan, the structure of this program is described.

2 Historical Background

Importance of optical science in the 21st century has been recognized in many countries and several official reports have been published [1, 2]. OECD has published a report on the emerging field developed by compact and high intensity short-pulse lasers [3]. Based on this activity, IUPAP has formed a Working Group “International Committee on Ultrahigh Intensity Lasers (ICUIL)” [4]. In parallel with this, “Asian Intense Laser Network (AILN)” has been started [5]. Also multinational cooperation with major laser laboratory participation has been started in EU as Laserlab Europe [6].

In Japan, many programs related to optical science are carried out under various schemes supported by MEXT, METI (Ministry of Economy, Trade and Industry), MIC (Ministry of Internal Affairs and Communications), and others. However, since optical science is a multidisciplinary field, it has been difficult to lay out a coordinated strategy on optical science. The Science Council of Japan has studied the importance of optical science, and has issued a Statement “Strengthening Optical Science and Technology for Creating New Fields” in 2005 [7].

Optical science has been included in the Third Science and Technology Basic Plan (FY 2006-2010), where construction of X-Ray Free Electron Laser has been started. In order to form stronger bases for research and education in optical science, MEXT has issued an “Interim Report for the Promotion of Photon Science and Technology” in 2007, in which the basic concept of “Photon Frontier Network” has been laid out [8].

3 Framework of Photon Frontier Network

The new program on optical science is composed of two parts [Fig. 1]. One is the “Photon Frontier Network” under MEXT, which is described in this article, and another is the Research Grants under JST (Japan Science and Technology Agency) on “Evolution of Light Generation and Manipulation”. These two programs are coordinated at the Board Meeting attended by the Program Officers responsible for implementing these programs, so that these programs contribute to form the basic structure of optical science and technology in Japan.

Photon Frontier Network, which is a 10-year program, is intended to (1) provide advanced facilities to general users, (2) develop advanced light sources, and (3) foster young researchers. This program is coordinated by Y. Kato as the Program Director (PD) and Y. Sano, S. Yagi and T. Yabuzaki as the Program Officers (PO). The JST Research Grants, which is a 8-year program, is composed of 2 categories “Team type research” and “Individual type research” coordinated by T. Ito and H. Masuhara, respectively.

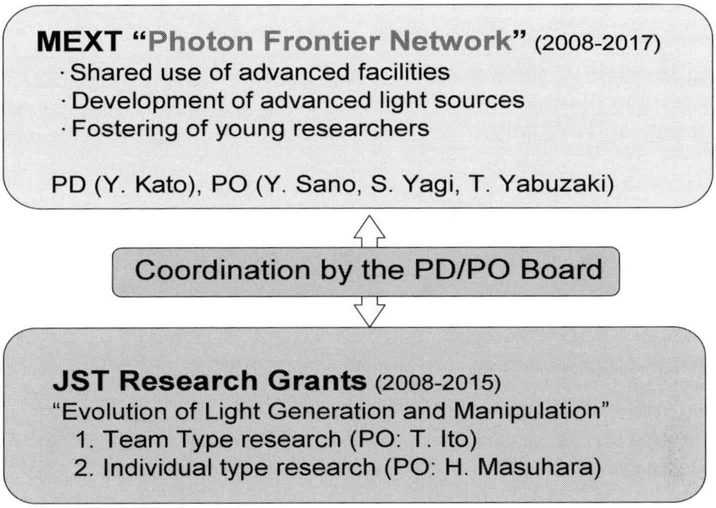

Fig. 1 MEXT "Photon Frontier Network" and JST Research Grants on "Evolution of Light Generation and Manipulation" started in 2008.

Photon Frontier Network is composed of two research consortia; "Advanced Photon Science Alliance" located in Kanto District and "Consortium for Photon Science and Technology (C-PhoST)" located in Kansai District [Fig. 2]. Advanced Photon Science Alliance is directed by M. Gonokami, with its center located at Photon Science Center of the University of Tokyo. C-PhoST is directed by R. Kodama, with its center located at JAEA Center for Photon Science and Technology. These two consortia are composed of several Core Organizations which are described in the following sections. At both of these consortia, it is planned that other organizations will join as Cooperating Institutes after a few years. Therefore it is expected that an extensive research/education network covering most part of Japan is formed in the field of optical science and technology. Collaboration with overseas programs in optical science is also planned to work with the international community.

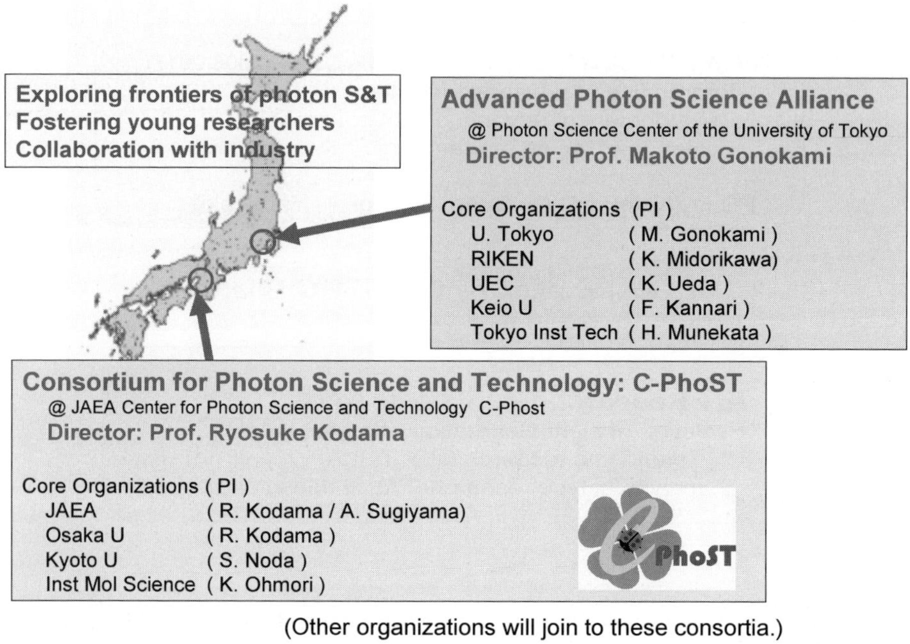

Fig. 2 Photon Frontier Network.

4 Advanced Photon Science Alliance

Advanced Photon Science Alliance is composed of 5 Core Organizations headed by Principal Investigators (written in parentheses): University of Tokyo (M. Gonokami), RIKEN (K. Midorikawa), University of Electro-communications (K. Ueda), Keio University (F. Kannari) and Tokyo Institute of Technology (H. Munekata).

In order to foster young researchers with strong research bases and broad scopes, Advanced Photon Science Alliance is operated under close partnership with the unique education programs which have been tested in these few years. One is CORAL (Consortium on Education and Research on Advanced Laser Science) headed by K. Yamanouchi of University of Tokyo, which is the education program for graduate students with participation of photonics companies as the lecturers to introduce forefront technologies in industrial fields. Another is Elementary Teaching Laboratory and Crisis/Limit Experience Program headed by H. Yoneda of UEC where graduate students plan and operate the education programs by themselves.

Advanced Photon Science Alliance is based on equal partnership of many scientists of various research fields belonging to the Core Organizations for achieving breakthroughs in science and technology which are difficult to implement individually. The research plan comprises several elements [Fig. 3]. The major theme is the collaboration between the frequency standard research [9] and the attosecond science research [10] through control of frequency and phase of the optical waves. Optical frequency standard with 10^{-18} precision generated by an optical lattice atomic clock will be sent through commercial optical fiber network to various users including RIKEN where attosecond and water-window radiation are generated as high order harmonics. Also planned is the development of fiber lasers and ceramic lasers [11] with high power and in UV regions by collaboration between the material science and the laser science researchers.

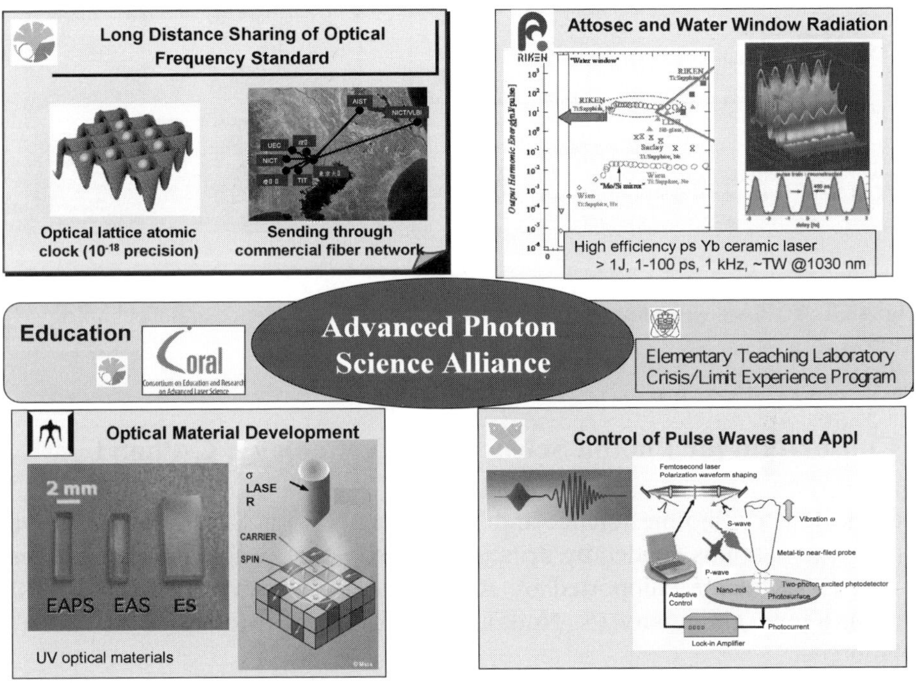

Fig. 3 Advanced Photon Science Alliance

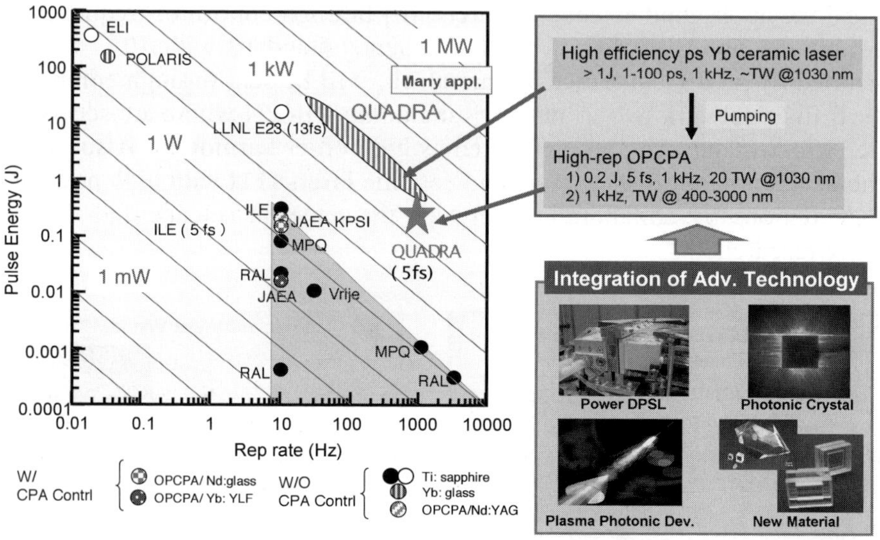

Fig. 4 QUADRA development at Consortium for Photon Science and Technology: C-PhoST

5 Consortium for Photon Science and Technology: C-PhoST

Consortium for Photon Science and Technology (C-PhoST) is composed of 4 Core Organizations headed by Principal Investigators (written in parentheses): JAEA (R. Kodama, supported by A. Sugiyama), Osaka University (R. Kodama), Kyoto University (S. Noda) and Institute for Molecular Science (K. Ohmori).

The major strength of this Consortium is the collaboration between the specialists in two fields: high power lasers and semiconductor lasers. Spatially coherent, high power laser diodes are being developed at Kyoto University as photonic crystal surface emitting laser diodes (PCSEL) [12]. This work will be coordinated with the high power laser research at JAEA Kansai Photon Science Institute, ILE Osaka University and IMS. It is planned that a high quality advanced laser system "QUADRA" will be developed through this collaboration, where LD-pumped high average power Yb ceramic laser (>1 J, 1 kHz repetition rate) will be used to pump OPCPA to generate ultrashort duration, high peak power laser (20 TW, 1 kHz) [Fig. 4]. QUADRA will be used to generate radiations from THz to gamma-ray regions, to develop plasma photonic devices [13] and to apply quantum control [14] to basic science and energy research.

Emphasis will be placed in the education to foster young researchers capable of taking leaderships in scientific projects through participation to the forefront researches taking place at C-PhoST and also participation to international collaboration activities. The education program will be implemented mainly by collaboration between Osaka University and Kyoto University. Also it is planned that "Optical Science Promotion Organization" will be established at Osaka University which will coordinate research and education in optical science carried out independently at various departments.

6 Conclusion

Scientific research is based on the initiative and originality of each scientist. Therefore it is necessary for scientists to keep focusing on the subjects they are interested in. Also completely new ideas may grow from interactions between the scientists working in different fields, where different views and new insights are obtained.

The two consortia of Photon Frontier Network are composed of very active researchers working in different fields (quantum optics, atomic and molecular physics, solid sate physics, plasma physics, x-ray science, laser engineering, etc.), and with different scales, from standard size laboratory scale to larger size facilities. It is hoped that intense interaction among these top scientists will result in new ideas and new fields which are not born from independent researches.

Also strong interaction with industry will enable working on various problems encountered in real worlds. This very active and interactive environment will be very effective to foster young researchers with broad scope and competitiveness.

Acknowledgement

We are indebted to Director S. Ohtake and T. Shimomura of Basic and Generic Research Division, the Research Promotion Bureau, MEXT for organizing the Photon Frontier Network Program. Also thanks are due to H. Hutchinson (Rutherford Appleton Laboratory, UK), W. Sandner (Max Born Institute, Germany), E. Moses and C. P. J. Barty (Lawrence Livermore National Laboratory, USA) and J. Dorfan (Stanford Linear Accelerator Center, USA) for providing useful information during the planning of this program.

References

1. National Research Council: *HARNESSING LIGHT: Optical Science and Engineering in the 21st century*, National Academy Press, USA, 1998
2. *German Agenda: Optical Technology for the 21st Century*, VDI-Technologiezentrum, Germany, 2000
3. OECD Global Science Forum: *Workshop on Compact High-Intensity Short-Pulse Lasers: Future Directions and Applications*, OECD, 2002
4. IUPAP Working Group ICUIL: The International Committee on Ultrahigh Intensity Lasers, http://www.icuil.org/, 2003
5. AILN: Asian Intense Laser Network, http://www.asianlasernet.org/ailn/, 2004
6. LASERLAB-Europe, http://laserlab-europe.net
7. The Science Council of Japan: Statement: *Strengthening Optical Science and Technology for Creating New Fields*, 2005
8. Special Task Force on Promotion of Photon Science and Technology, *Interim Report for the Promotion of Photon Science and Technology*, MEXT, 2007
9. Takamoto, M., Hong, F.L., Higashi, R. and Katori, H.: 'An optical lattice clock', *Nature*, 435, 321-324, 2005
10. Nabekawa, Y., Shimizu, T., Okino, T., Furusawa, K., Hasegawa, H., Yamanouchi, K. and Midorikawa, K.: 'Interferometric autocorrelation of an attosecond pulse train in the single-cycle regime', *Phys. Rev. Lett.*, 97, 153904, 2006
11. Lu, J., Yagi, H., Takaichi, K., Uematsu, T., Bisson, J.F., Feng, Y., Shirakawa, A., Ueda, K.I., Yanagitani, T. and Kaminskii, A.A.: '110 W ceramic $Nd^{3+}:Y_3Al_5O_{12}$ laser', *Appl. Phys. B*, 79, 25-28, 2004
12. Noda, S. et al.: 'Polarization mode control in two-dimensional photonic crystal laser by unit cell structure design', *Science*, 293,1123-1125, 2001
13. Kodama, R., et al.: 'Plasma devices to guide and collimate a high density of MeV electrons ', *Nature*, 432,1005-1008, 2004
14. Yokoyama, K. et al.: 'Optimal control of ultrafast selection', *J. Chem. Phys.* 120, 9446-9449, 2004

Part 2 – Transient Collisional X-Ray Lasers

Grazing Incidence Pumping (GRIP): Single- vs. Double-Pulse Arrangement

K.A. Janulewicz[1,2], C.M. Kim[1], H.T. Kim[1], J. Lee[1,2]

[1]Advanced Photonics Research Institute of Gwangju Institute of Science and Technology, Gwangju, 500-712 Republic of Korea
[2] School of Photon Science and Technology of Gwangju Institute of Science and Technology, Gwangju, 500-712 Republic of Korea

Abstract. There are two well established pumped schemes of X-ray lasers used in applications. The differences in the plasma kinetics and output of an X-ray laser pumped by a single profiled and double pump laser pulses in the grazing incidence pumping geometry are analysed. The differences caused by the structural features of the irradiating pulses were observed in both the experiment and numerical modelling. It was found that the structure of the profiled pulse is not the optimum one. The possible ways of the performance improvement are analysed by numerical modelling.

1 Introduction

There are two well established versions of grazing incidence pumping (GRIP) based either on a single profiled [1,2] or double-laser pulse [3-5] arrangement. Both these variants are sufficiently efficient and stable to be applied in practice [6]. It has already been shown that each of three components of the profiled pulse is very important for the scheme performance. However, profiling of the pump pulse used in the proof-of-principle experiment [1] was limited by the technological constraints put by the laser system itself. The pre-pulse, i.e. the first component of the shaped pulse [1,2] was generated in a regenerative amplifier and the main pulse was delayed by a fixed time gap determined by the round-trip time of the regenerative amplifier cavity. This time gap was equal to 4 ns. Within these experimental limitations some possible variations were tested. It was found that the best results were obtained when the pre-pulse and the main pulse were bridged with a 4 ns long flat-top pedestal originated mostly in the amplified spontaneous emission (ASE) of the power amplifiers [1,2].

It was obvious, that in spite of the stable performance the mentioned constraints could hinder exploiting the full potential of the scheme with a single profiled pump laser pulse. To understand better the physical mechanisms involved in the lasing action in both variants of the scheme [1-5] we conducted numerical simulations using hydrodynamics/atomic physics code

EHYBRID [7,8]. The simulations gave us data about behaviour of the most important plasma parameters as a function of space and time. These parameters include the plasma temperature, density, local gain coefficient and the average ionization stage. Taking into account a low level of the pre-forming laser energy the pre-plasma state is analysed in detail. The experimental conditions described in [1,2] were used in modelling.

2 Analysis of the experiment

The experiment was conducted at UQBF (Ultra-intense Quantum Beam Facility -100 TW titanium:sapphire laser) of Advanced Photonics Research Institute (APRI). Detailed description of the experiment was given in [1,2]. Two arrangements were tested in the experiment. The first, more traditional applied two pump laser pulses of different length to create and excite the active medium. The long pulse of 350 ps length was followed by a delayed in an adjustable way short, picosecond (8 ps length) pulse [2]. The experimental test established optimum delay to be at 300 ps and its changes towards the upper limit were relatively weak but with clear decreasing tendency. The reason for such behaviour is not fully clear. The delays up to 600 ps were tested. The preforming pulse impinged on the target normally to its surface bringing an energy about 300 – 500 mJ focused to a line with a width of 30-40 μm and a length of 7 mm. Strong lasing was observed with an estimated small-signal gain coefficient of 62.4 cm^{-1} and an effective gain-length product of 22.6. The output beam was regular even in the far-field zone and showed strong modulation of the beam intensity distribution caused most likely by the speckle effect.

The single profiled pulse was described and analysed in detail in [1]. It was found that the long ASE pedestal bridging the picosecond (8 ps) pre-pulse and the main pump pulse of the same length is crucial for the scheme work. However, also the pre-pulse emerged from the experiment as an important element of the profiled pulse influencing the spatio-temporal structure of the high-gain area [1]. Unfortunately, the mechanism of pulse shaping described in detail in [1] prevented it from changing the temporal gap between the pre-pulse and the main pulse. It was also found that the bridging pedestal should have a flat-top shape at a level about 5 orders of magnitude lower than the main pulse. This corresponds to a peak power of ~3 MW. Increase in the peak power level by one order of magnitude not only precluded increase in the output signal but nearly halved it. The total pump energy used in this part of the experiment was estimated to be between 1.2 and 1.5 J. The amplification characteristics gave a small-signal gain value equal to 76 cm^{-1} and the effective (including saturation effect) gain-length product of 28.2 [1]. Also in this case the beam shape was very regular in the near-field zone and showed good energy concentration in the far-field-zone. The scheme worked in a stable and reproducible way.

Both schemes demonstrated very good and comparable transverse coherence measured in a Young's double-slit interferometer. The single pulse arrangements seemed to demonstrate a reasonably high coherence even if due to the strong speckle effect the measurement result was very likely underestimated. The double pulse scheme proved the improved coherence level delivering ~3% of fully coherent photons in the output beam in comparison to the usually quoted value of 1% [9].

3 Modelling of the experiment

Modelling was conducted with the hydrodynamics/atomic physics numerical code EHYBRID dedicated to simulations of collisional X-ray lasers [7,8]. The code proved its reliability in some previous simulations of X-ray laser experiments [7,8,10]. However, its using here requires a comment. The pulses used in both modelled arrangements showed relatively low and in the case of the single pulse even very low level of the pre-forming energy. Ablation and plasma creation at these levels are very difficult to direct description due to scarce data for the equation of state. The used approximations introduce some level of uncertainty as far as the plasma dynamics is concerned. Comparison of the simulation results with the effects observed in the experiment suggests that the plasma expansion is noticeably underestimated while the plasma kinetics gives an accurate picture of the excitation process [10].

In modelling we concentrated on behaviour in space and time of four most important parameters of plasma used as the active medium: plasma density, electron temperature, average ionization stage and local gain coefficient. The pump laser beam irradiating target in any of the arrangements considered was focused to a line of 7 mm length and 30-40 μm width, in agreement with the experimental conditions. This gave a linear density of irradiation between 1.6 and 2 J/cm for the main pulse component in both arrangements. The pre-forming pulse in the double-pulse scheme delivered the linear density at a level between 0.4 and 0.7 J/cm. The performing part of the single profiled pulse included of very low energy about 20 mJ and this gave a density of 0.03 J/cm. The wavelength of the pump laser was equal to 800 nm with a length of the main (heating) pulse equal to 8 ps. The temporal gap between the pre- and main pulses was exactly 4 ns (peak-to-peak).

The simulations done for the double-pulse scheme with a delay of 350 ps (very close to the experimental one) gave the spatio-temporal dependences of the medium parameters presented in Fig. 1. The plasma state concluded from the plot in Fig. 1 is not in full agreement with the key criteria formulated by Pert in [7] as pre-conditions of efficient lasing in a collisional Ni-like scheme.

First of all, the plasma has no correct ionization stage as instead of expected or recommended abundance of Cu-like or Ni-like ions our average ionization stage after the pre-forming phase (at the onset of the main heating pulse) is at a level of Z^*=13-14 (Fig. 1d).

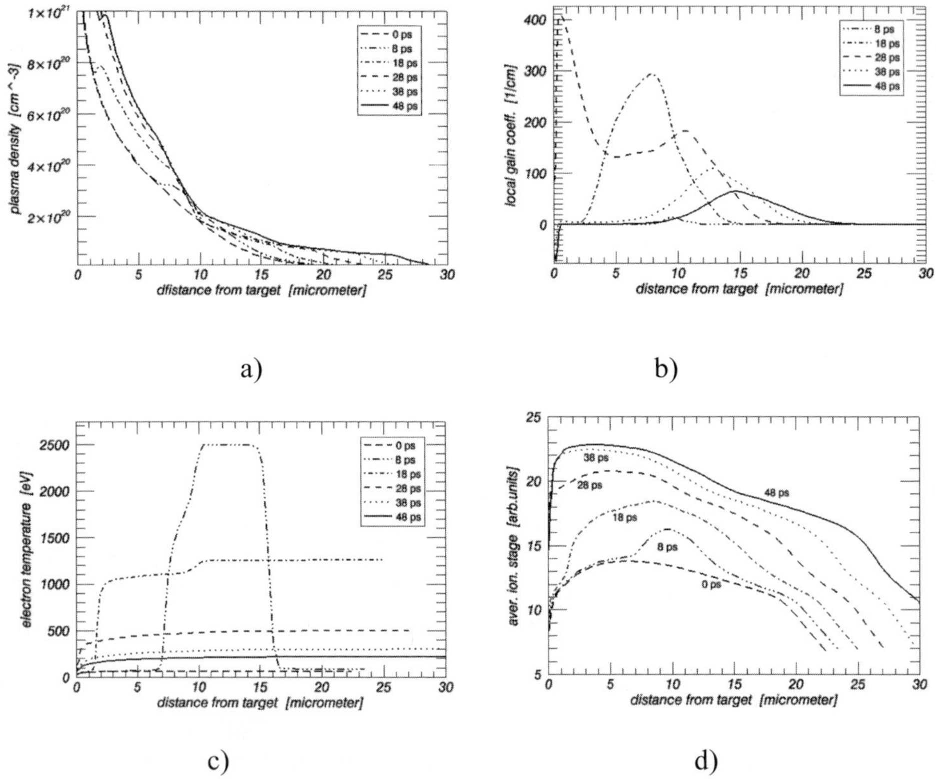

Fig. 1 Results of simulations done with the EHYBRID code for the experimental conditions in double-pulse pump arrangement. Plasma density, local gain coefficient, electron temperature and the average ionization stage are plotted in a), b), c) and d) figures respectively.

The required shallow density profile at a moderate density is achieved until at least 20 ps after the main pulse onset (Fig. 1a). On contrary, the plasma absorption is very efficient and the high temperature area appears in a controlled (pre-determined) distance from the target surface shortly after the onset of the heating picosecond laser pulse, as we can see from Fig. 1c. This area is well localized and the thermal energy deposited there flows in both directions, i.e. plasma plume outwards and towards the target surface. As a consequence, the plasma conditions do not fulfil also the last criterion requiring that at the conclusion of the main pulse, plasma has to be close to the optimum of density, temperature and ionization. For the sake of clarity, the conclusion moment in our modelling corresponds to the time point of 16 ps after the pulse onset.

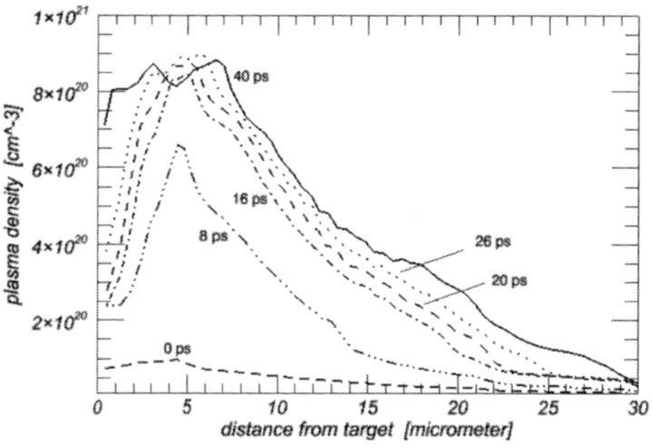

a)

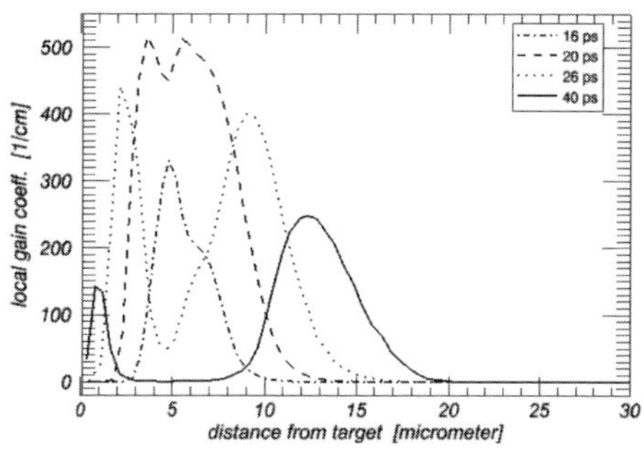

b)

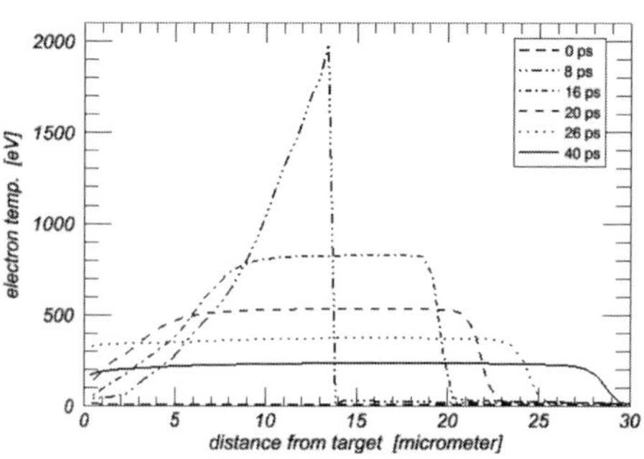

c)

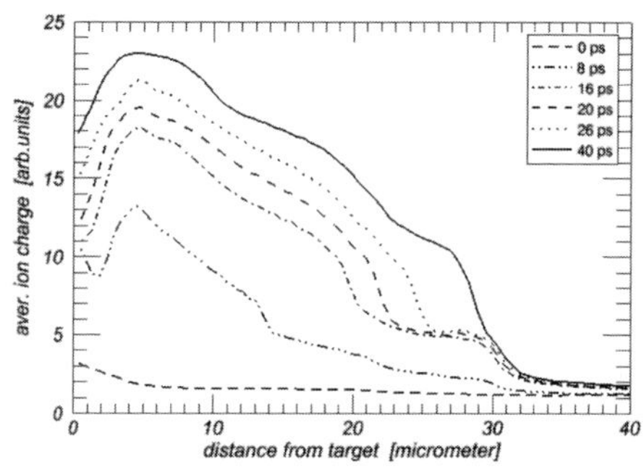

d)

Fig. 2 Plasma conditions and gain simulated with the EHYBRID code for the Ni-like silver X-ray laser pumped by a single profiled laser pulse.

These conditions result in the local gain coefficient drawn in Fig. 1b for different time points as a function of the distance from target. It is clear that high gain can be already obtained at about 18 ps after the main pulse onset. The required ionization stage is only slightly lower than the optimum. Also the density is not too high being about 4×10^{20} cm^{-3}. However, the density gradient is still high and above all the electron temperature at this stage of cooling and redistribution processes exceeds 1000 eV. These conditions used to vary very quickly and already 10 ps later (28 ps after the main pulse onset) a gain area split appears, being usually a signature of overionization. As the electron temperature reduces and spreads over the plasma plume in the region of high density the conditions for extreme high gain emerge in the high density area. The temperature is reduced there by efficient cooling to the bulk material and hence the ionization stage is close to the optimum leading to an extremely short-lived and high gain. Very high density with steep gradients (strong refraction) and the short time scale preclude using gain of this area. On the other end of the density range a clear and stable gain area appears for a long time at the density about the optimum value of 2×10^{20} cm^{-3}.

The single profiled pump laser pulse described in detail in [1] had totally different structure and the quantitative energetic relations between the pre-forming and heating components are also different. The result of simulations is shown in Fig. 2. It is immediately seen that the active medium development has totally different character. Peak gain is much higher than in the case of double-pulse arrangement and positioned closer to the target surface. Moreover, the profile of the plasma density distribution shows no zone with the relaxed density gradients, even if the average steepness seems to be significantly reduced. The origin of that can be deduced from the spatial and temporal changes of the electron temperature. It is seen in Fig. 2c that the spatial profile of the high-temperature region is strongly asymmetric. The energy of the main pulse is deposited in a broad area/volume due to high transparency (low absorption) of the preformed plasma. Significant part of the heating energy is deposited close to the critical surface in the area of higher density and the peak electron temperature is also strongly reduced. This effect changes the dynamics of the plasma during and after the heating process. The main heating pulse works also for increase of the ionization level. The way of the energy deposition pre-determines the plasma expansion (spatial changes of the electron density) in the phase after the main pulse conclusion. The shift of the plasma (electron) density from the target surface is caused by ionization of new external parts of the plume rather than by movement of the ionized plasma mass (compare Fig. 2a and Fig. 2d). There is no movement of the maximum ionization with time towards the high density and this causes so strong split of the gain area with the useful gain within the density close to 4×10^{20} cm^{-3} rather than to 2×10^{20} cm^{-3} considered as the optimum value in [7]. To identify the differences between both arrangements we shall look closer at the pre-forming phases of both schemes with so different input parameters.

3.1 Pre-forming phase

The pre-forming phase of the plasma development is terminated with the onset of the short heating pulse. While this time point could be changed in the case of the double pump pulse arrangement, the experimental conditions of pulse profiling put serious constrains on variations of the pre-pulse structure and its length. The delay between the short pre-pulse and the main pulse cannot be changed. Also relation between the amplitudes of these both components was correlated.

The plasmas generated in both cases were extremely different. The single-pulse-created plasma shows the electron temperature at a level of a few eV, the average ionization stage was equal to 2 across the whole plasma plume. The most dramatic difference between the preformed plasmas can be seen in the plasma electron density. The density is kept below 10^{20} cm^{-3} across the whole width of the plasma column being much broader than that in pre-forming by a long separated pulse. Similar difference is observed in Fig. 4 showing the electron temperature distribution for both pre-forming variants.

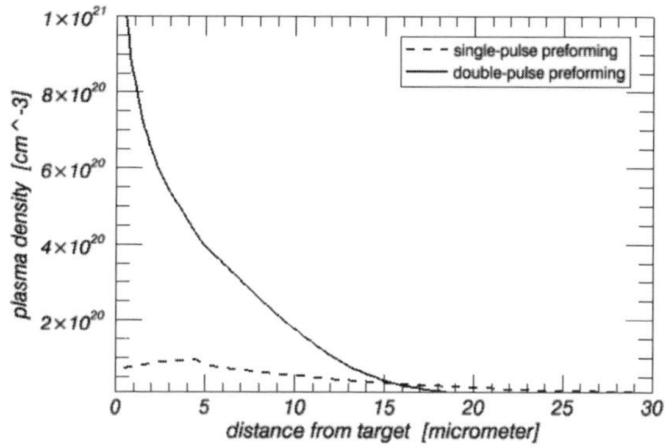

Fig. 3 Plasma density distributions for two different pre-forming arrangements in double- and single-pulse pump schemes.

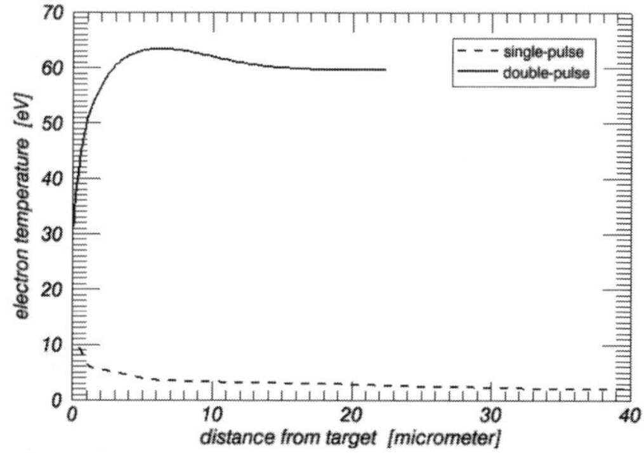

Fig. 4 Spatial distributions of the electron temperatures for both pump schemes at the moment of the main heating pulse onset.

It is seen that the maximum temperatures differ by more than 60 eV and while the pre-plasma created within the double-pulse scheme is most energetic in some distance from the target surface, the rarefied plasma created by the ASE pedestal in the single-pulse arrangement, extends over more than 100 micrometers from the target surface and has the maxima of the considered parameters exactly on the target surface. This explains also the strange behaviour (long tail on the target side) of the plasma temperature observed in the heating phase of the single-pulse and confirms that the plasma was too rarefied in the initial phase causing the deposition to be located in fact at the target surface. The plasma density along the path of the laser radiation was lower than the critical density and its very weak gradients reduced deflection of the pump laser beam. All other effects were the consequence of this one.

3.2 Plasma heating (excitation)

The heating pulse was assumed to be 8 ps long (FWHM) in both pump schemes with controlled rise time. The latter was defined as equal to the pulse width what is rather low value. Longer rise times were tested as well but no dramatic changes were observed. Heating within the double-pulse scheme corresponds well to general rules formulated in [7] while in the single pulse method the place of the energy deposition changes with time. This follows the build up of new absorption zone parallel to increased ablation/ionization with the increase in the delivered energy. As a consequence the maximum density achievable in this process does not exceed the value of 10^{21} cm^{-3} and the same is below the critical density. This maximum is shifted out of the target surface (see Fig. 2a). Moreover, the peak electron temperature is lower by about 20 % relative to the temperature observed at the same moment for the double-pulse arrangement. Created in this way deposition zone determines the sharp maximum of the ionization stage (Fig. 2d) resulting in strong overionization and split of the gain area (Fig. 2b). In spite of this, the local gain coefficient is very high and covers an area at least comparable with that in the double-pulse method.

4 Conclusions

In the experiment the new pump arrangement of collisional Ni-like X-ray laser based on a single profiled pump laser pulse proved to be a robust and reliable scheme. However, taking into account the value of the pump energy and the length of the silver target the output between 1 and 2 μJ, even if with improvement of the coherence level, has to be treated as a moderate one. Attempting to understand the most significant differences we modelled both pump processes. It was found that the single-pulse scheme does not fulfil the general rules treated as a safe path to efficient lasing. The responsibility for that is on the side of low flexibility of the pulse shaping procedure creating too weakly preformed medium. The main pulse has to work also for performing. Then the medium parameters are not optimal. The density and the temperature are a bit too high but in final effect give very high local gain. This positive side will be alleviated by non-relaxed high density area (density gradients) threatening efficient excitation process. In the experiment the gain-length products including saturation effect were exceptionally high for the single-pulse arrangement.

Acknowledgments

The work presented here was supported by the Ministry of Commerce, Industry and Energy of the Republic of Korea through the "Core Technology Development Program.", by the International Office of BMBF _(Grants Nos. KOR 06/014 and KAP 05/003) within the German-Korean Collaboration Program and by the School of Photon Science and Technology of the Gwangju Institute of Science and Technology.

References

1. Kim, H. T. *et al.,* "Demonstration of a saturated Ni-like Ag x-ray laser pumped by a single profiled laser pulse from a 10-Hz Ti:sapphire laser system", *Phys. Rev. A,* 77, 023807, 2008
2. Kim, H. T. *et al.,* " Characteristics of a Ni-like silver x-ray laser pumped by a single profiled laser pulse", *J. Opt. Soc. Am. B*, 25, B76, 2008
3. Rocca, J. J., Wang, Y., Larotonda, M. A., Luther, B. M., Berrill, M., Alessi, D., "Saturated 13.2 nm high-repetition-rate laser in nickel-like cadmium", *Opt. Lett.*, 30, 2581, 2005
4. Wang, Y., Larotonda, M. A., Luther, B. M. , Alessi, D., Berrill, M., Shlyaptsev, V. N., and Rocca, J. J., "Demonstration of high-repetition-rate tabletop soft-x-ray lasers with saturated output at wavelengths down to 13.9 nm and gain down to 10.9 nm", *Phys. Rev. A*, **72**, 053807, 2005
5. Kazamias, S., *et al.*,"Characterisation of a transient collisional Ni-like molybdenum soft-X-ray laser pumped in grazing incidence", *Phys. Rev. A*, 77, 033812, 2008
6. Brewer, C. A. *et al.*, "Single-shot extreme ultraviolet laser imaging of nanostructures with wavelength resolution", *Opt. Lett.*, 33, 518, 2008
7. Pert, G. J., "Optimizing the performance of nickel-like collisionally X-ray lasers", *Phys. Rev. A,* 73, 033809, 2006
8. Holden, P. B., Healy, S. B., Lightbody, M. T. M., Pert, G. J., Plowes, J. A., Kingston, A. E., Robertson, E. , Lewis C. L. S., and Neely, D., "A computational investigation of the neon-like germanium collisionally pumped laser", *J. Phys B*, 27, 341, 1995

9. Dunn, J. *et al.*, "Overview of Tabletop X-ray Laser Development at the Lawrence Livermore National Laboratory", in *X-Ray Lasers 2006, Proceedings of 10th International Conference, August 20–25, Berlin, Germany*, P.V. Nickles, K. A. Janulewicz, Eds., Springer, 2007
10. Janulewicz, K. A., Tümmler, J., Priebe, G., Nickles, P. V., "Plasma-kinetics perspective on a collisional Ni-like x-ray laser pumped by a single profiled laser pulse", *Phys. Rev. A*, 72, 043825~ 2005

An Improved Double-Pulse Non-Normal Incidence Pumping Geometry for Transient Collisionally Excited Soft X-Ray Lasers

Daniel Zimmer[1,2,3], Vincent Bagnoud[1], Boris Ecker[1,2], Udo Eisenbarth[1], Jamil Habib[3], Daniel Hochhaus[1,2], Dasa Javorkova[1], Sophie Kazamias[3], Thomas Kuehl[1,2], David Ros[3], Daniel Ursescu[4], Bernhard Zielbauer[3] and the PHELIX-Team[1]

[1] GSI Helmholtzzentrum für Schwerionenforschung, D-64291 Darmstadt

[2] Johannes Gutenberg-Universität Mainz, D-55099 Mainz

[3] Université Paris-Sud 11, F-91405 Orsay

[4] National Institute for Laser, Plasma and Radiation Physics, RO-77125 Bucharest

Abstract. In this manuscript an optimized pumping geometry for transient collisionally excited soft x-ray lasers is presented. In contrast to the standard scheme, where a nanosecond pre-pulse under normal incidence is assumed to provide the optimal plasma preparation and a picosecond pulse under grazing incidence performs the final heating- and excitation process, two picosecond pulses of equal duration under non-normal incidence are applied. Both pulses are produced in the front-end of the CPA pump laser. They are focused collinearly onto the target with the same spherical mirror under non-normal incidence, optimized for efficient traveling wave excitation for the main-pulse. X-ray lasing was achieved on Ni-like palladium (14.7 nm) at less than 500 mJ total pump energy on the target. This proves that this configuration is at least as favorable as the standard scheme, providing much simpler and more reliable operation. Moreover using this concept x-ray lasing was realized on Ni-like Samarium (7.3 nm).

1 Introduction

On the way to reaching shorter wavelength x-ray lasers a main obstacle was the high pump laser energy required for the preparation of the lasing medium. Over the last years great progress was achieved towards lowering the necessary energy. For this, two pulses in two independent beam-lines with different duration and at different angles of incidence were applied. This is a non-trivial complication, since the delivery of high-energy pulses require the use of large beam diameters. With the concept which is presented here the set-up is simplified tremendously. Since the double pumping pulse scheme uses only one beam-line to generate the x-ray laser gain medium, numerous advantages i.e. improved stability, reproducibilty and not least cost decrease are apparent.

In this article we present an improvement as compared to the pumping geometries of [1] and [2] by the investigation of XRL operation under optimized pumping parameters [3]. Since the pre-pulse is created in the front-end of the laser system and then, collinearly to the main pulse, propagated through the amplifiers and the compressor the complexity of the beam delivery is radically reduced while working with a comparable pumping efficiency. This yields a simplification of the XRL pumping scheme which is especially appealing to large aperture systems. On the one hand, the alignment of the XRL set-up is greatly simplified since only one beamline is used; but on the other hand one loses the ability to control independently the focus parameters and angle of incidence of each beam. Experimentally we observed that lasing of Ni-like palladium at 14.7 nm is achieved at a low pumping energy threshold without the tedious alignment required for the two beamline geometry thus confirming that the advantages of this pumping scheme compensate easily the inability to control the geometric parameters of each beam separately. Moreover the double-pulse non-normal incidence pumping enabled the demonstration of x-ray lasing on Ni-like Samarium at 7.3 nm at our installation.

2 Experiment

In the front-end of the driving CPA laser system the double pumping pulse is generated in a standard Mach-Zehnder type set-up. The stretched pulse is distributed into the two arms with an adjustable ratio via the combination of a wave plate and a polarizing beam splitter. One of the arms incorporates a delay line adjustable between 0 and 3 nanoseconds delay. The two pulses are then amplified through the chain of the CPA system which includes two regenerative Ti:Sapphire amplifiers and two Nd:Glass laser heads. After compression, the two short pulses are sent to the experiment chamber. In the pump configuration for the Pd XRL the two pulses have the same duration, which is similar to [1], but they can be adjusted between around 0.3 and 50 picoseconds by tuning the grating-to-grating distance of the compressor. For the pumping of the Sm XRL an additional adjustable compressor is installed in the pre-pulse arm, which allows us to produce a pre-pulse duration between 100 and 200 picoseconds. The pre-compressor set-up proved to be a good alternative to the similar concept of [4] which is using a stretcher module instead.

In the first experiment we investigated a nickel-like palladium $4d^1S_0 - 4p^1P_1$ transient collisionally excited soft x-ray laser at 14.7 nm by using a two-dimensional high-spatial-resolution diagnostics recording the XRL far-field. The experimental set-up in the target chamber is depicted in Fig. 1: The focusing system of [5] is used which produces a line focus with an intrinsic traveling wave speed of 1.2 c. The beam from the compressor is deflected by a flat mirror onto a spherical mirror with a focal length of 600 mm which is positioned off the normal incidence. The line focus on the Pd slab target was $5.5\,mm \times 50\,\mu m$ FWHM. Both pulses hit the target at the same grazing incidence angle Φ of 29 degrees, a value determined to be the optimal in the classical GRIP scheme for our pump laser. For the pump laser wavelength of 1053 nm the electron density at which the energy is absorbed amounts to $n_{e,abs} \approx 2.3 \times 10^{20}\,cm^{-3}$, following $n_e = n_c \cdot \sin^2\Phi$ with the critical density n_c.

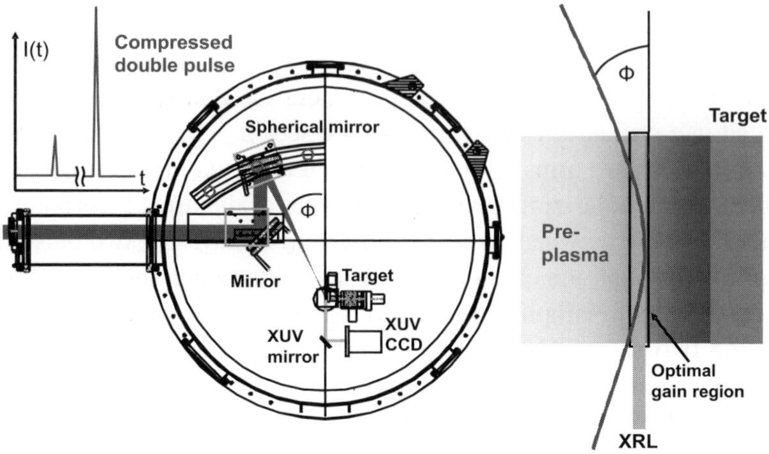

Figure 1. The experimental setup on the left is showing the beamline of the focussing system for the pump laser and the x-ray laser diagnostics. The insert to the right shows the schematic view of the non-normal incidence pumping scheme [3].

The insert in Fig. 1 shows geometrically the absorption of the two pulses on the target. The pre-pulse is hitting directly onto the target, creating an expanding plasma plume, in which the main pulse is refracted. In view of the formation of the pre-plasma, the fact that the pre-pulse has the same short time duration as the main pulse certainly introduces some difference to the typical scheme. It was assumed that thermalization should be reached before the main pulse is impinging onto the plasma, since a pulse separation of 1 ns was used.

In the experiment the intensity ratio between the two pulses was fixed to 1:4 and the level of other pre-pulses created in the chain by ASE and other effects was below 1:1000. The resulting irradiances on the target were $\sim 3.5 \times 10^{12}\, W/cm^2$ for the pre-pulse and $\sim 1.5 \times 10^{13}\, W/cm^2$ for the main pulse respectively. The quality and orientation of the line focus were checked with a microscope imaging its shape on a glass diffuser at the final target position. The XRL diagnostics consisted of a 2-D XUV far-field camera. The image was detected on a 16-bit back-thinned CCD camera after a deflection by a flat multi-layer mirror optimized for 14.7 nm, and a total propagation path of 30 cm. The resolution was limited by the 26 μm pixel size.

In the second experiment the nickel-like samarium $4d^1S_0 - 4p^1P_1$ transient collisionally excited soft x-ray laser at 7.3 nm was either recorded by a XRL far-field camera via a 45 degree XUV mirror optimized for that wavelength or by a grazing incidence reflection on a silver coated glas substrate which guided the XRL beam into a flat-field gold grating spectrometer with 1200 lines/mm. The experimental set-up in the target chamber is shown in Fig. 2: The focusing system produces a line focus with an intrinsic traveling wave speed of 1.5 c by using a combination of a 90 degree off-axis parabolic mirror and an on-axis spherical mirror. The beam from the compressor is deflected by a 30 cm diameter copper parabola with a focal length of 2 m via a flat folding mirror onto the 20 cm diameter spherical mirror with a focal length of 675 mm which is aligned off the normal incidence by 12 degrees. The line focus on the Sm slab target was $8\,mm \times 100\,\mu m$ FWHM. Both pulses hit the target at the same non-normal incidence angle of 50 degrees. The double-pulse delay was varied from 100 ps to 200 ps, values which are described as optimal in [6]. The intensity of the pre-pulse was changed between 5% and 50% of the total energy. The contrast level could not be measured better than 10^{-3} at the time of the experiment. The resulting irradiances on the target were $\sim 2.5 \times 10^{13}\,W/cm^2$ for the pre-pulse and $\sim 5 \times 10^{15}\,W/cm^2$ for the main pulse respectively.

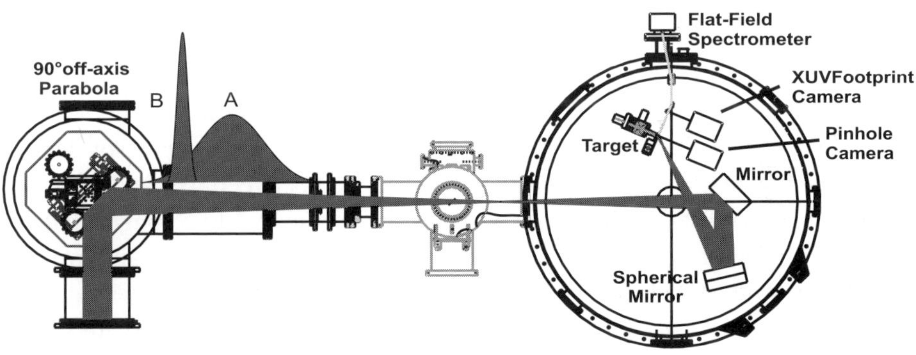

Figure 2. The experimental setup shows the beamline of the focussing system for the pump laser and the x-ray laser diagnostics. The temporal structure of the double-pulse is indicated above the beam.

3 Results

The palladium x-ray laser output is measured by integration of the plasma background corrected counts on the XUV CCD. The XRL intensity was recorded for pumping energies of 600 $\pm 10\%$ mJ. To obtain comparable results the output intensities are normalized to 600 mJ pump energy and used to investigate the influence of pulse delay and pulse duration. The dependence of the XRL output energy on the delay between the pulses is shown in Fig. 3(a). Strong XRL output could be observed up to 1100ps.

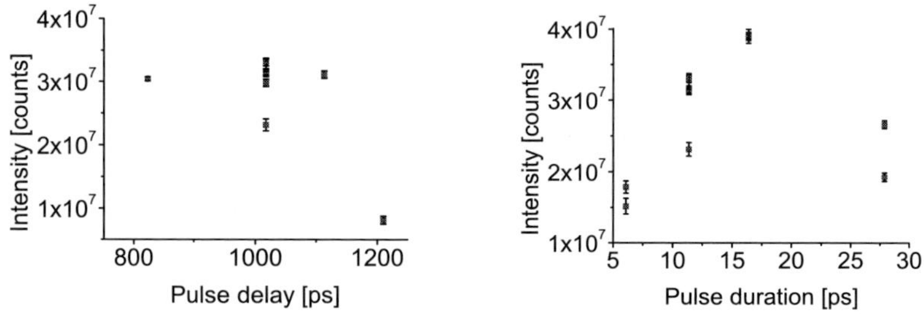

Figure 3. On the left: Dependence of the XRL intensity on the time delay of the pumping pulses with a pulse duration of 11 ps. On the right: Dependence of the XRL intensity on the time duration of the pumping pulses with a pulse delay of 1 ns [3].

The Pd XRL output for different pulse durations is shown in Fig. 3(b). This variation affected both pulses equally, since only the compressor settings were changed. The pulse duration was changed between 6 ps and 28 ps. A maximum is observed around 16 ps, indeed a factor of 4 to 5 longer than the typical value for optimized pumping in the standard scheme [7] and even longer than used in [1] and [2]. This effect could be related to a longer interaction time necessary to enable suitable pre-plasma conditions. In the scheme with two equally short pulses, the duration of the pre-pulse might be problematic in terms of the production of a homogenous plasma plume. This could be the reason for the untypically long optimal pulse duration.

A further improvement of the pumping efficiency might be achieved by using the additional compressor module in the pre-pulse arm of the Mach-Zehnder set-up to adjust the pre-pulse duration separately, as it was done for the pumping of the Sm XRL.

The output of the samarium x-ray laser at 7.3 nm was identified via the spectrum. Figure 4 shows the lineout of the spectrum of the region of interest. The peak of lasing at 7.3 nm is visible, but due to the strong x-ray background of the much hotter plasma the contrast to background is not as high as for the Pd XRL. An explanation could be that the pumping pulse parameters were not optimal for this configuration. Moreover the pump laser contrast of slightly below 1:1000 might have reduced the Sm XRL output as well, as reported by [6].

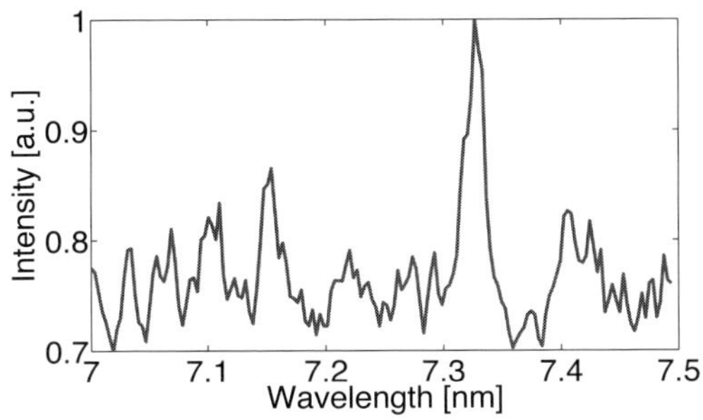

Figure 4. Lineout of the spectrum with the Sm x-ray laser line at 7.3 nm.

4 Conclusion

In conclusion, the improved double-pulse non-normal incidence pumping geometry for transient collisionally excited soft x-ray lasers provides a simple and efficient way to produce XRL output at close to 200 eV photon energy. Creating the double-pulses can be well achieved in a Mach-Zehnder like geometry. Using the single beam-line traveling wave focusing geometry the adjustment of the pump optics is straightforward and stable. In comparison to earlier work using similar double-pulse schemes the pumping energy for a reliable operation of a Pd-XRL is strongly reduced from above 1 J ([1]) to less than 500 mJ. The drastic improvement is attributed mainly to the optimized pumping parameters like GRIP angle, pulse separation, and pulse duration. In this experiment the highest Pd XRL intensity registered on the CCD camera corresponds to $\sim 0.3\,\mu$J of XRL energy at 700 mJ total pump energy. This is comparable with the best reported values e.g. in [7]. Therefore it opens the way to apply the concept to future routine operation at facilities like LASERIX [8]. The possibility for efficient operation at pulse durations above 20 ps can be used for a further reduction of pump laser requirements for applications [9]. The first result of the Sm x-ray laser pumping demonstrates the feasibility of applying the double-pulse pumping scheme for higher pumping pulse energies, which is still necessary for reaching XRL wavelengths close to the water window [10].

The work was supported through the Laserlab Europe Integrate Infrastructure Initiative.

References

[1] Balmer J 2007 SPPhy **115** 37
[2] Kim H T 2008 Phys. Rev. A **77** 023807
[3] Zimmer D 2008 Opt. Express. **16** 10398
[4] Kawachi T 2003 Appl. Opt. **42** 2198
[5] Neumayer P 2004 Appl. Phys. B **78** 957
[6] King R E 2001 Phys. Rev. A **64** 053810
[7] Wang Y 2005 Phys. Rev. A **72** 053807
[8] Kazamias S 2006 J. Phys. IV France **138** 13
[9] Kuehl T 2007 Laser Part Beams **25** 93
[10] Pert G J 2007 Phys. Rev. A **75** 063814

Generation of the Circularly Polarized X-Ray Laser Using the Pulse-Power Magnet

N. Hasegawa[1], T. Kawachi[1], A. Sasaki[1], H. Yamatani[1], A. Iwamae[1], M. Kishimoto[1], M. Tanaka[1], Y. Ochi[1], M. Nishikino[1], Y. Kunieda[1], H. Kawazome[1], K. Nagashima[1] and H. Yoneda[2].

[1]Quantum Beam Science Directorate, Japan Atomic Energy Research Agency, 8-1 Umemidai, Kizugawa, Kyoto, 619-0215, Japan.
[2]Institute for Laser Science, University of Electro-Communications, 1-5-1, Chofugaoka, Chofushi, Tokyo 182, Japan.

Abstract. We proposed the method of generation of the circularly polarized x-ray laser using the Zeeman splitting. External magnetic field of 20 T was applied to the gain medium plasma to separate the degenerated lines of nickel-like molybdenum x-ray laser. The splitting of the x-ray laser line was clearly obtained, and the strength of the magnetic field estimated from the quantity of the x-ray laser line splitting was quite higher compared with that of the external magnetic field. It implies that there might be alternative mechanism for enhancement of the magnetic field in the gain medium plasma.

1 Introduction

The polarized light sources are used for the analysis of the structure of many materials in the wavelength region of the visible and hard x-ray. In particular, the circular dichroism measurement is very useful to analyze the 3-D structure of the materials, such as classification of the optical isomer (chirality) [1]. The circular dichroism measurement is basically absorption spectroscopy, so the use of intense soft x-ray sources such as x-ray lasers (XRL) is desirable to improve the sensitivity of the measurement. In the wavelength region of visible and hard x-ray, transmission optics can be used as phase control devise [2] to obtain circularly polarized light. Whereas in the wavelength of the soft x-ray region, there is no appropriate transmission optics. Recently, the circularly polarized soft x-ray has been generated by as a combination of synchrotron and undulators [3], however still large facility is required. Thus the investigation of alternative compact method to obtain circularly polarized soft x-ray is quite important. In the following, we propose the new approach for the generation of circularly polarized XRL by use of the external magnetic field.

2 Circularly polarized XRL by use of the external magnetic field

We considered the transition of the XRL to generate the circularly polarized XRL. Figure 1 shows the Kastler diagram of the nickel-like XRL line. The XRL is generated between the *4d* ($J = 0$) and the partial *4p* ($J = 1$) levels. There are three magnetic sublevels with m_J = -1, 0, 1 in the lower level, and the XRL has three degenerated lines. The polarization of these lines are left-handed circular (σ'), linear (π) and right-handed circular polarization (σ). The direction of the radiation of the circular components and linear component are parallel and perpendicular to the quantization axis, respectively. If we take the quantization axis to the direction of the external magnetic field, the circularly polarized XRL can be extracted. If the strength of the magnetic field is large enough to separate each polarization component by the Zeeman effect, each circular polarization can be extracted separately after resolving with high-resolution spectrometer. The quantity of the Zeeman shift is proportional to the strength of the magnetic field (see in section 3.1).

The schematic figure of the method of the extraction of the circularly polarized XRL is shown in Fig. 2. In this setup (the direction of the magnetic field is parallel to the axis of longitudinal of x-ray laser medium), the left- and right-handed circularly polarized XRL can be propagated in the XRL medium. The advantage of this method is that the right- and left-handed circularly polarized XRL are obtained at the same time.

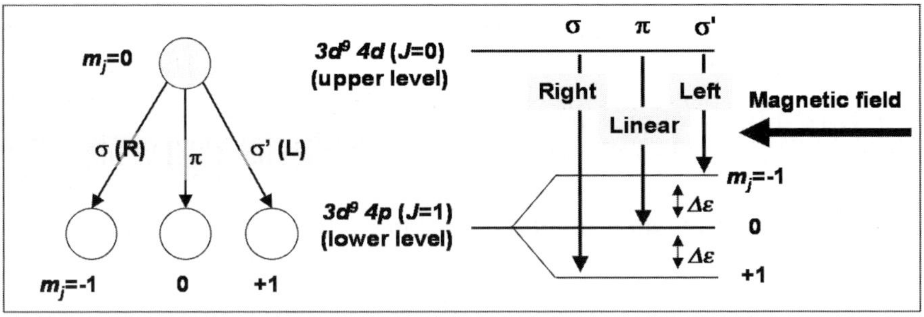

Fig. 1 Kastler diagram of the nickel-like XRL.

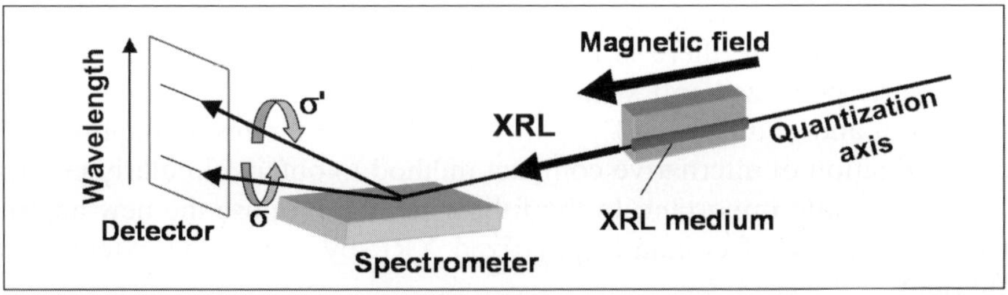

Fig. 2 Schematic diagram of the extraction of the circularly polarized XRL.

3 Experiment of the extraction of the circularly polarized XRL

In this study, we choose the Ni-like Mo XRL ($3d^94p\ ^1P_1 - 3d^94d\ ^1S_0$) as an example. The reasons for using this laser are as follows: the strong amplification has been obtained with the grazing incident pumping scheme (GRIP) [4] in previous work [5], and the wavelength can be covered by high-resolution spectrometer.

There were three steps to extract the circularly polarized XRL. First, we estimated the strength of the magnetic field required for the extraction of the circularly polarized XRL. Second, we made the pulse-power magnet system. Thirdly, we demonstrated the extraction of the circularly polarized XRL.

3.1 The strength of the magnetic field required for the extraction of the circularly polarized XRL

The Zeeman effect of the degenerated lines of highly charged ions such as XRL line can be treated by linear Zeeman effect and the quantity of the split is described as to be $2\Delta\varepsilon = \{J(J+1)\}^{1/2}\mu_B B$ ($= 1.6 \times 10^{-4}$ B eV), where μ_B is the Bohr magneton ($= 5.8 \times 10^{-5}$ eV/T), and B is the strength of the magnetic field. To extract the each circularly polarized XRL component, the Zeeman split value has to be larger than the spectral width of the XRL without the external magnetic field.

Figure 3 shows the spectral profile of the nickel-like molybdenum x-ray laser line taken by the high-resolution spectrometer, HIREFS [6], without the external magnetic field. Each dot and the solid curve show the experimental data point and the fitting curve with the Gaussian profile, respectively. The detail of the experimental setup is described in section 3.3. The spectral width of the XRL ($d\lambda_{XRL}$) was measured to be 0.00205 nm ($d\lambda_{XRL} / \lambda_{XRL} = 1.1 \times 10^{-4}$). Consequently the magnetic field required for the splitting of the circular polarization components was estimated to be 45 T. To obtain this value is not difficult by using pulse-power magnet system [7].

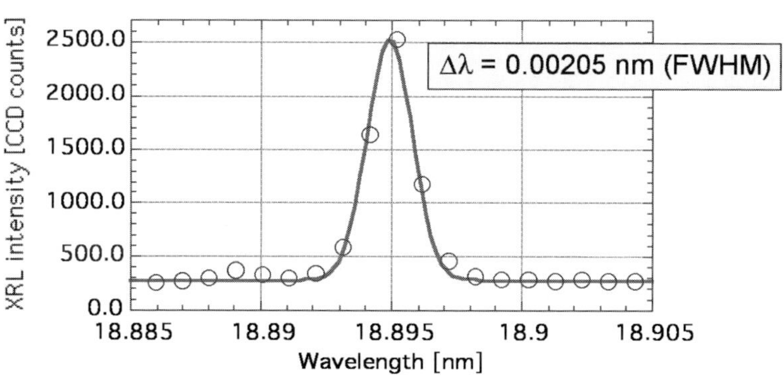

Fig. 3 The spectrum of the Ni-like Mo XRL.

3.2 The design of pulse-power magnet system

Figure 4 shows the scheme of the pulse-power magnet system. This system consists of five parts, DC power supply, capacitor, switching device, transmission line and solenoid coil. The coil, the transmission line and the capacitance of the system were designed to obtain the impedance matching between the coil and the capacitor. The impedance of this system was designed about 1 Ω. The inner diameter, length and the number of turn of the coil were 4 mm, 5 mm and 10, respectively. The electric current of 26 kA could produce 45 T magnetic field at the center of the coil. The width and thickness of the transmission line were 50 mm and less than 1 mm, respectively. The capacitance was 330 nF. As for the switching device, the laser triggered spark gap (LTSG) [8] was used. The characteristics of LTSG, i.e., tolerance for the large current and high voltage, fast switching (< 100 ns) and low jitter (< 10 ns) suited for the pulse power system. Nd:YAG laser at a wavelength of 532 nm was used for the trigger laser, and it was synchronized to the pumping laser of the XRL. The duration and energy of the trigger laser were 7 ns and 10 mJ, respectively. The magnetic field of 45 T could be obtained at the 30 kV charge.

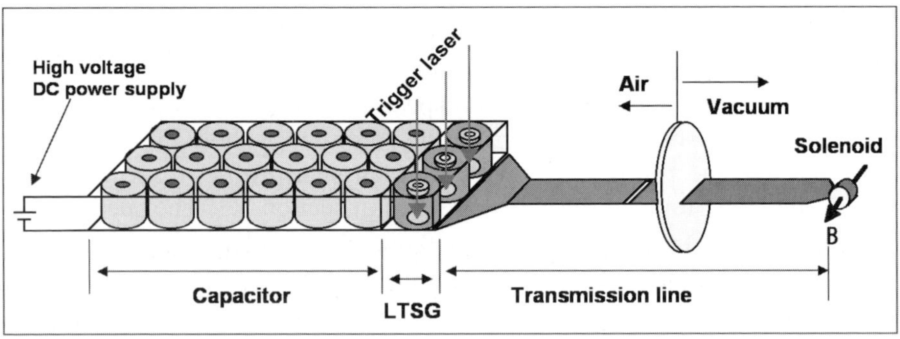

Fig. 4 Pulse-power magnet system.

3.3 Demonstration of the extraction of the circularly polarized XRL

We demonstrated the extraction of the circular polarized Ni-like Mo XRL. The experimental setup is shown in Fig. 5. The target was a thin rod (the cross section is 1 mm × 1mm and length is 30 mm) molybdenum to set at the center of the magnet coil. The peak value of the external magnetic field was estimated to be 20 T in the present experiment estimated from the temporal profile of applied voltage at the transmission line. Nd:glass laser at a wavelength of 1053 nm was weakly focused on the target surface under the grazing incident pumping (GRIP) configuration [4]. The grazing incidence angle was 14 degree, and the focal width and length were 70 μm and 5 mm, respectively. Nd:glass laser light consisted the pre-pulse and the main pulse. The total energy on the target was 12 J, and the energy ratio of these pulses was 1 : 4, respectively. The pulse separation was 2.0 ns, and each duration were 400 ps and 7 ps, respectively. The spectral profile of the XRL was measured by a high-resolution spectrometer, HIREFS made by Hettrick scientific [6]. The entrance slit position was 70 mm from the edge of the plasma. The far field image of the XRL on the slit was relayed to the detector position. The back-illuminated CCD (Princeton, PI-SX:1K) was used for the detector. The inverse linear dispersion of the spectrometer was determined by use of carbon Balmer α line (182.20 Å) and Ni-like Mo XRL line (188.95 Å) [9]. The inverse linear dispersion was 785 mÅ /mm on the CCD surface. This value together with the information of CCD pixel size, slit size (= 3 μm) and magnification of HIREFS (= 3.2) led to the resolution of the total system of 12.7 mÅ ($d\lambda / \lambda = 7 \times 10^{-5}$).

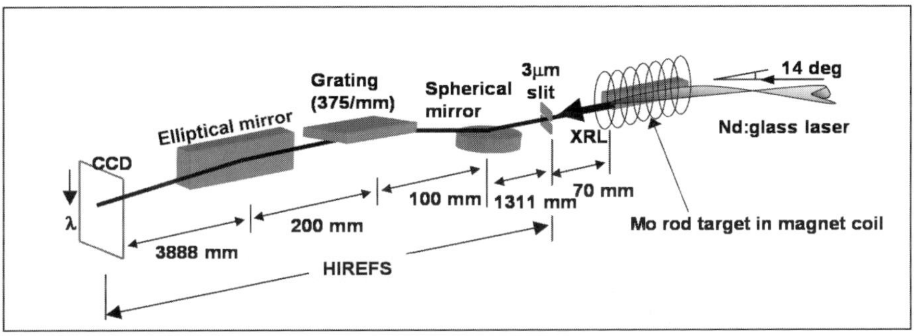

Fig. 5 Experimental setup for the extraction of the circularly polarized XRL.

Figure 6 (i) shows the obtained spectra of the Ni-like Mo XRL without and with the external magnetic field, respectively. As the difference between these spectra was clearly seen, and apparent split of the spectrum was obtained under the presence of the magnetic field.

Fig 6 (ii) shows the spectral profile of the XRL under the magnetic field. The notation (a),..(e) corresponds to the position in Fig. 6 (i). As the refraction angle increases, the separation becomes larger. If the separation of the spectrum was decided due to only the Zeeman effect, the strength of the magnetic field near the target surface was higher than that far from the target surface. The strength of the magnetic field in the gain medium estimated to be (a) 41, (b) 122, (c) 179, (d) 218, (e) 270 T at each position, and these quantities were quite higher compared with the expected external magnetic field. It implies that there might be alternative mechanism for enhancement of the magnetic field in plasma, such as the spatial compression of the magnetic flux due to the shock wave [10] occurred when the main pulse of the pumping laser was incident to the XRL medium. The detail of the mechanism is under consideration now.

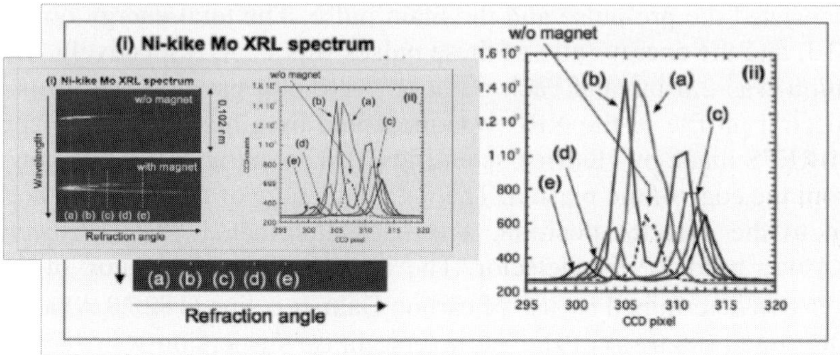

Fig. 6 (i) XRL spectra without and with the external magnetic field. (ii) Cross section of the XRL spectrum.

4 Summary

We proposed the method of the extraction of the circularly polarized XRL using the external magnetic field. From the measurement of the spectral width of the Ni-like Mo XRL (~ 0.002 nm), we estimated that magnetic field of ~ 45 T is required for the complete separation of the circular polarization components. Preliminary experiment was conducted under external magnetic field (~ 20 T), and the separation of the XRL line was obtained.

The estimated strength of the magnetic field from the split value of the spectrum of the XRL was quite higher than the expected value. This result implied there was alternative mechanism for enhancement of the magnetic field in plasma, such as the spatial compression of the magnetic flux due to the shock wave.

Acknowledgements

This work was supported by a Grand-in-Aid for Young Scientists (B) (No. 19740350) from the Japan Society for the Promotion of Science.

We would like to thank Prof. Hiroaki Nishimura, ILE, Osaka University for the use of high-resolution spectrometer HIREFS.

References

1. Fasman G. D. (Eds.) : 'Circular Dichroism and the Conformational Analysis of Biomolecules', Plenum Press, New York, 1996.
2. Hirano K, Okitsu K., Momose A. and Amemiya Y. : 'Recent Research Related to the Polarization and/or Phase of SR', Advances in X-Ray Chemical Analysis, Japan **33**, P. 25, 2001.
3. Tanaka M., Nakagawa K., Agui A., Fujii K., Yokoya A. : 'First Observation of Natural Circular Dichroism for Biomolecules in Soft X-ray Region Studied with a Polarizing Undulator', Phys. Scr. T115 , 873-876 (2005).
4. Keenan R., Dunn J., Shlyaptsev V. N., Smith R. F., Patel P. K. and Price D. F. : 'Efficient pumping schemes for high average brightness collisional x-ray lasers', Proc. SPIE Vol. 5197, 213, 2003.
5. Luther B. M., Wang Y., Larotonda M. A., Alessi D., Berrill M., Marconi M. C., Rocca J. J. and Shlyaptsev V. N. : 'Saturated high-repetition-rate 18.9-nm tabletop laser in nickellike molybdenum', Opt. Lett., **30**, 165, 2005.
6. Hettrick M. C., Underwood J. H., Batson P. J., and Eckart M. J. : 'Resolving power of 35,000 (5 mÅ) in the extreme ultraviolet employing a grazing incidence spectrometer', Appl. Opt. **27**, 200, 1988.
7. Kindo K. : '100 T magnet developed in Osaka', Physica B 294-295, 585, 2001.
8. Miyoshi Y., Hosokawa T. and Shintani M. : 'Breakdown of a Spark Gap in Air Triggered by a Laser', Jpn. J. Appl. Phys. **8**, 620, 1969.
9. Li Y., Nilsen J., Dunn J. and Osterheld A. L. : 'Wavelengths of the Ni-like $4d\ ^1S_0 - 4p\ ^1P_1$ x-ray laser line ', Phys. Rev. A, **58**, R2668, 1998.
10. Tomasel F. G., Shlyaptsev V. N., and Rocca J. J. : 'Enhanced beam characteristics of a discharge-pumped soft-x-ray amplifier by an axial magnetic field', Phys. Rev. A, **54**, 2474, 1996.

Gain Saturation of the Ni-like Antimony Laser at 11.4 nm in Grazing-Incidence Pumping Geometry

C. Imesch, F.Staub and J.E. Balmer

Institute of Applied Physics, University of Bern, Sidlerstrasse 5, 3012 Bern

Abstract. We report on gain-saturated operation of the 4d → 4p, J = 0–1, 11.4-nm soft-x-ray laser line in Ni-like antimony (Sb) at a pump energy of only 2.5 J. The driving laser used was a 1054-nm Nd:glass CPA laser system with a pulse duration of 7 ps (FWHM). The pump beam was focused with a tilted on-axis parabolic mirror in a grazing-incidence (GRIP) pumping configuration at an incidence angle of 45°. A fraction of 2.8% of the pump energy (~70mJ) was used for the prepulse, which was propagated along the same beam line as the main pulse and arrived at the target 4.4 ns before the main pulse.

1 Introduction

The GRIP scheme takes advantage of refraction to increase the path length of the pump beam in the gain region of the plasma, thereby increasing the fraction of pump energy absorbed in that region. Furthermore this scheme provides inherent travelling-wave excitation. The travelling wave is becoming more important when going to shorter wavelengths, because simulations predict that the gain lifetime decreases with decreasing wavelength.

There is a great interest to operate the x-ray laser in the gain-saturated regime. Gain-saturated operation is desired because it provides maximum extraction of the stored energy. However, if the gain lifetime is short compared to the temporal mismatch between the propagating pulse and the excitation, the x-ray laser does not reach the saturation regime, although a conventional saturation measurement would indicate saturation.

In the first part of this paper we describe our recent experiments aimed at achieving saturation of the Ni-like antimony line at 11.4 nm. As the starting point for the setup we used the settings used for the Ni-like tin (Sn) line (11.9 nm) gain saturation measurements. However, we increased the pump pulse duration from ~2 to ~7 ps (FWHM), to account for the fact that a longer length of the line focus was used. In the second part we discuss the gain-saturation measurements we performed to exclude the possibility that the finite gain lifetime causes a fake saturation.

2 Experimental setup

For our experiments we used the 1054-nm Nd:glass CPA laser system at the Institute of Applied Physics of the University of Bern. The laser is capable of delivering up to 15 J to the target, whereby the damage threshold of the compression gratings is the limiting factor of the system. In order to propagate this amount of energy across the gratings without damage, the beam diameter was expanded to 125 mm. In this series of experiments the pulse duration was ~7 ps (FWHM) after compression. The output beam was focused with a tilted parabolic on-axis mirror to a line focus on the target. The beam hit the target at an angle of 45°, which resulted in a mean intrinsic travelling-wave speed of 1.41c. Main pulse and prepulse were both propagated over this parabolic mirror. The FWHM of the length was measured to be ~19.6 mm and the width was ~50 μm (FWHM).

The x-ray laser emission was measured on-axis using a 1200-lp/mm, aberration-corrected, flat-field Hitachi grating. The spectrum was recorded by a phosphor screen imaged to a cooled CCD camera (Photometrics STAR I) having a pixel size of 23 μm.

In a first attempt we used a commercially available 25.5-mm wide thin foil of antimony as the target, which was glued onto a solid brass block. Although weak x-ray laser action at 11.4 nm was observed in this way, the main pulse energy had to be increased to ~10 J. It was obvious that the flatness of the target was a problem; in fact this was already observable by visual inspection. Better targets were subsequently obtained by coating a ~1-μm thick layer of Sb onto diamond-machined, 25-mm wide brass slabs. With these targets, intense x-ray laser output (about 20× more intense than before) was observed with a main pulse energy of ~2.3 J.

3 Experiment Results

Fig. 1 shows the 11.4-nm line intensity as a function of total energy incident on the target, whereby 2.8% of the total energy was assigned to the prepulse, which was 4.4 ns ahead of the main pulse. It is seen that measurable x-ray laser output starts at ~1.5 J of pump energy and then continually increases up to 2.5 – 3 J. At still higher pump energy the increase in x-ray laser output appears to diminish.

Fig. 2 shows the dependence of the laser intensity on the variation of the prepulse delay at fixed pump energy of 2.5 J. As in the case of Sn [6], strongest laser output was observed for delays greater than ~4 ns. This is in some discrepancy to the results of Refs. [3-5,7], where optimum delays below [3,4,5] or slightly above [7] 1 ns were reported. However, at least some of these experiments [3,4,5] had an additional, weak (some tens of mJ) second prepulse well before the main pulse (between 5 and 6.7 ns).

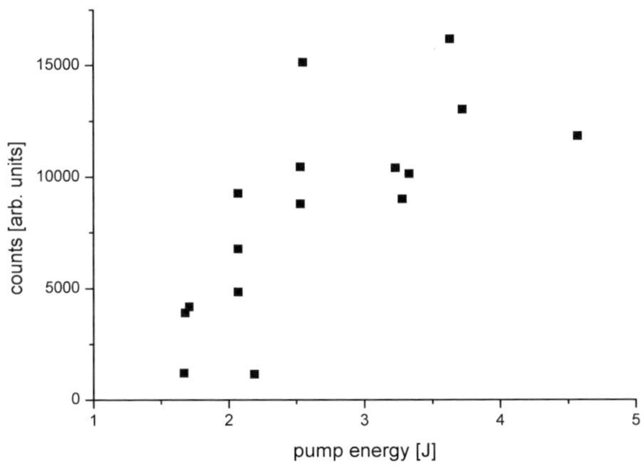

Fig. 1 Variation of the x-ray laser output vs. pump energy. 2.8% of the energy is in the prepulse that arrives 4.4 ns before the main pulse.

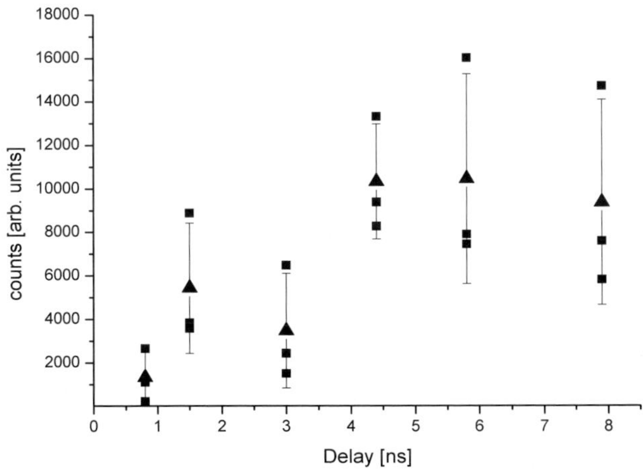

Fig. 2 Variation of the x-ray laser output as a function of the delay between pre- and main pulse. The pump energy is ~2.5 J whereof 2.8% are in the prepulse.

In order to determine the gain coefficient, the variation of the 11.4-nm laser output intensity with target length was measured for an incidence angle of 45°, the maximum possible angle accessible in our current setup. This angle had been found to be optimum in the case of Sn in previous work [6], and since Sn is the neighbour element to Sb in the periodic table, it was assumed to be at least close to optimum for Sb. The results are shown in Fig. 3. The 11.4-nm line intensity is seen to increase exponentially as a function of target length until it rolls off into saturation for target lengths above ~10 mm. At short lengths (4 and 5 mm) the intensity was below the threshold of the detector for some shots. More precisely, at 4 mm 5 of 8 and at 5 mm 6 of 7 shots showed measurable x-ray laser output. Therefore the averaged point at 4 mm in Fig. 3 tends to be overestimated. Nevertheless the gain measurement shows a clear roll-off effect, which is a sign of saturation. Whether or not this is caused by true gain saturation will be discussed in the next two sections.

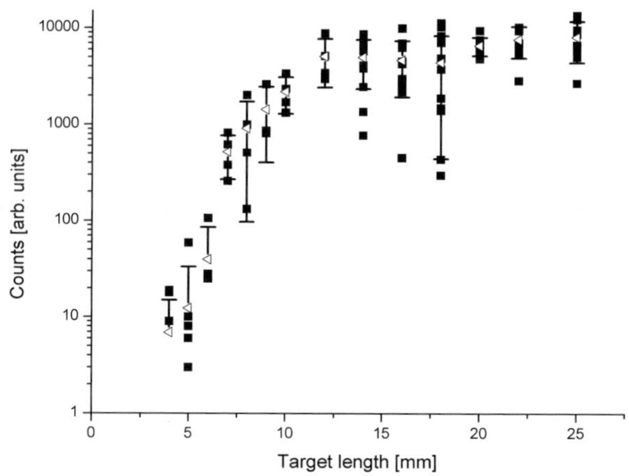

Fig. 3 Measurements of the output energy of the x-ray laser as a function of the target length.

4 Discussion of the roll-off effect

The roll-off of the gain curve cannot only be caused by gain saturation. Other reasons are:

- A spatially non-uniform gain profile caused by a non-uniform irradiance profile [8].
- A mismatch between the velocity of the x-ray pulse in the plasma and the excitation velocity of the gain (traveling-wave velocity mismatch) [9,10].

The first point should not be a problem in this experiment, because the roll-off appears long before the FWHM of the pump focus length is achieved. The second effect is dominating the roll-off if the temporal mismatch between the propagating pulse and the excitation becomes much greater than the gain lifetime.

Normally the roll-off is caused by a mix between saturation and the above described traveling-wave velocity mismatch. Because of that, different sets of the saturation intensity I_s, the 1/e-lifetime of the gain τ, and the small-signal gain coefficient g_0 may result in similar gain curves.

To find curves with similar appearance we need a criterion that connects all these curves. We use the fact that the transition from exponential to linear growth of the gain curve causes a peak in the second derivative of the output curve. We call this point the curvature transition point, described by the length and the fluence (L_T, E_T). By variation of the parameters (I_s, g0, τ) and leaving the transition point fixed we obtain curves with similar roll-off behavior.

Because the appearance of a roll-off is not a sufficient criterion for saturation, we need an additional indicator. Therefore we define the saturation fluence E_s as the fluence of the x-ray laser pulse, for which the intensity peak of the ASE pulse reaches the saturation intensity I_s. If the saturation fluence E_s of the x-ray laser is higher than the transition fluence E_T, the roll-off is not caused by gain saturation.

By assuming an exponential time decay of the gain [10] we get $E_s \approx \tau I_s/\ln(I_s/I0)$ in a first-order approximation. Additional numerical calculation shows that in the case of gain saturation the approximation for the transition point, $E_T \approx \tau Is/3$ is valid.

5 Data interpretation

We now return to the gain measurements of antimony and analyze if there is true saturation or only a roll-off effect caused by the finite lifetime of the small-signal gain coefficient g_0. In this process we assume that there is more than one parameter set that suitably fits the problem. To be sure that the antimony measurements show saturation, the saturation criterion has to be fulfilled by any of these possible parameter sets (I_s, g_0, τ).

To find these sets, we firstly consider only one fitting curve. For this curve we calculate the curvature transition length L_T and the fluence E_T. We then evaluate all the parameter sets which have the same transition point. Because the calculated transition point (L_T, E_T) is only an approximation of the true transition point, every fit curve does not have to go exactly through this approximated curvature transition point. Therefore we define a small area of transition points that give appropriate fitting curves. This area we approximate with a rhombus (see Fig. 4) with the side length $\Delta L_T = 1$ mm and height $\Delta E_T = 300$ (arb. units). The choice of a rhombus instead of a rectangle is made because the latter one allowed extreme (L_T, E_T) pairs (one has an extremely high, the other an extremely low value) which do not fit well the problem.

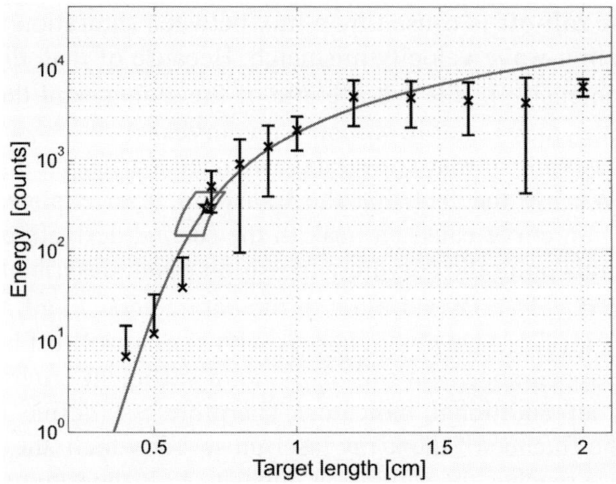

Fig. 4 Fit curve through the mean values of the measurement data. The fit parameters are $g0 = 40$ cm^{-1}, $\tau = 10$ ps and $r = I_0/I_s = 2.3 \cdot 10^{-7}$. The asterisk on the curve marks the transition point. The rhombus around this point identifies the area of transition points which generate suitable fit curves.

A curve having its transition point in this rhombus is shown in Fig. 4. The values for the sets of parameters (I_s, g_0, τ), having the curvature transition point inside this area, are g_0 between 32.5 and 47.5 cm^{-1}, τ between 7.5 and 27.5 ps, and $r=I_0/I_s$ between $1.2 \cdot 10^{-7}$ and $7.0 \cdot 10^{-7}$. For all the points in this area the fluence of the x-ray laser is higher than the saturation fluence E_s. Therefore the roll-off of all these fitting curves is caused by gain saturation. So we can say that the antimony x-ray laser at 11.4 nm is saturated even if we can determine the relevant parameters (I_s, g_0, τ) only within a certain range.

References

1. V.N. Shlyaptsev et al,, *Proc. SPIE*, Vol. 5197, 221-228, 2003
2. R. Keenan et al., *ibid.*, 213-220
3. M.A. Larotonda et al., *IEEE J. of Selected Topics in Quantum Electronics*, Vol.10 No. 6, 1363-1367, 2004
4. Y. Wang et al., *Phys. Rev. A* 72, 053807 1-7, 2005
5. J.J. Rocca et al., *Opt. Lett.* 30, 2581-2583, 2005.
6. M. Grünig, C. Imesch et al., *Optics Communications*, accepted for publication

7. J. Tümmler et al., *Phys. Rev. E,* 72, 037401 1-7, 2005
8. R. Tommasini and E. Fill, *J. Opt Soc. Am B*, Vol. 17 No.10, 1665-1670, 2000
9. R. Tommasini and E. Fill, *Phys. Rev. A*, 62, 034701 1-4, 2000
10. F. Strati and G.J. Tallents, *Phys. Rev. A*, 64, 013807 1-11, 2001

Temporal Coherence and Spectral Line Shape of a GRIP Transient X-Ray Laser

J. Habib[1,2], A. Klisnick[1], O. Guilbaud[1,2], D. Joyeux[3], B. Zielbauer[1,2], S. Kazamias[1,2], D. Ros[1,2], F. de Dortan[1] and M. Pittman[1,2]

[1] LIXAM, Laboratoire d'interaction du rayonnement X avec la matière, UMR 8624, Bat 350, Campus de l'université Paris XI, F-91405 Orsay Cedex France
[2] LASERIX, Centre Laser Université Paris-Sud, Campus de l'université Paris XI, F-91405 Orsay Cedex France
[3] Laboratoire Charles Fabry, Institut d'Optique, Campus de Polytechnique, rd 128, F-91127 Palaiseau cedex, France

Abstract. We have used a variable path-difference, wavefront-division interferometer to measure the temporal coherence of an unseeded Ni-like Mo GRIP X-ray laser. This quantity is inversely related to its spectral linewidth. We have investigated the role of several pump parameters on the inferred spectral width. Along this experiment, particular attention has been paid on the source stability.

1 Introduction

Recently, important progress have been achieved to turn soft x-ray lasers into table top, high repetition rate, fully coherent and intense short wavelength [1] sources. However, some aspects still need to be improved in order for these sources to support the comparison with X-FEL. Among them, the pulse durations still stay in the picosecond range, even in seeded mode [2, 3]. In this last case, the reason for the long pulse duration is related to the narrow spectral width of the x-ray laser amplifier. High Order Harmonic (HOH) seed has a broad spectrum, but its amplification in the x-ray laser plasma leads to a dramatic narrowing of its spectral width and to an important stretching of the pulse in the temporal domain.

In the context of seeded x-ray lasers, soft x-ray amplifiers generated in laser-produced solid target plasmas present some interest in terms of output energy (due to high saturation intensities) but they are also promising because of potentially important linewidth. Ion temperature and collisional broadening are expected to be higher than in OFI plasmas.

In this study, we report on a systematic study of the spectral linewidth of an ASE GRIP Ni-like Mo x-ray laser (λ =18.9nm) for different pumping conditions. The ultimate goal of this work is the maximization of the amplifier bandwidth at a given total pumping energy. Moreover the linewidth value is a critical parameter in the field of seeding x-ray laser because it will determine the coupling efficiency between the HOH pulse and the amplifier. The linewidth is deduced from the measurement of the temporal coherence of the source. The spectral width ($\Delta\lambda/\lambda < 10^{-4}$) is indeed for the moment too small to be resolved with a grating spectrometer. We used a dedicated interferometer [6, 7] to produce interference fringes with the x-ray laser beam. By following the fringe visibility as a function of the path difference between the interfering beams, we measure the temporal coherence. Only few measurements of the temporal coherence of x-ray laser sources have been reported. This kind of experiment requires (i) an excellent interferometer, (ii) a good stability of the source and (iii) a great number of shots (that is, a high repetition rate) to be able to make refined scans in path difference and systematic studies. In this work we benefitted from the 10Hz repetition rate of the LASERIX facility and paid particular attention to the stability of the source.

2 Experimental setup

The soft x-ray laser was generated using the laser front-end of the LASERIX facility which is a 10 Hz, multi-terawatt CPA Ti: Sapphire laser (λ_{IR}=800nm). The output beam (2J, 500ps) is divided into two parts with a beam splitter. The first part is kept uncompressed and is line-focussed on the molybdenum x-ray laser target using a combination of spherical and cylindrical lenses. The target length is 4mm and the line focus dimensions are 6mm x 80µm. A small prepulse can be added to the main pulse travelling on this beam, with variable energy and delay. The energy of the main long pulse is noted E_{ns} The second part of the laser beam is compressed to τ_{ps} =2 ps duration and is focussed on the target using a spherical mirror (4mm x 50µm). The grazing angle of the beam on the target is 20°. Optimal delay between the two pulses was found to be Δt = 400 ps.

Several diagnostics were used to monitor the energy, the source size and the beam uniformity of the x-ray laser. Figure 1 shows examples of the near-field (1a) and far-field (1b) distributions of XRL intensity obtained during the experiment. Near field images were obtained using an XUV aspherical mirror (f=500mm) forming the image of the exit aperture of the source on an XUV CCD camera with a magnification of 13. This device helped us to estimate the source size fluctuations that lead to spatial coherence variations. Far-field images were recorded on another XUV CCD camera placed after a multilayer mirror at a distance of 1.2m from the x-ray laser. With this device, any fluctuations of the beam uniformity and pointing direction can be monitored. Series of near-field and far-field images were acquired before each interferometric measurement.

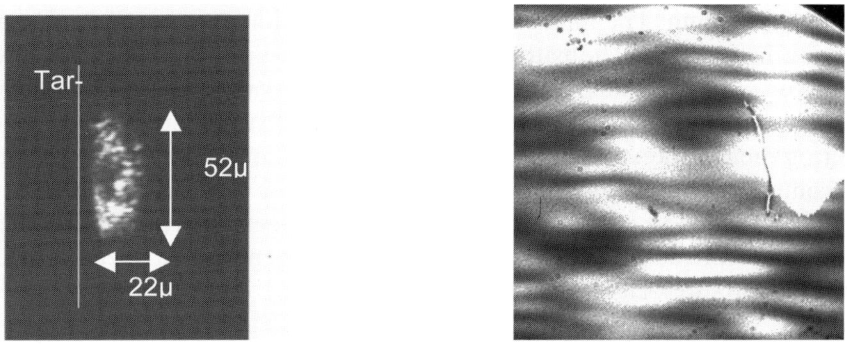

Fig. 1 (a) Near field image of the soft X-ray laser source generated with energies of E ns =500mJ in the long pulse and E ps = 800mJ in the short pulse. Short pulse duration τps is 2ps. The dimension of the exit aperture is 52μm vertical and 22μm horizontal. (b) Far field image of the x-ray laser beam in the same conditions. Field of view is 5 mrad x 5mrad.

The coherence time measurement relies on the determination of interference fringe visibility as a function of the path difference between the two interfering beams [7].

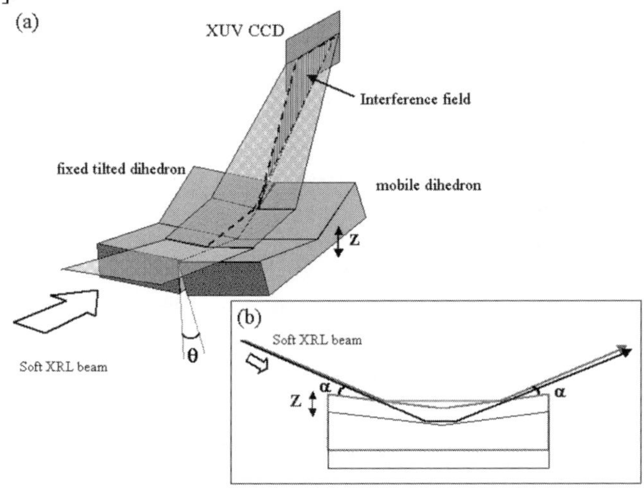

Fig. 2 (a) Fresnel interferometer for temporal coherence measurement. The insert (b) shows the propagation of the two beamlets reflected by the two dihedrons with one vertically shifted.

A schematic of the interferometer we used is presented in Figure 2a. This optical device is a variable path difference wavefront division interferometer developed by the Laboratoire Charles Fabry (Institut d'Optique).

This interferometer is achromatic and has a good transmission on a large range of wavelength. The incident beam is reflected into two beamlets that are interfering in far-field. The path difference between them can be changed by moving vertically one of the dihedron. The dihedral geometry (Fig. 2b) enables this operation without changing the transverse superposition of the interfering beams, keeping the spatial coherence conditions constant. The interferometer is 4m away from the source to ensure a high level of spatial coherence and a good intensity uniformity in the selected part of the beam. A multilayer mirror placed before the interferometer generates a spectral selection around the soft x-ray laser wavelength.

3 Source stability

The coherence time is determined as follows. The fringe visibility is measured as a function of the path difference. The temporal coherence length L_c, defined as the path difference for which the visibility is decreased by 1/e, is determined by fitting the experimental data with an analytical function. The coherence time is then L_c/c.

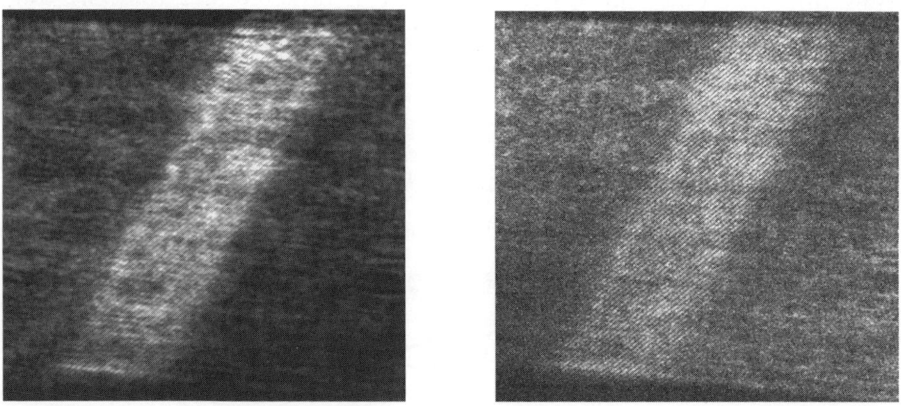

Fig. 3 Interferograms acquired for a path difference of zero (right) and 40000 λ (left).

In our setup, only positive path differences are investigated, as the evolution of the visibility should be symmetric. The zero path difference has been precisely set using a broadband spectrum source and checked with the IR laser strongly attenuated. The path difference between the two beamlets has been varied from 0 to 50000 λ by step of 5000 λ. For each step, more than 10 interferograms have been recorded. Because of the great number of shots required to make one coherence time measurement, the source stability is a crucial point to assess. Figure 3 presents two interferograms obtained for a path difference of zero and 40000 λ.

Different sources of fringe visibility fluctuations may be identified. First, the beam intensity spatial distribution changes from one shot to another. This may lead to a shot to shot fluctuation of the difference of intensity between the two interfering beamlets. Second, a similar effect may arise from x-ray laser beam direction fluctuations. Third, because the x-ray laser behaves like a spatially incoherent source, source size fluctuations will induce spatial coherence and hence visibility fluctuations. All these aspects have been controlled along the experiment. If the source size and pointing stability are remarkably stable, the beam non uniformity is important. However, the characteristic transverse length across which the beam is uniform is large enough to cover the two dihedrons lateral aperture and the field of interference. Finally different measurements of the coherence time for the same reference pumping conditions have been performed before each study. The measurements were found to be reproducible.

4 Experimental results

We have measured the coherence time for different pumping configurations. From the evolution of the fringe visibility it is possible to extract the source spectral profile by fitting the experimental data with analytical functions. A Fourier transform of the fitted function gives the spectral profile and the spectral line width.

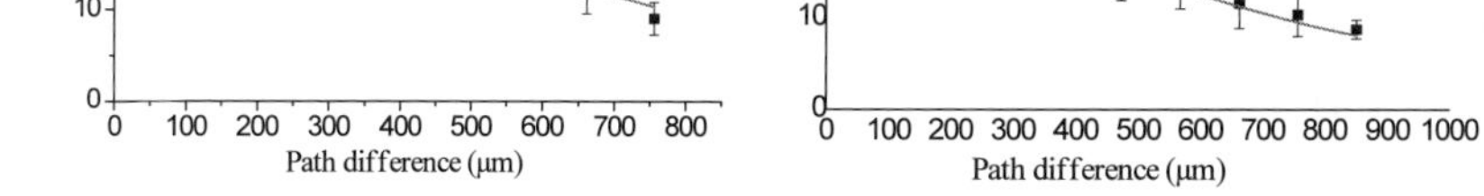

(a) **(b)**

Fig. 4 Effect of the long pulse energy on the temporal coherence. The graphs show the variation of the measured fringe visibility versus path difference. (a) The XUV laser is generated with **E_{ns}=500mJ** energy in the long pulse and a short pulse duration of **2ps**. The coherence is L_c =514±15μm and the spectral linewidth is **Δλ=2.2mÅ**. **b)** For **E_{ns}=300mJ,** the coherence length is L_c =569±22μm and the spectral linewidth is **Δλ=2mÅ**.

We have investigated the effect on linewidth of the long pulse energy E_{ns} , without any prepulse. Figure 4 presents the experimental results for E_{ns} = 500 mJ (4a) and E_{ns}=300 mJ. A small decrease of the linewidth is observed when decreasing the long pulse energy which might be associated to a decrease of the ion temperature in the gain zone. Hydrodynamic simulations of these two situations are underway to check this interpretaion.

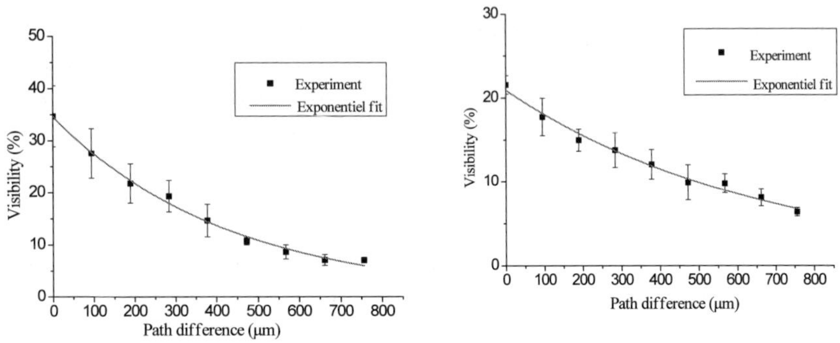

(a) **(b)**

Fig. 5 Effect of the short pulse duration. The graphs represent the variation of the measured fringe visibility versus path difference. The XUV laser is generated with E_{ns} =500mJ energy in the long pulse. a) Short pulse duration is 100fs. The coherence length is L_c =429±15μm and the spectral linewidth is Δλ=2.65mÅ. b) Short pulse duration is 4ps. The coherence length is l_c =671±33μm and the spectral linewidth is Δλ=1.69mÅ.

We have also studied the effect of the short pulse duration τ_{ps} on the spectral width of the x-ray laser. The evolutions of the fringe visibility as a function of the path difference are presented in Figure 5 for two different pulse durations, all other parameters being kept constant (E_{ns}=500 mJ). The corresponding values of the coherence time and spectral linewidth Δλ are summarised in Table 1. It can be seen that a longer pulse duration leads to a larger coherence time and hence to a smaller linewidth. The interpretation of this behaviour is still under discussion. One possibility could be that the decrease of the linewidth could be caused by the decrease of the pump intensity when the short pulse duration is increased.

Table 1 Measured values of the temporal coherence length, the coherence time and the spectral bandwidth for different pumping conditions.

	L_c (μm)	t_c (ps)	Δλ(mÅ)
E_{ns} =**300 mJ**, τ_{ps}= 2ps	569 ± 22	1.89 ± 0.07	2
E_{ns} =500 mJ, τ_{ps}=**2ps**	514 ± 15	1.71 ± 0.05	2.2
E_{ns} =500 mJ, τ_{ps}=**100fs**	429 ± 15	1.43 ± 0.05	2.65
E_{ns} =500 mJ, τ_{ps}= **4ps**	671 ± 33	2.23 ± 0.01	1.69

5 Conclusion

We have described in this paper the first systematic study of the spectral line width of an x-ray laser as a function of the pumping parameters. This work has been made possible by the high repetition rate of the source and by its stability. We have investigated the effect of the long pulse energy and of the short pulse duration. The induced linewidth variations are small but significant. Further data from other set of pumping conditions are still under analysis. This obtained results need to be interpreted using hydrodynamic simulations associated to a post-processor describing the evolution of the spectral profile during amplification in the plasma. Besides, simulations will help us to assess if the values and trends observed can be extrapolated to seeded operation.

References:

[1] Zimmer D. et al., this conference
[2] Rocca J. et al., this conference
[3] Guilbaud O. et al., this conference
[4] Celliers P. et al., 1995 Opt. Lett. 20 1907
[5] Smith R. et al., 2003 Opt. Lett. 28 1
[6] Klisnick A. et al., 2006 J.Q.S.R.T. 99, 370
[7] Guilbaud O. et al., 2006 Eur. Phys. J D 40, 125

Part 3 – High Repetition Rate X-Ray Lasers

High Coherence Injection-Seeded Table-Top Soft X-Ray Lasers at Wavelengths Down to 13.2 nm

J. J. Rocca[1], Y. Wang[1], F. Pedaci[1], B. Luther[1], M. Berrill[1], D. Alessi[1], E. Granados[1], M. Man Shakya[2], S. Gilbertson[2], Z. Chang[2]

[1] National Science Foundation Center for Extreme Ultraviolet Science and Technology Colorado State University, Fort Collins, CO 80523
[2] Kansas State University

Abstract. We discuss the demonstration of soft x-ray lasers with essentially full spatial and temporal coherence at wavelengths down to 13.2 nm by high harmonic seeding of soft x-ray plasma amplifiers created by irradiation of solid targets. These lasers are demonstrated to produce the shortest pulses obtained to date from plasma-based soft x-ray lasers. A pulse duration of 1.13 ± 0.47 ps was measured for a seeded Ne-like Ti plasma amplifier operating at 32.6 nm using an ultrafast streak camera.

1 Introduction

Gain-saturated high repetition rate table-top soft x-ray lasers (SXL) have been demonstrated at wavelengths as low as 13.2 nm in laser created plasmas [1, 2]. Recently a free electron laser based on a large electron accelerator was also demonstrated to operate in this region [3]. However, since in both cases the beam grows from spontaneous emission noise, its temporal coherence is limited by the emission of random uncorrelated phase wavetrains. The seeding of SXL amplifiers with high harmonic (HH) pulses [4-7] is capable of generating intense soft x-ray pulses with extremely high spatial coherence, low divergence, short pulsewidth, and defined polarization. Furthermore, the very narrow linewidth of the plasma amplifier can increase the temporal coherence of the seed pulse.

A first experiment demonstrated the amplification of HH pulses in a neon-like Ga plasma pumped by 600 J optical laser pulses [8]. However, the seed pulse was only by a factor of three. More recently, an optical field ionization SXL amplifier produced by femtosecond laser excitation of a Kr gas cell was seeded with the 25th harmonic of a Ti:Sapphire laser to generate saturated amplification in the 32.8 nm laser line of Ni-like Kr [4,9]. Our group demonstrated the saturated amplification of HH seed pulses in the 32.6 nm line of Ne-like Ti in a significantly denser transient collisional SXL plasma amplifier created by heating a solid titanium target [5]. Also recently, a Ne-like Mn SXL media was reported to amplify a HH seed from 4.7 pJ to 3 nJ [10]. The seeding of this type of higher density laser-heated solid target SXL amplifiers, that have an increased saturation intensity and broader laser linewidth, has the potential to lead to phase coherent lasers with higher intensities and shorter pulsewidths.

Herein we summarize results of the demonstration and characterization of injection-seeded lasers at wavelengths down to 13.2 nm. Highly coherent gain-saturated soft x-ray laser pulses were produced in dense laser-created plasmas, by amplifying high-harmonic seed pulses in the 18.9 nm, 13.9 nm and 13.2 nm transitions of nickel-like Mo, Ag, and Cd ions respectively [5-7]. These results, obtained using a table-top Ti:Sapphire pump laser extend to significantly shorter wavelengths and shorter pulsewidths the ability to generate bright phase coherent soft x-ray laser beams, offer new scientific opportunities for applications in small laboratory environments. Ultrafast streak-camera measurements of the pulsewidth of these injection-seeded solid-target soft x-ray plasmas amplifiers confirm they have sufficient bandwidth to generate pulses of ~ 1 ps duration. A pulse duration of 1.13 ± 0.47 ps was measured for a seeded Ne-like Ti plasma amplifier operating at 32.6 nm. This is the shortest pulse duration reported to date from a table-top soft x-ray laser amplifier. The result agrees with model simulations which suggest that intense femtosecond soft x-ray laser pulses could be obtained by injection-seeding a tailored plasma amplifier in which the gain is confined to a high density region.

2 Demonstration of injection seeded soft x-ray lasers at wavelengths below 20 nm

The results were obtained by injection seeding with HH pulses collisionally excited SXL amplifiers operating in the $4d^1S_0 \rightarrow 4p^1P_1$ transitions of nickel-like Mo, Ag, and Cd at wavelengths of 18.9 nm, 13.9 nm and 13.2 nm respectively [6,7]. The experiments were conducted using a 5 Hz repetition rate table-top pump laser using the experimental set-up schematically shown in Fig. 1. The HH pulses from the Ti:sapphire laser were injected into SXL amplifiers created by transiently heating a pre-created plasma from solid targets with intense optical laser pulses of ~ 0.9 J energy and 6.7 ps duration impinging at a grazing incidence angle of 23 degrees. Pulses produced by a single 815 nm wavelength table-top chirped pulse amplification Ti:Sapphire laser system consisting of three stages of amplification were used for both, generating the harmonic seed pulses and to pump the SXL amplifiers. The amplifiers consist of plasmas created by irradiating polished slabs up to 4 mm in length.

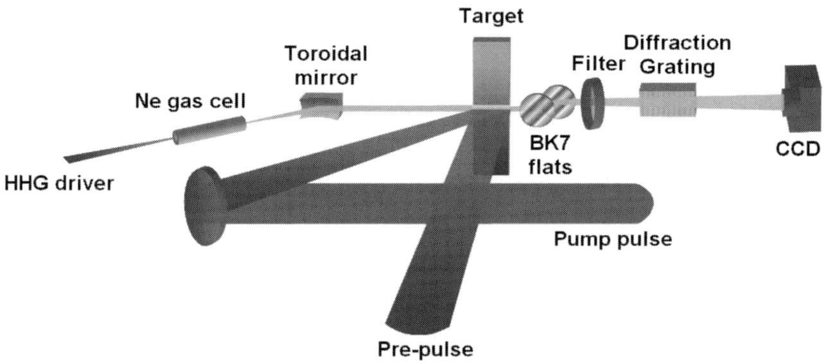

Fig. 1 Experimental setup of injection-seeded soft x-ray laser. Either a gas cell of a gas jet was used to generate the HH seed pulses.

The grazing incidence pumping geometry takes advantage of pump beam refraction to increase its energy deposition in the plasma region with optimum density for amplification [1,2,11,12]. The 20 mJ laser pulses used to drive the high harmonic generation were compressed to about 50 fs using a separate grating compressor, and focused on the input of either a Ne gas cell (for the Ni-like Mo and Ag experiments) or Ne gas jet (for the Ni-like Cd experiments). The high harmonic output of the gas cell was relay imaged onto a ~ 100 μm diameter spot at the input of the plasma amplifier using a gold-coated toroidal mirror designed to operate at a grazing incidence angle of 10 degrees. The wavelength of the selected harmonic order was made to overlap with that of the laser line tuning a thin etalon introduced in the first of the three multi-pass amplifiers of the Ti:sapphire laser system. Two BK7 windows positioned at grazing incidence and sets of either two or three thin film filters (0.3 μm and 0.5 μm thick aluminium for Mo, and 0.3 μm thick zirconium for Ag) were used to attenuate the straight light from the 815 nm beam used to generate the high harmonics. The output of the soft x-ray amplifier was dispersed with a variable space diffraction grating (1200 lines/mm) and was detected by a CCD.

The spectra in Fig. 2 illustrates the dramatic improvement in laser beam divergence, from about 10 mrad to 0.7 mrad, obtained by seeding a 2.5 mm long 18.9 nm nickel-like Mo amplifier with pulses from the 43th harmonic of Ti:sapphire. The energy of the injected seed pulse is amplified more than 400 times (Fig. 2d). The variation of the seed intensity and pulseshape as it propagates along the amplifier is initially governed by a dramatic narrowing of its bandwidth which is not supported by the much narrower amplifier linewidth, and later, by saturation of the gain. Similar results were obtained seeding the isoelectronic transition in nickel-like Ag at 13.9 nm with the 59th harmonic (Fig. 3). The most intense amplified pulses were measured to reach an energy of ~ 75 nJ. The fluence of these pulses significantly exceeds the computed saturation fluence of the 18.9 nm laser line of nickel-like Mo at these plasma conditions, $\sim 1.8\times10^{-3}$ J cm^{-2}. Since the Ni-like Ag amplifier was operated at a slightly smaller gain, saturation occurs later within the amplifier, at a length of ~ 3 mm (Fig. 3d). Strong amplification is observed over a narrow range of time delays , ~ 1-1.5 ps between the peak of the short pump pulse and the arrival of the seed pulse (inserts of Figs 2d). The experiments were successfully repeated for Ni-like Cd to produce a seeded beam at 13.2 nm.

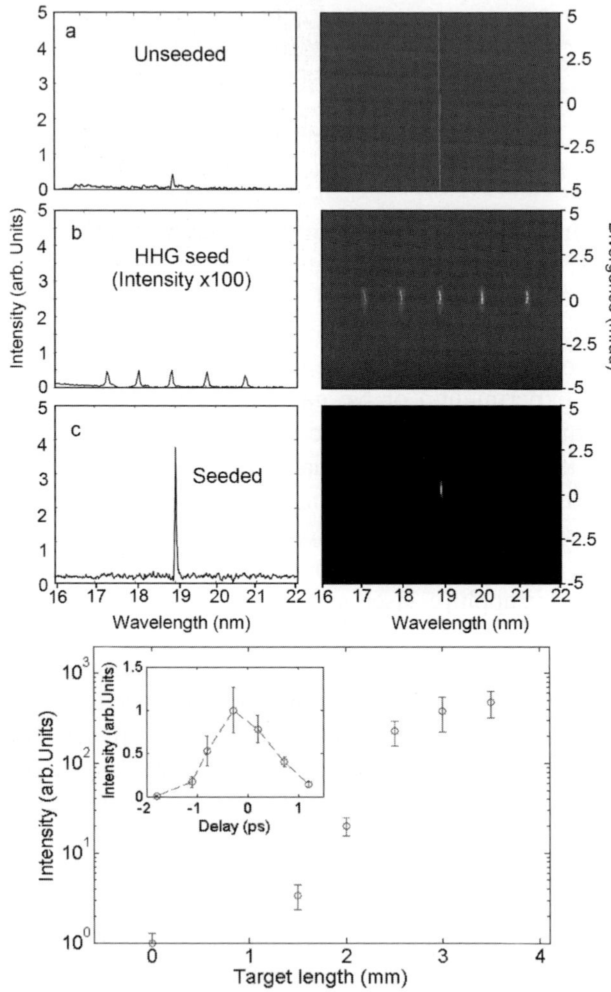

Fig. 2 Injection-seeded 18.9 nm Ni-like Mo laser characteristics. Spectra of a) unseeded soft x-ray laser, b) HH seed, c) seeded soft x-ray laser. The bottom fig. shows the intensity variation of the amplified seed as a function of amplifier length and time delay between peak of heating laser pulse and arrival of the seed pulse.

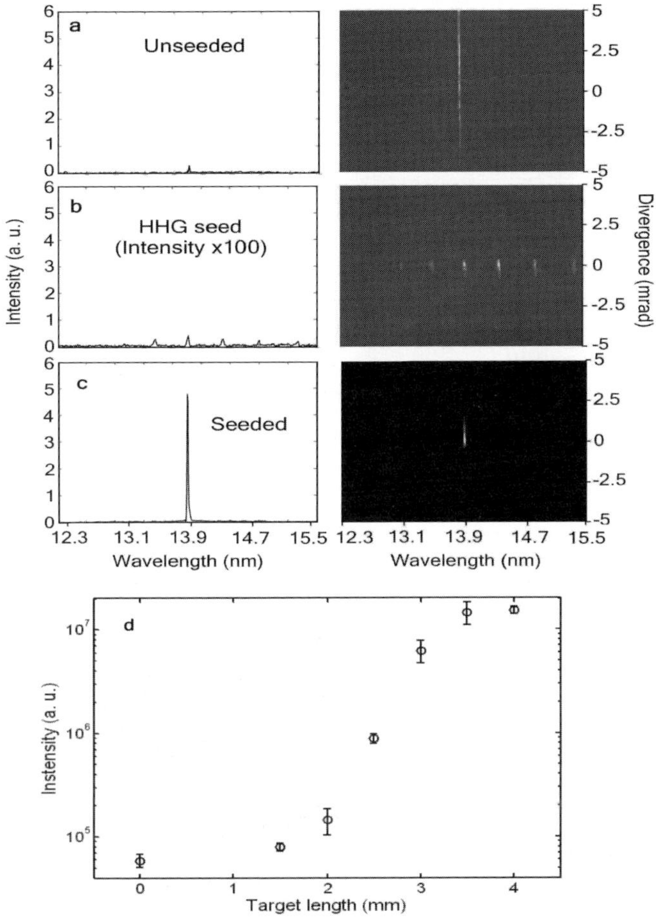

Fig. 3 Characteristics of the 13.9 nm seeded Ni-like Ag laser. a-c spectra illustrating the beam divergence and intensity of a) unseeded laser, b) HH seed, c) seeded soft x-ray laser. d) Variation of the amplified seed pulse energy as a function of plasma column length

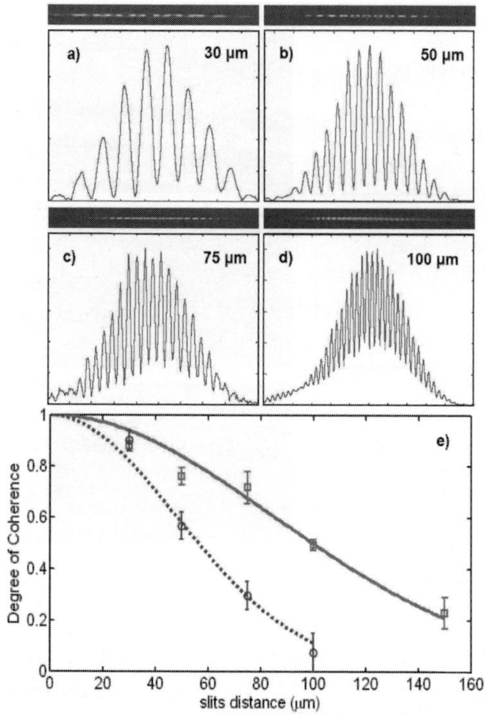

Fig. 4 a-d Two slit single shot interferograms and intensity lineout for the seeded Ni-like Mo 18.9 nm laser beam for different slit separations, and corresponding intensity lineouts; e) Degree of coherence $|\mu 12(\Delta x)|$ as a function of slit separation for the injection-seeded (solid) and unseeded (dashed) laser. The lines are Gaussian profiles for Rc=50 μm and Rc= 84 μm. The plots in e) represent a dramatic improvement of the spatial coherence by seeding, because the beam diameter of the seeded beam at the location of the measurement (80-100 μm) is an order of magnitude smaller than that of the unseeded beam (~ 1 mm).

3 Spatial coherence measurements

The dramatic improvement of the spatial coherence obtained by injection seeding was determined in a Young's double slit interference experiment comparing the fringe visibility as a function of slit separation for the unseeded and seeded lasers. Fig. 4 a-d show single shot interferograms and their profile lineouts for the seeded 18.9 nm nickel-like Mo laser. The interferogram corresponding to the 30 μm slit pair displays full fringe visibility. On the right hand side the visibility is degraded by uneven illumination of the slits. Fig. 4e shows the variation of the maximum measured fringe visibility as a function of slit separation for both the unseeded and the seeded nickel-like Mo lasers.

To characterize the spatial coherence length we use here the coherence radius R_c, within which the fringe visibility is larger than 0.61. Assuming a Gaussian profile, R_c of the unseeded laser is determined to be about 50 μm, only a very small fraction of the ~ 1 mm beam diameter at this location. This shows that only a very small part of the unseeded laser pulse energy is contained within the region where the beam is spatially coherent. In contrast, for the seeded laser R_c=84 μm, a value practically equal to the 80-100 μm beam diameter, indicative of a very high degree of spatial coherence thorough practically the entire beam. The degree of spatial coherence of the injection-seeded 13.9 nm nickel-like Ag and 13.2 nm Ni-like Cd lasers was measured to be practically identical to that of the seeded 18.9 nm Mo laser. This extraordinarily high degree of spatial coherence is accompanied by very high temporal coherence. The very narrow bandwidth of the amplifier ($\Delta\lambda/\lambda < 1 \times 10^{-4}$,) greatly limits any frequency chirp in the HH seed, effectively enabling the generation of a pulse with very high temporal coherence.

4 Pulsewidth measurements

We have also conducted measurement of the pulse duration of phase-coherent soft x-ray laser pulses from an injection-seeded plasma amplifier created irradiating a solid target [13]. Injection-seeding creates a fundamentally new regime for the generation of short soft x-ray laser pulses in which the pulse duration is independent of the gain duration. We used an ultrafast streak camera [14] to make single-shot measurements of soft x-ray pulses generated by seeding the 32.6 nm line of Ne-like Ti with the 25th harmonic of a Ti:Sapphire laser. The measured pulse duration of 1.13 ± 0.47 ps is to our knowledge the shortest soft x-ray pulse duration demonstrated to date from a plasma amplifier. This pulsewidth compares with a pulsewidth of ~ 5 ps obtained if the same plasma amplifier is operated in the unseeded mode [15].

The results are in good agreement with hydrodynamic/atomic physics model computations. The simulations suggest soft x-ray laser pulse a few hundred femtosecond in duration should be achieved by seeding an amplifier in which gain takes place in a higher density region where collisional broadening increases the amplifier bandwidth. For example, computations suggest gain-saturated laser pulses of ~ 290 fs duration could be obtained seeding a 3.5 mm long λ = 10.9 nm Ni-like Te plasma amplifier with an electron density of ~ 2×10^{21} cm^{-3} [13].

5 Conclusions

We have demonstrated phase coherent table-top SXLs with low divergence at wavelengths as short as 13.2 nm. A pulse duration of ~ 1 ps was measured for an injection-seeded 32.6 nm Ne-like Ti laser. It should be possible to obtain x-ray laser pulse a few hundred femtosecond in duration seeding higher density plasma amplifiers. These bright sources open the opportunity to realize application experiments requiring intense phase-coherent SXL light on a table-top.

Work supported by the Engineering Research Centers Program of the National Science Foundation under NSF Award Number EEC-0310717

References

1. Y. Wang *et al.*, "Demonstration of high-repetition-rate tabletop soft-x-ray lasers with saturated output at wavelengths down to 13.9 nm and gain down to 10.9 nm". Physical. Review. A **72**, 053807 (2005).
2. J.J. Rocca et al., "Saturated 13.2 nm high-repetition-rate laser in nickel-like cadmium", Optics Letters **30**, 2581, (2005).
I. R. Al'miev *et al,* "Operation of a free-electron laser from the extreme ultraviolet to the water window"., Phys. Rev. Lett. **99**, 123902 (2007).
3. Ph. Zeitoun *et al.*, "A high-intensity highly coherent soft X-ray femtosecond laser seeded by a high harmonic beam", Nature, **431**, 426 (2004).
4. Y. Wang *et al.*, "High-Brightness Injection-Seeded Soft-X-Ray-Laser Amplifier Using a Solid Target," Phys. Rev. Lett. **97**, 123901 (2006).
5. Y. Wang *et al.*, , "Phase coherent table top soft x-ray lasers at 18.9 nm and 13.9 nm by saturated amplification of high harmonic pulses in Nickel-like ions", Nature Photonics. **2**, 94-98 (2008).
6. F. Pedaci *et al* , "Highly coherent injection-seeded 13.2 nm table-top soft x-ray laser,", Opt. Lett., **33**, 491 (2008).
7. T. Ditmire, *et al.* "Amplification of XUV harmonic radiation in a gallium amplifier". *Phys. Rev. A* **51**, R4337-R4340, (1995).
8. T. Mocek, *et al.* "Absolute time-resolved X-ray laser gain measuremen"t. Phys. Rev. Lett. **95,** 1 73902, (2005).
9. N. Hasegawa *et al.*, "Direct measurement of the temporal profile of the amplification gain of the transient collisional excitation neonlike manganese x-ray laser medium", Phys. Rev. A, 76, 043805, (2007)
10. R. Keenan *et al.*, "High-Repetition-Rate Grazing-Incidence Pumped X-Ray Laser Operating at 18.9 nm", Phys. Rev. Lett. **94**, 103901 (2005).
11. B. M. Luther et al., "Saturated high-repetition-rate 18.9-nm tabletop laser in nickellike molybdenum", Opt. Lett. **30**, 165-167, (2005).
12. Y. Wang et al., "Measurement of 1 Picosecond Soft X-Ray Laser Pulses from an Injection-Seeded Plasma Amplifier," Phys. Rev. Lett, (submitted).

13. M. M. Shakya, Z. Chang, "Achieving 280fs resolution with a streak camera by reducing the deflection dispersion", Applied Physics Letters **87**, 041103 (2005).
14. M.A. Larotonda *et al.*, "Pulse duration measurements of grazing incidence pumped high repetition rate Ni-like Ag and Cd transient soft x-ray lasers," Optics Letters **31**, 3043 (2006).

Characterization of a Seeded Optical-Field Ionized Collisional Soft X-Ray Laser

J.P. Goddet[1], S. Sebban[1], O. Guilbaud[4], J. Gautier[1], Ph. Zeitoun[1], C. Valentin[1], F. Tissandier[1], T. Marchenko[1], G. Lambert[1], J. Nejdl[1], B. Cros[2], G Maynard[2], B.Robillard[2], S. Kazamias[4], K. Cassou[4], A. Klisnick[4], D. Ros[4], J. Benredjem[4], T. Mocek[3], M. Kozlová[3] and K. Jakubczak[3]

[1] Laboratoire d'Optique Appliquée, chemin de la hunière, 91128 Palaiseau
[2] LPGP, Université Paris-Sud, 91405 Orsay, France
[3] Institute of Physics, Department of X-Ray Lasers, Prague, Czech Republic
[4] LIXAM, "X-ray Lasers and Applications group", Université Paris-Sud, 91405 Orsay, France

Abstract. By seeding an optical-field-ionized population inverted plasma amplifier with the 25th harmonic of an infrared laser we have produced a compact, diffraction- limited and Fourier-limited laser beam in the soft x-ray spectral range. This laser beam is emitted within a cone of 0.7 mrad at a repetition rate of 10 Hz at a central wavelength of 32.8 nm, The beam exhibits a regular Gaussian spatial profile, and wavefront distortions smaller than $\lambda/17$. The measured coherence time of 5.5 ps is equal to the duration of the lifetime of the amplifying plasma which shows that this source has reached the Fourier limit.

1 Introduction

The high scientific activity related to the development of ultrafast, coherent soft x-ray sources is motivated by the large number of novel applications such as high-resolution microscopy, lithography, interferometry and holography that can be realized with such sources. Recently, fourth generation synchrotron sources have been designed to provide highly coherent, bright, soft x-ray beams, able to open up new opportunities in science (1). In parallel, dramatic advances in ultra-short pulse laser technology have made it possible to generate compact, high repetition rate, coherent soft x-ray sources. The most advanced sources are based on high-order harmonic generation (HHG) and plasma soft x-ray lasers (SXRL). HHG are produced by focussing a high intensity femtosecond laser pulse into a gas medium. Odd harmonics of the fundamental laser frequency (i.e. 3ω, 5ω, …) are produced in a directional, collimated beam with photon energy that can be extended up to 1 keV (2). HHG driven by current optical lasers exhibit high spatial coherence (3) with an average energy of a few hundreds of nanojoules at 30 nm (4-5). In contrast to HHG, SXRL provide much higher output energy per pulse and narrower linewidth (6).

The first laboratory SXRL at 20 nm was demonstrated nearly 20 years ago (7), and required a few kilojoules of laser energy for pumping. However, recent spectacular progress in this field has led to the development of a large variety of compact SXRL systems providing high energy, monochromatic coherent radiation down to 10 nm. The amplifying medium is a hot, dense, highly charged plasma column generated by the interaction of an intense infrared laser pulse with a solid or gas target (8-9) or by a high density capillary discharge in a rare gas (10).

Until now, in all the SXRL schemes operating at saturation, population inversion between the levels of the lasing ion is induced by electron collisional excitation, leading to high gain value at short wavelength. The short lifetime of the gain and the absence of high reflectivity soft x-ray optics, make the use of this SXRL amplifier in a complete laser cavity impossible. For this reason, SXRL emission generally results from the single-pass amplification of spontaneous emission. As a consequence, SXRL radiation is characterized by rather poor optical qualities in terms of beam profile homogeneity, spatial coherence, source size regularity and wavefront distortion, which prevents the use of SXRL for applications that require a highly intense, coherent, soft x-ray photon flux in a sub-micron spot size.

To fully explore the potential of plasma-based SXRL sources, improving their spatial beam quality is a crucial bottleneck to be overcome. The solution we have chosen to explore consists in seeding a soft x-ray amplifier with a HHG beam (11). This approach is a direct analogy of the "oscillator-amplifier" concept commonly used for infrared laser system, applied to the soft x-ray range. Here the HHG beam plays the role of an oscillator which is injected and amplified while propagating through a population-inverted SXRL plasma column amplifier. This emerging scheme (12-13) offers a prospect for compact SXRL chain combining both the high energy extracted from the SXRL amplifier as well as the high optical quality of the HHG seed.

Here we report an experimental demonstration of a compact, diffraction- and Fourier-transform-limited laser beam in the soft x-ray range. The beam was emitted by 32.8 nm Optical Field Ionised (OFI) SXRL amplifier seeded by a HHG beam. Thanks to the spatial filtering of the seed beam by the plasma amplifier, the 32.8 nm laser beam exhibits a regular Gaussian intensity profile with a divergence of 0.7 mrad. The wavefront distortions were measured to be smaller than $\lambda/17$ (1.9 nm) which demonstrates that this 32.8 nm laser beam is diffraction-limited. The temporal characterization shows that the measured coherence time is equal to the duration of the gain life time, i.e. 5.5 ps, which is the upper limit of the duration of 32.8 nm radiation. This consequently proves that this 32.8 nm seeded laser is Fourier-limited in the spectro-temporal domain. The laser line exhibits a Gaussian spectral profile with a full-width-at-half-maximum (FWHM) of only 3.61 mÅ.

2 Experimental set up

The experimental set-up is schematically illustrated in Figure 1. The experiment was performed using a 10 Hz, multi-terawatt Ti:sapphire laser system providing two independent 34 fs laser beams at a central wavelength of 815 nm. A first laser beam, containing about 10 mJ, was used to generate the seed HHG beam inside a 7 mm long gas cell filled with 30 mbar of Ar. We paid special attention to the 25th harmonic of the infrared laser which can be closely matched to the wavelength of the lasing transition of the SXRL amplifier. A grazing incidence toroidal mirror was implemented to image the output of the HHG source with a magnification of 1.5 at the entrance of the amplifier cell. The HHG seed spot was measured to be astigmatic with dimensions of about 50×100 μm (at 1/e2). A second beam delivers ~600 mJ on target and was used to create the OFI amplifying plasma column which drives the 3d94d (1S0)→ 3d94p (1P1) transition of the Kr8+ ion at 32.8 nm (14). The pump beam was circularly polarized and focused by a 1 m focal length spherical mirror to a spot diameter of 38 μm (at 1/e2) inside a 7.5 mm long gas cell filled with Kr.

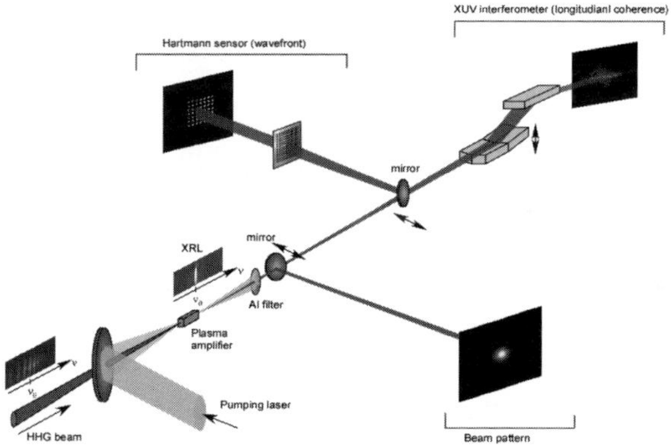

Fig. 1 Schematic description of the experimental arrangement.

Under these conditions, the amplifier exhibits remarkable properties, well suited for the amplification of a seed pulse. The amplifier has a circular aperture and the gain value does not significantly vary along and across the gain column (15). The transverse density gradient is weak enough to prevent the seed beam from serious refraction (i.e. deterioration) during amplification. Previous simulations and experimental observations indicated that the amplifier has rather sharp edges which is an important criterium to ensure the efficient selection of the amplified seed radiation, i.e. spatial filtering of the injected seed beam (16).

3 Experimental results

First, the 32.8 nm emission from the amplifier was optimized by varying its length, gas density, laser beam polarization, and focussing position. The most effective configuration was obtained for a pressure of 30 mbar, a cell length of 7.5 mm, and circular polarization. Second, the spatial and temporal overlapping between the HHG and SXRL beams was adjusted to produce the largest amplification of the seed radiation. The amplification started at a time delay between the SXRL plasma creation and HHG injection of 2 ps, reached a maximum (×180) at 3 ps, and lasted up to about 8 ps. This direct measurement of the time evolution of the gain of a SXRL amplifier (17) indicates that the actual duration of the amplified seeded pulse does not exceed 6 ps. Then, using several online diagnostics schematically shown in Figure 1, we characterized the intensity distribution, the wavefront, and the transverse and longitudinal coherence of the amplified 32.8 nm laser beam.

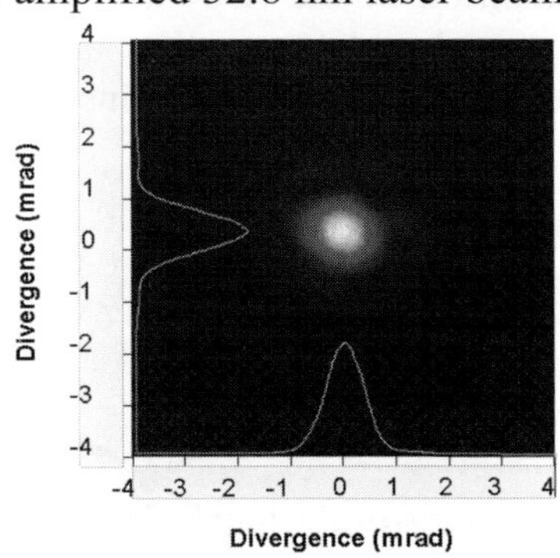

Fig. 2 Spatial profile of the seeded 32.8 nm laser beam.

The spatial profile of the seeded 32.8 nm laser was measured by placing a removable 45°, soft X-ray mirror made of Mo/B4C/Si tri-layers in the beam path, which redirected the beam towards a cooled, thin, back-illuminated charge-coupled device (CCD) camera. A 300-nm-thick aluminium filter was used to block the infrared beams. Figure 2 shows a typical profile of the generated seeded laser beam. The intense, monochromatic radiation is emitted in the direction of the HHG seed, being confined within a 0.7 mrad cone (FWHM). The spatial beam distribution is nearly perfectly Gaussian. Note that such a high beam quality has never been observed for other plasma-based SXRLs. In general, SXRLs exhibit highly contrasted intensity modulations (or speckles) forming a complex and irregular pattern (18), which is a direct consequence of low spatial coherence combined with the high temporal coherence of the amplified spontaneous emission (ASE) radiation (19). Here we take advantage of the fact that the HHG seed is partially coherent to avoid intensity modulations in the beam pattern. In addition, the circular aperture of the gas amplifier permitted to tailor the intensity distribution of the 32.8 nm radiation, resulting in the Gaussian beam profile.

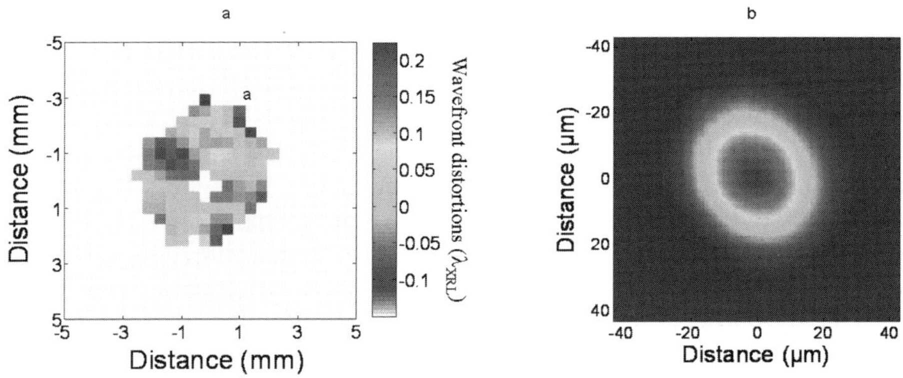

Fig. 3 (a) Measured wavefront and (b) reconstructed source size of the seeded 32.8 laser beam.

A crucial parameter to evaluate laser beam quality is the wavefront. The wavefront of the seeded 32.8 nm laser was measured using a soft x-ray Hartmann sensor with an accuracy better than λ/20 at 32.8 nm. In the Hartmann wavefront analysis a beam passes through a hole array and is projected onto a CCD camera that detects the beamlet sampled by each hole. The position of each individual spot centroid is measured and compared with a reference position. This enables the wavefront's local slope to be measured at a large number of points within the beam from which the wavefront can be reconstructed (20). Earlier investigations showed that neon-like Ar capillary discharge-driven SXRL at 46.9 nm exhibits wavefront distortion of 3×λSXRL, which was the only measurement performed for plasma-based SXRL source (21) so far. Figure 3(a) shows the reconstruction of the wavefront of the 32.8 nm seeded laser. The amplitude of the wavefront defects never exceeds 0.058×λSXRL (root-mean-square) which corresponds to beam wavefront distortions less than λ/17 (1.9 nm). According to the Marechal criterion, which states that a system is regarded as well corrected if the wavefront distortion does not exceed λ/14 (22), the generated seeded laser beam is clearly diffraction-limited. Note that the wavefront of the HHG seed beam was about λ/3 due to the severe astigmatism introduced by the toroidal mirror. The dramatic improvement of the wavefront of the 32.8 nm laser is a direct consequence of the spatial filtering by the amplifier. This measurement also made possible the reconstruction of the intensity distribution of the 32.8 nm laser beam at the exit of the amplifying plasma column which results from the convolution of the wavefront shape with the intensity distribution of the beam. As shown in Figure 3(b), the source shape is Gaussian with a diameter of 50 μm (at 1/e2) which is in reasonable agreement which previous simulations predicting that the diameter of the amplifying zone of the plasma is about 60 μm (16).

Due to the intrinsically narrow ion line width and the gain narrowing effect, SXRLs represent the most monochromatic sources in this spectral range. The spectral shape and line width of the 32.8 nm seeded laser has been inferred from the experimental measurement of the longitudinal coherence (19) by means of a wavefront division interferometer (Fig. 1). The SXRL beam was directed to the dihedron pair with a grazing incidence angle. After reflection on the dihedrons the incident beam was separated in two parts that converge and finally overlap in the far field, yielding interference fringes. As one of the dihedrons can be accurately translated in the vertical direction, a controlled delay (i.e. path difference) between the two interfering beams can be introduced. By following the decrease in fringe visibility with increasing delay, it is possible to reconstruct the spectral profile of the radiation with a resolution power better than 2×10^5.

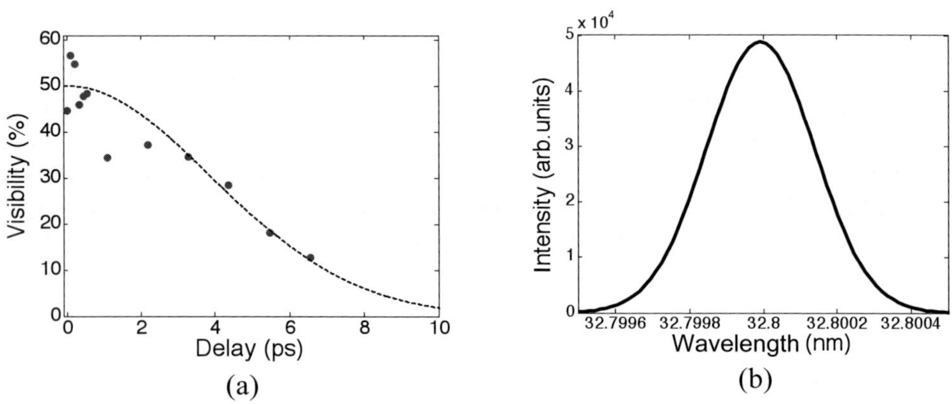

Fig. 4 (a) Variation of the measured fringe visibility as a function of the path difference. A decreasing Gaussian function was used to fit the data. (b) Reconstructed profile of the 32.8 nm laser line.

Figure 4(a) shows the measured fringe visibility as a function of the path difference. The experimental data were fitted with a Gaussian decreasing function from which the coherence time of 5.5 ps, defined as the path difference that decreases the maximum visibility by a factor 1/e, was inferred. The fact that this coherence time matches the measured life time of the SXRL gain medium, clearly demonstrates that the 32.8 nm seeded laser is longitudinally fully coherent. Assuming a Gaussian spectral profile, we have inferred a SXRL line width of Δν=1.1 x 1011 Hz, corresponding to Δλ = 3.61 mÅ. Simulations carried out using the COFIXE code predict an ion temperature of Ti=6 eV (23). Under these conditions the Doppler-broadening of the line is equivalent to 6.8 mÅ. Other calculations performed with the PPP code (24) showed a negligible effect of Stark broadening. The homogeneous broadening due to electron collisions has been calculated within the atomic model described in (16), leading to a value of 5 mÅ at 30 mbar. The effect of the amplification on the line shape has been calculated using the model described by J. Koch et al. (25), and assuming a uniform plasma column. For the amplification factor equal to the experimental value, the shape of the SXRL line is Gaussian with FWHM of 3.2 mÅ, in good agreement with the experimental results

4 Conclusion

By seeding of a laser created plasma amplifier we have demonstrated that it is possible to generate an intense soft x-ray beam having all the fundamental properties of common visible/IR/UV lasers. In a near future we anticipate a significant improvement of this concept, e.g. by using waveguiding technique to increase the length of the amplifier and thus boost up the SXRL output energy by at least one order of magnitude (15). Our measurements suggest that thanks to the perfect wavefront it should be possible to focus the 32.8 nm seeded laser into a near diffraction-limited spot, and thus achieve a soft x-ray intensity close to 1015 W.cm 2. The superb spatial beam quality and full longitudinal coherence make this source an excellent scientific tool for applications such as soft x-ray holography, phase contrast imaging, and microscopy.

References

1. Tesla FEL reports, http://flash.desy.de/reports_publications/index_eng.html /
2. J. Seres et al., Nature 433 , 596 (2005)
3. X. Zhang et al. , Opt. Lett. 29, 1357 (2004)
4. E. Takahashi et al., Phys. Rev. A 66, 021802 (2002)
5. J.-F. Hergott et al., Phys. Rev. A 66, 021801 (2002)
6. H. Daido, Rep. Prog. Phys. 65, 1513 (2002)
7. D.L. Matthews et al., Phys. Rev. Lett., 54, 110 (1985).
8. P.V. Nickles et al., Phys. Rev. Lett. 78, 2748 (1997)
9. S. Sebban et al., Phys. Rev. Lett. 86, 3004 (2001).
10. J.J. Rocca et al., Phys. Rev. Lett. 73, 2192 (1994)
11. T. Ditmire et al., Phys. Rev. A 51, 0R4337 (1995)
12. Ph. Zeitoun et al., Nature 431, 466 (2004)
13. Y. Wang et al., Phys. Rev. Lett. 97, 123901 (2006)
14. S. Sebban et al., Phys. Rev. Lett. 89 253901-1 (2002)
15. B. Cros et al., Phys. Rev. A 73, 033801 (2006)

16. J. Ph. Goddet et al., Opt. Lett. 32, 1498 (2007)
17. T. Mocek et al., Phys. Rev. Lett. 95, 173902 (2005)
18. S. Sebban et al., JOSA B 80, 195 (2003)
19. O. Guilbaud et al., Europhys. Lett. 74, 823 (5)
20. W.H. Southwell, J. OPt. Soc. Am. 70, 998 (1980)
21. S. Le Pape et al., Phys. Rev. Lett. 88, 183901-1 (2002)
22. A. Marechal, Rev. d'Optique, 26, 257 (1947)
23. G. Maynard et al., Contrib. Plasma Phys. 47, No. 4–5, 352 (2007)
24. B. Talin et al., Phys. Rev. A 51, 1918 (1995)
25. J.A. Koch et al., Phys. Rev. A. 50, 1877 (1994)

Investigation on the Spatial Properties of Silver X-Ray Laser Using GRIP Schemes

H. T. Kim[1], K. A. Janulewicz[1], C. M. Kim[1], I. W. Choi[1], J. H. Sung[1], T. J. Yu[1], S. K. Lee[1], T. M. Jeong[1], D. -K. Ko[1], P. V. Nickles[2], J. Tümmler[2], and J. Lee[1]

[1] Advanced Photonics Research Institute, GIST, Oryongdong 1, Bukgu, GwangJu, Korea.
[2] Max Born Institute, Max-Born-Strasse 2A, D-12489 Berlin, Germany.

Abstract. Spatial characteristics of Ni-like silver x-ray laser using grazing incidence pumping (GRIP) schemes were investigated. We succeeded in lasing from a nickel-like silver plasma in the GRIP geometry using both a double pump laser pulse as well as a single-profiled pulse.The transverse profile of the intensity distribution as well as the spatial coherence of both variants of Ni-like Ag x-ray laser were determined by far-field measurements and the Young's double slit interferometry. The x-ray laser pumped by a single-profiled pulse showed the output beam with a highly coherent part of a reasonable volume containing about 3 % of the total photon number.

1 Introduction

X-ray laser is a well-established source of coherent short-wavelength radiation. Recent progress in x-ray laser research has resulted in table-top size x-ray lasers requiring reduced pump energy of down to level of 1 J [1-4]. The table-top x-ray lasers have been tested in various practice-oriented applications such as nano-scale imaging [5], interferometry [6] and spectroscopy [7]. One of the important issues deciding about the progress in X-ray laser applications is improvement of the spatial quality and increase in the spatial coherence – the features of paramount importance for many applications.

In this paper, we report on the measurements of the spatial characteristics of Ni-like Ag X-ray laser pumped by using 10-Hz 100-TW Ti:Sapphire laser system. Recently, strong generation from Ni-like Ag X-ray laser has been achieved using both the conventional (traditional) double-pulse GRIP geometry as well as a variant of the GRIP scheme applying only single-profiled pump laser pulse [8]. Investigations of the spatial characteristics of Ni-like Ag X-ray laser output beam included spatial distribution of the radiation intensity and spatial coherence.

In sec 2, we briefly compare the double-pulse GRIP scheme with its variant using only single profiled pump pulse. The measurements of intensity distribution and spatial coherence of Ni-like Ag x-ray laser are described in sections 3 and 4, respectively. A summary of the results are is given in section 5.

2 Conventional GRIP vs. the GRIP geometry applying a single-profiled pulse

The GRIP scheme has opened a way to achieve saturated emission from x-ray lasers with pumping energy about 1 J [1-4]. The GRIP scheme has enabled an easy access to x-ray lasers and promoted new application areas because the necessary pumping energy for GRIP scheme could be achieved by commercial Ti:Sapphire laser systems. Since the GRIP geometry applies the transient inversion scheme, the conventional GRIP uses two pumping pulses: the pre-pulse and main heating pulse. The pre-pulse has usually duration of ns or sub ns and energy of few 100 mJ. The main pulse is mostly few picoseconds long with energy about 1 J. In this version of the GRIP geometry, the main pulse is obliquely (at a grazing angle) incident on a slab target to increase absorption of the pump energy in the plasma.

The main advantage of the GRIP scheme is reduction in the pump energy necessary to overcome lasing threshold. However, such a GRIP scheme forces using a disadvantageous system of two different beam lines. Since GRIP scheme conventionally uses a width of the focal line of few tens (30-50) of microns, perfect spatial overlapping of two focused beams is not a trivial task. In order to reduce the overlapping problem, it was attempted to irradiate the target just by obliquely incident single profiled main pulse. In final effect, saturation of Ni-like Ag x-ray laser has been achieved by using the new variant of GRIP geometry with the profiled pump laser pulse.

The experiments were performed at the Advanced Photonics Research Institute (APRI) 100 TW laser facility delivering driving pulses with 30-fs pulse duration and 3-J laser energy operating at 10 Hz repetition rate. For the double-pulse GRIP scheme, the stretched 350-ps laser pulse was divided before the pulse compressor. The uncompressed pulse of 0.5 J energy and 350-ps duration was focused on the target by spherical (f = 50 cm) and cylindrical (f= -20 cm) lenses and uased as a pre-pulse. The pre-pulse impinged on the target in the direction normal to its surface. The main pumping pulse was compressed to a length of 8 ps (full width at the half of maximum – FWHM). Such a laser pulse with energy of 1.5 J irradiated the target at an grazing angle of 18 degrees. Astigmatism of the focusing spherical mirror with a focal length of 75 cm was used to create a 7.2 mm long focal line on the target. The width of the line-focused beam was about 30-50 μm. The main pulse was 300 ps delayed relative to the prepulse. Silver slab target with width of 7 mm was used in all experiments. The X-ray radiation was recorded by a flat-field spectrometer (FFS) equipped with a cylindrical collecting mirror and a 512x2048 pixel CCD camera. The scattered light was blocked by Zr filters. A flat slab target was aligned to make the deflected X-ray laser beam propagating properly along the axis of FFS. For the single-profiled pulse pumping scheme, the temporal profile of main pulse was just modified, while the geometry and pulse energy of conventional GRIP scheme were kept constant. Details of the GRIP x-ray laser using single-profiled pumping pulse are descried in [8].

The output of the Ni-like Ag X-ray laser as a function the target length was measured in both cases, and the results are shown in Fig. 1. Each point of the scatter represents an average value and the error bars show deviation of the maximum and minimum values used in averaging. The experimental data of the X-ray laser measured output vs. the target length were carefully fitted using the modified Linford formula which takes into account the saturation effect [9]. In the case of double-pulse GRIP, the estimates based on the fitting procedure gave a small-signal gain coefficient of 60.4 cm^{-1} and an effective gain length product of 23.1. For the single pulse pumping, a small-signal gain coefficient of 76 cm^{-1} and an effective gain length product of 28.2 were extracted from the data. The X-ray laser output energy in the saturation regime was estimated to be about 1.5 μJ per pulse in both cases. Consequently, both GRIP geometries gave deep saturation of Ni-like Ag X-ray laser with exceptionally high gain-length product.

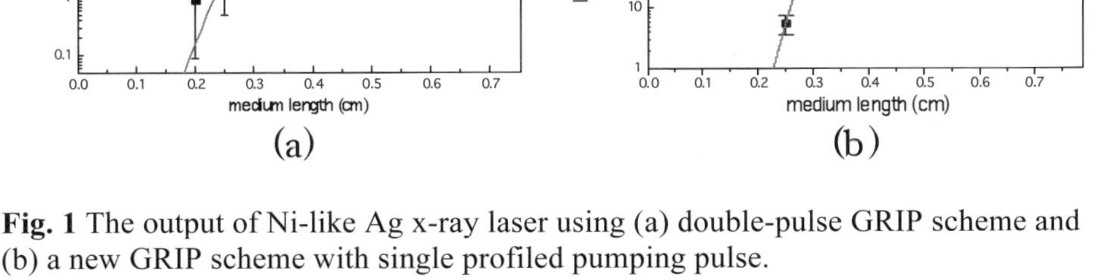

Fig. 1 The output of Ni-like Ag x-ray laser using (a) double-pulse GRIP scheme and (b) a new GRIP scheme with single profiled pumping pulse.

Simplification of experimental setup and stable operation by removing overlapping problem in GRIP scheme are the main advantages of the new pumping geometry. In order to confirm stable operation of the new GRIP scheme, the output of x-ray laser beam was monitored while keeping pumping conditions constant. The Ag target was refreshed by shifting the target before each shot. Figure 2 shows a level of the output signal of the x-ray laser pumped by single profiled pulse. The fluctuation of the x-ray laser output taken from two different series of successive shots was around 17 % in root mean square. Since the conventional GRIP geometry showed signal fluctuation above 20 %, the new variant of GRIP using single-profiled pulse improved stability of the x-ray laser output. It was ascribed to reduction of the overlapping effect. Jitter of timing in the amplification process in the pump laser system varying the temporal profile of the pumping pulse seems to be the main source of power fluctuation in the single pulse pumping scheme.

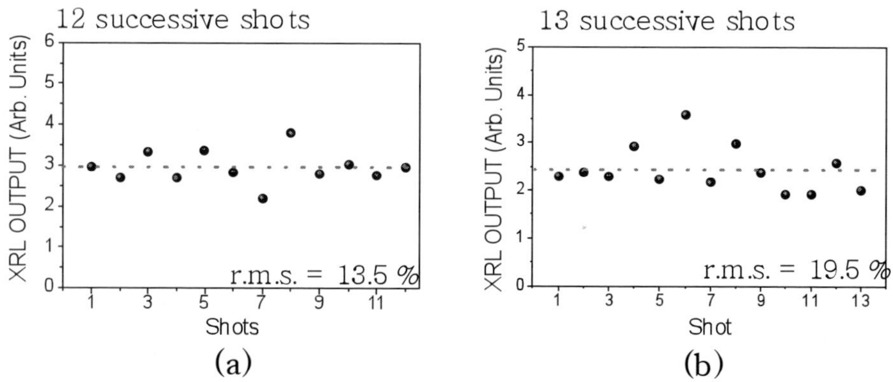

Fig. 2 X-ray output of two series of successive shots in the case of the GRIP scheme using single profiled pumping pulses.

3 Spatial distribution of the output beam of silver x-ray laser

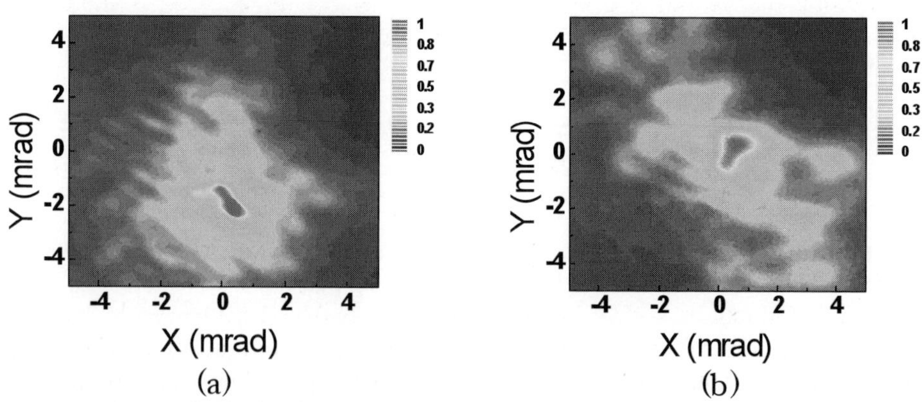

Fig. 3 Far field distribution of the x-ray laser output; (a) double-pulse GRIP and (b) single-profiled pumping pulse.

The output beam of X-ray lasers usually has irregular spatial intensity distribution caused by amplified spontaneous emission and high longitudinal coherence. In order to apply x-ray lasers to imaging techniques, knowledge of the beam quality of x-ray laser is of fundamental importance. The far-field patterns of Ni-like Ag x-ray lasers applying both pumping geometries were recorded. Figure 3 shows the footprints of x-ray laser beam taken at a distance of 55 cm behind the target emitting end. In both cases the beam divergence was well below 5 mrad. For the case of double-pulse GRIP scheme, the beam divergence was about 4 mrad as shown in Fig. 3 (a). The x-ray laser pumped in the modified GRIP geometry had a beam divergence about 3 mrad.

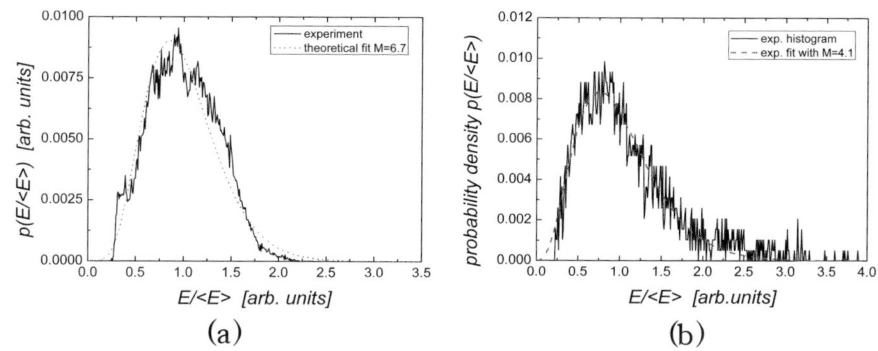

Fig. 4 Probability densities of the normalized integral intensity obtained from the recorded intensity spatial distributions (solid line) and fitting functions (dotted line) in the case of (a) conventional GRIP and (b) single profiled pumping pulse.

In the further steps of the analysis of spatial properties of x-ray laser beam, the number of spatial degree of freedom was calculated from the far-field intensity profile of the x-ray laser beams. Probability density function (p) of the normalized integral intensity (E) can be expressed by following formula in the case of unpolarized light [10].

$$p_E(E) \cong \left(\frac{2M}{\overline{E}}\right)^{2M} \frac{E^{2M-1} e^{-2M\frac{E}{\overline{E}}}}{\Gamma(2M)}, \quad (W^{3}\ 0) \tag{1}$$

Here, M is the number of degree of freedom, that means number of statistically independent radiation modes. Figure 4 shows fitting of the intensity histogram with Eq. (1) for the case of (a) double-pulse GRIP scheme and (b) GRIP with single-profiled pulse. The fits with Eq. (1) match quite well the intensity histogram of far-field patterns shown in Fig. 3. The number of independent mode was derived to be 4.1 and 6.7 for single-profiled pulse case and the conventional GRIP, respectively. The x-ray laser pumped by the modified GRIP arrangement shows a reduction in number of independent modes comparing to that of conventional GRIP. Consequently, x-ray laser using a new variant of GRIP with single profiled pulse has improved spatial quality of the beam. It is believed, that this improvement is caused by different plasma conditions and the gain structure strengthening spatial filtering condition in 7-mm long plasma medium.

4 Spatial coherence of silver x-ray laser

Young's double slit interferometer was used to investigate spatial coherence of the X-ray laser output. The interference fringes were obtained by insertion of a stainless plate with a set of double slits on the propagation axis of the X-ray laser. Firstly, we examine the spatial coherence of Ni-like Ag x-ray laser pumped in the double-pulse GRIP scheme. In this case, a double-slit set consisted of three pairs of the slits separated respectively by 60-μm, 100-μm, and 150-μm. The slits cut out in a 50-μm thick stainless steel plate by femtosecond laser pulses had 10-μm width and 2-mm length. The slits were positioned perpendicularly to the dispersion axis of the flat-field spectrometer. The centre of the double-slit separation was positioned on the axis of the X-ray laser beam 84 mm behind the target. Hence, the beam size (FWHM) at the slit plane estimated from the beam divergence was about 330 μm. The interferogram images from double slit are shown in Fig. 5.

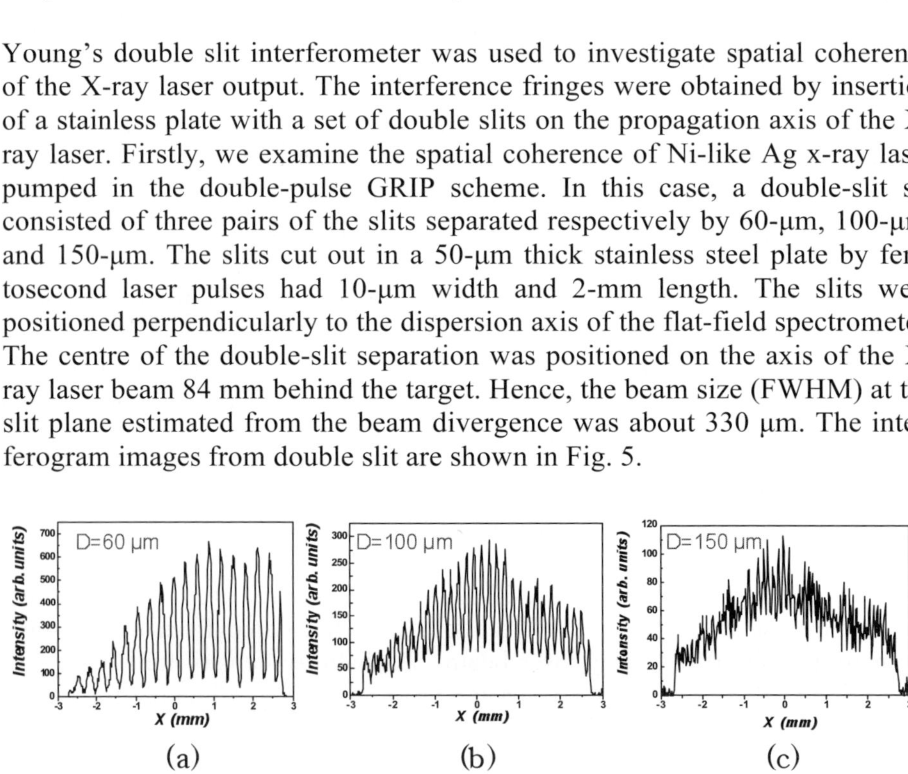

Fig. 5 Young's double slit interferograms of x-ray laser beam by nominal GRIP scheme with different slit separations of (a) 60 μm, (b) 100 μm and (c) 150 μm.

The spatial coherence of X-ray laser beam can be relatively easily determined from the visibility of interference fringes obtained in a Young's double slit interference experiment. The visibility of interference fringes is defined as $V=(I_{max}-I_{min})/(I_{max}+I_{min})$, where the I_{max} and I_{min} are the maximum and the adjacent minimum of the interference fringe at the centre of interferogram. When both slits are equally illuminated, the modulus of spatial coherence factor $|\mu|$ is equal to V. Thus, the visibility of the observed interference fringes can be considered as the spatial coherence factor because the slits are similarly illuminated in the experiment. For the circularly symmetric and uniform beam, the fringe visibilities for different slit separation can be described by the following equation [11]:

$$V(\Delta x) = 2J_1(\pi d_S \Delta x / \lambda z) / (\pi d_S \Delta x / \lambda z) \tag{2}$$

where the J_1 is first order Bessel function, d_S is the diameter of the equivalent incoherent source, Δx is slit separation, λ is wavelength and z is the distance from the source to double slit. Figure 6 shows the fringes visibility as a function of the slit separation in the case of the conventional GRIP scheme. The measurement points show the averaged visibility for the given slits separation and the error bar indicates the maximum and minimum visibilities for the given measurement point. The solid line is the fitting curve using Eq. (2) and the nonlinear least-squares method. From the fitting curve, we estimated the transverse coherence length L_C. The transverse coherence length L_C can be conventionally defined as the slit separation Δx where the $|\mu|$ decreased to 0.88. This criterion is most strict definition of highly coherent part of a partially coherent beam [11]. A transverse coherence L_C of 54 μm was estimated from the fitting curve. Since the beam diameter at the slit plane is 330 μm, the highly coherent part of the X-ray laser contains about 1.8 % of the total number of photons in the case of the double-pulse GRIP scheme, assuming that the x-ray laser beam has Gaussian profile.

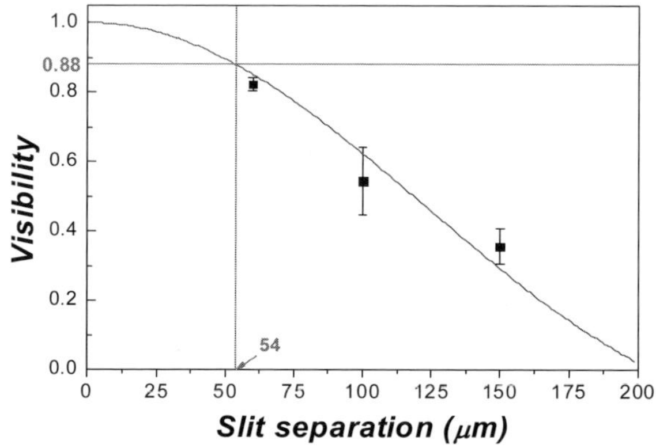

Fig. 6 Dependence of fringe visibility on slit separation in the case of nominal GRIP scheme. Solid line shows nonlinear least-squares fits to the averaged measured values.

The same method was used to measure spatial coherence of the modified GRIP arrangement as that used in the case of conventional GRIP scheme. Here, six different slit separations were used and the slits were positioned 76 mm behind the target end. The x-ray laser beam has diameter of 230 μm at the slit position resulting from a beam divergence of 3 mrad. Figure 7 shows the interferograms obtained with the x-ray laser beam using the single profiled pump laser pulse. It shows clear fringes even with a the slit separation of 160 μm.

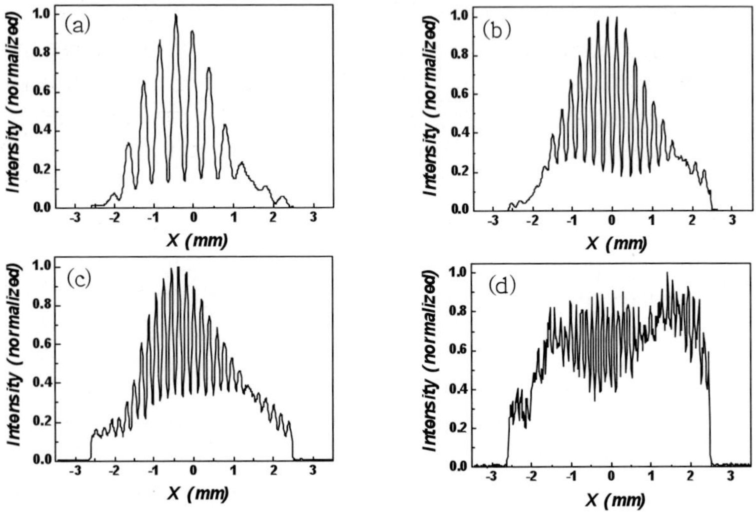

Fig. 7 Young's double slit interferograms of x-ray laser beam by single profiled pumping scheme with different slit separations of (a) 50 μm, (b) 90 μm, (c) 110 μm and (d) 160 μm.

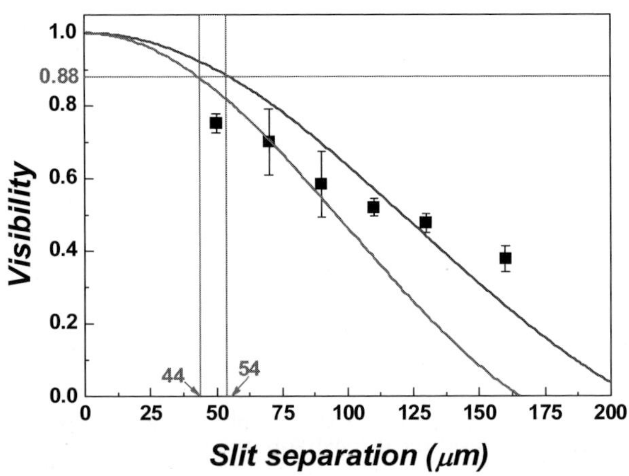

Fig. 8 Dependence of fringe visibility on slit separation in the case of single profiled pumping scheme. Solid line shows nonlinear least-squares fits to the averaged measured values.

Fitting the visibility vs. the slit separation using Eq. (2) was performed also in this case. The first three points of small slit distance show behaviour different from that of the last three points corresponding to larger slit distance. Performing separate fittings for each of the group it was found that the first three points indicate worse coherence than the last three points. Considering the coherence length can be positioned between the values estimated from these two fitting curves, it was found that this has to be in the range between 44 μm and 54 μm. Taking into account a beam diameter of 230 μm, the highly coherent part of the total number of photons is between 2.5 % and 3.8 % in the case of the GRIP geometry using single profiled pumping pulse. The x-ray laser beam using single profiled pump laser pulse has a better coherence about 3 % in comparison to that using nominal GRIP showing only about 2 % of fully coherent photons.

The percentage of the coherent photons in the emitted XRL beam was relatively higher than that recently reported for Ni-like Cd X-ray laser using also GRIP geometry [12]. This enhancement of coherence of X-ray laser can be mainly ascribed to the increased medium length. An Ag target nearly two-times longer than that used in the experiment reported in [12] was used in our experiment. This could improve modal selection of the propagated radiation and hence increase coherence of the beam. We think the further improvement of spatial coherence of the x-ray laser beam generated by new variant of the GRIP geometry can be achieved by modification of propagation conditions along the elongated plasma medium that can more efficiently filter the spatial modes. As shown in Sec. 3, the x-ray laser beam generated in this arrangement has smaller divergence and contains lower number of independent spatial modes than the beam obtained in the conventional double-pulse GRIP setup. It corresponds well to better coherence of x-ray laser beam measured in this pump geometry.

5 Summary

We investigated Characteristics of Ni-like Ag x-ray laser pumped in both the conventional double-pulse GRIP and the GRIP variant with single profiled pumping pulse were investigated with stress on the spatial dependences of the emitted radiation. Saturated generation from Ni-like Ag x-ray laser was obtained with small signal gain coefficients of 60.4 cm^{-1} and 76.0 cm^{-1} for conventional GRIP and single pulse pumping scheme, respectively. In order, Far-field distributions and Young's double slit interferograms were recorded to characterize spatial features of the generated from the x-ray laser beams. The x-ray laser beam obtained from the X-ray laser pumped by a single profiled laser pulse has better spatial characteristics than the conventional GRIP scheme as far as beam divergence, number of modes and coherence. Specifically, the x-ray laser beam of the new variant of GRIP geometry contains 3 % of highly coherent part among the total radiation that is sufficient to deliver within a single shot 10^9 fully coherent photons for applications even if few Mo:Si multilayer mirrors are used.

Acknowledgement

This work was supported by the Ministry of Knowledge and Economy of Korea through the Ultrashort Quantum Beam Facility Program.

References

1. Keenan, R. et al.: 'High-Repetition-Rate Grazing-Incidence Pumped X-Ray Laser Operating at 18.9 nm', *Phys. Rev. Lett.*, 94, 103901, 2005
2. Rocca, J. J. et al.: 'Saturated 13.2 nm high-repetition-rate laser in nickel like cadmium', *Opt. Lett.,* 30, 2581, 2005
3. Wang, Y. et al.: 'Demonstration of high-repetition-rate tabletop soft-x-ray lasers with saturated output at wavelengths down to 13.9 nm and gain down to 10.9 nm', *Phys. Rev. A,* 72, 053807, 2005
4. Tümmler, J. et al.: '10-Hz grazing–incidence pumped Ni-like Mo x-ray laser', *Phys. Rev. E*, 72, 037401, 2005
5. Sandberg, R. L. et al.: High numerical aperture tabletop soft x-ray diffraction microscopy with 70-nm resolution', *PNAS* 105, 24, 2008
6. Grava, J. et al.: 'Dynamics of a dense laboratory plasma jet investigated using soft x-ray laser interferometry', *Phys. Rev. E*, 78, 016403, 2008
7. Namba, S. et al.: 'Enhancement of Double Auger Decay Probability in Xenon Clusters Irradiated with a Soft-X-Ray Laser Pulse', *Phys. Rev. Lett.*, 99, 043004, 2007
8. Kim, H. T. et al.: 'Demonstration of a saturated Ni-like Ag X-Ray laser pumped by a single profiled laser pulse from a 10-Hz Ti:sapphire laser system', *Phys. Rev. A* 77, 023807, 2008
9. Lin, J. Y. et al.: 'Gain saturation of the Ni-like X-ray lasers', *Opt. Commun.,* 158, 55, 1998
10. Goodman, J. W.: *Statistical Optics*, Wiley Classics Library Edition, 2000
11. Born, M. and Wolf E.: 'Principles of Optics', Cambridge University Press, 1999
12. Liu, Y. et al. : 'Spatial coherence measurements of a 13.2 nm transient nickel-like cadmium soft X-ray laser pumped at grazing incidence', *Opt. Exp.*, 14, 12872, 2006

Spatial Filtering of High Order Harmonics by an OFI Plasma Amplifier

J.P. Goddet[1], S. Sebban[1], Ph. Zeitoun[1], J. Gautier[1], C. Valentin[1], F. Tissandier[1], T. Marchenko[1], G. Lambert[1], J. Nejdl[1], B. Cros[2], G Maynard[2], B.Robillard[2], T. Mocek[3], M. Kozlová[3] and K.Jakubczak[3]

[1] Laboratoire d'Optique Appliquée, chemin de la hunière, 91128 Palaiseau
[2] LPGP, Université Paris-Sud, 91405 Orsay, France
[3] Institute of Physics, Department of X-Ray Lasers, Prague, Czech Republic

Abstract. It is now possible to produce a 10 Hz soft x-ray laser beam having very high optical qualities. The solution consists in seeding an optical-field-ionized population inverted plasma amplifier with the 25th harmonic of an infrared laser. This concept was successfully realized in LOA in 2004 and an extensive investigation of the source has been recently performed. Indeed we demonstrate the first diffraction-limited laser beam in the soft x-ray spectral range. This laser beam at a central wavelength of 32.8 nm, emitted with a divergence of 0.67 mrad at a repetition rate of 10 Hz. The beam exhibits a regular Gaussian spatial profile, and wavefront distortions smaller than $\lambda/17$. A theoretical analysis of these results shows that this high beam quality is due to spatial filtering of the seed beam by the plasma amplifier aperture.

1 Introduction

Until now, in all the soft x-ray laser (SXRL) schemes operating at saturation, population inversion between the levels of the lasing ion is induced by electron collisional excitation, leading to high gain value at short wavelength [1]. The short lifetime of the gain and the absence of high reflectivity soft x-ray optics, make the use of this SXRL amplifier in a complete laser cavity impossible. For this reason, SXRL emission generally results from the single-pass amplification of spontaneous emission. As a consequence, SXRL radiation is characterized by an inhomogeneous beam profile [2], a low spatial coherence [3] and important wavefront distortions [4], which limits the use of SXRL for applications that require a highly intense, coherent, soft x-ray photon flux in a sub-micron spot size. To fully explore the potential of plasma-based SXRL sources, improving their spatial beam quality is a crucial bottleneck to be overcome. A promising solution to be explored consists in seeding a soft x-ray plasma amplifier with a High Order Harmonic (HOH) seed beam [5]. This Seeded soft x-ray laser (SSXRL) approach is a direct analogy of the "oscillator-amplifier" concept commonly used for infrared (IR) laser system, applied in the soft x-ray range. Here the HOH beam plays the role of an oscillator which is injected and amplified while propagating through a population-inverted SXRL plasma column. This emerging scheme [6-7] offers a prospect for compact SXRL chain combining both the high energy extracted from the plasma amplifier as well as the high optical quality of the HOH beam [8-9].

We report on the first observation of a diffraction-limited laser beam at 32.8 nm achieved by seeding an Optical Field Ionised (OFI) SXRL amplifier [10] with a HOH beam. The 32.8 nm SSXRL beam exhibits a circular energy distribution with a divergence of 0.67 mrad (half angular with at $1/e^2$) Wave-front distortions were measured to be smaller than λ/17 (1.9 nm). Numerical simulations show that the properties of the measured wavefront and energy distribution of the SSXRL result from spatial filtering of the seed beam in the amplifying plasma column.

2 Experimental setup

The experiment was performed at the Laboratoire d'Optique Appliquée (LOA) using a 10 Hz, multi-terawatt Ti:sapphire laser system providing two independent 34 fs laser beams at a central wavelength of 815 nm [11]. The first IR laser beam, containing about 10 mJ, was used to generate the HOH beam inside a 7 mm long gas cell filled with 30 mbar of Ar. The 25th harmonic of the IR laser was closely matched to the wavelength of the lasing transition of the SXRL amplifier. A grazing incidence toroidal mirror imaged the output of the HOH source with a magnification of 1.5 at the entrance of the amplifier cell. The second IR beam, circularly polarized, was delivering ~600 mJ on target. This pump beam, focused to a spot diameter of 38 μm (at $1/e^2$) inside a 7.5 mm long gas cell filled with Kr, was used to create the OFI amplifying plasma column which drives the $3d^94d\ (1S^0) \rightarrow 3d^94p\ (1P^1)$ transition of the Kr^{8+} ion at 32.8 nm [12].

3 Far field pattern

The spatial energy distribution of the SSXRL and HOH beams were measured after reflection from a removable 45°, soft X-ray mirror made of $B_4C/Mo/Si$ tri-layers, by a Charge Coupled Detector (CCD) camera. A 300 nm thin aluminium filter was used to block the IR beam. Figure 1(a) shows the energy distribution of the seed HOH. The HOH beam exhibits an astigmatic profile with horizontal and vertical divergence of 1.32 mrad and 0.48 mrad, respectively, due to a slight misalignment of the toroidal coupling mirror. This HOH beam is then injected into the 32.8nm amplifier, set for optimized parameters, i.e. filling pressure of 20 mbar, and cell length of 7.5 mm. The energy distribution resulting from the amplification of the HOH beam into the plasma column is dramatically improved as shown in Fig. 1(b). The intense monochromatic radiation is emitted in the direction of the HOH beam, and has a divergence of 0.67 ± 0.07 mrad ($1/e^2$). The spatial beam distribution is nearly Gaussian. Note that such a high beam quality has never been observed for other plasma-based SXRLs. In general, SXRLs exhibit highly contrasted intensity modulations (or speckles) forming a complex and irregular pattern [13], which is a direct consequence of low spatial coherence combined with the high temporal coherence of the amplified spontaneous emission (ASE) radiation [2]. Here the amplification of the partially coherent HOH beam strongly enhances the spatial coherence and results in a smooth beam pattern.

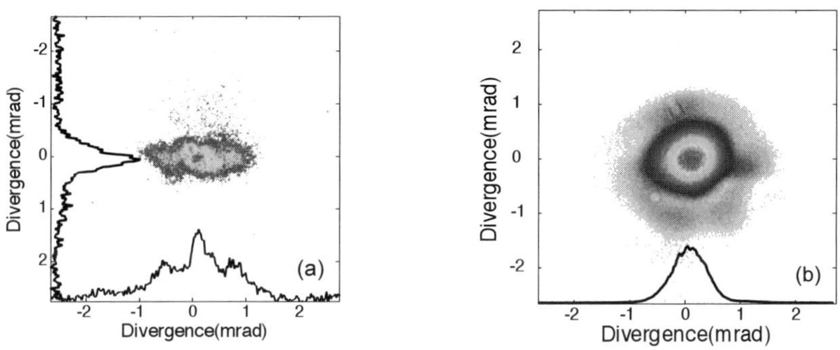

Fig. 1 Spatial profile of (a) the HHG and (b) the SSXRL beams.

4 Wavefront measurement

The wavefront of the SSXRL was measured with a soft x-ray Hartmann sensor [14]. In the Hartmann wavefront analysis a beam passes through a hole array and is projected onto a CCD camera that detects the beamlet sampled by each hole. The position of each individual spot centroïd is measured and compared with a reference position. The Hartman technique allows energy and phase to be measured simultaneously. The soft x-ray Hartmann sensor was made of a 100 microns thick, nickel plate placed at about 20 cm in front of a back-illuminated XUV CCD. The Nickel plate contained a matrix of 51×51 square holes (80 μm×80 μm and separated by 380 μm), over an area of 15×15 mm^2. This diagnostic permits to reconstruct the wavefront with a resolution of $\lambda/20$ at 32.8 nm [15]. A more accurate analysis of the spatial coherence can be obtained by combining intensity and phase measurements to get full information on the complex amplitude of the laser field. In the plane of the Hartmann detector, the complex amplitude of the laser field can be written as

$$E_H(x, y, z = z_H) = \sqrt{I_{SXRL}(\theta_x, \theta_y)} \exp\left[i \frac{\pi}{\lambda_{SXRL}} \left(\frac{x^2 + y^2}{z_H} + 2\delta(\theta_x, \theta_y) \right) \right] \quad (1)$$

In this equation, z_H is the distance between the detector and the entrance of the plasma, $\theta_x = x / z_H$ and $\theta_y = y / z_H$ are the beam divergences in the two directions perpendicular to the beam axis, I_{SXRL} is the laser intensity as reported in Fig. 1 and $\delta(\theta_x, \theta_y)$ represents the fluctuation of the wavefront; $\delta(\theta_x, \theta_y) = 0$ corresponds to a fully coherent, diffraction limited beam. Figure (2a) shows a typical example of the values of $\delta(\theta_x, \theta_y)$ for the seed HOH beam, measured by the Hartmann sensor. The HOH beam exhibits large phase fluctuations, with an average distortion of about $0.3\times\lambda$SXRL root-mean-square (rms). The wavefront fluctuations in the case of the SSXRL beam are shown in Fig. 2b. In comparison to the HOH case (Fig. 2a), the amplitude of the wavefront fluctuations of the SSXRL are much smaller, with a rms value of only $0.058\times\lambda_{SXRL}$, which corresponds to less than $\lambda/17$ (1.9 nm). According to the Marechal criterion, which states that a system is regarded as well corrected if the wavefront distortion does not exceed $\lambda/14$ [16], the generated SSXRL can be considered as a fully coherent diffraction-limited beam, which has never been demonstrated so far. Earlier works showed that neon-like Ar capillary discharge-driven SXRL at 46.9 nm exhibits wavefront distortion of $3\times\lambda_{SXRL}$, which is, to our date, the only measurement performed for plasma-based SXRL source [17] .

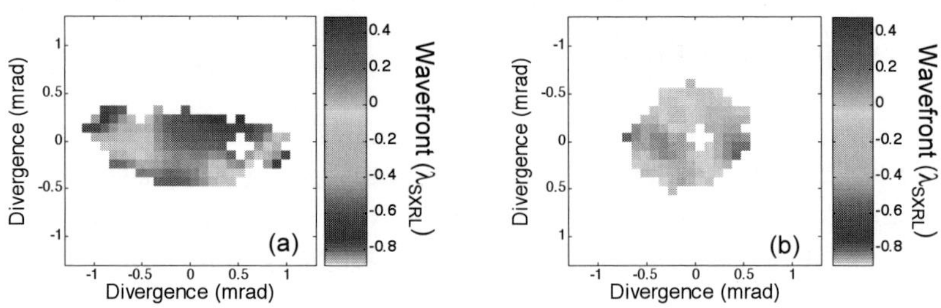

Fig. 2 Measured wavefront of (a) the HHG beam, and. (b) the seeded EUV laser at 32.8 nm.

5 Reconstructed sources

A simple theoretical analysis of the spatial coupling of the seed HOH beam into the plasma amplifier allows understanding how such a high beam quality can be achieved. As the intensity and the phase of the beams were measured using the Hartman sensor, the values of the complex amplitude of the electric field, Eq. (1), are known in the plane of the detector at each small hole position. Using a χ^2 minimization fit, these values are projected on a HG basis *(18)*, to make possible the reconstruction of the complex amplitude of either the HOH beam or the SSXRL one anywhere on the path of the beam.

The results of this analysis are presented in Figs 3. Figure 3(a) shows the HOH beam energy distribution reconstructed at the entrance of the amplifying plasma column, which is found to be oval with dimensions of 55×116 μm^2 (at 1/e^2). The SSXRL energy distribution at the exit of the plasma is shown in Fig. 3b. It has a more symmetrical shape with a surface of 58×77 μm^2 (at 1/e^2). These beam reconstructions indicate that, in terms of energy, the coupling between the HHG beam and the amplifier is well optimized as 60% of the incoming energy is amplified by the plasma. Thus the spatial filtering is quite efficient because the main part of the phase distortion comes from the outer part of the HHG beam.

Fig. 3 (a) Reconstructed profile of the HHG beam at the entrance of the amplifying plasma, and (b) reconstructed source size of the SSXRL beam.

6 Spatial filtering simulation

To analyze the importance of spatial filtering in the reduction of the wavefront distortions we used a simplified model in which the plasma is considered as a perfect amplifier, where the amplified output signal is a coherent reproduction of the input HOH beam. The plasma is assumed to be a cylinder of length L=7.5 mm and of radius R. The value for the Rayleigh length of the HHG beam is one order of magnitude larger than L, so that diffraction effect inside the plasma can be safely neglected. Therefore the complex amplitude for the SSXRL, at the exit of the plasma, takes the simple form

$$E_{SSXRL}(x,y,z=L)=\sqrt{G}E_H(x,y,z=0)H\left(R-\sqrt{x^2+y^2}\right), \qquad (2)$$

Where G is the amplifying factor, $H(u)$ is the heaviside step function and E_H is the complex amplitude defined in Eq. (1) for the HHG beam case. In Eq. (2), due to the large value of G and also to absorption by non-ionized gas, the amplitude of the field outside the plasma has been set to zero.

We see from Eq. (2) that, besides a constant factor, the influence of the plasma amplifier on the HHG beam is identical to the effect of a circular diaphragm. $E_{SSXRL}(x,y,z=L)$ was determined from Eq. (4), and projected again on the HG basis, in order to determine the complex amplitude of $E_{SSXRL}(x,y,z=z_H)$ in the plane of the Hartmann detector. These calculations were performed for plasma radius ranging from 15 μm up to 1000 μm. Figure 4 shows the distribution of energy as determined from $E_{SSXRL}(x,y,z=z_H)$ for 4 values of the plasma radius a) R= 100 μm, b) R= 50, c) R = 30 and d) R = 20 μm. As observed in Figure 4, the energy distribution of the SSXRL is strongly modified for plasma radius smaller than 100 μm. For the larger radius Fig. 4.a, the HOH beam size at the entrance of the plasma (Fig. 3a) is smaller than the plasma size, and the amplified beam shape is not modified. By decreasing the plasma radius down to 25 μm, the beam evolves from astigmatic to a perfect circular beam, as observed experimentally. For a plasma of 30 μm (Fig. 4c), the divergence of the SSXRL beam is of 0.46 mrad ×0.57 mrad, in reasonable agreement with our experimental measurements 0.67 mrad (Fig. 1b). For a plasma radius smaller than 25 μm (Fig. 4d), the divergence of the SSXRL beam increases due to diffraction.

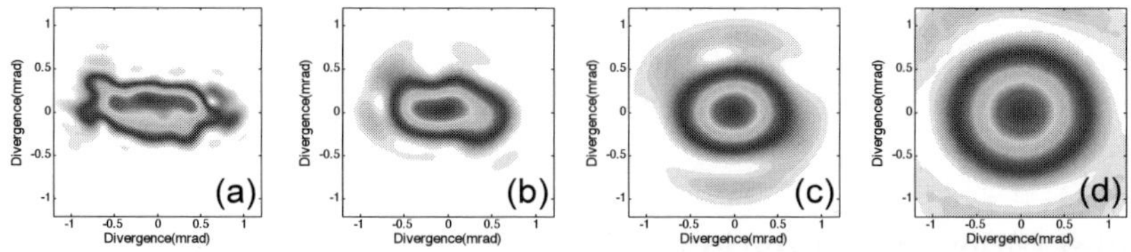

Fig. 4 2D reconstruction of the far-field pattern of the SSXRL beam by spatial filtering of the HHG beam trough a plasma aperture with variable radius: (a) 100 μm, (b) 50 μm, (c) 30 μm, (d) 20 μm.

The variation of wavefront distortion as a function of the plasma radius is plotted in Fig. 5. It shows that for our experimental conditions, the wavefront distortions of the SSXRL beam are significantly reduced for a plasma radius smaller than 60 μm and that the diffraction limit (λ/14) is reached when the amplifying plasma radius is smaller than 37 μm. According to these calculations our measurements of the wave front suggest that the HOH beam was filtered by a 30 μm radius plasma column; this value is in good agreement with the ones determined from the divergence of the beam and from the energy distribution (Fig. 3a). The agreement between experimental observations and this simple analysis implies that the plasma amplifier has a very low level of transverse dispersion, much lower than the dispersion of the intensity of the IR pump beam. This is due to the highly non-linear character of the OFI process, in which the lasing ion Kr^{8+} can be produced without significant changes of the plasma properties, while allowing nearly one order of magnitude variation for the IR intensity.

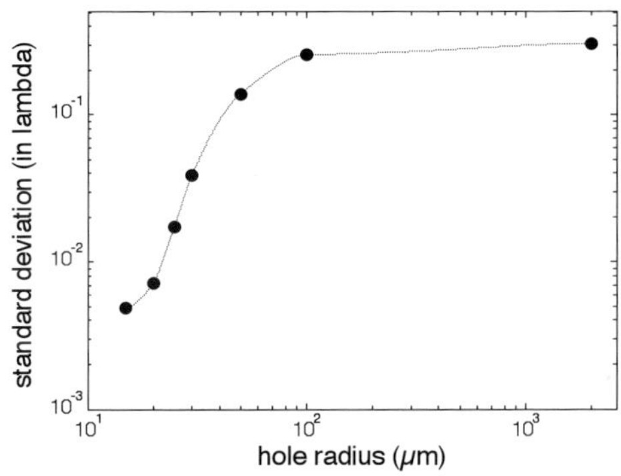

Fig. 5 Calculated wavefront of the of the SSXRL beam as a function of the plasma radius.

7 Conclusion

In conclusion, by seeding a laser created plasma amplifier we have demonstrated that it is possible to generate an intense soft x-ray beam having all the optical properties of common visible/IR/UV lasers. Our measurements suggest that thanks to the perfect wavefront it should be possible to focus the SSXRL beam into a near diffraction-limited spot, and thus achieve a soft x-ray intensity close to 10^{15} W.cm^{-2}. In a near future we anticipate a significant improvement of this concept, e.g. by using waveguiding technique to increase the length of the amplifier and thus boost up the output energy by at least one order of magnitude [19]. The excellent spatial beam quality, coupled to a high longitudinal coherence make this source an excellent scientific tool for applications such as soft x-ray holography, phase contrast imaging, and microscopy.

References

1. H. Daido, *Rep. Prog. Phys*. 65, 1513 (2002)
2. S. Sebban et al., *JOSA B 80*, 195 (2003)
3. J. Ph. Goddet et al., *Opt. Lett.* 32, 1498 (2007)
4. J.J. Rocca et al., *Phys. Rev. Lett.* 73, 2192 (1994)
5. T.Ditmire et al., *Phys. Rev*. A 51, 0R4337 (1995)
6. Ph. Zeitoun et al., *Nature* 431, 466 (2004)
7. Y. Wang et al., *Phys. Rev. Lett*. 97, 123901 (2006)
8. L.Le Déroff et al , *Phys. Rev. A*. 61, 043802 (2000)
9. K.Midorikawa et al., *Phys. Rev.A*. 66, 021802(R) (2002)
10. B. E. Lemoff et al., *Optics Letters* 19: 569-571 (1994)
11. M. Pittman et al., *Appl. Phys. B* 74, 529 (2002).
12. S. Sebban et al., *Phys. Rev. Lett*. 89 253901-1 (2002)
13. S. Sebban et al., *JOSA B* 80, 195 (2003)
14. P. Mercere et al. *Opt. Lett.* Vol. 28 (2003) 1534
15. W.H. Southwell, *J. OPt. Soc. Am.* 70, 998 (1980)
16. A. Marechal, *Rev. D'Optique*, 26, 257 (1947)
17. S. Le Pape et al., *Phys. Rev. Lett.* 88, 183901-1 (2002)
18. A.E. Siegman *Lasers*, University Science Books, Sausalito, California (1986)
19. B. Cros et al., *Phys. Rev. A **73***, 033801 (2006)

New Driver Laser System for Double Target X-Ray Lasers at JAEA

Y. Ochi[1], N. Hasegawa[1], T. Kawachi[1], M. Nishikino[1], M. Tanaka[1], M. Kishimoto[1], and T. Ohba[1]

[1]Japan Atomic Energy Agency

Abstract. We have developed a new driver laser system with repetition rate of 0.1 Hz for the double target x-ray laser at Japan Atomic Energy Agency. We named the new driver laser TOPAZ. TOPAZ delivers two beams with energy of > 5 J and duration of ~ ps. Ni-like Mo and Ag lasers at wavelength of 18.9 nm and 13.9 nm respectively have been successfully generated using TOPAZ.

1 Introduction

At Japan Atomic Energy Agency (JAEA), high quality x-ray laser (XRL) at the wavelength of 13.9 nm has been developed by means of the double target scheme [1]. The Ni-like silver laser is spatially full-coherent with diffraction limited beam divergence, 0.2 mrad. Using the double target XRL we already started a various of application studies such as holography, speckle measure measurement of condensed matter [2], interaction with cluster [3], XRL induced fluorescence [4], and so on. Because present driver system at JAEA adopts a Nd:glass rod amplifier as the power amplifier [5], generation of the x-ray laser is limited to be every 20 minutes due to the cooling time of the rod glass. In order to make more practical application studies, highly repeatable driver laser system is necessary. One of the candidates for the high-repetition driver laser is Ti:sapphire laser. By means of the gracing incident pumping scheme (GRIP) [6], 10-Hz XRLs have already been demonstrated [7-9]. The other approach is developing a new driver laser which can provide > 5 J and picoseconds pulse. Although GRIP using Ti:sapphire laser has an advantage in repetition rate, the XRL output energy is a few μJ at maximum and it is difficult to produce shorter wavelength XRL than 10 nm. Considering future extension to shorter wavelength XRL and higher energy XRL pulse, we decide to develop CPA laser with zigzag slab Nd:glass amplifiers. A name of the new driver system is "TOPAZ" (Twin OPtical Amplifiers using Zigzag slab). In this paper, a design of TOPAZ and results of XRL generation.

2 System Configuration of TOPAZ

TOPAZ bases on the CPA laser with zigzag slab Nd:glass amplifiers, which consists of the oscillator, pulse stretcher, OPCPA preamplifier, prepulse generator, zigzag slab Nd:glass power amplifiers, pulse compressor, and optics for producing a line focus on the target. Because two beam lines are necessary for the double target XRL, the laser light is divided into two beams after the OPCPA and following components are set up independently on each other.

The oscillator is a mode-locked Ti:sapphire laser (Spectra-Physics; TSUNAMI) pumped by 10-W diode-pumped solid state laser (Spectra-Physics; Millenia). The centre wavelength is 1053 nm and the spectral bandwidth is 20 nm in the full width at the half maximum (FWHM). The oscillation frequency is 80 MHz with typical power of 300 mW (~ 4 J / pulse). The pulse stretcher consists of a diffraction grating with 1740 grooves/mm, a spherical mirror with focal length of 1500 mm, flat mirror, and staring mirrors. A chirp of 250 ps/nm is generated by 4-path stretcher. After the stretcher, spectral bandwidth is limited to be 8 nm by the optics size. Output power is 100 mW (~ 1 nJ/pulse).

The stretched pulse is amplified by OPCPA. The pump source is 532-nm light from Q-switched YAG laser with seeder (Continuum; Powerlite Precision II) with 8-ns duration and 700-mJ energy. Total gain over 10^6 is obtained by using four BBO crystals at the pump intensity of 100 MW/cm^2. The output energy from the OPCPA is more than 10 mJ with the spectral bandwidth of 8 nm. A root mean square of energy fluctuation is 7.6%. The contrast ratio of the amplified laser pulse to the background is better then 10^4, shown in Fig. 1. Here the repetition rate is 10 Hz.

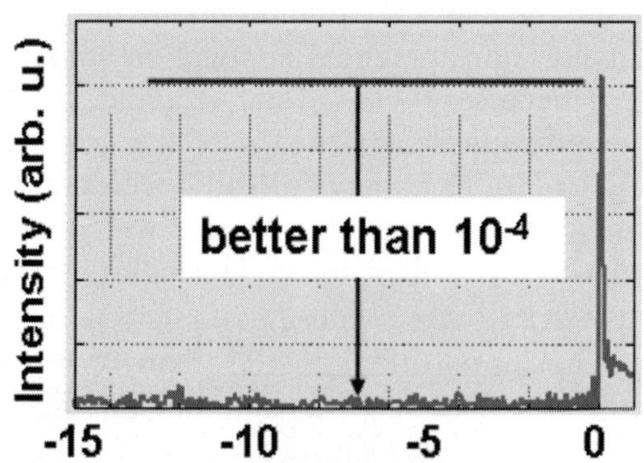

Fig. 1 Contrast of the amplified pulse by OPCPA.

After the OPCPA the laser pulse is divided into two beam lines for the double target scheme. Then each laser light is split into two pulses; one becomes prepulse and the other main pulse. Additional negative chirp is applied for the prepulse component by using grating pair in the compressor configuration. Pulse separation between the prepulse and main pulse is adjusted by a delay line in the mainpulse path. Then two pulses are combined with a beam splitter. Here the prepulse conditions, i.e. separation and intensity, of each beam line can be decided independently.

The main power amplifier consists of two Nd:glass zigzag slab amplifiers. The laser glass is a silica-phosphate with 1.0 wt.% Nd (Schott; APG-1). The first amplifier is used to amplify the laser light up to ~1 J with beam size of 10 x 10 mm^2 by using a laser glass of 15 x 17 mm^2 in cross-section and 99 mm in length. The laser glass is pumped by eight Xe flash lamps (four lamps on each side) from the side faces. The first amplifier is put into an image-relay cavity consisting of spherical lens pair and flat end mirrors. After the laser light is amplified up to ~ 1 J by multi-pass amplification in the cavity, it is extracted by means of polarization control using a pockels cell and a polarizer. Then the laser light is expanded to be 90 x 10 mm^2 by cylindrical lenses and goes to the second zigzag slab amplifier using larger laser glass of 110 x 17 mm^2 in cross-section and 205 mm in length. Here twenty four Xe flash lamps (twelve lamps on each side) are used for pumping. The stored energy density estimated from the small signal gain for each amplifier is 0.3 J/cm^3. Details of these amplifiers are described in Ref. 10. Figure 2 shows a preliminary result of amplification at 0.1-Hz repetition rate for 30 contiguous shots. The mean energy is 10.7 J with fluctuation of ~ 7% in the root mean square. Then the amplified laser light is reshaped by cylindrical lenses to be 80 x 80 mm^2 while going to the pulse compressor. The pulse compressor consists of three diffraction gratings with 1740 grooves/mm and flat mirrors set in a vacuum chamber [3].

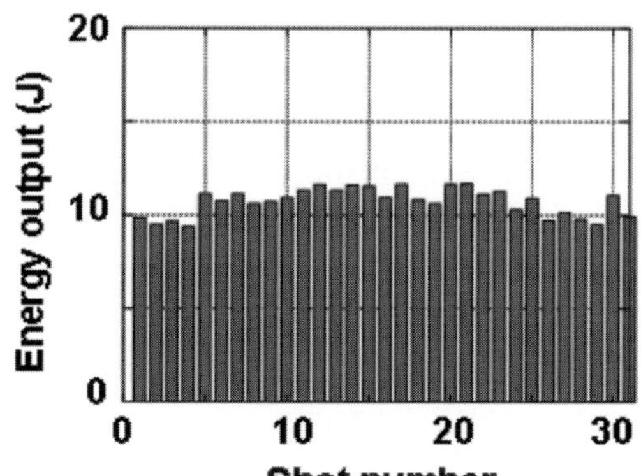

Fig. 2 Energy output and stability of TOPAZ.

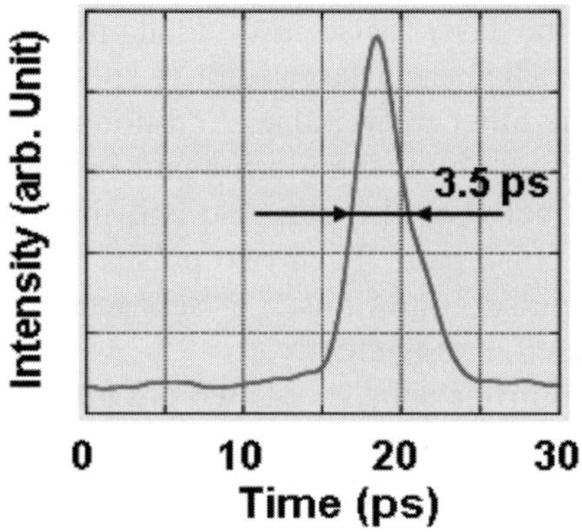

Fig. 3 Main pulse shape after compression.

Figure 3 shows the temporal shape of the compressed main pulse measured by a streak camera. Considering the instrumental resolution, the pulse duration is 3.5 ps at this time. It is noted that shorter duration, around 1-ps, can be possible by tuning the pulse compressor. Then the compressed laser lights are focused on the targets in a line shape by means of off-axis parabolic mirrors [3]. Throughput of the compressor and line focusing system is about 70%, therefore on-target laser energy is ~ 7.4 J.

3 XRL generation using TOPAZ

Using TOPAZ, we demonstrated XRL generation with Ni-like Mo (λ = 18.9 nm) and Ni-like Ag (λ = 13.9 nm). Pulse duration of the main pulse was 10 ps and 3 ps for Ni-like Mo and Ni-like Ag respectively. For the both case, prepulse of 400-ps duration was set at 2.5 ns before the main pulse. A lasing spectrum of the Ni-like Mo laser and a far field pattern of Ni-like Ag laser were clearly observed shown in Figs. 4. The XRL generation strongly depended on the prepulse condition. It should be noted that in case of the irradiation without prepulse, the XRL couldn't be observed. The optimization of the prepulse condition for XRL generation is still going on.

4 Summary

We developed new driver laser system, TOPAZ, for double target XRL at repetition rate of 0.1 Hz. TOPAZ provided high contrast laser light with energy of 7.4 J and duration of 3.5 ps. A prepulse of a few hundreds ps is added before the main pulse. Separation between the main pulse and the prepulse can be adjusted. Using TOPAZ, Ni-like Mo laser at the wavelength of 18.9 nm and Ni-like Ag laser at the 13.9 nm are successfully observed. Optimization of irradiation condition, especially prepulse condition, for the single and double target XRL generation is continued.

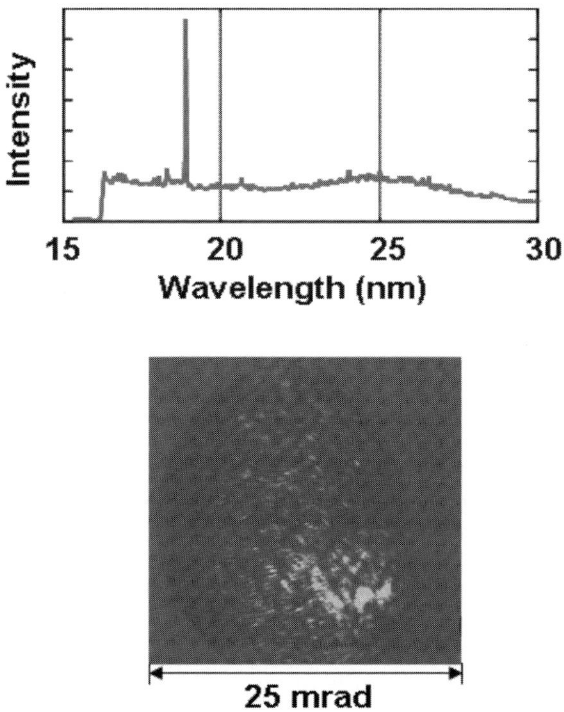

Fig. 4 Spectrum of Ni-like Mo XRL and far field pattern of Ni-like Ag XRL obtained by using TOPAZ.

Acknowledgement

This work was supported by KAKENHI 20740326.

References

1. Nishikino M., Tanaka M., Nagashima K., Kishimoto M., Kado M., Kawachi T., Sukegawa K., Ochi Y., Hasegawa N., and Kato Y.: 'Demonstration of a soft-x-ray laser at 13.9 nm with full spatial coherence', *Phys. Rev. A* 68, 0618021, 2003.
2. Tai R.Z., Namikawa K., Sawada A., Kishimoto M., Tanaka M., Lu P., Nagashima K., Maruyama H., and Ando M,: ' Picosecond View of Microscopic-Scale Polarization Clusters in Paraelectric BaTiO3', *Phys. Rev. Lett.* 93, 087601, 2004.

3. Namba S., Hasegawa N., Nishikino M., Kawachi T., Kishimoto M., Sukegawa K., Tanaka M., Ochi Y., Takiyama K., and Nagashima K.: ' Enhancement of Double Auger Decay Probability in Xenon Clusters Irradiated with a Soft-X-Ray Laser Pulse ', *Phys. Rev. Lett.* 99, 043004, 2007.
4. Tanaka M., Nishikino M., Yamatani H., Nagashima K., Kimura T., Furukawa Y., Murakami H., Saito S., Sarukura N., Nishimura H., Mima K., Kagamitani Y., Ehentraut D., and Fukuda T.: 'Hydrothermal method grown large-sized zinc oxide single crystal as fast scintillator for future extreme ultraviolet lithography', *Appl. Phys. Lett.* 91, 231117, 2007.
5. Kawachi T., Kado M., Tanaka M., Hasegawa N., Nagashima K., Sukegawa K., Lu P., Takahashi T., Namba S., Koike M., Nagashima A., and Kato Y.: 'Development of a pumping laser system for x-ray laser research', *Appl. Optics* 42, 2198-2205, 2003.
6. Luther, B.M.; Yong Wang; Larotonda, M.A.; Alessi, D.; Berrill, M.; Rocca, J.J.; Dunn, J.; Keenan, R.; Shlyaptsev, V.N.: 'High repetition rate collisional soft X-ray lasers based on grazing incidence pumping', IEEE Journal of Quantum Electronics, 42, 2006.
7. Tümmler J., Janulewicz K. A., Priebe G., and Nickles P. V.: '10-Hz grazing–incidence pumped Ni-like Mo x-ray laser', Phys. Rev. E 72, 037401, 2005.
8. Weith A., Larotonda M. A., Wang Y., Luther B., Alessi D., Marconi M. C., Rocca J. J., and Dunn J.: 'Continuous highrepetition-rate operation of collisional soft-x-ray lasers with solid targets', Opt. Lett. 31, 1994–1996, 2006.
9. Lindau F., Lundh O., Persson A., Cassou K., Kazamias S., Ros D., Pl´e F., Jamelot G., Klisnick A., de Rossi S., Joyeux D., Zielbauer B., Ursescu D., K¨uhl T., and Wahlstr¨om C.-G.: 'Quantitative study of 10 Hz operation of a soft x-ray laser—energy stability and target considerations', *Optical Express* 15, 9486-9493, 2007.
10. Ochi Y., Hasegawa N., Kawachi T., and Nagashima K.: 'Development of a chirped pulse amplification laser with zigzag slab Nd:glass amplifiers dedicated to x-ray laser research', *Appl. Optics* 46, 1500-1506, 2007.

Part 4 – Optical-Field-Ionised (OFI) X-Ray Lasers

Toward Ultraintense Compact RBS Pump for Recombination 3.4 nm Laser via OFI

S. Suckewer, J. Ren, S. Li, Y. Lou, A. Morozov, D. Turnbull, Y. Avitzour*

Princeton University, Princeton, NJ 08544, USA

Abstract. In our presentation we overview progress we made in developing a new ultrashort and ultraintensive laser system based on Raman backscattering (RBS) amplifier /compressor from time of 10th XRL Conference in Berlin to present time of 11th XRL Conference in Belfast. One of the main objectives of RBS laser system development is to use it for pumping of recombination X-ray laser in transition to ground state of CVI ions at 3.4 nm. Using elaborate computer code the processes of Optical Field Ionization, electron energy distribution, and recombination were calculated. It was shown that in very earlier stage of recombination, when electron energy distribution is strongly non-Maxwellian, high gain in transition from the first excited level n=2 to ground level m=1 can be generated. Adding large amount of hydrogen gas into initial gas containing carbon atoms (e.g. methane, CH_4) the calculated gain has reached values up to 150 – 200 cm^{-2} Taking into account this very encouraging result, we have proceed with arrangement of experimental setup. We will present the observation of plasma channels and measurements of electron density distribution required for generation of gain at 3.4 nm.

1 Introduction

The soft X-ray (SXL) and XUV lasers, due to their presently compactness, excellent beam quality and very reliable operation in wavelength range of 10 – 50 nm becomes practical tools for high resolution microscopy, micro-holography, very high plasma density measurements, applications to semi-conductor surface studies, and nano-lithography[1-4]. A key element here is the cost and availability of these devices, and intensive efforts was made to develop such compact soft x-ray lasers that their suitability for applications in academic and industrial laboratories became reality.

However right now, in our opinion (opinion also expressed at this conference), the crucial issue is development of a compact, repetition rate $\geq$ 1Hz X-ray laser in “water window’. This and applications of present SXLs should be “two legs” on which X-ray laser field would move towards wider recognition and broader implementations worldwide. We believe that now we have very good chances in succeeding of development recombination 3.4 nm laser.

Our recent computer modeling [5-7] provides us with good “receipt” about generating high gain in 2-1 transitions in CVI ions. We have now also appropriate laser for fast Optical Field Ionization(OFI) of carbon ions. This laser (~15 TW, 100 fsec Ti/Sapphire laser) can provide required intensity, according to results of calculations, of ~ 10^{19} Watt/cm^2 at the fundamental (0.8μm) and second harmonic (0.4μm) wavelengths. This Ti;Sapphire laser will be replaced in future by presently being developed at Princeton more powerful and more compact laser system base on Raman Backscattering (RBS) amplification and compression in plasma.

Beside X-ray laser development, ultrashort and ultraintensive pulse lasers become important for an increasingly large range of applications. The RBS amplification and compression for creation of ultrahigh intensity laser pulses may provide an alternative to the Chirped Pulse Amplification (CPA) method, which for two decades enabled tremendous progress in the generation of such pulses. The new RBS technique should be especially important for the development of 20-30 femtosecond pulses with intensities reaching 10^{20} W/cm^2 [8,9] and later 10^{21} W/cm^2 in quite compact, university size, systems. We also expect that the outcome from this effort will show a path towards RBS laser operating at intensities of 10^{24}-10^{25} W/cm^2 [10,11].

In next Section 2 we describe our recent (i.e. after Inter. X-Ray Laser Conference in Berlin, 2006) results on RBS amplification and compression, following by Section 3 on the most important results of calculations for development recombination 3.4nm laser and status of experimental setup.

2 Single and Double Pass Raman Backscattering Amplifier and Compressor

Compression of high laser energies through CPA [12] requires a large aperture diffraction grating, to avoid laser-induced damage. The above limits on CPA remain true, notwithstanding recent improvements through nonlinear optical techniques that have extended the range of frequencies, powers, intensities and pulse durations [13-15]. However, both CPA and derivative techniques [16] still rely upon time-stretching and time-recompressing of a short pulse. Hence, higher laser pulse energies can only be contemplated through larger diffraction gratings [17]. Plasma is clearly the medium of choice for laser pulses of extremely high power densities, and this is the basis of the ultraintense lasers we are in process of developing.

a. Enhancing the amplification and energy efficiency of the Raman Backscattering Amplifier and Compressor. Raman Backscattering Amplifier and Compressor scheme is based on the resonant interaction of two counter-propagating electro-magnetic waves, known as pump and seed, to convert energy stored in the long pump pulse to the short seed pulse. More specifically, in plasma the interaction of pump and seed, with frequencies of ω_{pump} and ω_{seed}, respectively, excites a longitudinal plasma wave (Langmuir wave), whose frequency ω_{pe} is determined by the plasma density. Under the resonant condition, $\omega_{pump} = \omega_{seed} + \omega_{pe}$, the plasma-mediated three wave interactions scatter energy from the pump to the seed. The output pulse undergoes simultaneous amplification and compression, and since plasma as the gain medium is impervious to optical damage, the short pulse can grow to extraordinary power.

We have demonstrated [8, 9] that by increasing the plasma amplification bandwidth and more efficient utilizations of the pump energy with a double pass design, the performance of Raman Backscattering Amplifier and Compressor can be significantly enhanced and the efficiency of energy transfer from pump pulse to amplified pulse can be very much improved. Figure 1 shows the experimental setup. We started from the first pass of amplification (removing dichroic mirrors M_1 and M_2 from the setup). The long pumping pulse (central wavelength 803nm, bandwidth of ~12 nm and pulse duration of ~ 20ps) enters the 2 mm long plasma channel from the right. The plasma channel is preformed by ionizing an ethane gas jet using a prepulse of 1064nm wavelength, ~6ns pulse duration and ~500mJ pulse energy. The pumping pulse interacts with the short seed pulse (central wavelength 878nm, with a bandwidth of ~9nm and pulse duration of ~ 500 fsec), which is injected from the left to the plasma channel.

The temporal overlap between the pump and the seed was initially roughly achieved through the pump and the seed cross signal in a second-harmonic-generation (SHG) crystal and then fine-tuned with the pump delay stage by looking for the maximum amplification during the experiment. The spatial overlap was first obtained by using diagnostic and imaging systems which consists of two He-Ne beams, alignment optics and a CCD camera. We then adjusted the pump reflecting mirror M_3 to tilt the injection of the pump in respect to the axis of the plasma channel. The plasma density profile had a plateau on its axis with increasing density along the radial direction due to expansion the plasma channel [8, 9]. Therefore, a plasma density gradient was introduced along the pump seed interaction path by letting the beam go slightly "off axis". Taking into account that as the laser pulse propagates through the plasma with off-axis geometry, its direction will be modified by the plasma gradient due to refraction, in the experiment we gradually adjusted and searched for the optimal pump direction and path in the plasma channel. The seed beam direction and path was adjusted accordingly (with seed reflecting mirror M_4) to ensure its best overlap with the pump for the optimal RBS amplification. The optimal RBS amplification was indicated by the maximum gain and the widest amplified seed spectrum, therefore the shortest seed pulse width.

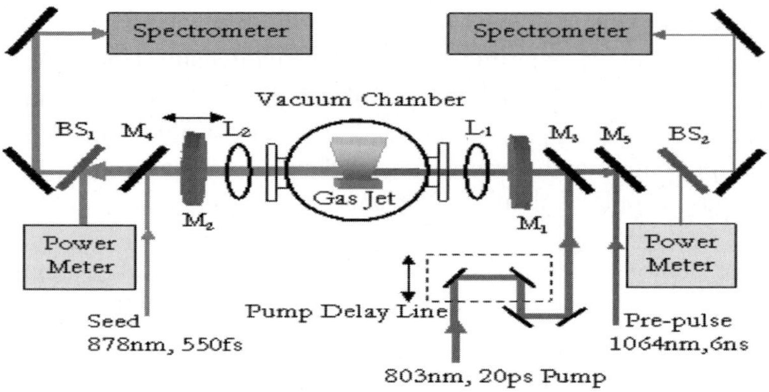

Fig. 1 Schematic diagram of the experimental set-up. Dichroic mirror M_1 has high reflectivity at 878nm (reflecting seed) and high transmission at 803 and 1064nm. Dichroic mirror M_2 has high reflectivity at 803nm (reflecting pump) and high transmission at 803 and 1064nm. M_3, M_5 and M_4 are reflecting mirrors for pump, prepulse and seed, respectively. L_1 and L_2 are the focusing lens and BS_1 and BS_2 are beam splitters.

Introducing a plasma density gradient along the path of pump and seed interaction imparts a larger amplification bandwidth for RBS. As the amplitude of the input seed grows during propagation in plasma, its bandwidth also increases and large bandwidth of the seed pulses result in breaking down of the resonant condition. Increased bandwidth of the plasma along the path of the interaction can enhance coupling between the pump and seed and leads to increased output seed energy. It is also noticed that the pump pulse was chirped, which comes from the procedure of lengthening the pump duration from its maximum compression (100fs, minimum chirp), to the desired pulse length of ~20ps. A chirped pump requires different resonant conditions along the plasma channel for different frequency component, and hence a properly introduced plasma density gradient can compensate for the pump chirp and enhance the RBS amplification.

After obtaining the maximum amplification for the first pass, we proceeded to the second pass by inserting a dichroic mirror (M_2) of high reflectivity at 803 nm and high transmission at 878 and 1064 nm between M_4 and L_2. Symmetrically, another dichroic mirror (M_1) of high reflectivity at 878 nm and high transmission at 803 and 1064 nm was inserted between M_3 and L_1. Setting up a second pass of RBS gain was based the observation that even for large amplification and compression in the first pass, only a small fraction of the total pump energy is being used. About 80 – 85% of the initial pump energy has passed through. Therefore a second round of amplification from reflecting the "unused" portion of the pump energy and the already amplified seed and having them interact again in the same plasma further improves the pumping efficiency. In the second pass, the temporal synchronization between pump and seed was adjusted by the horizontal position of the dichroic mirror M_2; and the spatial overlap was obtained with the assistance of the diagnostic and imaging systems.

Fig. 2 shows the output energy of the amplified seed in the first pass (squares points) and in the second pass (circle points) versus the temporal delay of the pump pulse. 0 delay means that pump and seed overlap throughout the whole plasma length. For both positive and negative delays, since the pump and the seed pass part of the plasma without interacting with each other, the gain drops down. The input pump energy in the first pass was ~ 87mJ and the input seed energy was ~16μJ. Taking into account plasma transmission of the pump beam (including pump energy transfer to the seed), and losses of the pump on optical surfaces while returning to the plasma for the second interaction, the input pump energy in the second pass was estimated only ~56mJ.

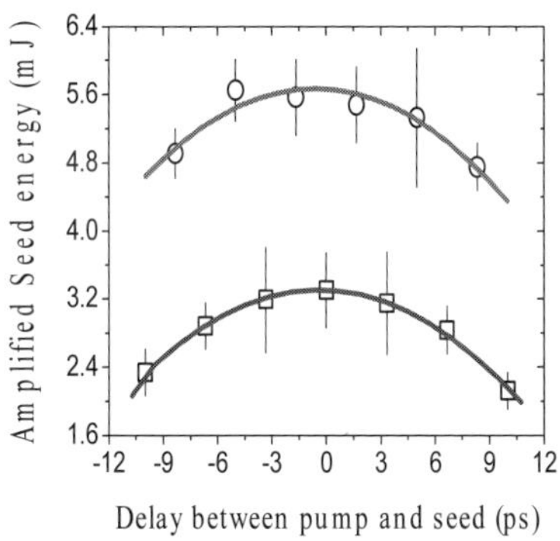

Fig. 2 Energy gain in the first (squares) and second pass (circles) with the input seed energy ~16μJ. The pump energy in the first pass was ~87 mJ, and in the second pass was ~56 mJ, which takes into account the pump losses while returning to the plasma for the second pass. The solid lines are curve fitting results to guide the eye.

With the maximum output energy shown in Fig. 2 ~5.6mJ from the second pass, and ~3.3mJ from the first pass, we measured a factor of 1.7 in increase of the pulse energy after the second pass. Since the pump energy in the second pass was slightly less than in the first pass, a new parameter which we called the "effective" gain length L_{eff} was introduced to better clarify the amplification process: $L_{eff} = L \times P_{pump}$, where L is the plasma length and P_{pump} is the pump power per square of beam diameter . Because pump pulse duration is the same for both passes and in the first approximation we can assume that also pump beam diameter is the same for the first and second pass, hence we can replace P_{pump} by pump energy E_{pump} in L_{eff} : L_{effl} = 2mm × 87mJ and L_{eff2} = 2mm × 56mJ. The ratio of the accumulated effective gain length after the second pass with that of the first pass (L_{efffl} + L_{eff2}) / L_{efffl} = 1.6, is not far from the measured energy output ratio of 1.7, indicating an almost linear grow of the output energy with the effective gain length.

Fig. 3 shows the measured output pulse width τ versus the output energy E. A least mean square error curve fitting of $E^a\tau$ = const indicated that a ≈1, which is consistent with theoretical prediction [10]

$$g_M \times \tau \sim \frac{\pi}{\sqrt{2\omega\omega_{pe}}} = const \qquad (1)$$

Here g_M is the maximum vector potential of the amplified pulse, and $g_M \propto (E/\tau)^{1/2}$. Using this relationship, we estimated the output pulse width after the second pass to be ~50fs based on the obtained second pass energy of ~ 5.6mJ.

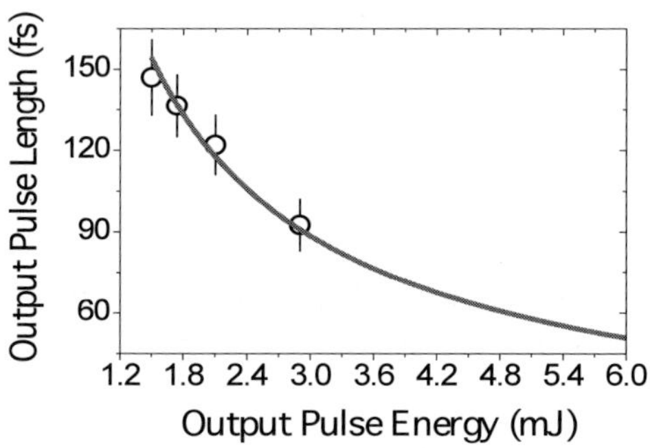

Fig. 3 The relationship between the measured output pulse width and the output energy, with the input seed energy ~16μJ.

b. Extending the region of resonance. With the gain increase from the second pass, the energy conversion efficiency from the pump to the seed reaches ~6.4%, a factor of more than 6 improvement compared to the best of our earlier result[18] .To further improve the efficiency and performance of the system, we were searching for the optimal operating conditions. Fig. 4 shows the amplified seed energy vs. the tilting angle of the pump beam in both x (left panel) and y (right panel) directions in the first pass, with pump energy ~100mJ and input seed energy ~16μJ. The best tilting angles were found to be ~5-6 degrees for x and y directions. It is interesting to note that these angles match the conditions in the Grazing-Incident Pumping scheme for soft x-ray lasers[19], where $\sin^2(\theta) = n_e/n_c$, θ is the grazing incident angle, n_e is the electron density and n_c is the critical plasma density.

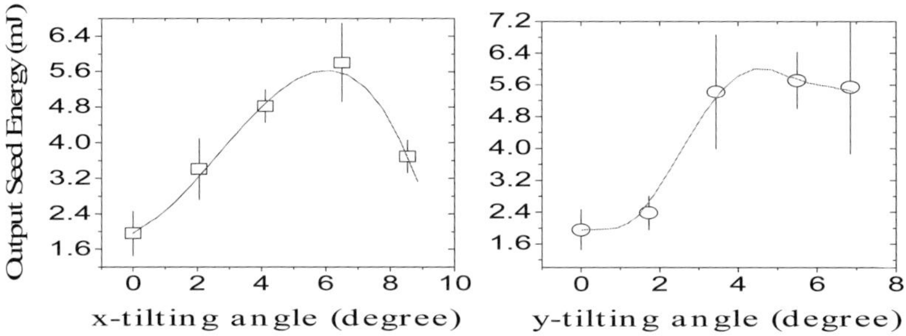

Fig. 4 The relationship between the tilting angle of the pump beam in both x (left) and y (right) directions in the first pass, pump energy ~100mJ and input seed energy ~16μJ.

We should also notice that all our previous results were obtained in a ~2mm plasma. As theory indicated, for a certain pump intensity and wavelength, the optimal plasma gain length is limited by instabilities such as Raman Forward Scattering and Modulational instability. Under the condition that the plasma wave breaks near its first maximum, the optimal plasma length is given by [10]

$$L_{amp} \sim 0.8 z_M^{2/3} \Lambda^{1/3} \tag{2}$$

Here z_M and Λ are the location of the first maximum in the amplified pulse and the number of exponentiations in the instability. For pumping intensity ~ 2×10^{14} W/cm^2 and pump wavelength $\lambda = 800$nm (these are typical parameters used in the experiment), L_{amp} ~ 3-6mm [10]. Recently, we demonstrated that a ~4mm plasma channel with the desired on-axis and radial density can be generated with 2 prepulses (Fig.5, also [9]), which should increase the gain length for RBS amplifier to further improve amplification and energy conversion efficiency.

c. To reach intensity of 10^{20} W/cm^2 in focus. In Fig.5 is shown a new setup for development RBS amplifier & compressor to reach intensity of 10^{20} W/cm^2 [9]. We will use newly installed 2 J Ti/Sapphire amplifier (~1.5 J after compressor) and a plasma channel that is 4 mm long and ~0.25 mm diameter for propagation the pump and seed with beam diameters ~0.20 mm, hence much larger than being use in recent experiments. Our extremely good recent results for intensity amplification and pulse compression [8], which are already close to the conditions required for generation very high intensity in focus, indicate that with more than order of magnitude higher pump energy our near future goal should be obtainable.

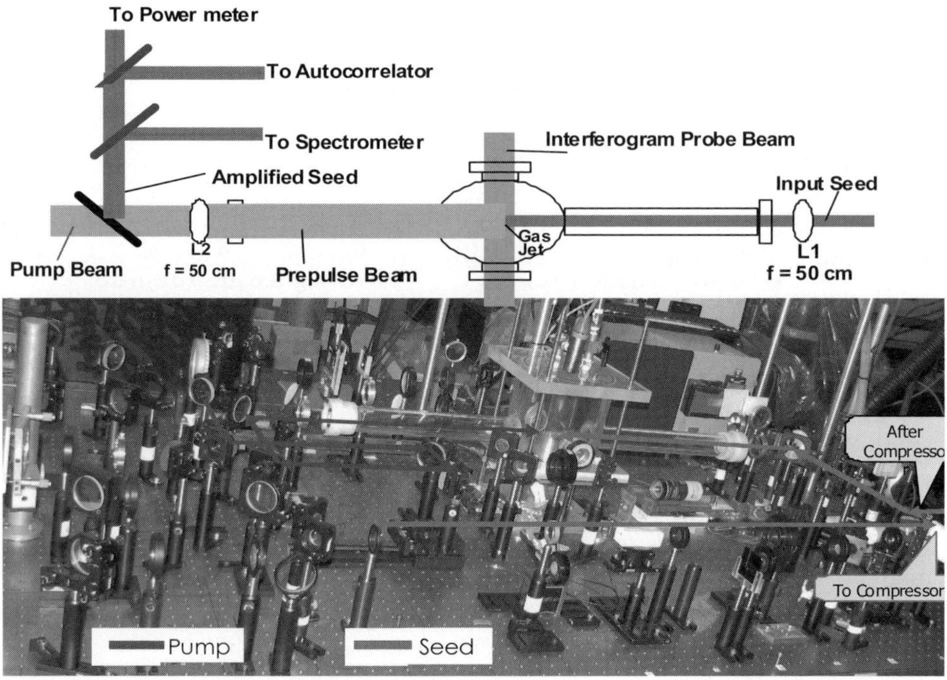

Fig. 5 Schematic and photo of a new setup for RBS laser with focus intensity of 10^{20} W/cm^2

To further improve the efficiency and performance of the system, we have concentrated our effort on generating good, 4 mm plasma channel [9]. In order to increase the plasma channel length from 2mm to 4mm in our new setup, the Rayleigh length of the ionization prepulse was increased by at least a factor of 2, and the prepulse energy was doubled to keep the intensity at the same level.

Additionally, in generating longer plasma, the temporal shape and spatial distribution of the prepulse might cause instabilities in the ionization process and affect the formation and longitudinal uniformity of the plasma channel. Therefore, our solution to this problem was to use two ionization pulses (two prepulses, prepulse1 and prepulse2). The foci of the two prepulse beams were adjusted and slightly separated (by ~0.5mm) along the axis of the plasma channel.

Fig. 6(a) shows the interferogram and Fig. 6(b) shows a 2-dimentional density profile of the plasma channel formed by two prepulses. The interferogram and the plasma density profile were obtained using a Mach-Zehnder interferometer. The two prepulses were both generated by two Nd:YAG lasers. Prepulse1 was ~ 10 ns with a focused spot size diameter ~ 65μm and ~300mJ beam energy and prepulse2 was ~ 6 ns also with a focused spot size diameter ~ 65μm and ~700mJ beam energy. The plasma was probed at a delay of ~22ns after prepulse1 and at a delay of ~16ns after prepulse2. It can be seen from Figs.6(a,b) that the plasma channel has a length of ~4mm with desired on-axis electron density and good its radial distribution for channeling laser beams. Obtained very good 4 mm plasma channel is very important result not only for our near future experiment but also for all our future experiments. Also very good plasma channel and relatively large diameters of the pump and seed will be very beneficial to obtain wave front amplified pulses close to diffraction limit (for experiment in [8] we measured amplified pulses to be near 2 – 2.5 diffraction limit even for much less favorable conditions for pulses propagation in plasma).

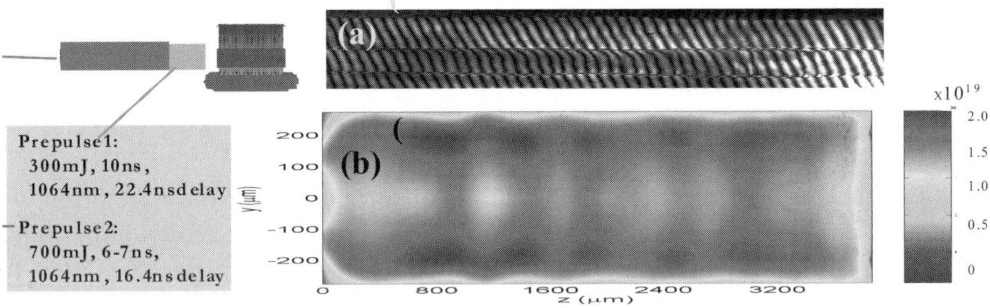

Fig. 6 Interferogram (a) and 2-dimentional density profile (b) of the ~ 4mm plasma channel for SRBS amplification (*note*: very good plasma uniformity along axis with appropriate density radial gradients for channeling of laser beams).

3 X-Ray Laser: Computer "Prescription" and Experimental Arrangement for Gain Generation in "Water Window" at 3.4nm

The main attraction of generating lasing in the transition to ground state of ions is its high quantum efficiency and favorable scaling to shorter wavelengths of the gain. It is possible to achieve lasing within the "water window" using relatively moderate energy (1 -2 J/pulse) lasers as pumps at repetition rate ≥ 1Hz. Our calculations [5-7] have shown the feasibility of achieving high gain with a recombination scheme in the 2→1 transition of CVI using ultrashort laser pulses (below100 fsec) with peak intensity in order of $I_p \sim 10^{19}$ Watt/cm^2. The ultrashort pulses are required in order to create strongly non-equilibrium plasma (strongly non-Maxwellian electron energy distribution) and generate minimum heat of the plasma during process of ionization, while atoms in lasing plasma column are totally stripped of electrons. These are crucial conditions for creation population inversion (hence a gain) in transition to ground states of hydrogen-like ions in general, and for 2-1 transition in CVI ions, in particular.

Presently the best way to satisfy above mentioned conditions is to use optical-field-ionization (OFI) [20-22] by means of ultraintense and ultrashort laser pulses. After the ions are fully stripped of their electrons they recombine by three-body recombination. The rate of such recombination is proportional to N_e^2 and $\mathbf{n}^4$ (N_e is electron density, and **n** is the principal quantum number of the ion state). Therefore the transition from C^{6+} to H-like CVI occurs primarily to the states with high n. Collisional transitions to level n = 2 occurs faster than to ground level n = 1, creating population inversion between levels n = 2 and n = 1.

a. Computer modeling results. We have developed an elaborate numerical model to characterize recombination gain in the 2 – 1 transition of C VI ions at 3.4 nm [7]. Recombination gain relies on having fully stripped C^{6+} ions in relatively cold plasma. The ionization mechanism that is used to achieve these fully stripped ions is tunneling ionization by ultrashort, ultraintense (with intensities in the range of $5x10^{18}-10^{19}$ W/cm^2) laser pulses. Due to the short pulse duration, minimal heating is produced during the ionization. However, when calculating the average energy that is absorbed during the ionization we found that the absorbed energy still corresponds to an electron temperature that would not allow for significant population inversion in the transition to ground state. Only by taking into account the actual phase-space distribution function of the plasma, including effect from the non-Maxwellian electrons distribution function after ionization, high gain is indeed feasible for the 2 – 1 transition[6].We have also shown that the gain can be enhanced and become less stringently-dependent on exactly matching the values of the experimental parameters, if quite large amount of hydrogen gas is added into the initial gas containing carbon atoms (e.g. methane, CH_4). In calculations hydrogen atoms density was as high as 95% of total gas density of potentially lasing medium for final electron density ~ 10^{20}/cm^3. Although diluting carbon density (gain G is proportional to the density of CVI ions) has negative effect on max gain for given plasma electron density, which is limited approximately to 10^{20}/cm^3 for lasing to ground state of CVI ions, nevertheless lower plasma temperature with such hydrogen concentration has much more positive effect on recombination rates, hence on gain.

The recombination was calculated by solving the rate equations, taking into account all the relevant atomic (ionic) processes. The values of the rate coefficients in the rate equations were obtained by integrating the cross-sections over the actual non-Maxwellian, time-dependent, electron distribution function that was obtained from the Fokker-Planck equations (Fig. 7). The results for gain G (in cm^{-1}) for CVI 2 – 1 transition for 2 different diameters of pumping laser beam (50 fsec,10^{19} W/cm^2) versus 3.4nm beam diameter as a function of time were obtained[7].The gains as high as G ≈150-200 cm^{-1} were calculated. From the experimental point of view, probably the most positive result is the one with 15 μm pumping beam diameter, for which the gain about $80 cm^{-1}$ is expected in up to 2.5 μm of X-ray beam diameter and lasting for ~0.6 psec.

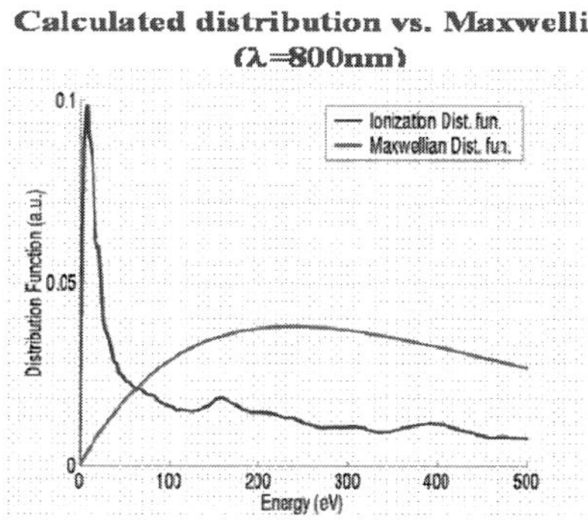

Fig. 7 Non-Maxwellian electron energy distribution function compared with Maxwellian distribution function for the same total energy for electrons. These distribution functions were calculated for the end of ionization time. Most of the electrons are concentrated in the lower energy region and participate in three-body recombination, whereas contribution of highly energetic electrons to recombination process is negligible.

b. Experimental Arrangement and Creation Plasma Channels. Our setup for development 3.4 nm laser is already constructed and we start using it for plasma channel creation (Fig, 8) and its density measurements relevant to gain generation in CVI ions. The experimental setup is shown in Fig. 9. In this setup we have relied on H. Milchberg and his group's pioneering works on "plasma fibers" for guiding high intensity laser beams through the plasma [23-26]. They have shown that combining axicon, which provides Bessel function type of elongated focus, with spherical lens can provides excellent plasma channel for X-ray laser development [27].

Fig. 8 Interferogram showing channel created in hydrogen plasma with electron density $\sim 10^{20}$ cm^{-3}, required for generation high gain at 3.4nm

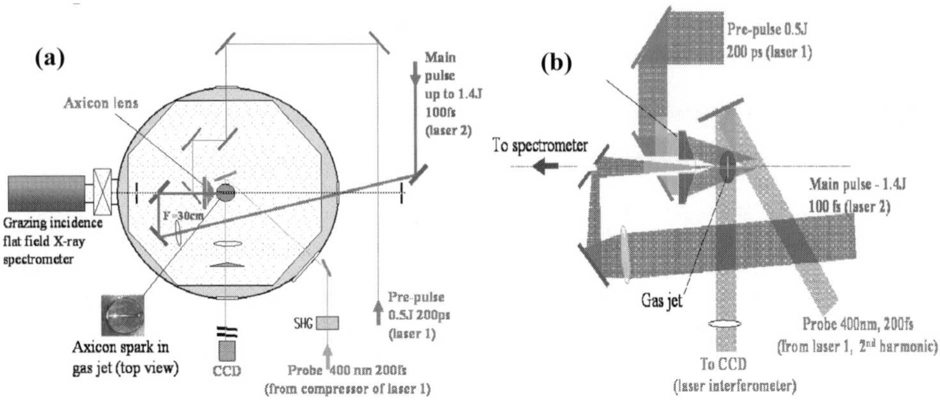

Fig. 9 Schematic of experimental arrangement with100 fsc ultrahigh intensity Ti/Sapphire laser

Acknowledgments

The authors are thankful to N.Fisch and J.Wurtele for very helpful discussions about RBS results, E. Valeo for help in computation of gain generation, and G.Pert for providing some of his papers prior to publications. N.Tkach provided very devoted technical support to both experiments.

This work was supported by NSF ECS, DOE/NSF -Plasma and DOE/NNSA grants.

References

*Present address: Texas University at Austin

1. Menoni, C. et al.,"Nano-Scale Imaging With Tabletop Soft X-Ray Laser: Sub-38 nm Resolution", Proc. 10th Intern. Conference on X-Ray Lasers, p.417 (Berlin, 2006)
2. Grava, J. et al., "Soft X-Ray Laser Interferometry of Colliding Plasmas", ibid, p471
3. Capeluto, M. et al., "Tabletop Nanopaterning Using Soft X-Ray Laser", ibid, p.491
4. Menoni, C. et al. This Conference
5. Avitzour,Y., Suckewer, S. and Valeo, E, Phys. Rev. E. **69** 046409 (2004)
6. Avitzour, Y., Ph.D. Thesis "Numerical Modeling of Recombination X-Ray Lasers in Transition to Ground State", Princeton University (2006).
7. Avitzour,Y., Suckewer, S., JOSA B **24** 819 (2007)
8. Ren, J., Cheng,W., Li, S. andSuckewer, S., Nature Physics **3**, 732 – 736 (2007)
9. Ren,J., Li,S., Morozov,A., Suckewer,S., Yampolsky, N.A. et al., Phys. Plasmas **15**, 056702 (2008)
10. *Malkin, V.M., Shvets, G. and Fisch, N.J., Phys. Rev. Lett. **82**, 4448(1999); also Malkin, V.M., Shvets,G., and Fisch, N. J., Phys. Rev. Lett. **84,** 1208 (2000)*

11. *Shvets, G., Fisch, N.J., Pukhov, A. and Meyer-ter-Vehn, J., Phys. Rev. Lett* ***81****, 48792 (1998); also V. M. Malkin and N. J. Fisch, Physics of Plasmas* ***8,*** *4598 (2001).*
12. Strickland, D. and Mourou, G., Opt. Commun. **56** 219 (1985)
13. Dubietis,A., Jonusauskas,G. and Piskarskas, A., Opt. Commun. **88** 437- 440 (1992)
14. Ross, I., Matousek,P., Towrie, A., Langley, M., Collier, J., Opt. Commun. **144,** 125 (1997)
15. *Ross, I., Collier, J., Matousek, P., Danson, C. et al., Appl. Opt.* ***39*** *2422 (2000)*
16. *Osvay, K., Kurdi, G., Klebniczki,J., Csatari, M. et al., Appl. Phys. B* ***74****, 1639 (2002); also Osvay,K. et al., Appl. Phys. Lett.* ***80****, 1704 (2002)*
17. *Stuart, B.C., Feit, M.D., Herman, S. et al., J. Opt. Soc. Am. B* ***13****, 459 (1996)*
18. Cheng,W., Avitzour,Y., Ping,Y., Suckewer, S., Fisch,N., Hur,M., Wurtele,J., Phys. Rev.Lett. **94**,045003(2005); also Ping,Y. et al., Phys. Rev. Lett. **92**,175007 (2004).
19. *Keenan, R., Dunn, J. Shlyaptsev,V., Smith, R.F., Patel P. and Price, D., SPIE* ***5197****, 213 (2003); also Keenan, R., Dunn, J., Price,D., Patel,P., Smith,R.F., Shlyaptsev,V., Phys. Rev. Lett.,* ***94*** *103901(2005)*
20. Keldysh, LV., Soviet Phys. JETP **20** 1307 (1965)
21. Perelomov,A.M, Popov,V.S.and Terent'ev,MV.,Soviet Phys.JETP **23** 924 (1965)
22. Burnett, N.H. and Corkum, P.B., JOSA B **6** 1995 (1989)
23. Durfee,C.G. and Milchberg, H.M., Phys. Rev. Lett. **71**, 2409 (1993)
24. Durfee,C.G. Lynch, J., Milchberg,H.M., Phys.Rev.E. **51** 2368 (1995)
25. Milchberg, H.M., Durfee, C.G., and McIlrath, T.J., Phys.Rev.Lett.**75**, 2494 (1995); alsoT. R. Clark and H. M. Milchberg, Phys. Rev. Lett. **78**, 2373 (1997)
26. Sheng,H., Kim, K.Y., Kumarappan,V., Layer, B.D., Milchberg, H.M., Phys.Rev. E.**72** 036411 (2005); also Kumarappan,V., Kim,K.Y., Milchberg,H.M., Phys.Rev.Lett. **94** 205004 (2005)
27. Milchberg,H.M., Durfee,CG., Lynch,J., JOSA B, **12** 731 (1995)

High Brightness Optical-Field-Ionization X-Ray Lasers Driven in Plasma Waveguides

M.-C. Chou[1], P.-H. Lin[2], R.-P. Huang[3], S.-Y. Chen[1,4], H.-H. Chu[4], J. Wang[1,4] and J.-Y. Lin[3]

[1]Institute of Atomic and Molecular Sciences, Academia Sinica, Taipei 106, Taiwan
[2]Department of Physics, National Taiwan University, Taipei 106, Taiwan
[3]Department of Physics, National Chung Cheng University, Chia-Yi 621, Taiwan
[4]Department of Physics, National Central University, Jhong-Li 320, Taiwan

Abstract. We experimentally demonstrate that the plasma waveguide can be efficiently produced in a pure Xe, Kr, and Ar cluster jet operated at a high backing pressure with axicon ignitor-heater scheme. The lasing photon number of Ni-like Kr laser at 32.8 nm generated in waveguide is dramatically enhanced by about three orders of magnitude in comparison to that without plasma waveguide, resulting in a photon number of 8×10^{10} and an energy conversion efficiency of 2×10^{-6} with a pump pulse of just 235 mJ. Due to the high atom density and long gain length provided by the plasma waveguide, Ne-like Ar laser at 46.9 nm is achieved in OFI plasma channel at the first time and simultaneous lasing in two ion species is also generated in a 10-mm gas jet with Kr and Ar mixtures. With sufficient x-ray photons, we show that x-ray digital holographic microscopy can be achieved in single-shot measurement providing a spatial resolution of < 500 nm and a temporal resolution of < 10 ps.

1 Introduction

Since the first collisionally excited optical-field-ionization (OFI) x-ray laser was demonstrated by Lemoff et al. for the 5d-5p transition of Pd-like xenon at 41.8 nm [1], OFI collisional-excitation x-ray lasers pumped by femtosecond high-repetition-rate lasers have been shown to be a promising scheme that meets the requirements of practical applications. Saturated lasing of the same spectral line with an output of 5×10^9 photons/pulse using 330 mJ pump pulses was achieved by Sebban et al. [2]. Later, strong lasing at 32.8 nm for the 4d-4p transition of Ni-like krypton with 3×10^9 photons/ pulse has also been demonstrated using 760 mJ pump pulse by the same group [3]. All these results were achieved by focusing an intense, circularly polarized laser pulse into a gas cell filled with pure xenon or krypton. In view of the versatility of gas jets, our group demonstrated nearly saturated x-ray lasing in xenon and krypton clustered gas jets with outputs of 2×10^{10} x-ray photons/pulse and 1×10^{9} photons/ pulse, respectively [4,5].

In OFI x-ray lasers, longitudinal pumping is the preferred choice to meet the requirements of high-intensity pumping and travelling-wave pumping. A serious problem for longitudinally pumped gas-target x-ray lasers is the limited gain length caused by ionization-induced refraction [2,4]. Implementation of a waveguide to maintain a small beam size over a long distance is the best solution to increase the length of gain region for the x-ray lasing. Recently, enhancement of OFI x-ray lasing in preformed plasma waveguides driven by discharge was reported. The gain media were a capillary tube filled with xenon and buffer gas [6] or a multi-mode capillary waveguide [7]. Up to 1.5×10^{11} photons/ pulse was produced by using a pump energy of 1 J. Here we report the demonstration 400-fold enhancement of the OFI krypton 32.8-nm laser in a gas jet by an optically preformed plasma waveguide. The large gain-length product also resulted in lasing at an additional line around 33.5 nm. Optically preformed plasma waveguide in a gas jet [8–10] is most favorable among all the guiding methods reported because it allows guiding of the pump pulse and the lasing x-ray pulse simultaneously and damage-free long-term highrepetition-rate operation for practical applications. With the assistance of plasma waveguide, the high-threshold low-gain transition in Ne-like Ar ions at 46.9 nm is also achieved in OFI plasma channel at the first time, demonstrating the capability of guiding the pump beam at high intensity in the waveguide. Besides, taking advantage of the high atom density and large gain length provided by the plasma waveguide, simultaneous lasing at 32.8 and 46.9 nm in two ion species is generated in a 10-mm gas jet with Kr and Ar mixtures.

The difficulty in the development of x-ray imaging has been the light-source requirements. Bright, fully coherent illumination is necessary, which generally has meant the use of large-scale, limited-access facilities such as third-generation synchrotron radiation sources [11], or x-ray free electron lasers [12]. In this paper, we also report the demonstration of flash digital Fourier holographic microscopy with single exposure pulse from a tabletop optical-field-ionization Ni-like Kr x-ray laser at 32.8 nm [13]. An imaging x-ray concave mirror is added between CCD and sample to overcome the limit of the pixel size of CCD and increase the effective numerical aperture of the imaging system simultaneously. The amplitude and phase contrast images of sample are obtained by numerically solving Fresnel integral for a certain propagation distance via 2D fast Fourier transforms. For a 1024×1024 hologram, the reconstruction process takes only 1 second in a fast PC. In addition to the capability of high-speed imaging and post-processing, this technique also provides the way, like optical microscope, to study large objects and their details with adjustable field of view. By translating the object away from or toward the x-ray concave mirror, one can observe the sample in large scale and then zoom in to a small region of interest.

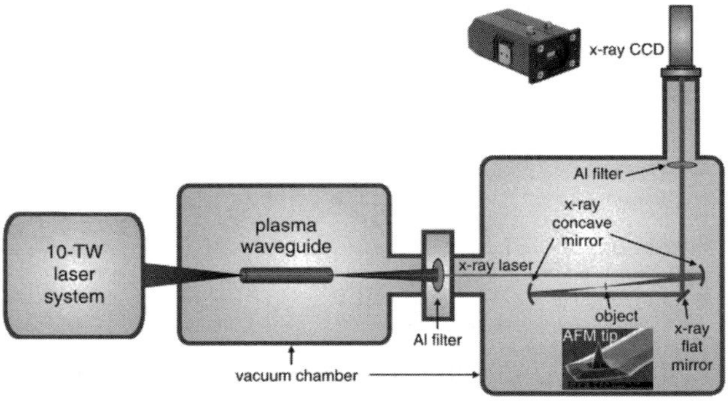

Fig. 1. Schematic diagrams of the experimental setup of soft x-ray digital holographic microscopy.

2 Experimental Setup

The experimental setup was similar to that described in [13]. A 10-TW, 45-fs, 810-nm, and 10-Hz Ti:sapphire laser system based on the chirped-pulse amplification technique was used in this experiment. The pump pulse was focused by an off-axis parabolic mirror of 30-cm focal length onto a cluster jet. The focal spot size of the pump pulse was 10-μm diameter in full width at half maximum (FWHM) with 85% energy enclosed in a Gaussian-fit profile. A quarter-wave plate was used to change the pump polarization. The clustered gas jet used for this experiment is produced from a slit nozzle and a pulsed valve. The gas-jet profile has a flat-top region of 8 mm length and a boundary of 500 μm length at both edges along the long axis. The preformed plasma waveguide was produced in pure krypton gases or the mixture of krypton and hydrogen gases with the axicon ignitor-heater scheme. The ignitor was a compressed 45 fs pulse, and the heater was the uncompressed beam with a pulse duration of 160 ps. They were focused by an axicon of 30◦ base angle to a line focus of >10-mm length. The longitudinal intensity distribution of the line focus can be approximately controlled by a beam expander installed before the axicon. A relay-imaging system was used to measure the injection beam profile at the end of the plasma waveguide to verify the guiding of the injection beam. The on-axis diagnostic was a flat-field spectrometer (FFS) made of an aperiodic grazing-incidence grating with an average groove density of 1200/mm and a back-illuminated 16-bit soft x-ray CCD camera. Mach-Zehnder interferometry with a probe pulse passing transversely through the cluster jet was use to monitor the beam propagation and waveguide formation.

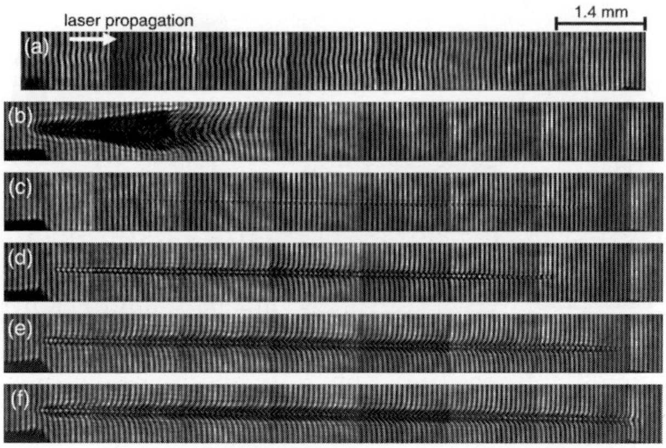

Fig. 2. Interferograms of the plasma taken at 10 ps after the pump pulse passed through the gas jet. (a) Using only the pump pulse with a focal position at 2.75 mm behind the entrance of the gas jet. Krypton atom density = 8×10^{17} cm^{-3}. (b) Using only the pump pulse with a focal position at 500 μm behind the entrance of the gas jet. Krypton atom density = 1.6×10^{19} cm^{-3}. (c) Using only a 45-mJ ignitor pulse . (d) Using only heater pulse. (e) Using a 45-mJ ignitor pulse and a 225-mJ heater pulse 200 ps after. There is no pump pulse and the interferogram is taken at 2.5 ns plus 10 ps after the heater pulse. (f) Using the pump pulse guided by the waveguide shown in (e). Krypton atom density = 1.6×10^{19} cm^{-3}.

The experimental setup of x-ray digital holographic microscopy is shown in Fig. 1. A 0.25-μm-thick aluminum filter was used to eliminate the copropagating driving laser light. The x-ray holographic microscopy consisted of three Mo/B_4C/Si multilayer mirrors with a reflectivity of 19% at 32.8 nm and a 16-bit back-illuminated x-ray CCD camera. The first x-ray concave mirror of 30-cm focal length was used to collect and focus the 32.8-nm beam and the second x-ray concave mirror of 20-cm focal length was used as an imaging lens. The 45° flat mirror was used to redirect the x-ray beam to the x-ray CCD camera. The holograms were recorded by the x-ray CCD camera with a 1024 × 1024 array of 13-μm pixels. The sample was placed between the two x-ray concave mirrors and the inherent magnification of this imaging system was 7.6. An AFM cantilever and a carbon-foil mesh were served as the test objects. The physical characteristics of the AFM cantilever are as follows: cantilever length 125 μm, width 30 μm, thickness 4 μm, tip height 10—15 μm. The carbon foil is a square mesh made of a two-dimensional array of 7 μm × 7 μm holes and 2-μm bars with 20-nm thickness. The objects were mounted on a motorized translation stage moving parallel to x-ray beam propagation.

3 Results and Discussion

3.1 OFI laser driven in plasma waveguide

In order to increase the lasing signal by increasing the atom density and extend the length of the gain region simultaneously, a plasma waveguide was implemented to overcome the adverse effect of ionization-induced refraction. The optically preformed plasma waveguide was produced by using the axicon-ignitor-heater scheme [10]. A short intense ignitor pulse ionizes the neutral gas by multiphoton ionization to provide seed electrons. After several hundred picoseconds, a succeeding long, high energy heater pulse heats up the plasma efficiently via inverse bremsstrahlung heating and further ionizes the gas by electron collision. After an adequate delay the plasma electron density in the encircling outer region becomes larger than the on-axis density and thereby a plasma waveguide capable of guiding a laser pulse is produced. Figure 2(a) shows the interfeorgram taken at 10 ps after the pump pulse has passed through the 10-mm Kr gas jet without preformed plasma waveguide. The maximum x-ray lasing of Ni-like krypton at 32.8 nm was observed at an atom density of 8×10^{17} cm^{-3} and a pump focal position of 2.75 mm behind the entrance of the gas jet as the optimal balance between pump beam convergence and ionization-induced refraction was achieved to produces the longest gain region and the shortest x-ray reabsorption region. The number of photons at the 32.8-nm lasing line was 2×10^{8} which is 5 times lower that that produced from a 5-mm gas jet due to a longer absorption length. Figure 2(b) shows the interferogram of the plasma taken at the same condition as in Fig. 2(a) except that the krypton atom density is raised to $1.6\times10^{19}cm^{-3}$ and the pump focal position was moved to 500 μm behind the entrance of the gas jet. Severe ionization-induced refraction drastically reduces the length of the gain region, and thus no x-ray lasing is observed at this condition. The plasma produced by only ignitor or heater pulse is shown in Fig. 2(c) and 2(d), respectively. It was found that plasma waveguide cannot be generated if only the ignitor or the heater was used. The guiding capability of the plasma waveguide was verified by imaging the beam size of the probe beam at the entrance and the exit of the plasma waveguide. Instead, a uniform plasma waveguide extended to ~9 mm at a 40-μm diameter was observed when both ignitor and heater were deployed as shown in Fig. 2(e). The interferogram of the plasma waveguide taken at 10 ps after the pump pulse has passed through the gas jet is shown in Fig. 2(f). The 235-mJ pump pulse was circularly polarized with a focal position of 500 μm behind the entrance of the gas jet and a delay of 2.5 ns after the heater. The guided beam size was measured to be <15 μm (FWHM) with more than 50% of the energy in the vacuum focal spot transmitting through the preformed waveguide. With the plasma waveguide the x-ray lasing photon number is dramatically enhanced by a factor of 400 compared to that with

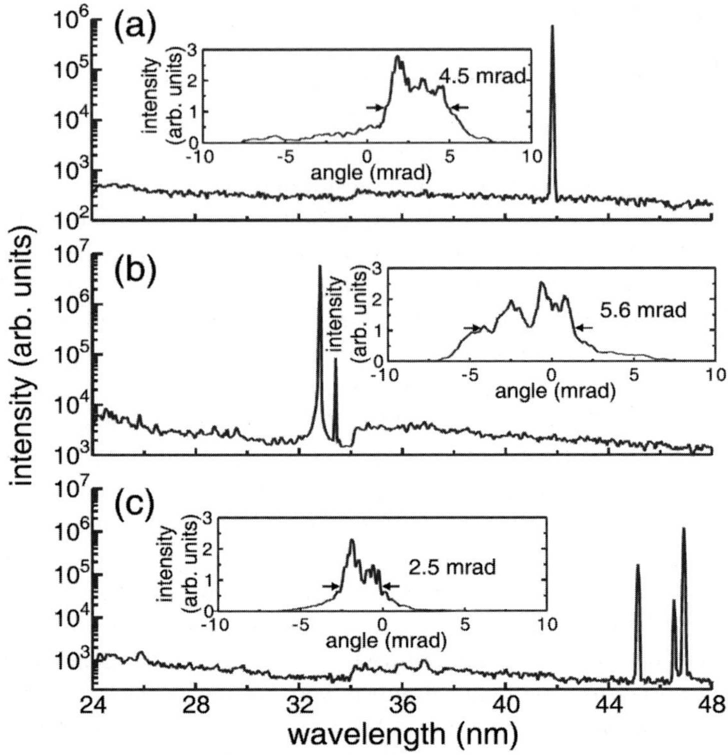

Fig. 3. X-ray spectra for an atom density of (a) $N_{Xe} = 4\times10^{19}$ cm^{-3}, (b) $N_{Kr} = 1.6\times10^{19}$ cm^{-3}, and(c) $N_{Ar} = 2.7\times10^{19}$ cm^{-3}. The heater energies were 180, 225, and 325 mJ, and the heater-pump delays were 5.5, 2.5, and 1.5 ns, respectively. The ignitor energy was 45 mJ and the ignitor-heater separation was 200 ps. Insets show the angular distributions of the 41.8 nm Xe^{8+}, 32.8 nm Kr^{8+}, and 46.9 nm Ar^{8+} lasing lines, respectively.

only the pump pulse at the optimal condition. The output reached 8×10^{10} photon/pulse with 10% fluctuation and the x-ray beam divergence was decreased to 5.6 mrad in FWHM with 20% fluctuation. The beam divergence was close to the aspect ratio of the waveguide. There was no lasing signal with only the waveguide forming pulses. The increase of on-axis plasma electron density and the strong x-ray lasing observed when the pump pulse is turned on reveals that the x-ray lasing results from optical-field ionization of the gas driven by the pump laser pulse. The spectrum and angular distribution of Ni-like Kr laser generated in the plasma waveguide are shown in Fig. 3(b). It is interesting to note that another lasing line at around 33.5 nm was also observed in the case of pure krypton plasma waveguide with an output of 4.4×10^{8} photon/pulse. It was found that the output of the 33.5-nm lasing line is always about two orders of magnitude smaller than that of the 32.8 nm lasing line for various atom densities, pump polarizations, pump energies, and heater energies. This seems to indicate that the 33.5-nm lasing line comes from a transition from the same upper level to another satellite lower level close to that of the primary lower level. It may be attributed to the $3d^9 4d\ ^1S_0 - 3d^9\ 4p\ ^3D_1$ transition. The first time observation of this satellite lasing line verifies a high gain length of such an x-ray laser scheme.

Under the same configuration of the pump and waveguide-forming pulses, high-threshold OFI collisional excitation Ne-like Ar laser at 46.9 nm was also achieved for the first time as shown in Fig. 3(c). The output photon of Ar laser was 3.4×10^9, and the beam divergence was 2.5 mrad at an atom density of 2.7×10^{19} cm^{-3}. This result suggests the presence of plasma waveguide is beneficial to x-ray laser generations required high pumping intensity. In addition to the lasing transition at 46.9 nm in Ne-like Ar ions, two other lasing lines at 45.1 and 46.5 nm were also observed and designated as $3d\ ^1P_1 - 3p\ ^1P_1$ and $3d\ ^1P_1 - 3p\ ^3P_1$ line respectively. These two lasing lines are predicted to have lasing gains when self-photo-pumping mechanism is considered [14]. This shows even low-gain lasing lines can be amplified significantly in a plasma waveguide which provides a longer gain length. Similarly strong lasing for Xe^{8+} ions at 41.8 nm was also achieved as shown in Fig. 3(a). The output photon of Pd-like Xe laser was 2.7×10^9, and the beam divergence was 4.5 mrad at an atom density of 5.1×10^{18} cm^{-3}. With a mixed-gas (Kr:Ar = 1:1) plasma waveguide x-ray lasing in Ni-like Kr at 32.8 nm and Ne-like Ar at 46.9 nm were obtained simultaneously under adequate conditions. The output photon number for Kr and Ar x-ray lasers was about 8×10^8 each. Simultaneous x-ray lasing in multiple ion species may become a power tool in plasma nonlinear optics.

3.2 X-ray digital holographic microscopy

As shown in Fig. 1, the imaging x-ray mirror produces a magnified image of the testing object, and the inherent magnification of the imaging system was 7.6 on CCD. By translating the specimen further toward the imaging mirror, the image plane will locate behind CCD to form a larger magnified object and a smaller field of view. The image recorded on CCD can be regarded as the hologram produced by the magnified object and the transmitted x-ray beam. In order to record sub-micron holographic images with a single picosecond x-ray exposure, the specimen was illuminated with a focused x-ray laser beam for producing stronger scattering signal and thus high-contrast holograms. Since two concave x-ray mirrors were arranged in a confocal geometry, the untouched beam was collected and collimated by the imaging mirror to form the reference beam on CCD. With such arrangement, the digital reference beam can be assumed as a plane wave without the need to know the exact parabolic phase function of the divergent reference beam. Image reconstruc-

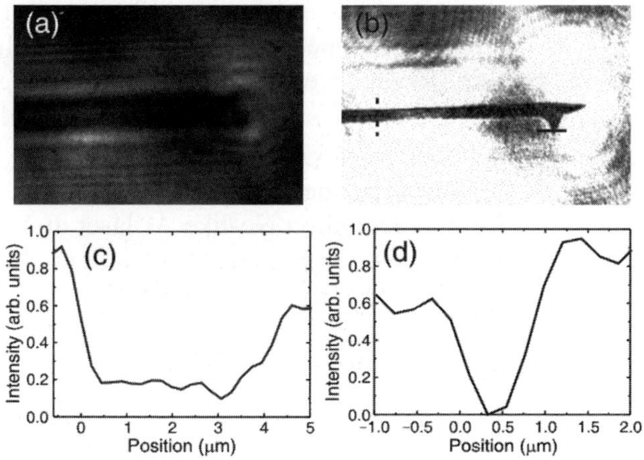

Fig. 4. X-ray hologram (a) and reconstructed amplitude image (b) of the AFM cantilever with a system magnification of 29.7×. (c) shows the line scan of the cantilever thickness at the dashed line shown in (b). (d) shows the intensity profile of the AFM tip at the solid line shown in (b).

tion was based on Fresnel-Kirchhoff diffraction integral of the recorded hologram with a propagation distance to the magnified image plane (focusing plane). A simple double fast Fourier transform algorithm treating the diffraction integral as the convolution function was written in MATLAB to simulate the amplitude and phase contrast images in 1 second. Figure 4(a) shows the single-shot AFM cantilever hologram recorded by x-ray CCD with a system magnification of 29.7. The spatial non-uniformity of the flash hologram is attributed to large speckles of the x-ray laser beam. With a proper choice of the focusing distance between the hologram and reconstruction image plane, the amplitude image of the AFM cantilever is obtained and shown in Fig. 4(b). A line scan across the AFM cantilever of 4-μm thickness is labeled in dashed line and plotted in Fig. 5(c). It is shown that the measured thickness of the cantilever fits the manufacture's specification well but the boundary sharpness of both sides is quite different due to 7.5-μm optical path difference resulting from the various widths of the cantilever. By changing the reconstruction distance, we can perform numerical focusing on the other side of the cantilever and bring the side with shape edge out of focus. This indicates digital holography enables microscope to numerically focus on 3D objects without mechanical movements while attains the same resolution as obtained from direct imaging using the same imaging optics. The lateral resolution of x-ray DHM is determined by measuring the edge response of the AFM tip. An intensity profile of the tip image is shown in Fig. 4(d). The spatial resolution of the reconstruction image derived from the edge responses of AFM tip is ~ 480 nm which also agrees well with that attained from the lineout of the AFM

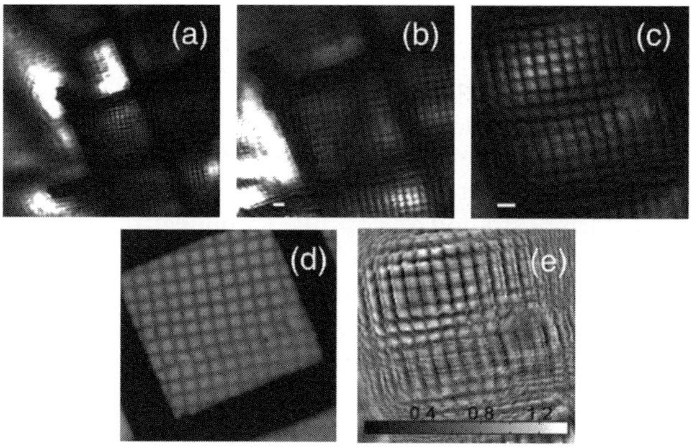

Fig. 5. Reconstructed intensity images of the carbon foil with magnifications of 17.9(a), 28.7(b), and 54.6(c), respectively. The scale bar in each figure is 20 μm. The optical microscope image of the carbon foil is shown in (d). (e) shows the reconstructed phase image of the carbon foil with a magnification of 54.6.

cantilever. The spatial resolution of x-ray DHM is limited by the numerical aperture (NA) of imaging optics and the wavelength of the illumination source. The NA in our system was around 0.05, corresponding to a spatial resolution of 400 nm in theory. The discrepancy may come from the aberration of imaging optics due to off-axis incidence of 32.8-nm laser on the x-ray concave mirror. In addition to single-shot measurement, we also performed x-ray DHM by accumulating data from 10 laser shots. Although multiple exposures can produce more uniform and better contrast holograms, but the resolution of the reconstructed image decreases as a result of the mechanical vibrations.

Figure 5 shows the reconstructed intensity images of a 20-nm carbon-foil mesh with various system magnifications. As can be seen, x-ray DHM demonstrates the flexibility for observing specimen in a large field of view and studying the details of the sample at high resolutions without complicate alignments and movements. The broken and bending mesh bars on the carbon foil caused by the sample cutting process are clearly seen in the reconstructed images and show good agreements with the observation using optical microscope shown in Fig. 5(d). The main advantage of the holography is the preservation of full phase information of the testing object within the two-dimensional hologram. The phase contrast image of the carbon foil at a magnification of 54.6 is shown in Fig. 5(f). The non-uniformity of the phase pattern mainly comes from the distortion (bending) of the sample as some parts of the image are not in focus. The retrieved phase difference caused by the carbon mesh is 0.8–1.2 rad, corresponding to the thickness of 19–28 nm by taking account of the refractive index of the carbon foil at 32.8 nm. This is close to the manufacture data of 20 nm.

References

1. Lemoff, B. E *et al.*: 'Demonstration of a 10-Hz femtosecond-laser-driven XUV laser at 41.8 nm in Xe IX', *Phys. Rev. Lett.* 74, 1574–1577, 1995
2. Sebban S. *et al.*: 'Saturated amplification of a collisionally pumped optical-field-ionization soft X-ray laser at 41.8 nm', *Phys. Rev. Lett.* 86, 3004–3007, 2001
3. Sebban S. *et al.*: 'Demonstration of a Ni-like Kr optical-field-ionization collisional soft X-ray Laser', *Phys. Rev. Lett.* 89, 253901, 2002
4. Chu H.-H. *et al.*: 'Collisional excitation soft x-ray laser pumped by optical field ionization in a cluster jet', *Phys. Rev. A* 71, 061804(R), 2005
5. Chou M.-C. *et al.*: 'Experimental investigation of the parameter space for optical-field-ionization cluster-jet x-ray lasers', *Phys. Rev. A* 74, 023804, 2006
6. Butler A. *et al.*: 'Demonstration of a Collisionally Excited Optical-Field-Ionization XUV Laser Driven in a PlasmaWaveguide', *Phys. Rev. Lett.* 91, 205001, 2003.
7. Mocek T. *et al.*: 'Dramatic enhancement of xuv laser output using a multimode gas-filled capillary waveguide', *Phys. Rev. A* 71, 013804, 2005.
8. Durfee III C. G., Lynch J., and Milchberg H. M.: 'Development of a plasma waveguide for high-intensity laser pulses', P*hys. Rev. E* 51, 2368–2396, 1995
9. Volfbeyn P., Esarey E., and Leemans W. P.: 'Guiding of laser pulses in plasma channels created by the ignitor-heater technique', *Phys. Plasmas* 6, 2269–2277, 1999
10. Xiao Y.-F. *et al.*: 'Efficient generation of extended plasma waveguides with the axicon ignitor-heater scheme', *Phys. Plasmas* 11, L21–L24, 2004
11. Miao J.W., Charalambous P., Kirz J., and Sayre D.: 'Extending the methodology of X-ray crystallography to allow imaging of micrometre-sized non-crystalline specimens' *Nature*, 400, 342–344, 1999
12. Chapman, H.N. *et al.*: 'Femtosecond diffractive imaging with a soft-X-ray free electron laser', *Nat. Phys.* 2, 839–843, 2006
13. Chou M.-C. *et al.*: 'Dramatic enhancement of optical-field-ionization collisional-excitation x-ray lasing by an optically preformed plasma waveguide', *Phys. Rev. Lett.* 99, 063904 , 2007
14. Nilsen J., 'Lasing on the 3d→3p neonlike x-ray laser transitions driven by a self-photo-pumping mechanism', *Phys. Rev. A* 53, 4539–4546 (1996)

Temporal Coherence and Spectral Linewidth of a Seeded Soft X-Ray Laser Pulse

O. Guilbaud[1] , J.P. Goddet[2] , S. Sebban[2] , D. Joyeux[3] , D. Ros[1] , J. Gautier[2] , K. Cassou[1], S. Kazamias[1], A. Klisnick[1] , J. Habib[1] , P. Zeitoun[2] , D. Benredjem[5] , S. de Rossi[3] , G. Maynard[4] , B. Cros[4] , A. Boudaa[4] , D. Phalippou[3] and A. Calisti[6]

[1] LIXAM, Laboratoire d'interaction du rayonnement X avec la matire, UMR 8624, Bat 350, Campus de l'Universite Paris Sud, F-91405 Orsay Cedex France
[2] LOA, Laboratoire d'optique applique, ENSTA, Chemin de la Hunire, 91761 Palaiseau Cedex, France
[3] LCFIO, Laboratoire Charles Fabry de l'Institut d'Optique, UMR 8501, Bat 503, Campus de l'Universite Paris Sud, F-91405 Orsay Cedex, France
[4] LPGP, Laboratoire de Physique des gaz et plasma, UMR8578 BAT 210, Campus de l'Universite Paris Sud, F-91405 Orsay Cedex, France
[5] Laboratoire Aime Cotton, CNRS, BAT 505, Campus d'orsay, 91405 Orsay Cedex, France
[6] Physique des interactions ioniques et molculaires, Universit de Provence, Facult de Saint-Jerome, 13397 Marseille Cedex, France

Abstract. We present in this paper the first measurement of the spectral profile of a seeded soft x-ray laser. Using a varying path difference interferometer the temporal coherence of a seeded OFI x-ray laser has been experimentally determined, leading to a coherence length of 5ps of the order of the gain lifetime. The measured bandwidth is of order 3.4 mÅ which is in good agreement with the prediction of a theoretical model presented here.

1 Introduction

The principle of seeding a soft x-ray laser by a high order harmonic pulse has recently been demonstrated in both OFI [1] and [2] (with solid targets) soft x-ray lasers. The promises of this scheme are progressively confirmed by different works investigating the amplification factors, the divergence and most of all, the full spatial coherence[2][8] of the beam and its wavefront [8]. Another promising aspect of seeding a soft x-ray laser is the potential to obtain pulses reaching exactly the Fourier limit. The XUV pulse duration will then be determined by the amplifier bandwidth. This bandwidth also controls the coupling efficiency between the seed and the amplifier. The knowledge of this parameter is therefore of great interest. Soft x-ray laser linewidth measurement is a difficult task because very high spectral resolutions are required. Interferometric methods are a promising way to achieve this goal. They rely on the measurement of the source coherence time which is inversely proportional to the source linewidth. We present in this paper the first experimental measurement of the linewidth of a seeded soft x-ray laser emitting at λ =32.8nm. The x-ray laser amplifier was generated using the OFI scheme. The temporal coherence was measured using an achromatic varying path difference interferometer. The linewidth is deduced from this measurement. Effect of the amplification length on the linewidth has been investigated and compared to the prediction of a simulation.

2 Experimental device

The soft x-ray laser amplifier considered here is a Ni-like Kr OFI soft x-ray laser (λ = 32.8nm). The experimental arrangement is similar to the setup described in the reference [1]. The main part of the energy of the LOA 2 J Ti-Saph laser is compressed at 30 fs and circularly polarized before being focussed by a f=1 m spherical mirror in a gas cell filled with krypton. A small fraction of the laser energy is dedicated to the generation of High Order Harmonic (HOH) in argon. The HOH beam is focussed at the entrance of the OFI gas cell by a toroidal grazing incidence mirror. When the harmonic and OFI amplifier are spatially, temporally and spectrally overlapped, a dramatic reduction of the divergence of the x-ray laser beam is observed on a far field XUV camera. This better collimation of the beam is associated to the full spatial coherence of the amplified harmonic. Besides, this phenomenon is also associated to a strong increase of the x-ray laser line level on the spectrometer due to a better collection of the laser output energy.

During the experimental campaign we worked with a constant harmonic seed energy. The amplification factor of the seeded laser is defined as the ratio of the seeded x-ray laser output energy devided by the harmonic energy. This factor has been optimised for a Krypton pressure of 30 mbar, a delay between the main IR laser and the HOH seed of 3ps, and a cell length of 7.5 mm. In these conditions the maximum amplification factor is of 180. In the following parts, results presented were obtained with this set of experimental conditions.

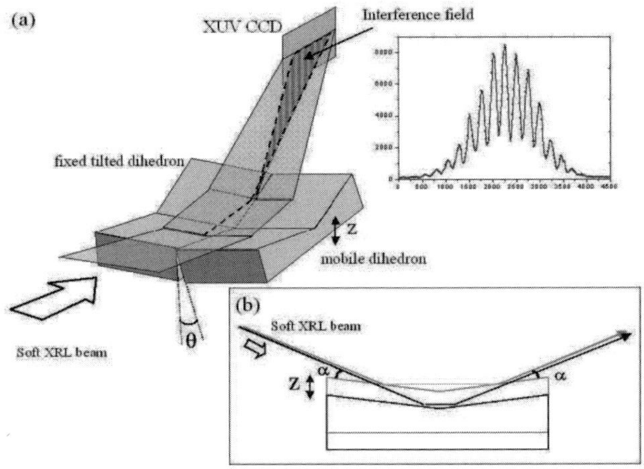

Fig. 1 (a) Schematic overview of the XUV interferometer used for longitudinal coherence measurement. (b) One of the dihedrons can be translated vertically(z) to introduce a controlled path difference between the two X-ray laser interfering half-beams, without changing the geometry of their transverse superposition.

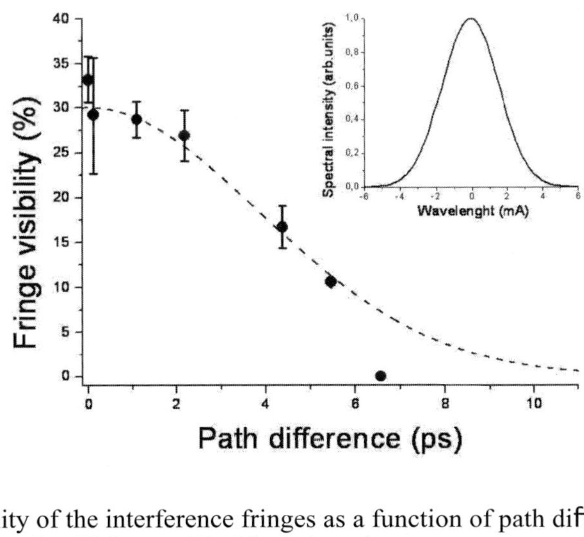

Fig. 2 Visibility of the interference fringes as a function of path difference for an amplifying length of 7.5 mm. The black dotted curve represents a Gaussian fit of the experimental data. The insert shows the reconstructed spectral profile deduced from the fit.

The seeded laser beam propagates towards a variable path difference interferometer. A detailed presentation of its principle can be found in [3] and in [4]. This device, represented in Figure 1, is composed of two dihedrons, tilted towards each other with a small angle. The incoming beam is separated in two beamlets that converge after reflection on the dihedrons. In their overlapping region, interference fringes can be recorded using a XUV CCD camera. One of the dihedrons can be translated vertically in order to introduce a path difference between the interfering beams. The geometry of the dihedrons (Figure 3b) ensures a constant lateral overlapping geometry, and thus constant spatial coherence conditions when changing the delay. By following the evolution of the fringe visibility as a function of the delay, it is possible to reconstruct the spectral profile of the incoming radiation through a Fourier transform.

3 Results

An experimental visibility scan obtained for a cell length of L=7.5 mm is presented in Figure 2. The experimental data have been fitted by a Gaussian function from which a coherence time of 5.4± 0.3ps was inferred. Assuming, as a first approximation, a Gaussian spectral profile, these would lead to a spectral width of $\Delta\lambda$ = 3.4± 0.2 mÅ. It has been also observed that the value of the coherence length and hence of the spectral line width does not change significantly when the amplifier length is changed from 3.5 mm to 7mm. The spectral width values for L=3.5 mm, L=5 mm and L=7.5 mm are plotted in Figure 3. Below 3.5mm the signal was too weak to perform any accurate measurement.

4 Discussion

A modelling of the spectral profile of the seeded x-ray laser as a function of amplifying length has been performed. The spectral shape of the harmonics has been modelled by a Gaussian shape with a width equal to the experimental value measured on a transmission spectrometer. The gain of the OFI amplifier has been set to 70 cm^{-1} close to the experimental value obtained in [5]. The homogeneous broadening of the x-ray laser amplifier bandwidth has been calculated with a collisional-radiative model that takes into account the non-maxwellian nature of the electron energy distribution [5]. The calculated homogeneous linewidth is of 5mÅ. We have also introduced an inhomogeneous contribution mainly dominated by the Doppler effect. Simulations with the PPP [6] code have shown that the ionic Stark effect can be neglected. The

Doppler broadening arise from the rapid heating of the ions (few hundreds of fs) that follows the ionization by the optical field. Theoretical works have indeed shown that the OFI plasma is produced in a state where important ion correlations exist [9]. The plasma relaxes rapidly in an uncorrelated form with a characteristic time equal to the plasma period. For a gas pressure of 30 mbar the temperature reached by the ions after this relaxation is of 6 eV. Finally the ratio between the HOH intensity and saturation intensity has been set so that the theoretical amplification factor is equal to the experimental factor. Gain narrowing and saturation re-broadening have been taken into account following J. Koch et al. [7].

Numerical results from this simulation are displayed in figure 3. The black line represents the linewidth as a function of the amplifier length. Black squares represent experimental data. For an amplification length of 7.5 mm, a theoretical linewidth of 3.6 mA in good agreement with experiment. Good agreement is also found for other amplification lengths. The simulation shows that the bandwidth decreases very slowly with the plasma length L when the saturation is reached, that is for L greater than 3mm. This behaviour is consistent with the experimental observations.

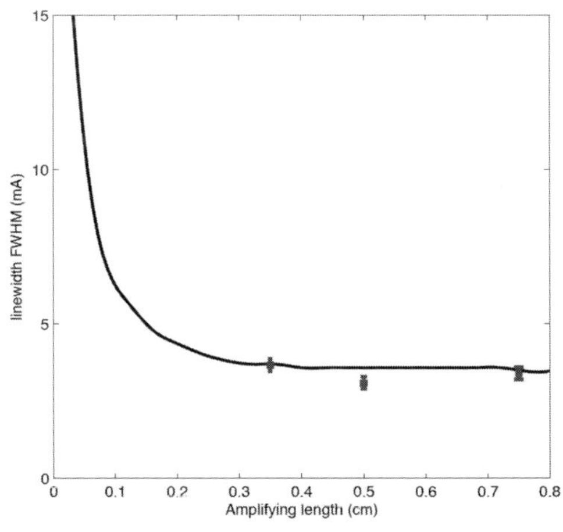

Fig. 3 Spectral linewidth (FWHM) as a function of the plasma amplifier length. The black line represents the evolution calculated with conditions described in the text. The experimental values are represented by the black squares.

5 Conclusion

We measured for the first time the coherence time of a seeded soft x-ray laser. The spectral profile reconstructed from the experimental values is a Gaussian function and the linewidth is of $\Delta\lambda$ =3.4 ± 0.2 mÅ for an amplifier length of 7.5 mm. If the pulse is assumed to be Fourier-limited, this spectral profile would correspond to a Gaussian shape pulse with a duration (at half maximum) of 4.8ps. This long duration results from the very narrow linewidth of the OFI amplifier. Shorter pulse duration might be expected from solid target plasma amplifiers due to higher ion temperature.

Acknowledgments

We warmly thank Jerome Guigand, Eric Bouisset and Jean-Claude Lagron for their precious help. The support of the technical staff at LOA is gratefully acknowledged. The multilayer optics used in the experiments were manufactured at LCFIO by the CEMOX facility, under the supervision of F. Delmotte.

References

[1] Zeitoun Ph, Faivre G, Sebban S, Mocek T, et al. 2004 Nature 431 466
[2] WangY, Granados E, Larotonda M, Berrill M, Luther B, Patel D, Menoni C and Rocca J 2006 Phys. Rev. Lett. 97 123901
[3] Guilbaud O, Klisnick A, Joyeux D, Benredjem D, Cassou K, Kazamias S, Ros D, Phalippou D, Jamelot G and Moller C 2006 Eur. Phys. J. D 40 125
[4] Habib J. et al., in this conference
[5] Cros B, Mocek T, Bettaibi I, Vieux G, Farinet M, Dubau J, Sebban Sand Maynard G 2006 Phys. Rev. A 73 033801
[6] Talin B, Calisti A, Godberg L, Stamm R, Lee R, Wand Klein L 1995 Phys. Rev. A 51 1918
[7] Koch J, MacGowanB, DaSilva L, Matthews D, Underwood J, Batson B, Lee R, London R and Mrowka S 1994 Phys. Rev. A 50 1877
[8] Goddet J P, et al., in this conference
[9] Maynard G, Lambert F, Andreev N, Robillard B, Boudaa A, Clerouin A, Cros B, Lenglet A, Mocek T, Sebban S 2007 Contrib. Plasma Phys. 47 352

Part 5 – Theory and Simulations

The Scaling of Recombination Following Tunnel Ionisation and its Suitability for Generating X-Ray Laser Gain

G J Pert†

† Department of Physics, University of York, Heslington, York YO10 5DD, UK

Abstract. An investigation of the scaling of ATI ionisation and cascade recombination is used to identify possible conditions for generating population inversion and gain on the Lyman α transition of hydrogeic ions.

1 Introduction

It has been proposed earlier that Lyman α transitions of hydrogen-like ions may be suitable candidates for relatively short wavelength laser action by using cold electrons released during tunnelling ionisation by linearly polarised light. To minimise the electron temperature, and thereby maximise populations in a recombination cascade, a background environment of hydrogen is used with the lasant at low concentration. Following ionisation, the electron distribution is relaxed to an approximate Maxwellian by inverse bremsstrahlung, electron-electron relaxation and three body recombination.

Using relatively simple models, we investigate the atomic number scaling of the necessary primary experimental variables required to obtain population inversion. This identifies typical values for laser intensity, electron density and lasant concentration, and thus conditions under which gain might be generated.

We confirm these conclusions by the use of more detailed numerical simulations in carbon, where several different physical phenomena involved can be modelled consistently. A number of deficiencies of the model are identified and their potential effects discussed. We note that the carbon system has also recently been investigated by Avitzour and Suckewer [1]

2 Tunnel Ionisation Scaling

Using the Ammosov, Delone, Krainov [2] formula for the probability of ionisation of a hydrogenic ion of charge Z under a strong laser field, an estimate of the ionisation probability in a linearly polarised field of intensity I_0 is given by

$$\overline{p} = 1.05 \times 10^{13} Z^2 \alpha^{-1/2} \quad s^{-1} \tag{1}$$

where

$$\alpha = \frac{2}{3} Z^3 \frac{E_{at}}{E_0} \approx 1.25 \times 10^8 Z^3 I_0^{-1/2} \tag{2}$$

where E_{at} is the atomic and E_0 the peak electric fields. Hence the intensity required for full ionisation in a time of $\approx$50 fs of ions of atomic number $Z \sim 6$ is about

$$I_0 \approx 1.4 \times 10^{14} Z^6 \quad \text{Wcm}^{-2} \tag{3}$$

and typically $\alpha \approx 10$.

The corresponding peak electron quiver energy (twice the ponderomotive energy) is

$$\varepsilon_q \approx 1.87 \times 10^{-13} \lambda^2 I_0^2 \approx 26.1\lambda^2 Z^6 \quad \text{eV} \tag{4}$$

where λ is the laser wavelength in μm. The average free electron energy is ε_q/α

This simple high field, tunnelling model is valid only if the Keldysh parameter [3]

$$\gamma = \sqrt{\frac{V_i}{\varepsilon_q}} \approx 0.52 \lambda^{-1} Z^{-2} << 1 \tag{5}$$

3 Recombination Scaling

To generate a population inversion we require the upper states of the ion to be populated by a recombination cascade, in which electrons diffuse through the upper states in response to both excitation and de-excitation collisions. On balance however the latter dominate so that the electrons move progressively into the lower states. This process maintains a near Boltzmann distribution of the populations in the upper states. However as the electron energy gap increases towards lower states, excitation rates can no longer approximately balance de-excitation to maintain approximate equilibrium, and the electrons fall towards the ground state. The level at which this transition occurs is known as the ‘bottleneck’. Clearly the collision rate across the bottleneck determines the cascade population rate, and hence the recombination rate of the upper laser state, $n = 2$ [4].

In contrast since radiative transitions are strongest for large energy transitions, the ground state, $n = 1$, is preferentially populated by direct radiative radiation and to a lesser extent by radiative decay from the upper excited states.

Under plasma conditions of electron density n_e and temperature T_e the two rates scale as

$$\begin{array}{lcl} \text{Cascade rate} & \sim & n_e^2\ Z^3\ T_e^{-9/2} \\ \text{Radiative rate} & \sim & n_e\ Z^2\ T_e^{-1/2} \end{array}$$

Clearly there is a limiting condition for the cascade rate to exceed the radiative, i.e. for the population rate of the upper state to exceed that of the lower, and hence for an inversion to form. Therefore for inversion we require

$$T_e \quad < \quad T_{lim}(n_e) \tag{6}$$

The form of the collision rates implies a characteristic scaling with Z within which the recombination process is self-similar [4], namely

$$n_e \quad \sim \quad Z^7 \qquad \text{and} \qquad T_e \quad \sim \quad Z^2 \tag{7}$$

Thus for a given temperature $T_e \sim Z^2$ the limiting electron density scales $n_{e\,lim} \sim Z^7$, plotted subsequently in fig.2. Hence we observe that the required density to generate inversion requires high electron density for high atomic number ions. Typically for carbon, $Z = 6$, we require $n_e \sim 10^{20}$ cm^{-3}.

At such high density, depression of the ionisation level [5] becomes important.

$$n_{lim} \quad \approx \quad 1.1 \times 10^4\ Z^{1/2}\ n_i^{-1/6} \tag{8}$$

where n_i is the ion density. Since this model implies the development of a cascade equilibrium amongst the high lying states, these scalings are strictly only valid if the depression is weak, i.e. n_{lim} is sufficiently large. In fact for the cases considered here $n_{lim} \approx 13$ and ionisation depression leads to a small deviation from the perfect scaling.

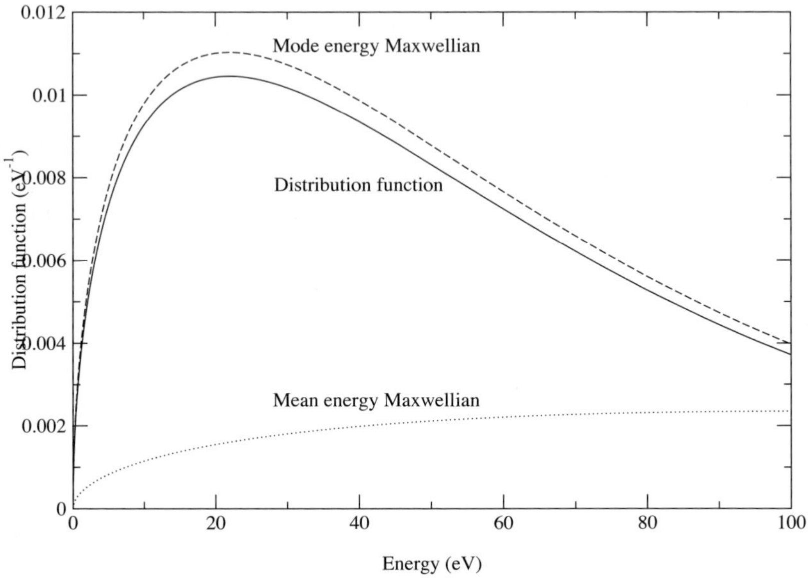

Figure 1. Plot of the electron distribution compared with that of a Maxwellian with the mode temperature and one with the same mean energy.

4 Electron distribution following ionisation

Whilst the laser beam is irradiating the gas, ionisation progressively strips the ions releasing electrons with a distribution strongly peaked about near zero energy. However different stages of ionisation give rise to electrons of markedly differing mean energies, depending on the ionisation stage. Thus the distribution is formed with a high energy tail, which will be exacerbated by re-scattering. Once released the electrons will gain additional energy during collisions with ions (inverse bremsstrahlung). Since the electron quiver speed is much larger than the thermal, all electrons will have approximately the same absorption coefficient. Furthermore the distribution due to inverse bremsstrahlung alone will have an approximately Maxwellian form.

Thus we find that at the conclusion of the laser pulse, the electron distribution is approximately Maxwellian, but with a substantial non-Maxwellian tail at high energy. The slow electrons may be accurately described by a Maxwellian with temperature defined by the mode (peak) of the distribution. The mode temperature is much less than that given by the mean electron energy (fig.1).

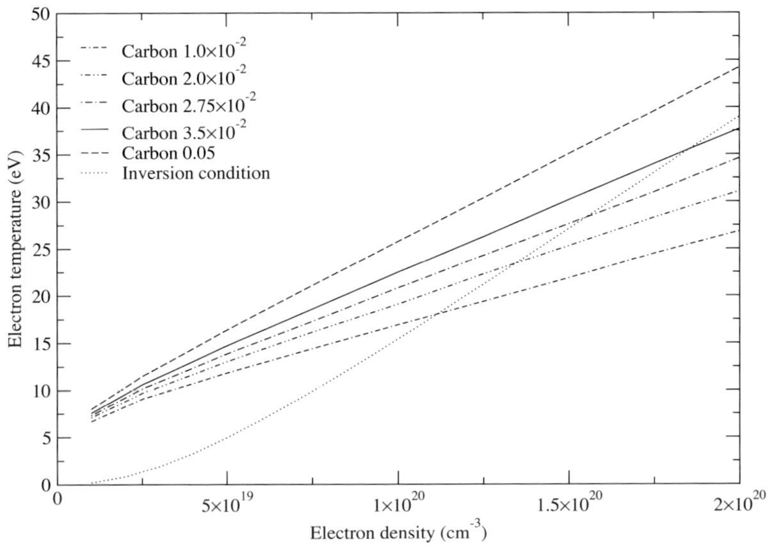

Figure 2. Plots of the mode temperature as a function of electron density for different carbon concentrations. Also plotted is the inversion limit. Pulse time-to-peak and half width 50fs and intensity 6.5×10^{18} W cm^{-2}

The cascade recombination rate is dominated by the slow electrons and therefore by the mode temperature.

The mode temperature for a pure carbon plasma is much too large for any population inversion to be generated. It is therefore appropriate to dilute the carbon with hydrogen to reduce the ATI energy, and thereby the mode temperature. To achieve ionisation low carbon concentrations of less 3.5×10^{-2} are necessary. The required electron density is provided by a suitably large hydrogen density.

Fig. 2 shows a nearly linearly scaling of the mode temperature with electron density. This is primarily due to strong field inverse bremsstrahlung absorption with a small contribution from electron-electron relaxation. The basic ATI contribution is associated with the temperature at low density. Thus we may write

$$T_e \approx (A + B(c)\, n_i) \approx a\, Z^6 + b(c)\, Z^{13} \tag{9}$$

The terms respectively represent the ATI and inverse bremsstrahlung component. Note the latter is dependent on lasant concentration c through the term

$\overline{Z^2}$ in the inverse bremsstrahlung coefficient. This useful relation allows the matching of the mode temperature to that required by recombination if inversion is to be generated.

5 Development of the population inversion

Numerical simulation of the full recombination mode accompanied by ATI tunnelling ionisation (without re-scattering), inverse bremsstrahlung and electron-electron relaxation is accomplished using a Fokker-Planck model [6, 7].

These calculations showed that for electrons with an initially Maxwellian distribution defined by the mode temperature in a pure carbon plasma

1. Ionisation depression decreased the populations of the higher levels, $n = 3$ and 4, more strongly than lower levels as their equilibrium (Saha) populations were decreased by the reduction of the ionisation energy. The reduced population inversion of the levels $n = 1$ and 2 was little changed.

2. Three body re-heat, whereby the recombining electrons heats the spectator electron, leads to a relatively small temperature rise as the recombination of high lying levels only is involved with collisional energy exchange. As a consequence, only the populations of high lying levels were markedly changed, being reduced by the temperature increase. In this case the population inversion was little changed.

3. Electron relaxation played little or no role.

These calculations confirmed that the simple scaling model of Pert [4] could be used with reasonable accuracy to predict the conditions for the onset of inversion if the temperature following ionisation were known.

6 ATI ionisation and recombination

The inclusion of the electron distribution following simple tunnel ionisation is modelled using the model and showed results in line with the preceding discussion.

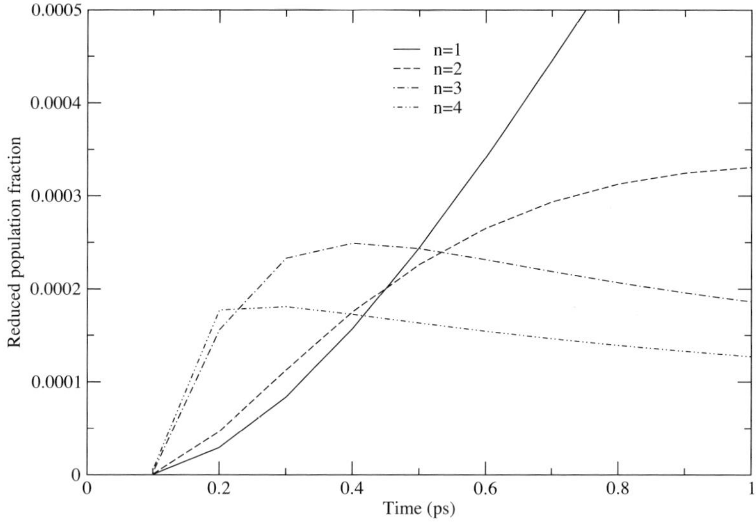

Figure 3. Plots of the reduced population fraction of the levels $n = 1 - 4$ following ionisation. Pulse time-to-peak and half width 50fs and intensity 6.5×10^{18} W cm^{-2}

Fig.3 shows the development of the populations of the levels $n = 1 - 4$ following ionisation by a laser pulse of 0.25μm wavelength and 6.5×10^{18} Wcm^{-2} intensity in a Gaussian pulse of 50fs duration (half-width). The gas was a mixture of 95% hydrogen and 5% carbon. A number of interesting features can be seen.

- A small population inversion is rapidly formed once the laser pulse is fully switched off at 100fs. reaching a maximum of about 0.0001 reduced population fraction after about a further 300fs and lasting for a further 300fs. This occurs at a slightly higher carbon fraction than predicted by the simple model based on the mode temperature.
- There is very little population growth until the the laser is finally switched off as the upper states are rapidly re-ionised by photo-ionisation.
- After an initial growth the populations of the higher excited states, $n = 3$ and 4, reach a maximum and then decrease. This is ascribed to the increasing mode temperature associated with electron re-heat. The effect on the lower levels, $n = 1$ and 2, however is small.

In principle we may calculate the gain using this value of the population inversion. However to do so would require an accurate value of the line width. This is unlikely to be known unless the nature of the gas used is specified due to the Coulomb explosion of the carbon ions if the carbon is initially in molecular form, e.g methane, and the Stark broadening at the high electron density.

7 Limitations of the model

The conditions under which population and inversions and gain can be achieved in carbon are quite severe and it is therefore important to examine the limitations of the relatively simple model we have used. These stem from the nature of both the ionisation process and the state of the plasma.

The tunnelling model. The applicability of the tunnelling model is established by the condition that the Keldysh parameter, γ eqn.(5), is small. For the higher stages of carbon this condition is well established. However at the intensities required for full ionisation of hydrogen, eqn.(3), this is not the case. Thus the hydrogen will be ionised in the multiphoton ionisation regime rather than the tunnelling. The initial electron distribution will therefore be in error. However this is not a serious error provided the hydrogen is ionised near threshold. Since the residual energy will be less than the photon energy., i.e. a few eV., and small compared to inverse bremsstrahlung absorption energy, the latter being comparable with the estimated initial mode energy.

A more serious limitation of the model may be the omission of rescatttering immediately following ionisation when the electron is still correlated with the ion. If the electron is released at the appropriate phase, returns to the ion and is back-scattered the electron may have energy $5\varepsilon_q$. Fortunately most electrons are not back-scattered and may not have the required phase [8]. None-the-less about 50% of the electrons are likely to possess additional thermal energy beyond that predicted by the simple tunnelling model. This difficulty is avoided if the hydrogen, which provides most of the electrons, is ionised in the multi-photon regime.

However this additional heating is not a factor in the multiphoton limit. Since the electrons originating from the hydrogen and the low ionisation stages are the only ones contributing to the recombination, the others being ineffective due to their high temperature, we can mitigate against this problem by ensuring that the hydrogen ionisation occurs at low intensity near threshold, possibly using a low intensity pre-pulse to accomplish this.

The plasma conditions We require a dense low temperature to achieve inversion. Typically we have electron density $n_e \approx 2 \times 10^{20}\text{cm}^{-3}$ and temperature $T_e \sim 25\text{eV}$. Under these conditions the electron-electron strong coupling parameter

$$\Gamma_{ee} = \frac{n_e^{1/3} e^2}{\varepsilon_0 k T_e} \sim 4 \tag{10}$$

The dilute approximation is thus only approximately valid. Some measure of correlation may therefore be important in three body collisions, which is not taken into account. However as the interactions are accounted for by a detailed balance based on electron-ion ionising collisions, the error is probably small.

The electron-ion strong coupling coefficient is also of order 1, but the effect of ion-electron correlations is probably weak in this context.

At high densities and low temperatures electron degeneracy should also be considered, since it would lead to serious overestimates of the three body rate. The Fermi energy is given by

$$E_F = \frac{1}{2m_e}\left(\frac{3}{8\pi}n_e\right)^{2/3} h^2 \tag{11}$$

For $n_e \approx 2 \times 10^{20}\text{cm}^{-3}$ $E_F \approx 0.125\text{eV}$. Consequently if $T_e = 25\text{eV}$, the system is not degenerate.

Laser conditions At the required laser intensity of 6.5×10^{18} Wcm^{-2} the ponderomotive force is much stronger than the thermal pressure. Consequently we may expect the electrons may be significantly displaced, and as a result gain energy from the work done on them by the force. In the wake field limit this will lead to a group of relativistic electrons. The condition for avoiding the most serious form of this behaviour is that the laser pulse duration must be long compared to the plasma wavelength c/ω_p where ω_p is the plasma frequency. At an electron density $n_e \approx 2 \times 10^{20}\text{cm}^{-3}$, this corresponds to a wavelength of about $2.2\mu\text{m}$ and a pulse length of about 10fs.

The electrons may also be displaced in the transverse direction across the focal spot. The displaced electrons generate a space charge field opposing the force resulting from to the gradient in the ponderomotive force, which, if severe, will lead to channelling. The electrons relaxing back into the channel recover the work done by the ponderomotive potential as kinetic energy. If the channelling is strong, this may give rise to appreciable heating. We may estimate this effect by balancing the ponderomotive force in a laser spot of radius R against the space charge field following an electron displacement D. The energy gain per electron W is estimated from the work done by the ponderomotive force. Thus we obtain

$$W \approx \frac{n_e\, e^2}{\varepsilon_0} D^2 \approx \frac{\varepsilon_0\, (I/c)^2}{n_e\, (n_c e^2)^2\, R^2} \tag{12}$$

where n_c is the critical density. The maximum allowable value of W places a lower limit on R and an upper limit on D. This effect therefore identifies a minimum allowable value of the focal spot diameter, typically in excess of 20μm.

8 Conclusion

It is shown that it is possible to design systems to generate population inversion on the L_α of hydrogenic carbon provided care is taken to avoid all possible sources of unnecessary heating. In particular it may be advisable to use a separate low intensity pre-pulse to ionise the hydrogen in the multi-photon regime. Since a typical system is likely to involve a methane-hydrogen mixture at high pressure, whose ionisation will lead to a Coulomb explosion heating the carbon ions, we have not attempted to estimate the gain in this work.

References

[1] Avitzour Y and Suckewer S *J.Opt. Soc. Am* B **24** 819–828

[2] Ammosov M V Delone N B and Krainov V P *Sov. Phys.-JETP* **64** 1191

[3] Keldysh L V *Sov. Phys.-JETP* **20** 1307–1314

[4] Pert G J 2007 *Phys. Rev.* E **76** 056404

[5] More R M 1986 *Adv.At.Mol.Phys* **21** 305

[6] Pert G J 2001 *J. Phys. B: At. Mol. Opt. Phys.* **34** 881–908

[7] Pert G J 2006 *Phys. Rev.* E **73** 066401

[8] Paulus G, Becker W, Nicklich W and Walter H *J. Phys. B: At. Mol. Opt. Phys.* **27** L703–708

Advances in Understanding the Anomalous Dispersion of Plasmas in the X-Ray Regime

Joseph Nilsen and K. T. Cheng

Lawrence Livermore National Laboratory, Livermore, CA 94551

Walter R. Johnson

University of Notre Dame, Notre Dame, IN 46556

Abstract. Over the last several years we have predicted and observed plasmas with an index of refraction greater than one in the soft X-ray regime. These plasmas are usually a few times ionized and have ranged from low-Z carbon plasmas to mid-Z tin plasmas. Our main computational tool has been the average atom code AVATOMKG that enables us to calculate the index of refraction for any plasma at any wavelength. In the last year we have improved this code to take into account many-atomic collisions. This allows the code to converge better at low frequencies.

In this paper we present our search for plasmas with strong anomalous dispersion that could be used in X-ray laser interferometer experiments to help understand this phenomena. We discuss the calculations of anomalous dispersion in Na vapor and Ne plasmas near 47 nm where we predict large effects. We also discuss higher Z plasmas such as Ce and Yb plasmas that look very interesting near 47 nm. With the advent of the FLASH X-ray free electron laser in Germany and the LCLS X-FEL coming online at Stanford in another year we use the average atom code to explore plasmas at higher X-ray energy to identify potential experiments for the future. In particular we look near the K shell lines of near solid carbon plasmas and predict strong effects. During the next decade X-ray free electron lasers and other X-ray sources will be available to probe a wider variety of plasmas at higher densities and shorter wavelengths so understanding the index of refraction in plasmas will be even more essential.

1 Introduction

Since the earliest days of lasers, optical interferometers have been used to measure the electron density of plasmas [1] using the assumption that the index of refraction of the plasma is due only to the free electrons and is therefore less than one [1-2]. With this assumption the electron density of the plasma is directly proportional to the number of fringe shifts in the interferometer. Over the last decade many interferometers [3-8] have been built in the soft X-ray wavelength range of 14 to 72 nm (89 to 17 eV). The experiments done with these sources all assume that only the free electrons contribute to the index of refraction. In the next few years, interferometers will be built using the X-ray free electron lasers, which will extend lasers to even shorter wavelengths [9].

In the last several years interferometer experiments [4-6] of Al plasmas, Ag and Sn plasmas [10], and C plasmas [11-12] observed fringe lines bend in the opposite direction than was expected, indicating that the index of refraction was greater than one. Analysis of the experiments showed that the anomalous dispersion from the resonance lines and absorption edges of the bound electrons have a larger contribution to the index of refraction with the opposite sign as the free electrons [12-15]. Since the original analysis [13] of the experiments with Al plasmas we have developed a new tool AVATOMKG [16] that enables us to calculate the index of refraction for any plasma at any wavelength. This tool is a modified version of the INFERNO average atom code [17] that has been used for many years to calculate the absorption coefficients for plasmas.

In this work we search for other materials that can be used to create plasmas with index of refraction greater than one at X-ray laser energies. We discuss how neutral Na vapor and singly-ionized Ne plasma look to be promising candidates to use in interferometer experiments based on the Ne-like Ar X-ray laser [18] at 26.44 eV or 47 nm. We also discuss higher Z plasmas such as Ce and Yb plasmas, which look very interesting near 47 nm. With the advent of the FLASH X-ray free electron laser in Germany and the LCLS X-FEL coming online at Stanford soon we look at the K shell lines of near solid carbon plasmas and predict strong effects.

2 Analysis of Interferometer Experiments

When the electron density is much less than the critical density, as is typical for laser produced plasmas, the traditional formula for the index of refraction of a plasma due only to free electrons is approximated as $n = 1 - (N_{elec} / 2N_{crit})$ where N_{elec} is the electron density of the plasma and N_{crit} is the plasma critical density. For a uniform plasma of length L the number of fringe shifts observed in an interferometer equals $(1 - n) L / \lambda$ or $(N_{elec} L) / (2 \lambda N_{crit})$. For a non-uniform plasma one does a path length integral. The fringe shifts are referenced against a set of reference fringes in the absence of any plasma and the formula assumes that the interferometer is in a vacuum. When analyzing an experiment one counts how far the fringes have shifted compared with the reference fringes taken with no plasma and converts this into electron density. For the 46.9 nm Ne-like Ar X-ray laser the number of fringe shifts in the interferometer is $(N_{elec}\ L) / (4.8 \times 10^{18}\ cm^{-2})$ and the critical density is $5.07 \times 10^{23}\ cm^{-3}$.

To understand the contribution of the bound electrons we look at the relationship between the absorption coefficient and the index of refraction. The total absorption coefficient $\alpha = N_{ion}\ \sigma = (4\ \pi\ \beta) / \lambda$ where N_{ion} is the ion density of the plasma, λ is the wavelength, σ is the absorption cross-section, β is the imaginary part of the complex index of refraction n* defined by $n^* = 1 - \delta - i\beta$. The real part of the index of refraction $n = 1 - \delta$. The Henke tables [19] tabulate the dimensional-less optical constants f_2 and f_1 for neutral materials. These coefficients are related to δ and β by $\delta = f_1\ N_{ion} / (2\ N_{crit})$ and $\beta = f_2\ N_{ion} / (2\ N_{crit})$. From the total absorption cross-section σ we determine the optical constant f_2. We then derive the optical constant f_1 as a function of photon energy E using the Kramers-Kronig dispersion relation [20] by taking the principal value of the integral

$$f_1(E) = Z_{nuc} + \frac{2}{\pi} P.V. \int_0^{\infty} \frac{f_2(\varepsilon)\, \varepsilon\, d\varepsilon}{E^2 - \varepsilon^2}$$

where Z_{nuc} is the atomic number of the element. For neutral materials the oscillator sum rules insure that f_1 goes to zero at zero energy and Z_{nuc} at infinite energy. For an ionized plasma with average ionization Z* then $f_1 = Z^*$ at $E = 0$.

In the absence of any bound electrons f_1 is equivalent to the number of free electrons per ion. This means that we can replace N_{elec} with $f_1\ N_{ion}$ in the formula for the number of fringe shifts. The free electron approximation for the index of refraction is true when f_1 is equal to Z*.

The number of fringe shifts observed in the experiment for a real plasma is now equals $(f_1\ N_{ion}\ L) / (2\ \lambda\ N_{crit})$. Taking the ratio of f_1 to Z* gives the ratio of the measured electron density to the actual electron density. When the ratio is negative, the index of refraction is greater than one and the fringes bend the opposite direction than expected in the interferometer.

3 Finding Anomalous Dispersion in Ne and Na Plasmas

For the interferometer that uses the Ne-like Ar X-ray laser at 46.9 nm (26.44 eV) we begin our search for plasmas that might have a large anomalous dispersion near 26.44 eV by first looking at the Henke data to find materials with absorption edges in the 20 – 30 eV range. Since Ne and Na both looked like interesting materials with L shell edges at 21.6 and 30.6 eV, respectively, we extrapolated the Henke data to estimate the f_1 versus photon energy for neutral Ne and Na. Finding negative f_1 values for neutral materials is usually a good clue to find negative f_1 in plasma that are only a few times ionized.

Neutral Ne plasma is very opaque to the 26.44 eV X-rays since they are above the L-shell absorption edge at 21.6 eV. However for single ionized Ne the L-edge moves to 41 eV and the plasma becomes much more transparent. Using an ion density of 10^{20} cm^{-3} at a temperature of 4 eV the average atom code calculates a $Z^* = 1.05$, which is close to singly-ionized. Figure 1 shows the optical constant f_1 versus photon energy for the Ne plasmas. For the case $Z^* = 0$, shown by the dotted line, the Henke data has been extrapolated to lower energy. At an energy of 26.44 eV $f_1 = -1.4$ for the neutral Ne but this case would be highly absorbing. For the singly ionized case with $Z^* = 1.05$, shown by the solid line, f_1 is now –4.4 at the photon energy of 26.44 eV. This is due to being on the low energy side of the strong 2s – 2p absorption line, which is shifted by +4.18 eV in the calculation so that the line agrees with the experimentally measured value of 26.86 eV [21]. While not shown in the figure, as one continues to ionize Ne to doubly-ionized the strong 2s-2p line moves to lower energy at 25.33 eV and the net result is f_1 reaches a positive value of 3.9 at the 26.44 eV photon energy. In a real experiment using the Ar X-ray laser one could observe the neutral Ne gas go from opaque to transparent as the gas is ionized and furthermore watch the fringe shifts go from negative to positive as Ne is ionized from singly to doubly ionized. At some point there would be no fringe shift when there was the right mixture of singly and doubly ionized Ne. One could even envision using the ionization state of Ne as a switch that could move the Ar X-ray laser from one direction to another.

Na has a simpler situation than Ne because neutral Na has its L-edge at 30.6 eV and this moves to higher energy as Na is ionized so Na is transparent to the Ar X-ray laser photons for neutral as well as partially ionized ions. The doubly ionized case has the spectrum shifted by 5.5 eV so that the 2s-2p line agrees with the experimentally measured value of 32.7 eV [22]. Figure 2 plots the optical constant f_1 versus photon energy for the two cases with $Z^* = 0$ (Henke data) and $Z^* = 2.0$ (average atom code with temperature of 7.3 eV). This yields $f_1 = -0.87$ for neutral Na and 1.44 for doubly-ionized Na at 26.44 eV. For singly ionized case the average atom code estimates a value of –0.15 for f_1 at this energy. This suggests that an experiment with neutral Na gas would observe the anomalous dispersion and the fringe shifts would go through a zero point near singly ionized Na and eventually look quite normal as one approached doubly-ionized Na.

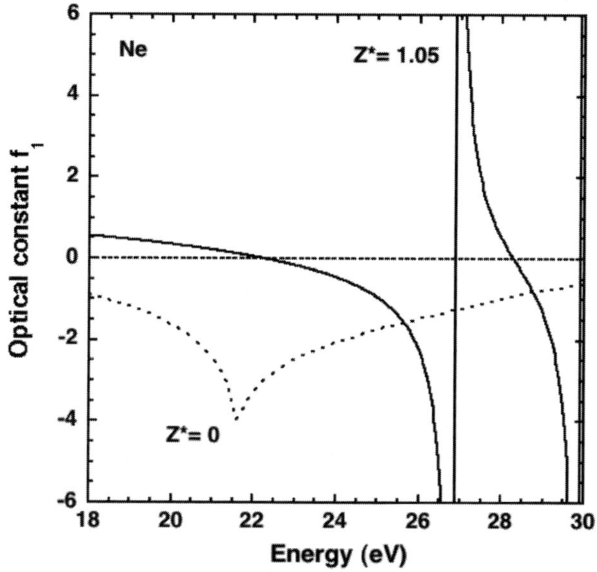

Fig. 1 Optical constant f_1 versus photon energy for Ne plasmas. The dotted line is from the Henke data. The solid curve is calculated by AVATOMKG code for Z* = 1.05. The dashed line is a visual aid for $f_1 = 0$.

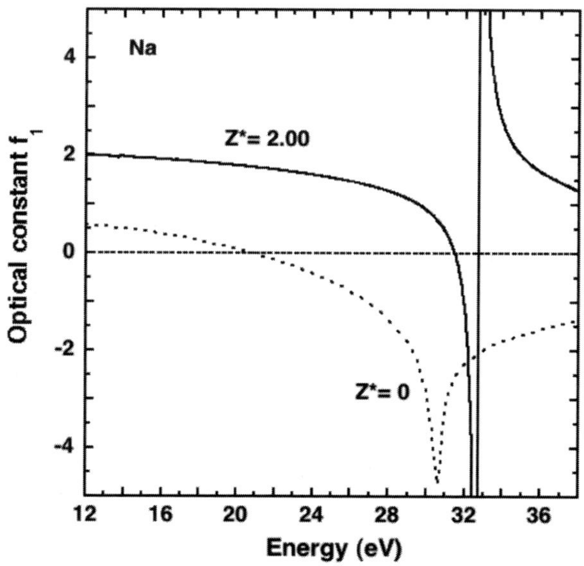

Fig. 2 Optical constant f_1 versus photon energy for Na plasmas. The dotted line is from the Henke data. The solid curve is calculated by AVATOMKG code for Z* = 2.0. The dashed line is a visual aid for $f_1 = 0$.

4 Finding Anomalous Dispersion in Ce and Yb Plasmas

In addition to low Z materials such as Ne and Na we looked for higher Z elements that could have anomalous dispersion near 26 eV. Two materials that looked interesting were Ce (Z=58) and Yb (Z=70). For the case of neutral Ce there are strong O-edges [19] for the 5p electrons at 19.8 and 17.0 eV that will move to higher energy as the outer electrons are ionized. In particular Ce^{4+} looked very interesting because it is a closed Xe-like core with closed 5s and 5 p sub-shells. The ionization energies for Ce^{0+} to Ce^{3+} are 5.54, 10.85, 20.198 and 36.758 eV, respectively. To estimate the size of the optical constant f_1 we first used the multi-configuration Dirac-Fock (MCDF) code from Grant [23] to calculate the oscillator strengths of ground state transitions in Ce^{4+}. This resulted in five absorption lines whose wavelengths have been measured and documented by NIST[24]. We adjust the position of the calculated lines to agree with the measured line positions. Using the oscillator strength we calculate a value for the absorption coefficient f_2 and use the Kramers-Kronig dispersion relation to calculate the optical constant f_1, shown in Fig. 3. Also shown is the case for neutral Ce extrapolated from the Henke tables. The dominant transitions are two 5p – 5d lines at 25.672 and 31.046 eV with oscillator strengths of 0.256 and 9.31, respectively, and a 5p-6s line at 30.799 eV with oscillator strength of 0.40 which results in f_1 = -17 at 26.44 eV. In an experiment we would expect to observe an index of refraction larger than one and a large anomalous dispersion.

The second material that looks interesting is Yb. Neutral Yb has O-edges [19] for the 5p electrons at 24.1 and 30.3 eV and an optical constant f_1 = -6.6 at 26.44 eV. As we ionize Yb^{2+} has a completely closed N-shell and closed 5s and 5 p sub-shells. Using the MCDF code to calculate the oscillator strengths of the ground state transitions we predict several strong 5p – 5d lines at 29.2 and 36.1 eV. Using the same procedure described above we then calculate the optical constant f_1 as a function of photon energy, as shown in Fig. 4, for Yb^{2+}. We predict f_1 = -5.2 at 26.44 eV so we would expect an X-ray laser interferometer at 26.44 eV to observe an index of refraction larger than one for this plasma.

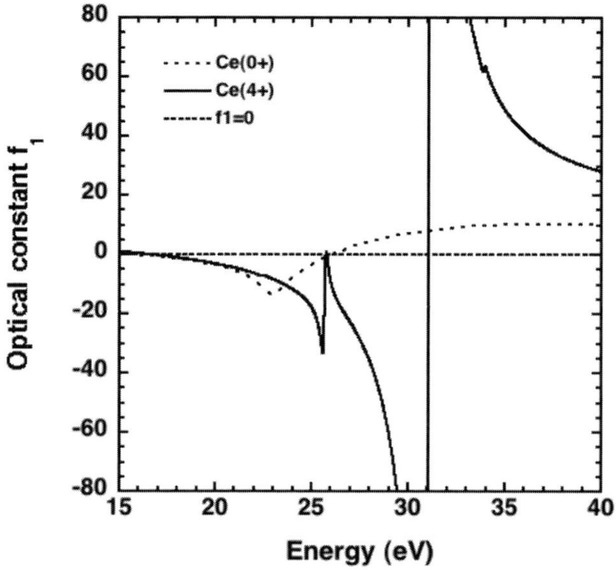

Fig. 3 Optical constant f_1 versus photon energy for Ce plasmas. The dotted line is from the Henke data. The solid curve is calculated for Ce^{4+}. The dashed line is a visual aid for $f_1 = 0$.

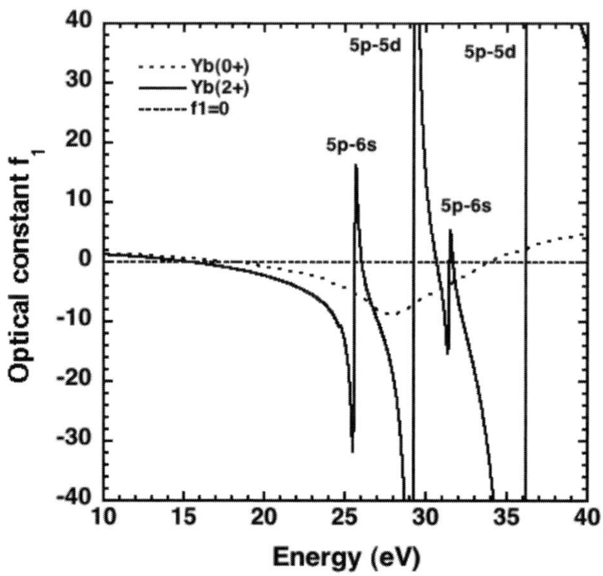

Fig. 4 Optical constant f_1 versus photon energy for Yb plasmas. The dotted line is from the Henke data. The solid curve is calculated for Yb^{2+}. Thc dashed line is a visual aid for $f_1 = 0$.

5 Modeling of Carbon Plasmas at Higher Energy

With the advent of the FLASH X-ray free electron laser in Germany and the LCLS X-FEL coming online soon at Stanford we look near the K shell lines of near solid carbon plasmas and predict strong effects near the K-shell absorption edge. We have previously modeled C plasma and predicted and observed strong anomalous effects near 26.44 eV [11,12].

The AVATOMKG code is used to predict the optical constant f_1 versus energy for C at density 0.2 g per cc, temperature of 12 eV, and resulting $Z^* = 1.91$, as shown in Fig. 5. Also shown is the Henke data for neutral C. For the doubly-ionized C one predicts that f_1 is less than zero for energies from 256 to 273 eV, which means that the index of refraction is greater than one over this range and one would observe anomalous effects. With the AVATOMKG code we are prepared to calculate the optical properties of many different plasma over the entire X-ray regime that is available to laboratory X-ray lasers as well as the new X-FEL facilities.

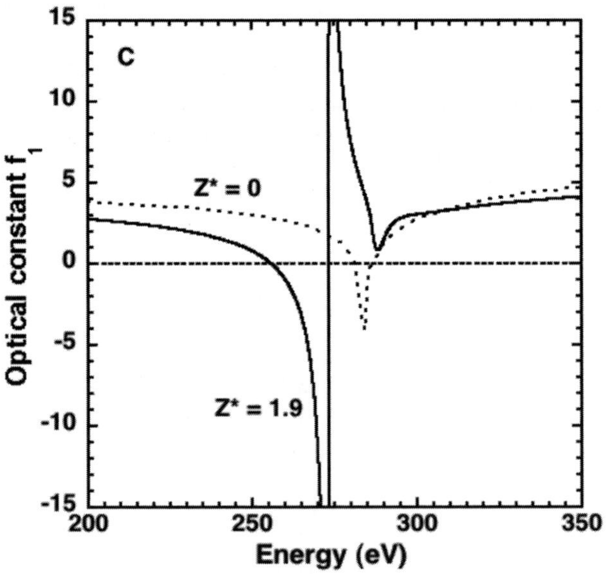

Fig. 5 Optical constant f_1 versus photon energy for C plasmas. The dotted line is from the Henke data. The solid curve is calculated by AVATOMKG code for $Z^* = 1.9$. The dashed line is a visual aid for $f1 = 0$.

6 Conclusions

For decades the analysis of plasma diagnostics such as interferometers have relied on the approximation that the index of refraction in plasmas is due solely to the free electrons. This makes the index of refraction less than one. Recent X-ray laser interferometer measurements of Al, Sn, Ag, and C plasmas at wavelengths ranging from 13.9 to 46.9 nm observed anomalous results with the index of refraction being greater than one. The analysis of these plasma show that the anomalous dispersion from both the resonance lines and absorption edges due to the bound electrons can have the dominant contribution to the index of refraction.

To understand how general this anomalous index of refraction effect is we searched for plasmas that should have an index of refraction greater than one in this soft X-ray regime. We present calculations of neutral Na vapor, singly-ionized Ne plasma, Ce^{4+} and Yb^{2+} plasma that predict an index of refraction greater than one at the 46.9 nm (26.44 eV) wavelength of the Ne-like Ar X-ray laser. With the advent of the FLASH X-ray free electron laser in Germany and the LCLS X-FEL at Stanford we look at the K shell lines of near solid carbon plasmas and predict strong anomalous effects with the index of refraction greater than one from 256 to 273 eV.

During the next decade X-ray free electron lasers and other sources will be available to probe a wider variety of plasmas at higher densities and shorter wavelengths so it will be even more essential to understand the index of refraction in plasmas. X-ray laser interferometers may become a valuable tool to measure the index of refraction of plasmas in the future.

Acknowledgements

Work performed under the auspices of the US Department of Energy by the Lawrence Livermore National Laboratory under Contract DE-AC52-07NA27344. The work of one author (WRJ) was supported in part by NSF Grant No. PHY-0456828.

References

1. G. J. Tallents, J. Phys. D. **17**, 721 (1984).
2. H. R. Griem, *Principles of Plasma Spectroscopy*, (Cambridge University Press, Cambridge, 1997) p. 9
3. L. B. Da Silva et al., Phys. Rev. Lett. **74**, 3991 (1995).
4. H. Tang et al., Appl. Phys. B **78**, 975 (2004).
5. J. Filevich et al., Phys. Rev. Lett. **94**, 035005 (2005).

6. J. Filevich et al., J. Quant. Spectrosc. Radiat. Transfer **99**, 165 – 174 (2006).
7. J. Filevich at al., Opt. Lett **25**, 356 (2000).
8. D. Descamps et al., Opt. Lett. **25**, 135 (2000).
9. A. Meseck et al., Nucl. Inst. And Meth. A **528**, 577 (2004).
10. J. Filevich et al., Phys. Rev. E **74**, 016404 (2006).
11. J. Filevich et al., Laser and Particle Beams **25**, 47 (2007).
12. J. Nilsen et al., "Understanding the anomalous dispersion of doubly-ionized carbon plasmas near 47 nm," HEDP (in press 2008) DOI:10.1016/j.hedp.2008.05.003
13. J. Nilsen and J. H. Scofield, Opt. Lett. **29**, 2677 (2004).
14. J. Nilsen and W. R. Johnson, Applied Optics **44**, 7295 (2005).
15. J. Nilsen, W. R. Johnson, C. A. Iglesias, and J. H. Scofield, J. Quant. Spectrosc. Radiat. Transfer **99**, 425 – 438 (2006).
16. W. R. Johnson, C. Guet, and G. F. Bertsch, J. Quant. Spectrosc. Radiat. Transfer **99**, 327 – 340 (2006).
17. D. A. Liberman, JQSRT **27**, 335 (1982).
18. J. J. Rocca et al., Phys. Rev. Lett. **73**, 2192–2195 (1994).
19. B. L. Henke, E. M. Gullikson, and J. C. Davis, ADNDT **54**, 181 - 342 (1993).
20. L. D. Landau and E. M. Lifshitz, *Electrodynamics of Continuous Media*, (Pergamon, New York, 1984) pp. 280 - 283
21. W. Persson, Phys. Scr. 3, 133 (1971).
22. T. Lundstrom and L. Minnhagen, Phys. Scr. 5, 243 (1972).
23. I. P. Grant et al., Comput. Phys. Commun. **21**, 207 (1980).
24. NIST web site at http://physics.nist.gov/PhysRefData/ASD/index.html

Recent Developments on Seeded or Unseeded Transient X-Ray Lasers

A. Klisnick[1], O. Larroche[2], F. De Dortan[1], J. Habib[1,3], O. Guilbaud[1,3], S. Kazamias[1,3], D. Ros[1,3] , B. Zielbauer[1,3]

[1] LIXAM, Laboratoire d'Interaction du rayonnement X Avec la Matière, UMR 8624, Bat 350, Campus de l'Université Paris XI, F-91405 Orsay Cedex, France
[2] CEA-DIF, Bruyères –le-Châtel, 91297 Arpajon Cedex, France
[3] LASERIX, Centre Laser Université Paris-Sud, Campus de l'université Paris XI, F-91405 Orsay Cedex France

Abstract. The COLAX code was used to investigate the role of several parameters on the spatial and temporal behaviour of the output pulse of a transient X-ray laser seeded by a harmonic pulse. The importance of the collisional linewidth on the duration of the output pulse is underlined.

1 Introduction

Seeding high order harmonics radiation in a transient X-ray laser plasma has been successfully demonstrated recently [1, 2, 3]. This opens the route to fully coherent, high-brightness sources that should be extended towards higher energy, shorter wavelength and shorter pulse duration. In order to support the experimental efforts towards these goals we have upgraded the COLAX code [4], so as to include seeding by a femtosecond harmonic pulse, as well as travelling wave pumping geometry. The Maxwell-Bloch model used in COLAX is well suited to the description of the amplification of a femtosecond pulse in a narrow bandwidth lasing medium, as is the case for X-ray lasers. We have shown that the amplification of the harmonic seed in the X-ray laser plasma is a dynamical process [5], that involves the temporal response of the atomic polarisation on the lasing ions to the injected field. The harmonic pulse hence generates a wake that follows by a few picoseconds and is amplified while propagating along the plasma length. In this paper we will discuss the role of the main parameters that modify the spatial and temporal behaviour of the output amplified seed, namely the time or level of injection. Further, the collisional spectral width of the lasing transition will be shown to play a crucial role in the duration of the output pulse. This simulation study is linked to the experimental work currently carried out at the LASERIX facility, aimed at seeding a GRIP transient X-ray laser [6].

2 Simulations of a seeded X-ray laser with the COLAX code

2.1 The Maxwell-Bloch model

The COLAX code, initially described in [4], is based on the resolution of Maxwell-Bloch equations to calculate the temporal and spatial evolution of the electric field associated with the amplified radiation. The code was recently upgraded to include the presence of a high-order harmonic radiation pulse injected at the entrance plane of an X-ray laser (XRL) plasma. The code is 2D (x, y) and only the linear component in the z direction is considered. The Maxwell equation, written in the paraxial slowly varying envelope approximation, describes the time and space evolution of the field travelling in either the forward (+) or backward (-) y direction:

$$\frac{\partial E_{\pm}}{c \cdot \partial t} \pm \frac{\partial E_{\pm}}{\partial y} = \frac{ic}{2\omega} \frac{\partial^2 E_{\pm}}{\partial x^2} + \frac{i\omega}{2c}(\varepsilon_R E_{\pm} + 4\pi P_{\pm}) \tag{1}$$

Here $\varepsilon_R + 1$ is the dielectric constant for the free electrons and $P_{\pm}$ is the atomic polarisation density on the lasing transition.

Equation (1) is solved together with the equation describing the temporal evolution of the polarisation density:

$$\frac{\partial P_{\pm}}{\partial t} = -\gamma P_{\pm} - i\omega D E_{\pm} + \Gamma_{\pm} \tag{2}$$

Here γ is the polarisation dephasing factor due to both elastic and inelastic collisions on bound electrons. This quantity is related to the collisional width of the lasing line. It will be shown below that it plays a crucial role in their temporal behaviour of the amplification of the harmonic seed. D is the reduced, non-dimensional population inversion that is taken at the quasi-steady state equilibrium, while saturation is taken into account. Finally Γ is a source term accounting for the contribution of spontaneous emission to polarisation.

It should be noted that keeping a time-dependent treatment for the polarisation is required here as the spectral width of the injected radiation, namely the harmonic pulse, is much larger than the width of the XRL gain, so that the adiabatic approximation, where dP/dt is set equal to zero and P is proportional to the electric field E, is not valid and should not be used. Using this time-dependent treatment we have shown [5] that the presence of the femtosecond harmonic pulse induces a dynamical response of the lasing plasma leading to the formation of a wake following the harmonic pulse.

2.2 The input data

The COLAX code is designed to be used as a postprocessor of the EHYBRID hydrocode to provide a detailed description of the amplification of harmonics pulse in conditions close to experimental ones. However for the present study we have used a simpler model case for the input data that describe the state of the plasma in which the harmonic pulse is injected. The use of such a model case gives the possibility to investigate separately the role of several seeding and plasma parameters and to get a better insight on the main trends of the spatial and temporal behaviour of the seeded XRL output. Further such a study is also useful to test the capability and the reliability of the code.

The values that were considered as input data are consistent with the conditions of maximum gain in a Ni-like silver XRL plasma that is calculated from a detailed EHYBRID simulation. The local gain has a Gaussian distribution in both space and time, with a maximum value of 50 cm^{-1} and a spatial width of 40 μm, a temporal width of 20 ps. The saturation intensity is 10^{10} W/cm^2 and the γ factor, which appears in Eq. (2) is 10^{12} s^{-1}. The electron density decreases exponentially with the transverse coordinate x whereas the electron temperature is constant in space and time at 400 eV.

The plasma length is 4 mm and we consider that this plasma is heated by the short pulse in a travelling-wave geometry along the longitudinal coordinate y. This geometry implies that the history of the gain coefficient at a given longitudinal position y is similar to the one at y = 0 but delayed by the time it takes to the short pulse to travel from zero to this position. The resulting spatial distribution of gain at a given time is shown in Fig. 1: the region where the gain is significantly greater than zero is limited in both the transverse (x) and longitudinal (y) directions. This gain region travels along the plasma length (from left to right in Fig. 1) at the speed of light.

Finally a harmonic pulse is injected at the input plane of the plasma at y = 0. This harmonic field has a Gaussian space and time envelope with a duration of 30 fs and a maximum intensity of 10^9 W/cm^2, that is one tenth of the saturation intensity.

From the calculated evolution of the electric field as the harmonic pulse travels along the plasma, we can calculate the evolution of the field intensity flowing through the output plane as a function of time and of the transverse position. Such a representation is interesting because it could be directly compared to a time-resolved near-field image that would be obtained in an experiment by setting the slit of a streak camera at the image plane of a device imaging the exit plane of a seeded XRL plasma. In the following all the results of COLAX that will be discussed will deal with the time-resolved distribution of intensity at the output plane.

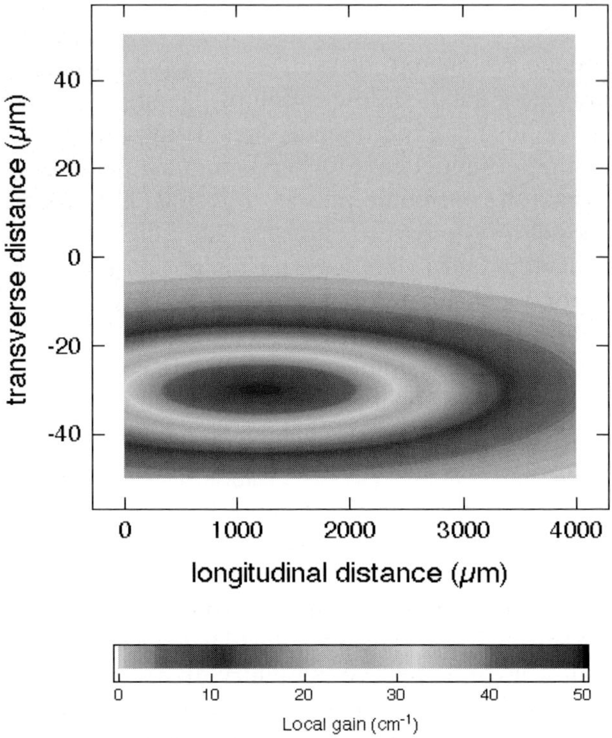

Fig. 1 Spatial distribution of the local gain at a given time during the propagation of the seed in the longitudinal direction. The limited size of the gain zone in the longitudinal direction is due to the presence of a travelling wave irradiation.

2.3 Seeded versus unseeded (ASE) output

In this section we compare the main features of the output intensity with and without seeding. Fig. 2.a shows the calculated distribution of output intensity versus time and space for the unseeded case, i.e. when the output intensity is only due to amplification of spontaneous emission (ASE). What is remarkable here is the presence of small-scale structures which are inherent to the ASE regime, due to a limited coherence of the amplified radiation. Such small-scale structures were also observed experimentally in the near-field pattern of a molybdenum XRL that was recorded recently at the LASERIX facility [6, 7].

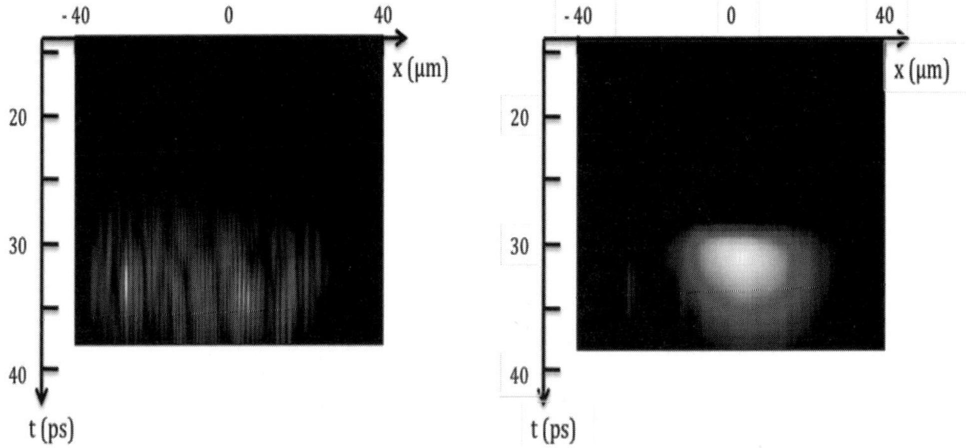

Fig. 2 Distribution of the output intensity versus time and transverse position predicted by COLAX. (a) unseeded case; (b) seeded case. Note that the color scale is not the same in the two images as the calculated signal is much stronger in case (b).

Fig. 2.b shows the distribution of intensity with the same temporal ad spatial scales as in Fig. 2.a, but here a seed pulse was injected. Note that the color scale is different in the two images as the ASE signal in Fig. 2.a is much weaker than the amplified seed in Fig. 2.b. It can be seen that the seeded output pulse is both smoother and smaller than the ASE one, whereas it has a similar duration of the order of 5 ps. Correspondingly, in the far field, COLAX predicts that the amplified seed has a much narrower divergence (about 1/10 mrad) than the ASE beam, in agreement with experimental observations [2].

3 Role of the time and level of injection

3.1 Role of time of injection

The time of injection was varied between 10 ps and 23 ps, where 0 ps is the start of the simulation and 20 ps is the time when the local gain is maximum at the entrance plane y =0. The results are summarised in Fig. 3 which shows the calculated maximum output intensity as a function of the time of injection, compared with the ASE level (dotted horizontal line). One can see that at the optimal injection time the maximum intensity of the amplified harmonics is 11 times the ASE one, and about 4 times the saturation intensity which was used for those calculations.

On the other hand the value of the optimal time of injection is 17 ps, i.e. about 3 ps before the time when the gain is maximum at the injection plane.

This temporal shift is directly related to the dynamical response of the plasma to the incident radiation. The radiation which is amplified is the wake following the femtosecond harmonic pulse by the same temporal delay.

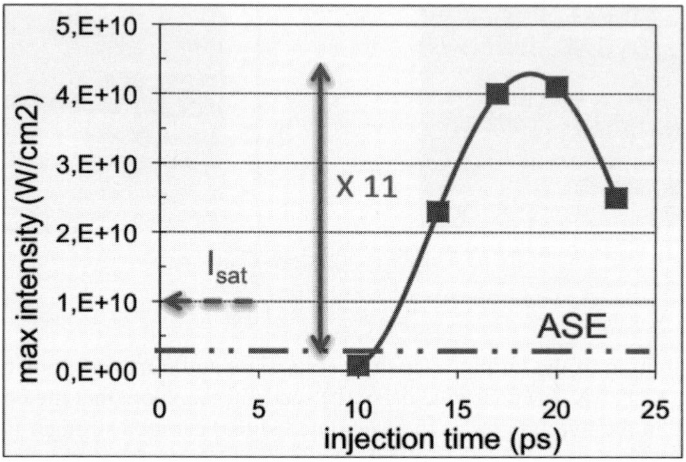

Fig. 3 Calculated maximum output intensity for increasing values of the time of injection, compared with the level of ASE intensity (horizontal dotted line) and with the saturation intensity I_{sat}.

3.2 Role of level of injection

The time of injection was then kept constant at 17ps and the level of injection was varied. The results are summarized in Fig. 4 which shows the calculated maximum output intensity for increasing values of the injected intensity, between 10^8 to 5.10^9 W/cm^2, or between 1/100 and 1/2 of the saturation intensity.

One can see that the curve which fits the calculated data is far from a linear increase, which would be expected at least for the lowest injected intensities that are well below the saturation intensity. However, due to the large gain length value of 20 which was used for these calculations, all the output intensities in the explored domain are above the saturation intensity of 10^{10} W/cm^2 . Because of saturation the slope is lower for the largest injected intensities. As a result the output intensity is more sensitive to small variations of the injection level in this part of the curve. Finally the highest injected intensities correspond to situations where the gain is strongly depleted by saturation in the region where the amplified intensity is maximum.

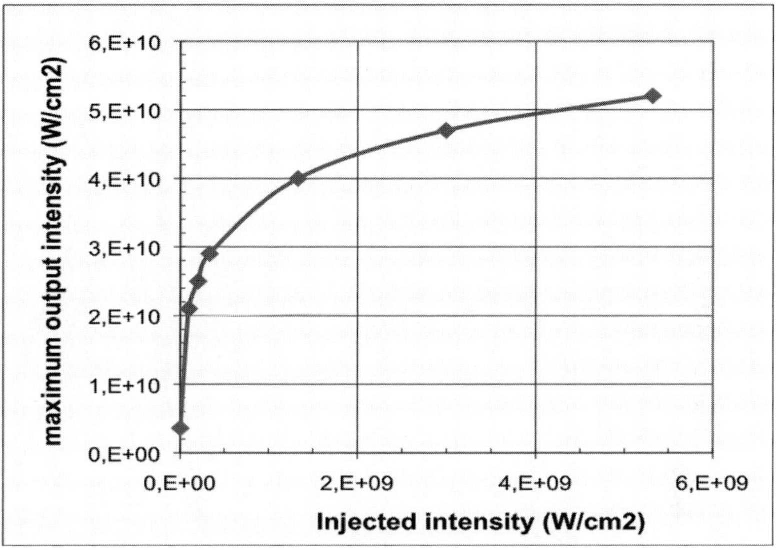

Fig. 4 Calculated maximum output intensity for increasing values of the injected intensities. Saturation intensity: 10^{10} W/cm^2

4 Duration of the amplified HHG

In this section we investigate the role of two parameters that might influence the duration of the amplified seed pulse. Fig. 5 shows the temporal history of the output pulse in three cases. In the first case the duration of the local gain was taken at 20 ps FWHM. The corresponding output pulse has a duration of about 5 ps. When the gain duration is reduced to 10 ps FWHM, this does not change much the duration of the output pulse which is barely reduced to 4 ps. However if we now come back to a gain duration of 20 ps and increase the factor γ by a factor 5 we see a dramatic reduction of the duration of the output pulse which is now of about 1ps.

It should be reminded that the factor γ is the polarisation dephasing rate, which is linked to the collisional linewidth of the lasing transition. This factor hence plays a crucial role in the temporal behaviour of the amplified harmonic and its value has to be modelled and evaluated precisely.

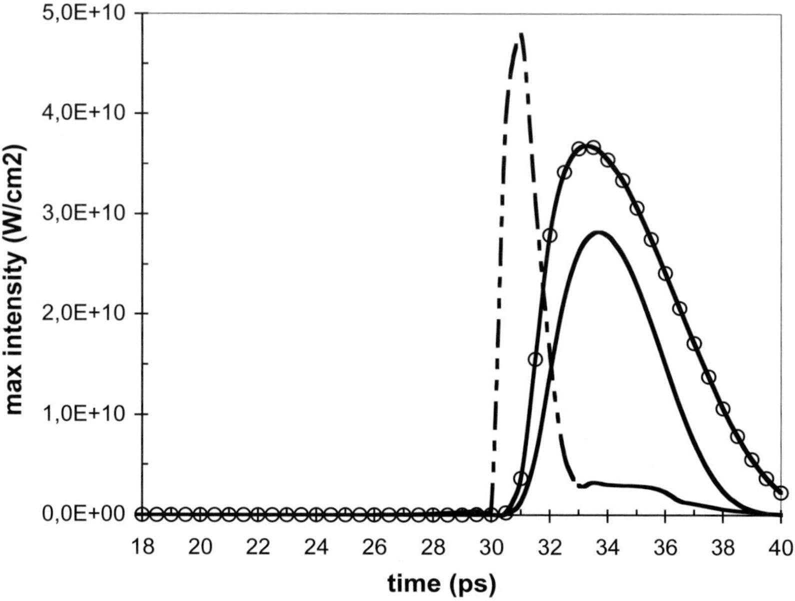

Fig. 5 Temporal history of the amplified seed pulse predicted by COLAX in three cases: (___) duration of gain 20 ps FWHM, $\gamma = 10^{12}s^{-1}$; (-o-) duration of gain 10 ps (FWHM) , $\gamma = 10^{12}s^{-1}$; (---) duration of gain 20 ps, $\gamma = 5.10^{12}s^{-1}$

5 Conclusions and future work

In this work we have used the COLAX code to investigate the role of several parameters that influence the temporal and spatial behaviour of the amplification of a harmonic seed in an XRL plasma.

We have varied the time of injection and we have shown that the optimal time is shifted by about 3ps with respect to the time of maximum gain at the injection plane. This corresponds to the delay between the harmonic pulse and the wake which is formed as a response of the medium to the incident radiation.

The output intensity increases very rapidly with the level of injection when this level is much lower than the saturation intensity (i.e. $I_{inj} \sim I_{sat} /100$). The increase gets lower when the level of injection approaches the saturation intensity.

Finally we have shown that the duration of the output pulse is predominantly controlled by the polarisation dephasing rate, the factor γ, which is related to the collisional linewidth. A precise knowledge of this quantity is hence required to ensure reliable predictions. Recent measurements of this quantity are reported in companion papers [7, 8].

The COLAX code will be further improved by including a time-dependent treatment for the excited level populations involved in the population inversion. A second step will be to include inhomogeneous Doppler broadening as in the current version only homogeneous broadening is taken into account.

References

1. Ph. Zeitoun et al *Nature* 431, 426, 2004
2. T. Mocek et al., *Phys. Rev. Lett.* 95, 173902, 2005
3. Y. Wang et al., *Phys. Rev. Lett.* 97, 123901, 2006
4. O.Larroche et al., *Phys. Rev. A* 62, 043815, 2000
5. I. AlMiev et al., *Phys. Rev. Lett.* 99, 123902, 2007
6. D. Ros et al., *This Conference*
7. J. Habib et al., *This Conference*
8. O. Guilbaud et al., *This Conference*

Influence of the number of atomic levels on the modelling of collisional X-ray lasers

F. de Dortan[1,2], M. Busquet[3], A. Bar-Shalom[3], M. Klapisch[3], J. Oreg[3], B. Rus[2], M. Kozlova[2] and J. Nejdl[2]

[1]LIXAM, Laboratoire d'interaction du rayonnement X avec la matière, UMR 8624, Bat 350, Campus de l'Université Paris XI, F-91405 Orsay Cedex, France
[2] Department of X-ray Lasers, Institute of Physics / PALS Centre, Prague 8, Czech Republic
[3] ARTEP, inc. Ellicott City, MD21024, USA

Abstract. The number of atomic excited levels considered in codes modelling X-ray lasers is generally very limited. A few dozens or even two or three are considered while thousands times more do exist. We compute populations, gain and spectra with tens of thousands of levels out of equilibrium to model a Zinc Ne-like X-ray laser. The maximum gain, its thermodynamic conditions as well as the transverse spectrum compare well with the experiment. The gain is marginally reduced for stationary plasmas when the number of levels increases. But the radiative losses are strongly enhanced for thick plasmas. For transient schemes, some neighbouring levels are strongly coupled with the lasing levels and will modify the solution of the Maxwell Bloch equations in case of seeding with high harmonics.

1 Introduction

In the plasma generating an X-ray laser, the temperature reaches a few hundred eV and the electronic density some 10^{20} to 10^{21} e^-/cc. In these conditions, many electrons may be excited simultaneously in each atom and the number of possible excited levels in each ion may be of a few thousands. The modelling of these X-ray lasers requires an hydro radiative computation to follow the plasma through time and space, a detailed atomic description to calculate the gain and a ray tracing code to evaluate the divergence and possibly the coherence of the beam. For time limiting reasons, the number of levels taken into account must be small to solve the equation of populations in each hydrodynamic cell, usually a few dozens [1, 2]. In computations studying the amplification of a High Harmonic in the lasing plasma and its coherence, only the upper and lower levels of the laser transition are usually considered [3, 4].

It has been noticed that the hydro-radiative codes (cited above) lead to very high gains (close to 1000/cm for Ni-like transient lasers by example) while the maximum gains observed in experiments reach approximately 70/cm in the same conditions [5]. Multiple reasons may damper the theoretical gains as time averaging, isotopic composition of the lasing media and the subsequent shift, hyperfine structure splitting [6] or use of a more precise model for the calculation of collision strengths [7]. We assume that the number of atomic levels taken into account in the lasing ion may also have some influence. A large number of levels deplete the low lying levels in favour of the high lying ones to keep constant the total population of the lasing ion. Regarding the modelling of High Harmonics amplification, we assume that some neighbouring levels of the lasing ones are strongly coupled to them and therefore should be considered in the dynamic resolution of the populations of the lasing levels.

We present in the second paragraph our modelling of a Zinc Ne-like laser. We use the SCRIC [10] code to compute the maximum possible gain, the transverse emission spectrum and compare it with the experimental results obtained on the PALS laser in the third paragraph. In the fourth paragraph we look for the levels strongly coupled with the lasing ones by comparing the rate coefficients of a Ni-like Silver transient X-ray laser obtained with the EHYBRID code [1].

2 Theoretical modelling of a Ne-like Zinc laser plasma

At the PALS laser, the time duration of the Ne-like Zinc X-ray laser is approximately 150ps, so we assume that the plasma and the populations are stationary. This strongly limits the computing time. We select the configurations necessary for the calculation on energy basis: energies of configurations are lower than 7 times the electronic temperature. The optimum temperature is close to 450 eV so configurations below 3000 eV are kept. In thermodynamic equilibrium this allows to keep all configurations and levels whose populations are above $9\times10^{-4}g$ where g is the degeneracy ; out of equilibrium – *i.e.* for low densities – this lower limit is even smaller. We assume that in these conditions, we keep a good accuracy on the calculation of populations below electronic densities of 10^{21} e^-/cc. For Ne-like ions, the configurations found with SCDF [8] are $1^2 2^8$, $1^2 2^7 3,4,5,6,7^1$, $1^2 2^6 3^2$ and $1^2 2^6 3^1 4^1$ where n^i refers to all possible electronic distributions of i electrons in the sublayers of layer n. This means 1 non relativistic configuration for the ground stage (the first group of configurations), 50 non relativistic configurations for the singly excited configurations (the second group of configurations) and 54 doubly excited configurations (the two last groups of configurations) or 241 singly excited detailed levels and 3784 detailed levels when all levels (singly and doubly excited) are included. Finally, 5 ions are included in the calculation from 18+ to 22+ and they account for 495 non relativistic configurations.

All levels, energies (parametric potential, relativistic) and cross sections (DW) have been computed with the HULLAC [9] suite of codes which makes 16384 detailed levels, approximately 10 million lines, 14 million collisional ionisations, 14 million photoionisations, 29 million collisionnal excitations, 80,000 autoionisations and the same number of reverse processes.

The SCRIC [10] collisional-radiative solver computes the populations of these levels but also the detailed absorption and emission spectra of homogeneous plasmas in limited time. This allows determination of the optimum thermodynamic conditions to maximise the gain. We finally characterise experimental lasing plasmas by comparing their transverse X-ray emission spectra with synthetic ones. Lines are computed with a Voigt profile including Doppler, natural and collisional broadening, for each line.

3 Comparison with experiment

The Zinc Ne-like laser currently working in PALS shows experimental gains around 7/cm for electronic densities close to $3x10^{20}e^-/cc$ [11]. The lasing ionization is around 20+, so the ionic density equals $1.5x10^{19}$ at/cc.

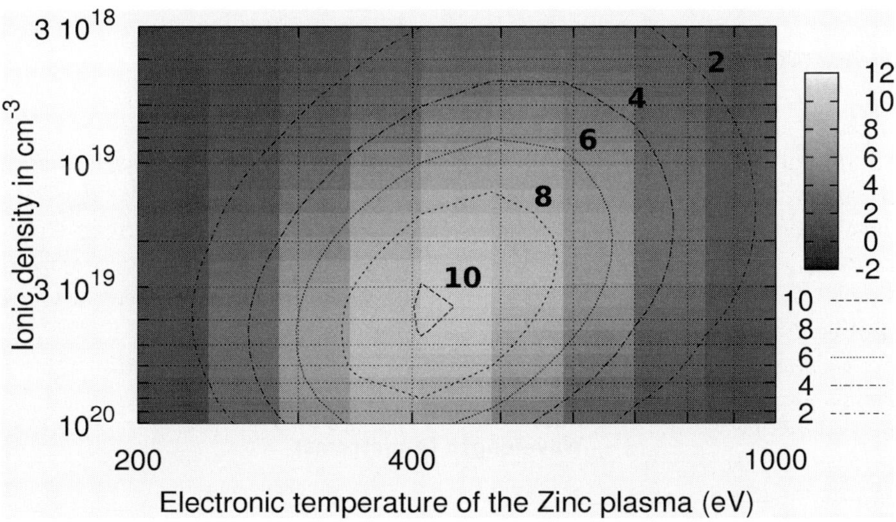

Fig. 1 Synthetic gain (cm^{-1}) of a 3p-3s 21.2 nm transition of a Ne-like Zinc homogeneous plasma in regard of electronic temperature and ionic density.

We present in figure 1, the theoretical gain obtained with the SCRIC code. The maximum gain is 10.5 / cm and obtained for 420 eV and $3.5x10^{19}$ at/cc which corresponds to an electronic density of $7x10^{20}$ e^-/cc. Such a density can't be reached with the PALS laser whose wavelength is 1.315 micron and critical density 6.510^{20} e-/cc. But at $3x10^{20}$ e^-/cc, the synthetic gain is maximum at 420 eV and reaches 8 between 420 eV and 550 eV, within 15% of the experimental value. Larger gains could be reached with lower wavelength. Concerning the number of levels included, it appears that in the optimum lasing conditions the population of highly excited levels remains very low and doubly excited levels account for only 2% of the total population of ion Zn^{20+}. This ratio increases for higher temperatures and / or densities but remains low (5% at 700 eV, $8x10^{19}$ at/cc and a 2/cm gain). Therefore, the gain is also little modified by the very large increase in the number of levels. However, a higher proportion of very excited levels -and the subsequent decrease of gain- may occur for temperatures highly above the stationary optimum gain conditions. These conditions are reached in transient X-ray lasers where the high gains come from the decoupling between ionization and temperature.

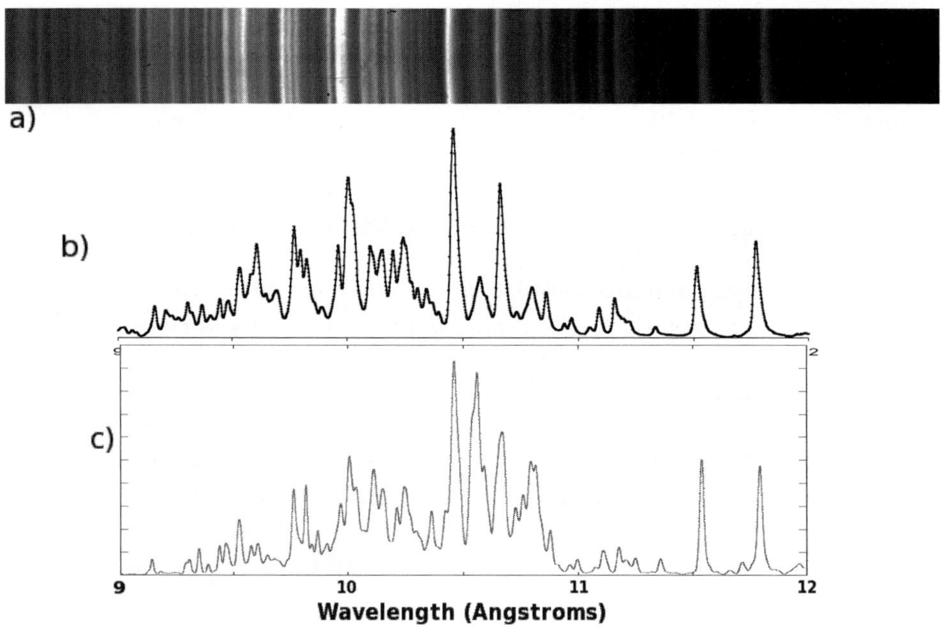

Fig. 2 Experimental [12] and synthetic transverse emission spectra of a Zinc X-ray laser plasma in optimum lasing conditions. a) Experimental, time integrated b) experimental, space and time integrated ; c) synthetic 420 eV, $2x10^{19}$ at/cc, $4x10^{20}$ e^-/cc and 150 microns thick ; an artificial 3eV broadening has been added to compare with experimental spectra.

In figure 2 we compare the experimental [12] and calculated X-ray transverse spectra obtained for the optimum lasing conditions mentioned above. We notice good convergence even if two discrepancies remain at 10.6 and 10.8 Angströms corresponding to an overestimation of Na-like Zinc population. This allows to confirm the thermodynamic conditions already mentioned : Te ~ 420 eV and Ne ~ $3x10^{20}$ to $4x10^{20}$ e^{-}/cc. Concerning the highly excited levels, even if their population is small, their contribution to the spectrum is high – all the background – because of the number of the satellite lines and of the thickness and opacity of the plasma which limits the height of the resonant lines. A comparison can be made with a Titanium X-ray laser synthetic spectrum containing only resonant lines in [13].

In these conditions, the total internal energy of the column of plasma (3cm * 150 microns diameter) is close to 35 Joules (~60% thermal energy, ~40% ionization + excitation energy) to be compared with the 450 Joules of the pumping laser. This means that the pumping laser energy is either not absorbed or radiated and this paves the way to new schemes with better laser absorption or less radiative losses.

4 Strongly coupled levels in High Harmonic amplification

When the lasing plasma is seeded with a high harmonic, the calculation of the amplification relies on the Maxwell Bloch equations which couple the electric field with the populations of the lasing levels [3]. The resolution of these equations shows, in specific situations, large modulations in the amplitude of the field, known as Rabbi oscillations [4]. These correspond to large variations of the populations of the lasing levels. When the lasing levels become strongly depleted, they may also deplete some neighbouring levels strongly coupled with them. These last levels, retroactively, won't be able anymore to repopulate the lasing levels and the global solution of the Maxwell-Bloch equations will be modified. For this reason their individual inclusion in the Maxwell-Bloch equations will improve the results as has already been observed for the different magnetic sublevels of the lower lasing level [14].

We identify in table 1 all the levels strongly coupled with the lasing levels of a Ni-like Silver transient laser. We use the EHYBRID code to model the plasma and the atomic physics [1]. These levels account for more than 10% of the maximum populating rate of the lasing levels and more than 3% of their total populating rate. The rates have been computed for the maximum gain (of 430/cm) obtained during a double pulse at 1.06 micron: 600ps at $5x10^{11}$ W/cm^2 followed by 0.4ps at $5x10^{14}$ W/cm^2 after 850ps. The lower lasing level is strongly populated by the ground level for approximately 1/3rd, by the upper lasing level for 1/3rd and by 7 other levels for approximately 1/3rd also. These last 7 levels are little populated (of the same order of magnitude as the lasing levels) consequently they will be rapidly depleted if the lower lasing level is depleted and will strongly retroact on it. The upper lasing level is populated for half by the ground level and just less than half by a more excited level $3d^9 4f$ (3/2, 5/2) J=1 which is populated similarly to the lasing level. Therefore it will also strongly retroact on the upper lasing level in case when they are both depleted. The ground level is very populated compared to the lasing levels and its population will not vary during the oscillations of the lasing populations.

Finally, a more complete treatment of the Maxwell Bloch equations should include the previously mentioned excited states, plus the lasing levels and the global populating/depopulating rates (condensing all the other populating/depopulating rates from the other levels including ground). In case of strong constraints on computing time, an improvement would be obtained by including only the $3d^9 4f$ level and possibly other levels beginning in descending order of populating rate of the lower lasing level.

Table 1 Maximum populating rates of the lasing levels in a transient Silver Ni-like X-ray laser (result obtained with the EHYBRID code). The lasing levels are in bold.

Level	Population/ ion pop.	Populating rate low. lasing lev.	Populating rate up. lasing lev.
Ground	0.642	29.10 %	49.80%
(5/2, ½) J=2	0.00592	5.45%	
(3/2, ½) J=2	0.00532	3.14%	
(5/2, 3/2) J=1	**0.00889**		**4.83%**
(5/2, 5/2) J=1	0.00500	7.97%	
(5/2, 5/2) J=2	0.00469	3.27%	
(3/2, 3/2) J=1	0.00273	3.04%	
(3/2, 3/2) J=2	0.00431	3.25%	
(5/2, 5/2) J=0	**0.0170**	**33.00%**	
(3/2, 5/2) J=1	0.0344	3.86%	44.20%
Total population of the ion Ag 19+	1.	100% =	100% =

5 Conclusion and perspectives

We have noticed that the gain of a stationnary plasma is modified by a few percents by the number of levels increase but that the radiative losses are strongly enhanced. The determination of the thermodynamic conditions of the experimental plasma is improved when a large set of atomic configurations is used. At high temperatures in transient X-ray lasers, the number of atomic levels considered in the simulation may reduce more the calculated gain and this still has to be checked. In transient Ni-like lasers the $3d^9 4f$ (3/2, 5/2) J=1 level is strongly coupled to the upper lasing level and a group of six less excited levels is also strongly coupled to the lower lasing level. Their influence on the dynamics of the amplification of a high harmonic in the lasing plasma has to be computed.

Acknowledgement

The author is indebted to G. Pert of the University of York for the lending of his EHYBRID code and J. Dubau of LIXAM, University of Orsay for his help on this code and his precious advices. This work benefited from support of the Czech Science Foundation grant No. 202/08/1734 and from the Czech Republic's Academy of Sciences project KAN300100702.

References

1. Blanc S. P. L. and Sauerbrey R. : 'Spectral, temporal, and spatial characteristics of plasma-induced spectral blue shifting and its application to femtosecond pulse measurement', *J. Opt. Soc. Am. B,* 13, 72-88, 1996
2. Pert, G. J. : 'Algorithms for the calculation of Ionization in rapidly Changing Plasmas', *J. Comput. Phys*., 27, 241-155, 1978
3. Larroche, O., Ros D., Klisnick A., Sureau A., Möller C. and Guennou H. : 'Maxwell-Bloch modelling of X-ray-laser-signal buildup in single- and double-pass configurations', *Phys. Rev. A*, 62, 043815, 2000
4. Robillart B. et al. : 'Modelling of an ultra-short x-ray laser pulse amplification through an optical-field-ionised gas using a Maxwell-Bloch treatment', This publication
5. Rocca, J. J. , Wang Y., Larotonda M., Luther B., Berrill M., Alessi D., Weith A., Marconi M., Patel D., Menoni C. S., Shlyaptsev V. N., Dunn J., Liu Y. and Attwood D. T. : 'Advances in high repetition soft X-ray lasers : lasing down to 10.9 nm and high brightness operation of a seeded soft X-ray amplifier using a solid target' P 149-159 in *Proceedings X-Ray Lasers 2006*, Springer Berlin, 2007
6. Dubau, J., Zabaydullin O., Möller C. and Benredjem D. : 'Copper neon-like x-ray laser modelling including hyperfine structure levels' P 347-354 in *Proceedings X-Ray Lasers 2004*, Institute of Physics Publishing, 2005
7. Aggarwal K. M. and Keenan F. P. : 'Effective collision strengths for transitions in Ni XIX', *Astron. Astrophys*., 488, 365-374, 2008
8. Ankudinov A. L., Zabinsky S. I. And Rehr J. J. : 'Single configuration Dirac-Fock atom code', *Comput. Phys. Comm*., 98, 359-364, 1996
9. Bar-Shalom A., Klapisch M. and Oreg J.: 'HULLAC, an integrated computer package for atomic processes in plasmas', *J. Quant. Spectrosc. Radiat. Transfer*,

71, 169-188, 2001 ; Busquet M., Bar-ShalomA., Klapisch M. and Oreg J. : 'An improved version of the HULLAC code', *Journal de Physique IV*, 133, 979-975, 2006

10. F. de Gaufridy de Dortan: *SCRIC, un code pour calculer l'absorption et l'emission detaillees de plasmas hors equilibre inhomogenes et etendus*, CEA Report no 6115, Editor CEA, 2006
11. Rus B., Mocek T., Prag A. R., Kozlova M., Jamelot G., Carillon A., Ros D., Joyeux D. and Phalippou D. : 'Multimillijoule, highly coherent x-ray laser at 21 nm operating in deep saturation through double-pass amplification', *Phys. Rev. A* 66, 2002, 063806
12. Nejdl, J., Rus B., Kuba J., Mocek T., Kozlova M., Polan J. and Homer P. : 'Electron temperature measurements of model Ne-like X-ray lasers plasmas from resonnance KeV spectra' P113-119 in *Proceedings X-Ray Lasers 2006*, Springer Berlin, 2007
13. Demir, A., Akman E., Bilikmen S., Demir P., Ince S., Kacar E., Yurdanur E. and Yaltkaya S. : 'Longitudinally pumped Ne-like titanium X-ray laser simulation with a post-processor code coupled to EHYBRID' P341-346 in *Proceedings X-Ray Lasers 2006*, Springer Berlin, 2007
14. Sureau A.& Holden PB.'From amplification of spontaneous emission to saturation in x-ray lasers : a Maxwell Bloch treatment', *Phys. Rev. A*, 52, 3110-3125, 1995

Modelling of Capillary Z-Pinch Recombination Pumping of Hydrogen-Like Ion EUV Lasers

P. Vrba[1], N. A. Bobrova[2], P. V. Sasorov[2], M. Vrbova[3] and J. Hubner[1]

[1] Institute of Plasma Physics AS CR, Prague 8, Czech Republic
[2] Institute of Theoretical and Experimental Physics, Moscow, Russia
[3] CTU Faculty of Biomedical Engineering, Kladno, Czech Republic

Abstract. We report here new results of simulations of Z-pinch recombination pumping for capillary filled by boron and compare them with previous nitrogen results. In both cases the same capillary radius and current pulse shape are taken into account. Evaluated pressure optimized gains for boron and nitrogen are 1.04 cm^{-1} and 0.11 cm^{-1}, respectively. Influence of possible wall ablation is also analyzed.

1 Introduction

Capillary pinching discharge has been demonstrated as a very successful method for pumping of compact and efficient soft X-ray lasers. Namely, pinching plasma column inside argon filled capillary has been shown to be a very efficient way to get lasing at 46.9 nm [1]. In this case the active medium was formed during the pinch compression stage when collision excited neon-like argon ions Ar^{8+} were created. There is significant interest in extending capillary discharge pumped lasers to shorter wavelengths. Recombination pumping scheme leading to the population inversion on the Balmer alpha transitions of low Z- elements is an alternative. The primary pumping process is three-body collision recombination, taking place in non-stationary under-cooled plasma created during the pinch expansion stage.

To understand the recombination pumping process quantitatively and to design optimum parameters of a new experimental device we perform computer modelling [5, 8]. Optional experimental parameters – capillary radius, current pulse shape and filling atom density – are varied.

We have done modelling of a fast electrical discharge in non-ablative nitrogen filled alumina capillary to estimate lasing at 13.4 nm with hydrogen-like nitrogen atoms [8]. We have found out rather high values of slope and amplitude of current pulse passing through the capillary, which are required to get successful lasing [5]. For example, in CAPEX-U experiment, where initial current slope is about 1.10^{12} A/s current amplitude I_{max} = 50 kA and capillary radius R_0 = 0.16 cm [7], the evaluated gain $G_{max} \sim 0.1$ cm^{-1} is too small to be measured. To get measurable gain it is very important to increase the current slope preferentially two times and slightly increase the current amplitude (to have 60 – 70 kA).

We expect that lasing on Balmer alpha quantum transition of hydrogen-like ions with lower atomic number Z is less demanding from the point of view of necessary current pulse amplitude and slope values. The influence of the wall ablation on the evolution of plasma quantities and achievable gain, expected for high currents has been investigated.

2 Computer Modelling of Laboratory Experiments

It is presumed that the electric current passing through the capillary has a shape of damped sinus and the discharge system is specified by the following five parameters: Capillary radius R_0, peak value of the current pulse I_{max}, initial current slope $dI/dt_{|t=0}$, damping time t_1 and initial atom density N_0. Simulation for any shot is performed in the following tree steps: (1) Radial and time dependences of mass plasma density $\rho(r,t)$, plasma electron temperature $T_e(r,t)$, plasma ion temperature $T_i(r,t)$ and plasma electron density $N_e(r,t)$ are evaluated by means of the MHD code NPINCH [2], (2) time dependences of ionization fractions and energy level populations for lithium-, helium- and hydrogen-like ions are evaluated by means of the kinetic code FLY [3] and (3) gain factor is assessed according to [4].

We investigate the discharge dynamics in a capillary pre-filled by partially ionized gas. During the discharge, the degree of plasma ionization is quickly changed due to of increasing electron temperature. In the plasma parameter range under consideration, the following dissipative processes are included: electron thermal conductivity, Joule heating, Nernst and Ettinghausen effects, the radiation losses, and ion viscosity. It is also important to incorporate the degree of ionization both into the equation of state and into the dissipation coefficients. We use the approximation of two-temperature (ion and electron) one-fluid magneto hydrodynamics.

The radiative properties of Z-pinch plasma in non-stationary and non-equilibrium state are described using kinetic code FLY [3] as a post processor. The code provides detailed information on the ionized and excited stages of ions. It enables us to study the evolution of lithium-like, helium-like and hydrogen ion stages of filling gases.

3 Boron Pinching Plasma

Boron plasma behaviour has been modelled for the capillary radius $R_0 = 1.6$ mm and the current pulse specified by $dI/dt_{|t=0} = 10^{12}$ A/s, and $I_{max} = 50$ kA, i.e. for current pulse shape, which was achieved with the CAPEX experiment [7]. The time dependence of the approximated current for this experiment is shown as **Fig. 1(a)**. The non-ablative capillary is presumed. In the simulations we assume free boundary conditions (Lagrangian boundary $r = R_0$). Boron is assumed to be slightly ionized at t = 0 ns.

Trajectories of the plasma boundary elements are shown in the r-t plane for three different initial electron density values see **Fig. 1(b)**. Expansion of the plasma during first ~20 ns is relevant to the chosen boundary condition. The overall picture of the discharge can be described as follows: the current pulse heats the plasma and creates the azimuthally component of the magnetic field, thus leading to the pinching of the plasma. The pinch time increases with increasing initial atom density.

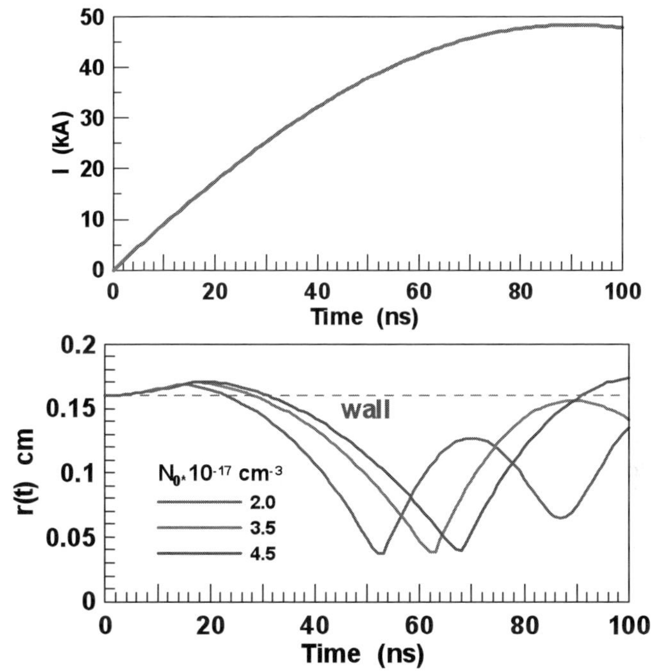

Fig. 1 **(a)** Shape of the current pulse, **(b)** Movement of the outer plasma element for three various atom densities N_0

3.1 Radial and Time Evolution of Plasma Quantities

Temporal evolutions of the calculated radial distributions of plasma parameters inside the channel are seen from **Fig. 2**. The first, the second, and the third frames show the distribution of the mass density compression ratio $\rho(r,t)/\rho_0$, the electron density $N_e(r,t)$, electron temperature $T_e(r,t)$ and the fourth frame shows the distribution of plasma axial current density $j_z(r,t)$. We can see the propagation of a compression shock wave from the walls toward the channel axis and its reflection from the region near the axis. A general picture of the motion of boron plasma under the action of self-consistent magnetic field is similar to the dynamics of Z-pinches. A cylindrical shock wave propagates in the discharge plasma and is reflected from the neighbourhood of the axis and turns into a diverging shock wave. The role played by magnetic pressure in the plasma acceleration toward the axis is important up to the time of the maximum plasma compression. It is interesting to point out that the highest mass compression and the highest electron densities are achieved at the capillary axis. But, it is not the case for the electron temperature and current density.

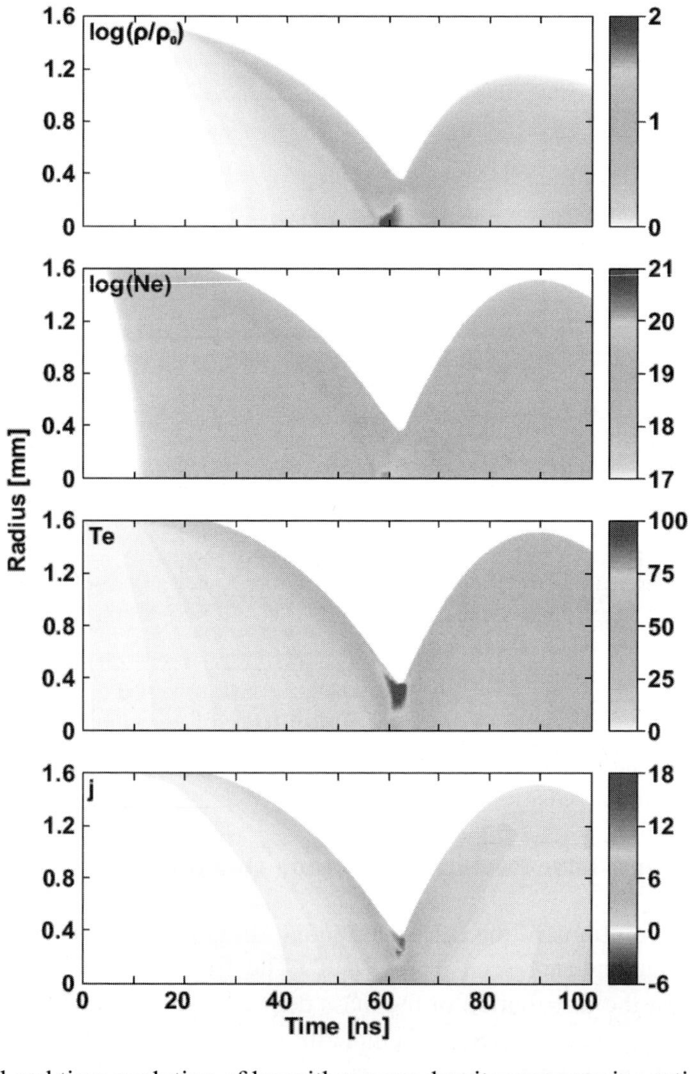

Fig. 2 Radial and time evolution of logarithm mass density compression ratio, logarithm electron density $N_e(0,t)$ in cm^{-3}, electron temperature $T_e(0,t)$ in eV and current density $j_z(0,t)$ in MA.cm^{-3} for initial boron atom density $N_0 = 3.5\ 10^{17}$ cm^{-3}

This situation differs from those for filled capillaries with ablating walls [11], where the redistribution of the electric current between the filling and ablated wall plasmas strongly influences the plasma dynamics.

3.2 Boron Ion Abundances, Population of Energy Levels and Gain on Capillary Axis

Kinetics of boron ions on the axis for initial atom density $N_0 = 3.5\ 10^{17}\ cm^{-3}$ is seen from **Fig. 3**. The ion abundances are quickly changed during the pinch collapse. Predominant density of the fully ionized atoms is achieved at the pinch time (around 62 ns) when the temperature is about 70 eV.

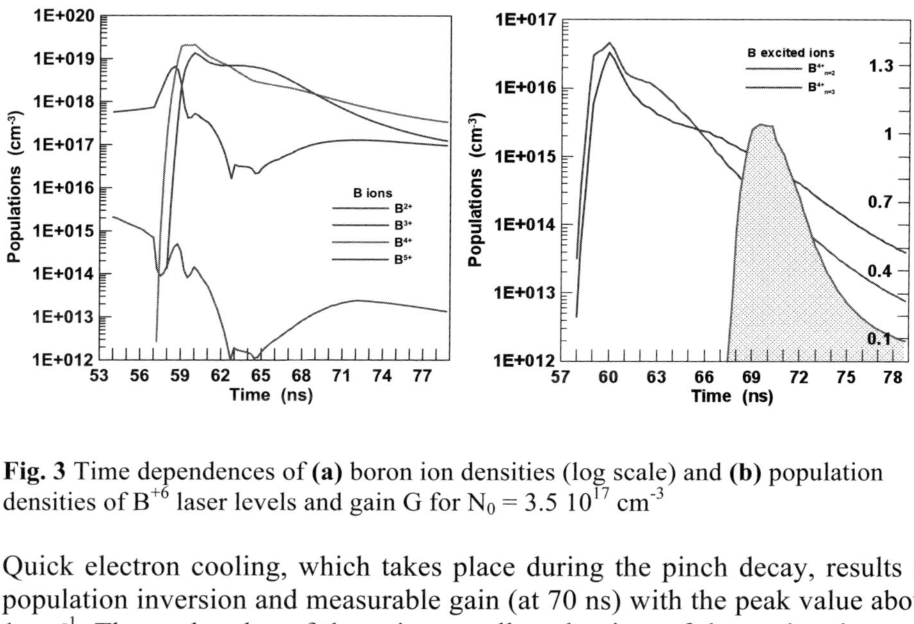

Fig. 3 Time dependences of **(a)** boron ion densities (log scale) and **(b)** population densities of B^{+6} laser levels and gain G for $N_0 = 3.5\ 10^{17}\ cm^{-3}$

Quick electron cooling, which takes place during the pinch decay, results in population inversion and measurable gain (at 70 ns) with the peak value about 1 cm^{-1}. The peak value of the gain as well as the time of the peak value vary with variations of the initial atom densities.

3.3 Off Axial Gain

Following the plasma parameters along the plasma trajectories, the off axial gain may be also evaluated. The values of electron N_e and temperature T_e along the selected plasma tube are used as input data for the FLY code. Gain along the plasma tube is evaluated in the same way as along the capillary axis. Representative results of radial-time evolutions are seen from the two-dimensional diagrams for three selected initial atom densities (**Fig. 4**). If the initial pressure of boron vapours is increased, the pinch time is more delayed and the radial and time evolution changes its character. For lower pressures the peaks of the gain are situated on the axis and the created active medium have shapes of cylinder, whereas for greater pressure the peak value of the gain is not found on the capillary axis and laser active medium has a shape of an annulus.

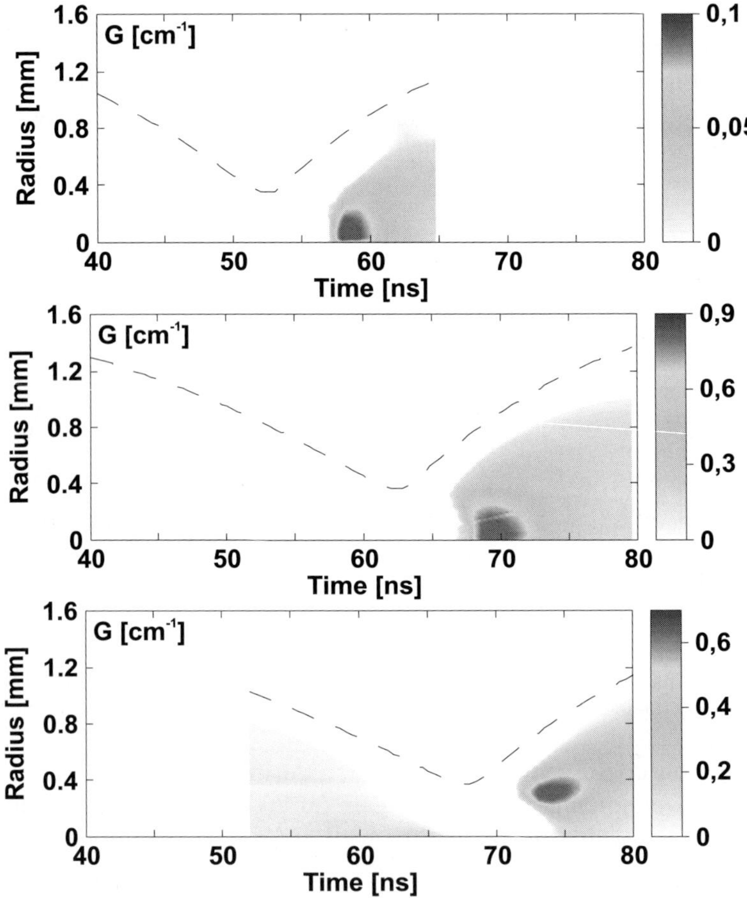

Fig. 4 Radial and time evolution of the gain evaluated for various initial boron atom densities **(a)** $N_0 = 2.0\ 10^{17}\ cm^{-3}$, **(b)** $N_0 = 3.5\ 10^{17}\ cm^{-3}$, **(c)** $N_0 = 4.5\ 10^{17}\ cm^{-3}$

4 Influence of Capillary Wall Ablation on Plasma Evolution

All the results presented above were obtained under the presumption that material ablation during the first pinch may be neglected. It is valid, if ceramic capillaries and low current peaks are discussed.

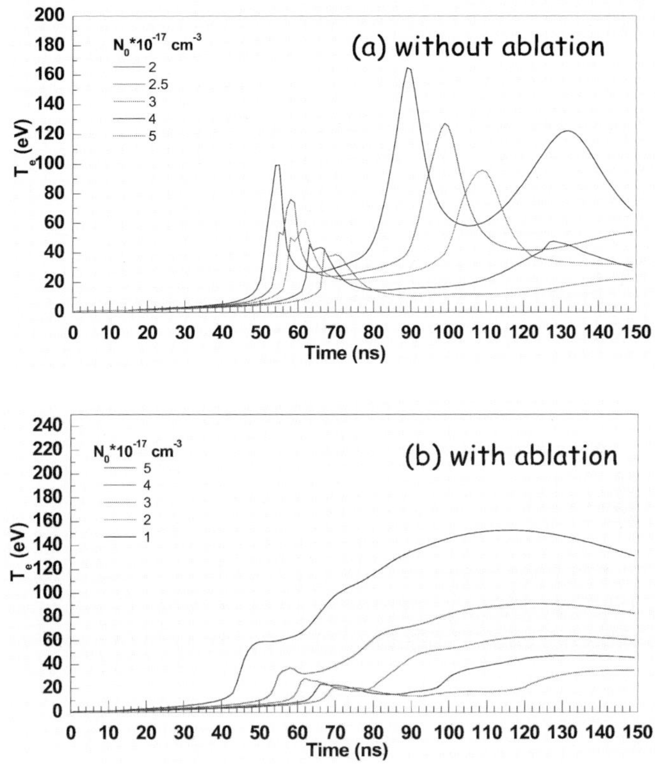

Fig. 5 Time dependences of electron temperature on the axis for various boron atom densities **(a)** without ablation, **(b)** with ablation

Our estimations show that in considered range of parameters the ablation of the ceramic capillaries can be neglected. Nevertheless, the accuracy of our estimations is not very high, and there exists the possibility of the wall ablation before the pinching of plasma in the capillary. To demonstrate importance of the ablation for the plasma behaviour we have performed simulations, where the plasma-wall interaction was modelled by considering the evaporated material of the wall as a cold neutral gas of high density and of sufficiently high total mass. In this case amount of wall material involved into the discharge is limited only by the energy flux delivered by thermal conduction that is only available for production of relatively hot plasma from the cold gas. Results comparing the two limiting cases (without ablation and with overestimated ablation) are presented (**Fig. 5**). We can see that process of ablation diminishes considerably plasma quantities at the time of pinching and the cooling effect becomes very slow. Calculated gains for such discharges are very low. We believe that such low parameters would be typical for plastics capillary and not for the ceramic ones.

5 Conclusion

Efficient amplification of spontaneous emission at boron Balmer alpha (26.23 nm) may be achieved with lower current peaks and lower current slopes than for nitrogen (13.38 nm). The achievable gain is one order higher than that with nitrogen, if $dI/dt_{|t=0} = 10^{12}$ A/s, and $I_{max} = 50$ kA. In the case of boron filling the electron plasma temperature required to get the dominating abundance of fully ionized atoms is reasonably lower than in the case of nitrogen. Preliminary appraisal of wall ablation prognosticates a serious obstacle on the way to capillary recombination pumping of higher Z hydrogen-like ions. The ablated material has a serious cooling effect; the electron temperature on the axis becomes lower than in the case without wall ablation.

Acknowledgement

This research has been supported by the Czech Science Foundation – project No. 102/07/0275 and by the Ministry of Education Youth and Sports in the frame of Research Centre Program – project No. LC528.

References

1. Rocca, J. J.: *Review of Scientific Instruments* **70**, 3799-3822, 1999
2. Bobrova N.A. et al.: *Plasma Physics Reports,* **22,** 387-402, 1996
3. Lee R. W. and Larsen J. T.: *J.Q.S.R.T.*, **56**, 535-556, 1996
4. Elton R. C.: *X-Ray Lasers*, Academic Press, New York, 1990
5. Vrba P., Vrbova M.: *Proc.of SPIE* Vol. **6702**, 67020W 1-8, 2007
6. MacFarlane J.J.: *Comput. Phys. Commun.* **56**, 259-278, 1989
7. Jancarek A.. et al: *Springer Proc. in Physics* **115**, Springer, Berlin, 687-682, 2007
8. Vrba P. et al.: *Central European Journal of Physics* **3**, 564-580, 2005
9. Vrbova M. et al: 33rd IEEE ICOPS Conf. Record-Abstracts, Michigan, 216, 200610. Vrba, P. et al.: *Proc.of* 28th ICPIG, Prague, 1205-1208, 200711. Vrbová M. et al: AIP CP641 X-Ray Lasers, 139-144, 2002

Propagation of a High-Harmonic Pulse Through a Population-Inverted Medium

Chul Min Kim, Karol A. Janulewicz, Hyung Taek Kim,
Do-Kyeong Ko and Jongmin Lee

Advanced Photonics Research Instittute,
Gwangju Institute of Science and Technology,
261 Cheomdan-gwagiro, Buk-gu, Gwangju 500-712, Korea

Abstract. We theoretically investigate the propagation of a high-harmonic pulse through a population-inverted medium. By using a Maxwell-Bloch model, the evolution of the 59th harmonic pulse of 820 nm laser through a dominantly nickel-like silver plasma resonant at 13.9 nm is simulated. The characteristics of ultrashort pulse amplification are anlayzed by considering the role of input pulse duration and lasing medium lifetime. The understanding of the amplification mechanism obtained in this way will be useful in developing an ultrashort coherent x-ray source which combines the merits of high harmonics and x-ray lasers.

1 Introduction

X-ray lasers based on plasma generated by high-power optical lasers have been developed for several decades. Nowadays, they deliver soft x-ray or extreme ultraviolet (e.g. 13.9 nm) pulses having few-μJ energy, picosecond duration, and monochromaticity ($\triangle\lambda/\lambda \sim 10^{-5}$), at the repetition rate of the driving optical lasers. Although it has partial coherence, more than 90 % of energy is incoherent due to the intrinsic randomness in the generation process [1]. To increase coherent energy, injection of high harmonics, which has almost full spatial coherence [2], has been investigated by several groups, and improvement of coherence and amplification has been demonstrated [3, 4]. Sophisticated numerical simulations incorporating plasma dynamics have also been attempted [4, 5].

In this paper, we address the fundamental physics of the harmonic injection, namely, the propagation of a femtosecond (fs) pulse through a lasing medium with picosecond (ps) lifetime. In the first approximation, this process can be investigated by concentrating on the coherent interation of field and matter without detailed plasma dynamics. To simplify the situation further, we assume ideal traveling wave amplfication in which both the driving laser pulse and harmonic pulse propagate at the same speed: the harmonic pulse encounters fresh population-inverted ions on its propagation.

This paper is organized as follows. In section 2, the physical model is described. The characteristics of ultrashort pulse amplification is discussed in section 3. In section 4, conclusion is made.

2 Physical model

To describe the pulse propagation in a lasing medium, we use a Maxwell-Bloch model, in which the density matrix equations for matter and the wave equation for field are solved self-consistently [6]. The matter consists of open two-state atoms modelling Ag^{19+} $4d(^1S_0) - 4p(^1P_1)$ transition. The harmonic pulse is assumed to be polarized along z-direction, and, thus, only the transition without change in magnetic quantum number ($m = 0 \rightarrow m = 0$) is considered.

The density matrix equations for the atoms are given as follows.

$$\begin{aligned}
\dot{\rho}_{ba} &= -\left(i\omega_{ba} + \gamma_{ba}\right)\rho_{ba} + iV_{ba}\left(\rho_{bb} - \rho_{aa}\right) \\
\dot{\rho}_{bb} &= -\gamma_b \rho_{bb} - 2V_{ba}\mathrm{Im}\left\{\rho_{ba}\right\} \\
\dot{\rho}_{aa} &= -\gamma_a \rho_{aa} + 2V_{ba}\mathrm{Im}\left\{\rho_{ba}\right\} + \gamma_{br0}\rho_{bb}
\end{aligned}$$

where $V_{ba} =< b|z|a > E(\tau) = z_{ba}E(\tau)$. b and a refer to the upper and lower states, respectively. The polarization is obtained with $P(\tau) = -2n_i z_{ba}\mathrm{Re}\{\rho_{ba}\}$. Various decay constants are obtained from either simulations or experimental data quoted in the literature: $n_i = 8.4 \times 10^{16}$ cm^{-3}, $\gamma_a = 2.3 \times 10^{12}$ Hz, and $\gamma_b = 8.5 \times 10^{11}$ Hz form EHYBRID [7]; $\gamma_{br0} = 5.9 \times 10^{10}$ Hz and $z_{ba} = 0.27$ atomic units from MCDFGME [8]; $\gamma_{ba} = 2.7 \times 10^{12}$ Hz, corresponding to $\triangle\lambda/\lambda = 4 \times 10^{-5}$, and $n_e = 3.0 \times 10^{20}$ cm^{-3} from typical experimental results [1].

The general form of the wave equation is as follows.

$$\nabla^2 E(z,t) - \frac{1}{c^2}\frac{\partial^2 E(z,t)}{\partial t^2} = \mu_0 \frac{\partial^2 P(z,t)}{\partial t^2}$$

Considering the assumption of ideal travelling wave amplification, we describe the pulse propagation in moving frame ($\tau = t - z/c$). Furthermore, the medium is optically rare for 13.9 nm radiation, and we can use slowly-evolving-wave approximation, i.e. slowly-varying-envelope approximation in moving frame [9]. With these approximations, the wave equation in Fourier frequency domain is as follows.

$$\frac{\partial \tilde{E}(z,\omega)}{\partial z} = i\left(\frac{\mu_0 c\omega}{2}\right)\tilde{P}(z,\omega)$$

The contribution by the free electrons, $\tilde{P}_e = -n_e\tilde{E}/\omega^2$, is added to $\tilde{P}(z,\omega)$. The density matrix equations and wave equation constitute the description of the numerical model.

Note that the dominant parts of physical quantities are oscillating at the resonance frequency. We can use additional assumptions: rotating wave approximation and on-resonance condition [6]. With these approximations, the equations are so reduced that only the envelopes are involved [10].

$$\begin{aligned}
\dot{\sigma}_{ba} &= -\gamma_{ba}\sigma_{ba} + iV_{baC}\left(\rho_{bb} - \rho_{aa}\right)\\
\dot{\rho}_{bb} &= -\gamma_b\rho_{bb} - 2\mathrm{Im}\left\{V_{baC}^*\sigma_{ba}\right\}\\
\dot{\rho}_{aa} &= -\gamma_a\rho_{aa} + 2\mathrm{Im}\left\{V_{baC}^*\sigma_{ba}\right\} + \gamma_{br0}\rho_{bb}\\
\tfrac{\partial E_C(z,\tau)}{\partial z} &= -i\tfrac{\mu_0 c\omega_{ba}}{2}P_C(z,\tau)
\end{aligned}$$

where $\sigma_{ba} = \rho_{ba}\exp\left(i\omega_{ba}\tau\right)$. The subscript C refers to the complex envelope obtained from the analytic signal of the original quantity [11]. We use this set of equations for simulation.

The lasing medium has the gain of $g_0 = 225\ \mathrm{cm}^{-1}$, which is the highest local gain from EHYBRID calculation. In the current model, setting $\rho_{bb} = 0.59$ and $\rho_{aa} = 0.42$ gives the gain value.

3 Characteristics of ultrashort pulse amplification

In the amplification of a pulse, the duration of the input pulse and the lifetime of the medium are the two temporal factors which determine the temporal profile of the output pulse. Furthermore, they also determine the energy extraction efficiency of the amplifying medium, as shown below.

The temporal profile of the output pulse as a function of propagation length is shown in figure 1 for the case of 25-fs input harmonic pulse. At the entrance ($z = 0\ \mu$m), only the 25-fs pulse is present. As it propagates, the harmonic itself is almost intact, but it initiates the lasing medium to radiate. Because the lifetime of the medium ($1/\gamma_{ba} \sim 370$ fs) is much longer than the

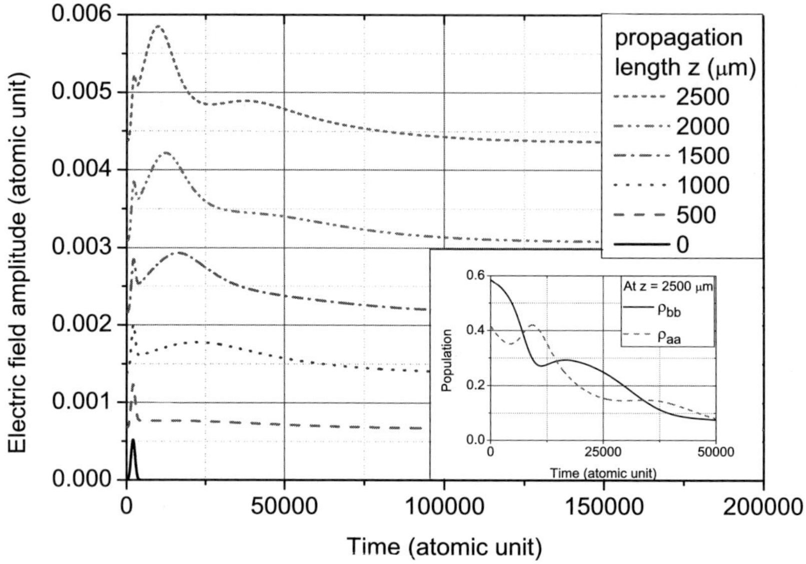

Figure 1. Temporal profile of an initially 25-fs pulse at various propagation length. (The inset shows the temporal evolution of population at $z = 2500\ \mu$m)

harmonic pulse duration, the newly generated radiation trails the input pulse: this phenomenon is also observed in the sophisticated simulations involving plasma dynamics [4, 5].

Upon further propagating, the trailing radiation becomes so strong that stimulated emission begins to affect population transfer between the upper and lower states, i.e. Rabi oscillation occurs. The modulation in the falling part of the pulse at $z = 2000,\ 2500\ \mu$m is the signature of Rabi oscillation; if ρ_{bb} becomes smaller than ρ_{aa} due to the oscillation, the medium becomes absorptive, and the field intensity decreases until ρ_{bb} becomes equal to or larger than ρ_{aa}. The corresponding population evolution is shown in the inset of figure 1. Note that the duration of the output pulse is as long as the lifetime of the medium even though the input pulse duration is much shorter than the lifetime.

The opposite case in which the input pulse is longer than the lifetime time is shown in figure 3. In this case, the input pulse experiences a significant decay of population inversion within its own duration. As a result, only

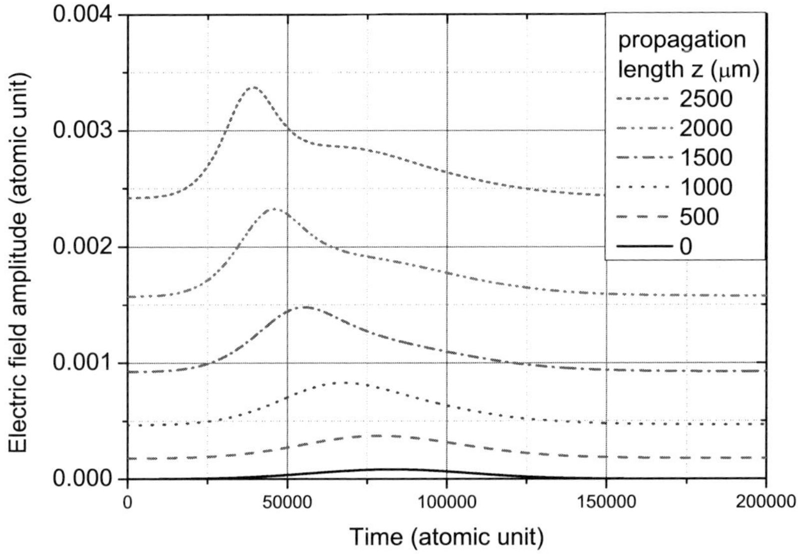

Figure 2. Temporal profile of an initially 1-ps input pulse at various propagation length.

the leading part is constantly amplified while the falling part is kept almost intact. As the pulse propagates, it suffers broadeinig because the leading part is elongated towards the temporal origin due to the continuous amplification. Emergence of Rabi oscillation is also observed at $z = 2000,\ 2500\ \mu$m.

The relative magnitudes of the input pulse duration and the lifetime of medium determine the energy extraction efficiency, as shown in figure 3. When the propagation length is short ($z < 1000\ \mu$m), the pulse of which duration is the closest to the lifetime is the most efficient because the most part of the pulse is amplified over the entire lifetime. The shorter pulses do not fully exploit the gain in lifetime interval, and the longer pulse are amplified only in the leading part. However, for long propgation length ($z > 2500\ \mu$m) at which the extracted energy overwhelms the input energy, the pulses of which duration is shorter than the lifetime are the most efficient, and the output energy values of these pulses converge to one another; once the input pulse duration is shorter than the lifetime, the energy extraction efficiency in saturation regime is independent of pulse duration. For these

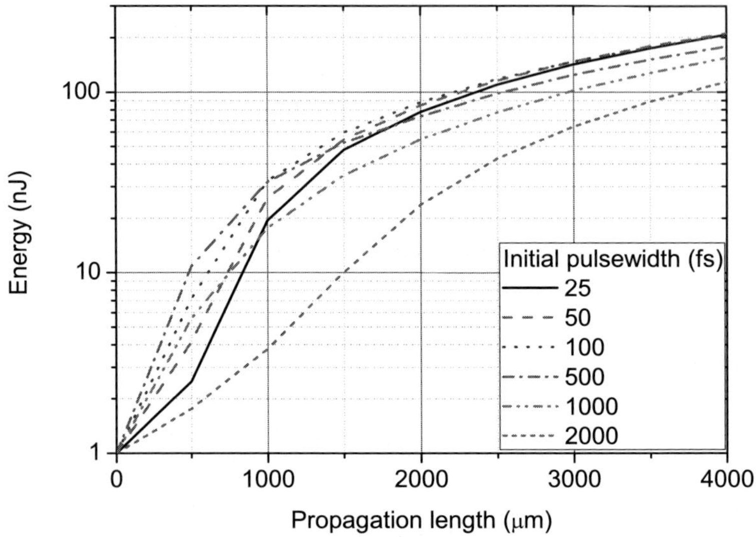

Figure 3. Energy of output pulses of various pulsewidth as a function of propagation length.

pulses, as shown in figure 1, the whole energy is kept within the lifetime of the medium, which maximizes energy extraction. For longer pulses, the part outside the lifetime interval is not amplified, and the efficiency reduces.

4 Conclusion

By using a Maxwell-Bloch model, we investigated the propagation of a harmonic pulse through an x-ray lasing medium. The characteristics of ultrashort pulse amplification was discussed by considering the input pulse duration and the medium lifetime. From this study, it is concluded that, once the input pulse duration is shorter than the medium lifetime, the duration and energy of the output pulse in saturation regime does not depend on the input pulse duration. In this case, the output pulse duration is close to the medium lifetime. Therefore, to decrease the output pulse duration, we should decrease the medium lifetime, e.g. by increasing free electron density and, thus, enhancing collisional dephasing. This result will be informative for designing high-harmonic-seeded x-ray lasers.

In the current model, we treated the lasing medium as a pure amplifier incapable of amplified spontaneous emission and considered only the coherent interaction. For analysis of coherence property and more realistic description of pulse propagation, the spontaneously emitted incoherent radiation should be properly incorporated, e.g., as suggested in [12]. At the same time, $m = \pm 1$ states of the lower level and the two polarizations of the x-ray pulse should be considered because spontaneous emission is unpolarized.

Acknowledgements

This work was supported by the Ministry of Knowledge and Economy of Korea through the Ultrashort Quantum Beam Facility Program. We appreciate Prof. G. J. Pert for providing EHYBRID code and invaluable discussion, and Dr. P. Indelicato for providing MCDFGME code.

References

[1] Jaeglé P 2006 *Coherent Sources of XUV Radiation* (New York: Springer)

[2] Pfeifer T *et al* 2006 Rep. Prog. Phys. **69** 443–505

[3] Zeitoun Ph. *et al* 2004 Nature **431** 426–429

[4] Wang Y *et al* 2008 Nature Photonics **2** 94–97

[5] Al'miev I R *et al* 2007 Phys. Rev. Lett. **99** 123902

[6] Yariv A 1989 *Quantum Electronics (3rd ed.)* (New York: John Wiley & Sons)

[7] Pert G J 1983 J. Fluid Mech. **131** 401–426

[8] MCDFGME (MultiConfiguration Dirac Fock and General Matrix Elements program, release 2005) written by J. P. Desclaux and P. Indelicato (http://dirac.spectro.jussieu.fr/mcdf)

[9] Brabec T and Krausz F 1997 Phys. Rev. Lett. **78** 3282–3285

[10] Armandillo E and Spalding I J 1975 *J. Phys. D: Appl. Phys.* **8** 2123–2135

[11] Born M and Wolf E 2005 *Principles of optics (7th ed.)* (Cambridge: Cambridge University Press)

[12] Larroche O *et al* 2000 Phys. Rev. A **62** 043815

Modeling of an Ultra-Short X-Ray Laser Pulse Amplification Through an Optical-Field-Ionized Gas Using a Maxwell-Bloch Treatment

B Robillart[1], G. Maynard[1], B.Cros[1], A.Boudaa[1], J.Dubau[2], S.Sebban[3], and JP.Goddet[3]

[1] LPGP, Université Paris-Sud 11, 91405 Orsay, France
[2] LIXAM, Université Pars-Sud 11, 91405 Orsay, France
[3] LOA, ENSTA, 91120 Palaiseau

Abstract. It has been recently demonstrated experimentally that seeding a high-harmonic pulse into an Optical-Field-Ionized gas can generate a coherent soft x-ray laser beam of up to 1μJ. In order to analyze the physical processes involved in the amplification of the x-ray laser pulse through the plasma amplifier a 3D numerical code named COFIXE_MB has been developed using a Maxwell-Bloch treatment. It brings detailed information about the x-ray pulse evolution, especially regarding the fast evolution of the pulse temporal profile and the spatial filtering of the wave front structure by the amplifier.

1 Introduction

Experiments results on soft X-ray laser (SXRL) obtained by seeding a high order harmonic (HOH) beam into an Optical-Field Ionized (OFI) gas have been recently reported [1]. These experiments performed using the "salle Jaune" laser equipment of LOA, ENSTA have demonstrated that with a few Joules pumping infra red (IR) beam of sub-picosecond duration focussed inside a krypton gas cell it is possible to obtain a fully coherent SXRL at 32.8 nm with up to 1 μJ of energy.

Efficiency of this SXRL depends on many parameters such as delay between HOH and IR beams, polarization and intensity of the IR beam, pressure and length of the Kr gas cell. Theoretical support is therefore deeply needed in order to understand the influence of each of these parameters. Moreover previous experiments [2] have shown that using capillary tubes, it is possible to guide the high intensity IR up to several centimetres. For such large lengths, the seeded SXRL will be highly saturated, therefore it is important to predict its properties.

The main objective of modelling is to get informations in terms of temporal evolution of the beam and also of its spatial properties, in particular concerning the structure of the SXRL wave front. To get it, two constraints have to be satisfied. First, the duration of the injected HOH (~ 30 fs) must be much shorter than the typical collision times in the plasma amplifier, related to line broadening and electronic excitation/ionization rates. The adiabatic approximation, usually used in laser physics, in which it is assumed that time evolution of the laser beam is much slower than plasma evolution at the microscopic scale, cannot be therefore used in our case. The time evolution of the laser field and of the populations of the atomic excited states leading to amplification has to be treated simultaneously. Second, to allow direct comparison with the experimental results concerning the wave front structure of the SRXL it is necessary to perform a full 3D calculation at least for studying transport and amplification of the X-UV signal. In particular, it will allow to take into account small miss-alignment between the HOH and the IR beam which experimentally is difficult to avoid.

A numerical code has been constructed to satisfy the former two constraints by solving Maxwell-Bloch equations in 3D following Kärtner MIT lecture course [3]. The numerical method used differs from Larroche et al. approach [4,5]. It is clear that approximations are needed in order to reduce the complexity of the problem ; this is described in section 2. In section 3 we present results obtained with our code for temporal and phase evolution of the SXRL.

2 Modeling

To describe the properties of the plasma amplifier we introduced two main approximations that are relevant to OFI process in gas. First, the plasma is rather uniform and macroscopic expansion can be neglected during gain amplification. We have therefore assumed that the plasma properties can be defined locally, neglecting any gradient effect. Second, tunnelling can be considered as instantaneous compare to the period of the IR field, moreover the duration of the IR pulse (35 fs) is much smaller than characteristic time of 1 ps for the plasma evolution, so that properties of the plasma created through OFI is supposed to depend on the maximum intensity of the IR field but not on the shape of the envelope of its amplitude.

Using these two approximations, the modeling of OFI x-ray laser amplification can be done through 3 independent numerical codes :

The first one concerns the infrared laser propagation through a gas contained in a cell or capillary tube calculated in a 2D cylindrical geometry within the paraxial approximation [2]. It provides a 2D map for the maximum values of the IR intensity at each point inside the plasma.

The second code solves a kinetic model in order to determine the temporal evolution of a uniform gas interacting with an IR beam.

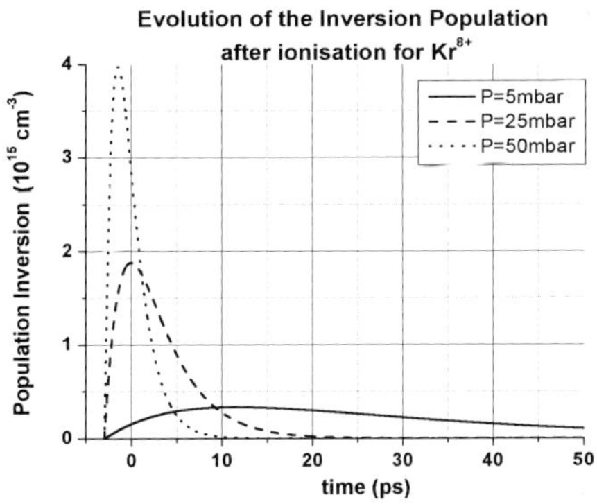

Fig.1 Evolution of the population inversion after ionization for Kr^{8+}

The initial free electron distribution function (FEDF) generated by the OFI process is highly in a non-equilibrium state. The relaxation of the FEDF is calculated through Fokker-Planck equation [6] while atomic data are determined from the SUPERSTRUCTURE [7] code and electron-ion collision strength are derived within the Distorted-Wave approximation [8]. More details on this second code are given in [2]. All the needed data entering the Maxwell-Bloch equations (see below) are stored and then parameterized in terms of the gas pressure and of the values of the maximum intensity of the IR beam. A typical example of results is reported in Fig. 1 for the population inversion between the two levels of the SXRL lasing line at 32.8nm in krypton. In this figure, it can be seen that density of excited states is nearly proportional to density while duration of the gain is inversely proportional to density. The duration of the gain is fixed by density reduction of the lasing ions due to ionization through electron collisions.

The third part of the calculation is performed by the COFIXE_MB code devoted to solving Maxwell-Bloch equation (MBE) in a time evolving plasma described by the resulting data of the two previous codes. To reduce the computational work, the MBE is solved within an effective two levels model. In a 3D ρ, z, ϑ cylindrical geometry MBE equations reads as

$$\begin{cases} \left\{\dfrac{\partial}{\partial z}+\dfrac{1}{c_0}\dfrac{\partial}{\partial t}\right\}A=\dfrac{i}{2k}\left[\dfrac{\partial^2}{\partial^2\rho}+\dfrac{1}{\rho}\dfrac{\partial}{\partial\rho}+\dfrac{1}{\rho^2}\dfrac{\partial^2}{\partial^2\vartheta}\right]A+D-\eta A & (1)\\ \dfrac{\partial}{\partial t}D=-\dfrac{1}{T_2}D+\alpha AW & (2)\\ \dfrac{\partial}{\partial t}W=-\dfrac{W-W_0}{T_1}-\left(AD^*+A^*D\right). & (3) \end{cases}$$

In these equations, $A(\rho,z,\vartheta)$ is the field amplitude of the x-ray laser, $D(\rho,z,\vartheta)$ and $W(\rho,z,\vartheta)$ are the dipole moment and the population inversion for the two lasing levels respectively, $W_0(\rho,z)$ is the values of W without an SXRL laser beam, T_1 and T_2 figure as respectively the energy and phase relaxation times of the lasing line, α is an atomic constant and $\eta(\rho,z,\vartheta)$ represents an attenuation factor, which role is important outside the ionised part of the target gas. Both W_0, T_1 and T_2 are deduced from the kinetic and the IR propagation codes. As our calculations are devoted to the amplification of an injected HOH signal, the contribution of spontaneous emission can be safely neglected. To numerically solve Eqs. (1-3), the complex scalar fields amplitude is written as $A=\sum_{\ell} A_{\ell}(z,\rho)\exp(i\ell\vartheta)$ with similar expressions for D and W, each mode function $A_{\ell}(z,\rho)$ being projected on a rectangular grid. In our 3D calculations, typical values are 40 angular modes, 75 grid points for ρ values in the transverse plane and 10 000/cm point for ☒ along the beam axis.

3 X-ray Amplification Results

3.1 Temporal profile and spectrum evolution

COFIXE_MB can bring information about the temporal and spectrum profile evolution of a resonant x-ray radiation field. To identify more clearly the temporal effects we present here results obtained for a 1d propagation through a homogeneous amplifier.

In Fig. 2, is represented the time evolution of the SXRL field amplitude for short plasma length. The non-adiabatic character of the amplification is clearly exhibited in this figure. The peak on the left of the figure represents the injected HOH signal with duration of 35 fs which is much less than the relaxation rates T_1 and T_2 entering the MBE Eqs. (2-3), which at the considered pressure of 25 mbar are around few ps. At the shortest length of 0.3 mm, we can observe an exponential decay of the signal behind the pulse. Making reference to the damp oscillator model, we can state that the damping value is above the critical damping value so that there is no oscillation. For larger lengthes 0.5 and 1 mm, the tail is much more amplified than the initial peak yielding to an exponential growth of the X-UV signal whereas the response to the initial peak leads to a linear increase of the amplification factor.

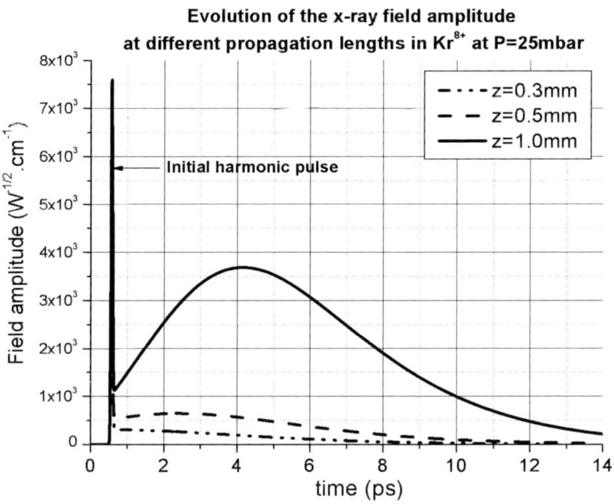

Fig. 2 Temporal profile of the x-ray pulse field amplitude at short propagation length (<1mm)

At larger propagation length (Fig.3a), there is a change of evolution, especially when the pulse duration gets limited by the gain duration traduced in Fig.1. Saturation of the gain has been reached at this point.

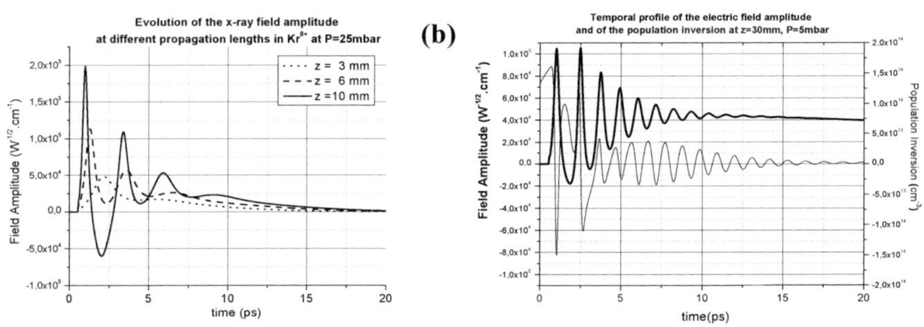

Fig. 3 (a) Temporal profile of the x-ray pulse field amplitude at long propagation length (3-10mm) at P=25mbar for Kr^{8+} **(b)** Oscillating temporal profile compared to population inversion temporal profile for Kr^{8+}, at z=30mm and P=5mbar.

As the pulse propagates and the SXRL intensity increases, the electric field amplitude starts to oscillate (Figs 3.a and b) with a strong peak (Fig.3.a) just behind the initial HOH one, indicating that the damping becomes larger than the critical damping value of the related oscillator. In fact, those oscillations are related to Rabi oscillations $\Omega_{rabi} = \sqrt{2\alpha.I}$ that are characteristic to dipolar transition between 2 levels [3,9]. As the Rabi frequency increases with the SXRL intensity, the frequency of the damped oscillator increases as well for a fixed value of the damping. Above a given threshold for the intensity, the period of the oscillation becomes smaller than the damping time, and oscillations become visible.

The setting of the Rabi oscillations is due to a non equilibrium process generated by the formation of the first intense peak. As the pulse propagates, it grows by taking energy from a depopulation of the upper lasing level of the amplifying ions. When the peak field becomes strong enough, population inversion gets negative after a short time and the medium starts to absorb incoming radiation until population inversion recovers a positive value. Actually, this process reproduces over the time and population inversion clearly oscillates (Fig.2a) under the influence of a strong field. The direct consequence is an induction of the same oscillations to the amplifying signal field amplitude.

However, those oscillations always decay and set the field amplitude and the population inversion to a stationary regime. The characteristic relaxation time of this decay can be determined as a function of the energy- and phase-relaxation times T_1 and T_2.

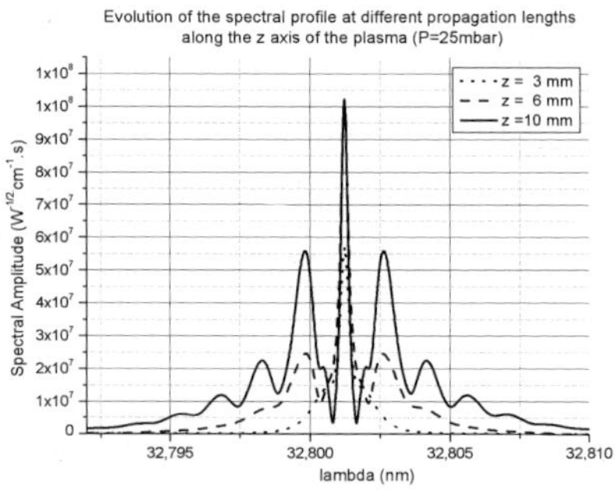

Fig. 4 Spectral profile of the x-ray field amplitude at different propagation lengths

Seen in the Fourier space, at early amplification, the x-ray transition line has a Lorentzian profile. Then, after saturation has occurred, sidebands related to the Rabi oscillations appear aside the main peak (Fig.4).

3.2 Transverse map of the outgoing x-ray signal energy

Using full 3D calculations COFIXE_MB can be used to analyse the spatial profile of the SXRL in the plane transverse to the beam axis, yielding information on the beam divergence and on the intensity and phase fluctuations, which are important quantities for application purposes. The initial 2D values of the HOH amplitude at the entrance of the plasma is calculated from experimental data extracted from a wave sensor detector. From the COFIXE_MB code, we determine at the exit of the plasma the complex amplitude of the SXRL, which is further more propagated up to the detector for comparison with experimental data. A typical example is reported on Figs. 5(a-c), for the intensities at the entrance and exit of the plasma and at the detector

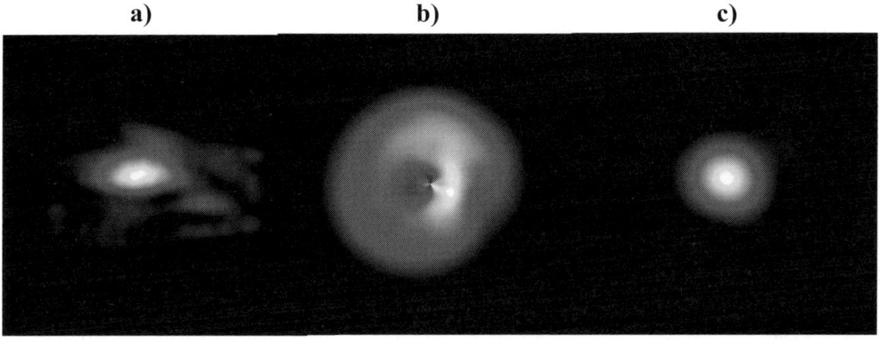

Fig. 5 (x,y) Footprint of the SXRL intensity in plane perpendicular to the beam axis, as determined by the COFIXE_MB code; a) : at the entrance of the OFI plasma; b) at the exit of the OFI plasma and c) at the detector position 4m away from the plasma. Note that each figure has a different scaling length

We can observe strong variations between these figures. The plasma amplifier first acts as a spatial filter, eliminating the intensity fluctuation of the HOH signal that are outside the plasma cylinder, so that Figure 5.b has better defined boundaries. Close to the plasma axis, the intensity of the IR beam is slightly to high leading to an over ionised plasma with a small gain so that the SXRL has a ring shape at the plasma exit (Fig. 5.b). In our simulation this is partially due to the fact that the IR beam is assumed to have a prefect cylindrical symmetry leading to a strong maximum of the intensity close to the axis. After propagation up to the detector, 4 m from the plasma, this central hole in the SXRL yield fluctuation only well outside the main central spot, so that it can easily be withdrawn by a diaphragm. In Fig. 5.c, the main spot has a nearly perfect Gaussian shape, with very little phase fluctuation and a radius of about 2 mm. In accordance with the experimental result, these calculations demonstrate that the HOH signal amplified by an OFI plasma results in an intense diffraction limited beam of low divergence.

4 Conclusion

We presented results of numerical modeling for the amplification of a high order harmonic pulse by a plasma created by optical field ionisation. Using approximations that are appropriate to OFI case we were able to construct a full 3D time dependent description of the amplification process. Regarding time evolution of the SXRL, we found that for large enough plasma length at which SXRL intensity is above a given threshold, time oscillations start to grow, leading to sidebands in the frequency spectra. These oscillations are expected to play a major role when infrared beam is guided over several centimetres. Concerning the spatial structure of the SXRL, results show that for an optimum value of gas pressure, at which the amplification factor is maximum, the plasma amplifier acts as an efficient spatial filter. Propagation of the amplified X-ray laser pulse outside the plasma for a few meters leads to an intense central spot with an almost Gaussian wavefront.

References

[1] J.P. Goddet et al. Optics Letters **32**, 1498 (2007) ; see also S. Sebban et al this proceedings
[2] B. Cros et al. Phys. Rev. A **73**, 033801 (2006)
[3] F. Kärtner, Ultrafast Optics, MIT Open Courseware (Spring 2005)
[4] O. Larroche et al. Phys. Rev. A **62**, 043815 (2000)
[5] I.R. Al'miev et al. Phys. Rev. Lett. **99**, 123902 (2007)
[6] G.J. Pert J.Phys. B **34**, 881 (2001)
[7] W. Eissner, M. Jones, H. Nussbaumer Comput Phys. Commun. **8**, 270 (1974)
[8] W. Eissner, Comput. Phys. Commun. **114**, 295 (1998)
[9] A.E. Siegman *Lasers*, University Science Books, ISBN 0-935702-11-3 (1986)

Effects of Inhomogeneous Incident Line Focus on 2D Hydrodynamic Behaviour of X-Ray Laser Plasma on Slab

T. Cheng[1,2], Y. J. Li[1,2], L. M. Meng[1,2], J.Zhang[1]

[1] Laboratory of Optical Physics, Institute of Physics, Chinese Academy of Sciences, Beijing 100190, China
[2] China University of Mining and Technology, Beijing 100083, China

Abstract. Two dimensional hydrodynamic behavior of x-ray laser plasma on slab under non-equilibrium condition was studied in a novel way. Changes of the characters of plasma with time were investigated. It is found that the inhomogeneity of plasma along the line focus can be relaxed by the hydrodynamic motion. Usually the process takes more than 1ns for temperature to get uniform and several hundreds picoseconds for electron density. Hydrodynamic oscillation of plasma due to the non-uniformity of line focus was obtained.

1 Introduction

After the demonstration of Ne-like x-ray laser using exploding foil target in 1985 [1], the soft x-ray laser research field experienced a rapid development. In 1987, slab target was firstly introduced into the experiment by Lee and they successfully reduced the power demand of laser to 10^{12}W/cm^2 [2]. However in laser produced plasma high gradient narrower the gain region. The amplification of spontaneous radiation was hindered and it is hard to get a saturation gain of x-ray laser. Tremendous success was achieved after the proposal of pre-pulse technique [3] applying which gain saturation of x-ray laser has been demonstrated in neon-like Zn, Ge, Se and etc. in large numbers of experiments. In 1996 short pulse was firstly tried and improved the efficiency greatly [4]. Energy requirement was dramatically reduced to 40J. The next year transient scheme [5] was demonstrated in the laboratory and then the use of short pulse as pumping pulse was widely adopted in nickel-like or neon-like transient x-ray laser scheme [6]. Recently a new pumping geometry based on the studying of plasma parameter on slab named Grazing Incidence Pumping (GRIP) which makes full use of refraction in the plasma brings another development into table-top x-ray laser [7].

In most recent plasma x-ray laser experiments, pre-pulse technique and slab dominate the experimental design. Effort is exerted on studying the relations between the output of x-ray laser and pre-pulse, main pulse, also the delay time [4,8]. Understanding of pre-pulse technique in one dimension is that the pre-plasma could relax the density and temperature gradient of plasma in the delay time by expansion offering a wider gain region. In previous work what we concerned is the character of plasma in the expanding direction while assuming a uniform distribution along the line focus in calculation or experimental designing. However in practical experiment the focal line is not ideal. The non-uniformity will introduce non-equilibrium to the plasma and then affects the obtainment of x-ray laser. It is quite necessary to study the behaviour of plasma under such non-equilibrium conditions.

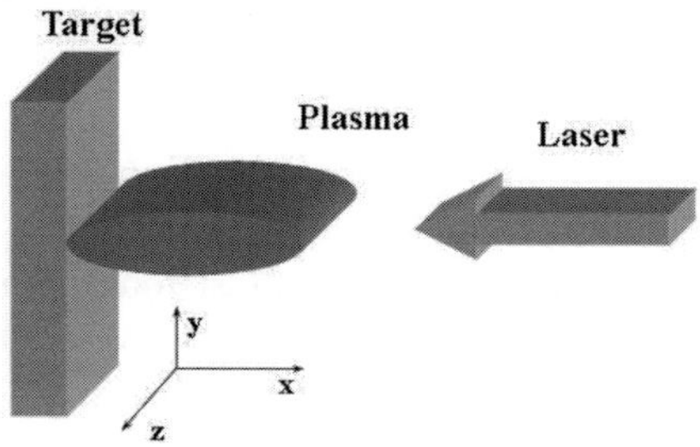

Fig. 1 Experimental configuration

Based on the similarity equation [9], we developed a method to study large scale (cm) hydrodynamic behaviour of plasma in two directions under the non-equilibrium by combining the 1D solution [9] for x direction (vertical to slab as shown in Fig. 1) with numerical calculation for z direction (along the focal line in Fig. 1). Temporal and spatial evolutions of plasma under non-equilibrium conditions are studied in this work.

2 Inhomogeneity of line focus

In the experiments applying pre-pulse technique, the quality of line focus is important. However whatever optical apparatus are chosen, non-uniform intensity distribution is unavoidable.

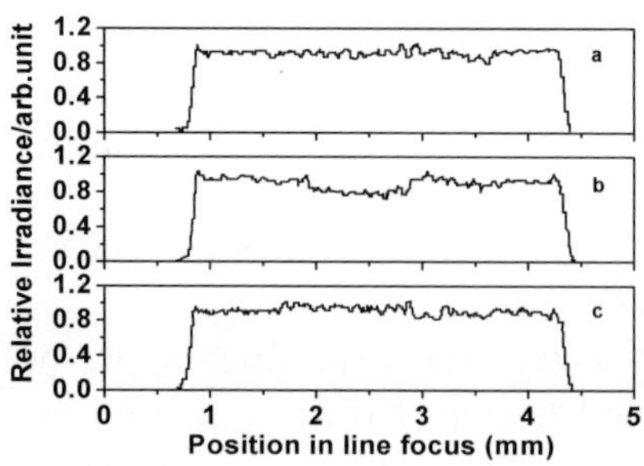

Fig. 2 Focal line sampling on "ShenGuang II" laser facility

Figure 2 shows the intensity distribution in z direction. Horizontal axis denotes the position in the line focus and vertical axis the intensity of incident laser, arbitrary units. After optimization the unevenness of incident laser is still obvious as Fig. 2 shows. Generally the range of non-uniformity of intensity is between $\pm 15\%$ of the average value [4], which will bring noticeable inhomogeneity to the plasma. As studied previously, expansion in x direction will relax the gradient and lead to a more uniform plasma. In the same way, hydrodynamic motion will also affect the plasma distribution in z direction. However detailed process has not been studied systematically.

3 Model equations

Considering the diffusion and viscosity, hydrodynamic evolution of plasma can be described as follows:

$$\frac{\partial \rho}{\partial t} = -\rho \frac{\partial \upsilon_x}{\partial x} - \upsilon_x \frac{\partial \rho}{\partial x} - \rho \frac{\partial \upsilon_z}{\partial z} - \upsilon_z \frac{\partial \rho}{\partial z}$$

$$\frac{\partial \upsilon_x}{\partial t} = -\upsilon_x \frac{\partial \upsilon_x}{\partial x} - \upsilon_z \frac{\partial \upsilon_x}{\partial z} - \frac{1}{\rho}\frac{\partial p}{\partial x} - \frac{1}{\rho}\overline{\nabla}_x q$$

$$\frac{\partial \upsilon_z}{\partial t} = -\upsilon_x \frac{\partial \upsilon_z}{\partial x} - \upsilon_z \frac{\partial \upsilon_z}{\partial z} - \frac{1}{\rho}\frac{\partial p}{\partial z} - \frac{1}{\rho}\overline{\nabla}_z q$$

$$\frac{\partial T}{\partial t} = -\upsilon_x \frac{\partial T}{\partial x} - \upsilon_z \frac{\partial T}{\partial z} - \frac{1}{c_v}\frac{p}{\rho}\left(\frac{\partial \upsilon_x}{\partial x} + \frac{\partial \upsilon_z}{\partial z}\right) + \frac{1}{c_v} H + \frac{1}{\rho c_v}\frac{\partial}{\partial z}\left(K_0 T^{\frac{5}{2}} \frac{\partial T}{\partial z}\right) \tag{1}$$

Where υ_z, ρ, T, L, p, K_0 denotes velocity in z direction, density of plasma, temperature, characteristic scale and coefficient of heat transport respectively. In the equation c_V and H are the same as described in reference [9]. Where q is equal to $C_0^2 L_0{}^2 \rho[\overline{\nabla} \cdot \vec{\upsilon}]^2$, C_0 is coefficient for viscosity and $L_0{}^2 \sim \Delta x^2 + \Delta y^2$. In contrast with the expansion in x direction, evolution in z direction can be considered as subsidiary course. Considering the self-similar model we simplified the equations (1) to describe the plasma as follows.

$$\frac{\partial \rho}{\partial t} = \frac{1}{L}\frac{dm}{dt} - \frac{L'}{L}\rho - \rho \frac{\partial \upsilon_z}{\partial z} - \upsilon_z \frac{\partial \rho}{\partial z}$$

$$\frac{\partial \upsilon_z}{\partial t} = -\upsilon_z \frac{\partial \upsilon_z}{\partial z} - \frac{1}{\rho}\frac{\partial p}{\partial z} - \frac{1}{\rho}\nabla_z q$$

$$\frac{\partial T}{\partial t} = -\upsilon_z \frac{\partial T}{\partial z} - \frac{1}{c_V}\frac{p}{\rho}\left(\frac{1}{t} + \frac{\partial \upsilon_z}{\partial z}\right) + \frac{1}{c_V} H + \frac{1}{\rho c_v}\frac{\partial}{\partial z}\left(K_0 T^{\frac{5}{2}} \frac{\partial T}{\partial z}\right) \tag{2}$$

$$\frac{\partial \rho}{\partial t} = -\frac{L'}{L}\rho - \rho \frac{\partial \upsilon_z}{\partial z} - \upsilon_z \frac{\partial \rho}{\partial z}$$

$$\frac{\partial \upsilon_z}{\partial t} = -\upsilon_z \frac{\partial \upsilon_z}{\partial z} - \frac{1}{\rho}\frac{\partial p}{\partial z} - \frac{1}{\rho}\nabla_z q$$

$$\frac{\partial T}{\partial t} = -\upsilon_z \frac{\partial T}{\partial z} - \frac{1}{c_V}\frac{p}{\rho}\left(\frac{1}{t} + \frac{\partial \upsilon_z}{\partial z}\right) + \frac{1}{\rho c_v}\frac{\partial}{\partial z}\left(K_0 T^{\frac{5}{2}} \frac{\partial T}{\partial z}\right) \tag{3}$$

Where $\frac{dm}{dt}$ is ablation rate [10] as a function of position and time. Equations sets (2) and (3) can effectively model the evolution when the pulse is on and after the pulse respectively.

4 Comparison with 1D similar solution

In order to verify the validity of our 2D model, we compared the numerical results under special conditions with 1D similar solution.

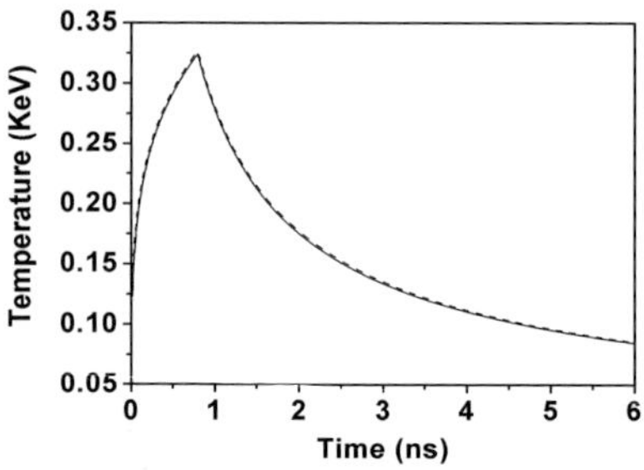

Fig. 3 Electron temperature versus time

With assumption that the distribution of pulse intensity is homogenous along the line focus, taking parameters as intensity of pre-pulse is 7×10^{11}W/cm^2 with pulse duration of 800ps on Pd target. Figure 3 displays the evolution of temperature given by our 2D model (solid line) and 1D analytical model (dash line) respectively. One can find that the results are obviously consistent with each other perfectly which certified the accuracy of our 2D model.

5 Results discussion

5.1 Evolution of temperature

In x-ray laser plasma, ionization degree is mainly determined by the temperature. Only properly ionized plasma can be the effective medium. Therefore uniform distribution of temperature is important for the x-ray laser. Referring to the practical measurement, we adopted 12% of the average intensity as the range of inhomogeneity in the calculation and introduced scrambling spatial intensity distribution of incident laser pulse.

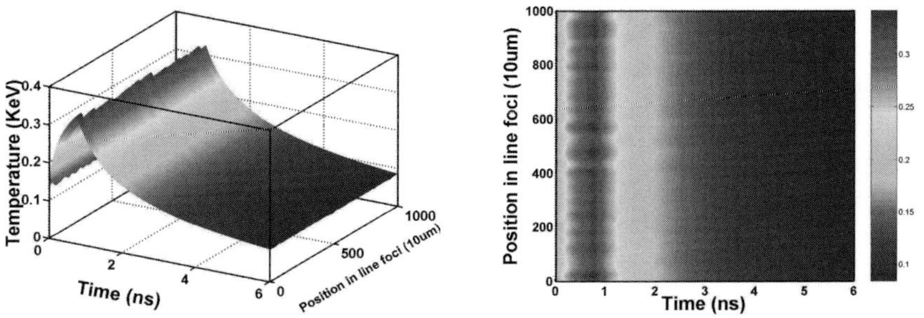

Fig. 4 Temporal evolution of plasma temperature distribution in z direction Average intensity of laser is 0.7TW/cm^2 and duration is 800ps, Ni-like Pd.

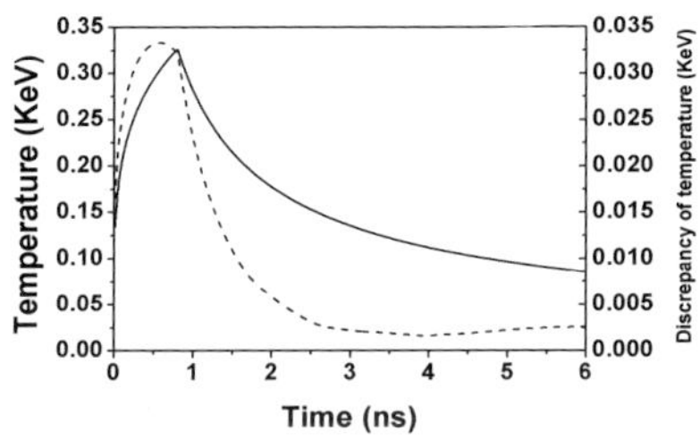

Fig. 5 Average temperature and discrepancy in z direction. Solid line is scaled to the left and dashed line the right.

Figure 4 display the temporal evolution of plasma temperature along focal line. At 800ps, the temperature reached its peak value and it is apparent that distribution of temperature is inhomogeneous. In Fig. 5 dashed line shows that the discrepancy decreases with the time after the pre-pulse which might be due to both the drop of temperature and the tendency to homogeneity in z direction. Solid line presents the temporal evolution of electron temperature showing that after the pre-pulse the temperature drops gradually. Contrasting dashed line with solid line, one can easily find that the former drops much more rapidly which could confirm our prediction that hydrodynamic behaviour will bring homogeneity to the temperature distribution.

During the period when pulse was still on, discrepancy of temperature increased faster than the average value which is due to the continual interaction between the uneven focus line with slab and plasma. At time when the pulse was turned off, discrepancy is larger than 33eV. At 2.5ns the distribution of temperature was more uniform with discrepancy smaller than 2eV. As given in the work [6], delay time is set as 1.34ns which gave a good output of x-ray laser. According to the dashed line when 1.34ns after the peak of pre-pulse the discrepancy along z direction is lower than 6ev which verified that the conditions adopted in experiment [6] was optimal.

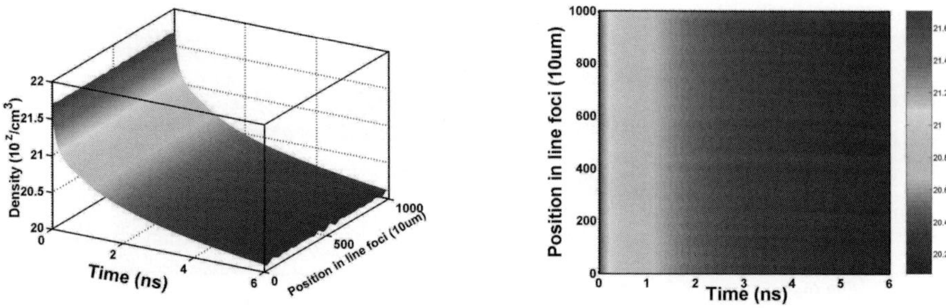

Fig. 6 Temporal evolution of electron density distribution in z direction. Intensity of laser is 0.7TW/cm^2 and duration is 800ps, Ni-like Pd.

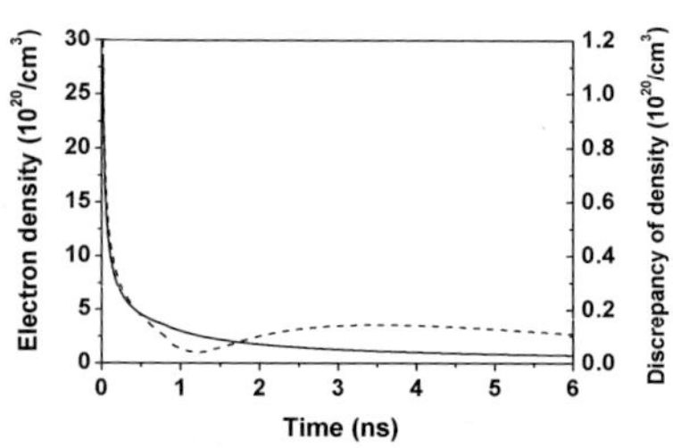

Fig. 7 Electron density and discrepancy in z direction. Solid line is scaled to the left and dashed line scaled to the right

5.2 Evolution of electron density

Figure 6 presents the evolution of electron density. One can find that the density distribution also trends to be uniform after the pulse. Dashed line in Fig. 7 shows that discrepancy decreased faster than the drop of average density from the time 400ps before the end of pre-pulse. At time pre-pulse was turned off, discrepancy of density was neither the maximum nor the minimum, different from that of temperature with the maximum value. The discrepancy kept decreasing till 1.2ns to a very small value, less than $4\times 10^{18}/cm^3$ comparing to the average density $2.6\times 10^{20}/cm^3$. This could be considered as the uniformity of density distribution. Compared with dashed line in Fig. 5, the interval needed to get so called uniform state is different. It was 400ps for density while 1.7ns for the temperature.

In Fig. 5 and Fig. 7, dash lines which denote the discrepancy of density or temperature have rising tails showing that after certain time the discrepancy get large again. The average temperature or electron density decreases with time, therefore the rising tails imply there might be hydrodynamic oscillation inside the plasma.

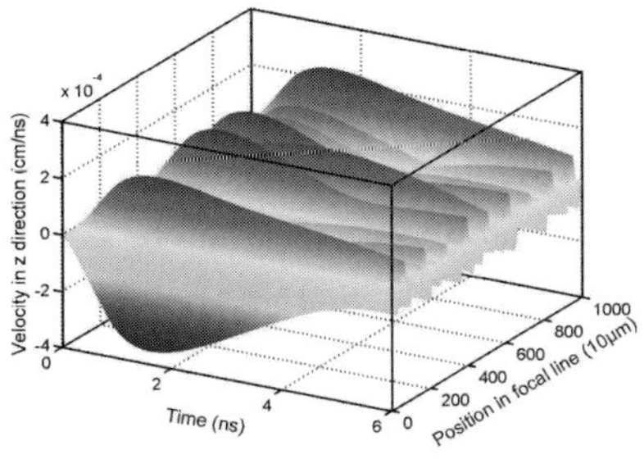

Fig. 8 Temporal of velocity in z direction

5.3 Oscillation of the plasma

At 1.2ns the velocity is large and with a non-uniform distribution shown in Fig. 8, which is due to the non-uniform plasma produced by the non-uniform incident laser. As time passing, orientation of velocity altered and the value reduced.

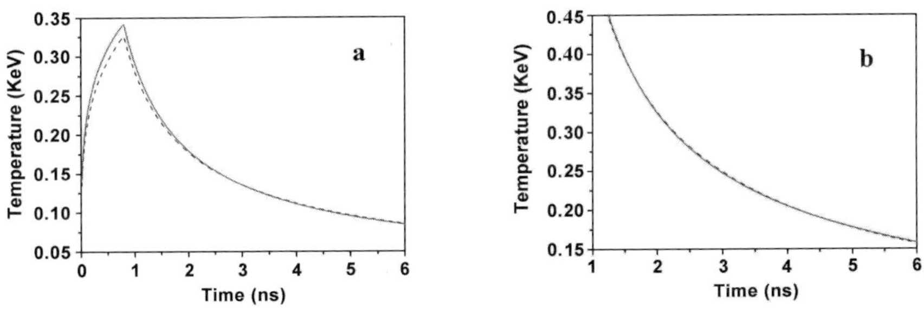

Fig. 9 Average temperature of plasma and temperature at fixed position in the line focus. Pulse duration is 800ps with intensity is 0.7TW/cm^2 for a, and 2.0TW/cm^2 for b, Pd target.

Figure 9 displays the temporal evolution of temperature, in which solid line shows the temperature at fixed point in line focus and the dashed line the average temperature. When the intensity is 0.7TW/cm^2, the temperature at fixed point is first above the average and then below it. When enlarging the intensity to 2TW/cm^2 we can observe a complete quiver in the same period.

This hydrodynamic oscillation was attributed to the non-uniform laser line focus. Inhomogeneous distribution of plasma density and temperature form the non-uniform distribution of pressure in z direction, and then drive the hydrodynamic motion. Like the swing of pendulum, when the distribution gets uniform but velocity remains thus oscillation happened. Comparing the conditions in Fig. 9a and Fig. 9b, we can find when the average intensity is higher the frequency of oscillation is larger. It can be explained by that higher intensity enlarge the velocity thus shorten the intervals needed.

6 Conclusions

In this work we developed a method of studying plasma under non-equilibrium to two dimensions basing on the similarity equations. We presented the evolution of plasma of large scale (cm) under the non-uniform incident pulse. Calculation show the temperature and electron density distribution varies a lot along focal line at the end of pulse, but such non-uniformity relaxes in the delay time. Tendency of electron density to uniform is faster than that of temperature. We obtained large scale oscillation in z direction resulted from the inhomogeneity of plasma. Such hydrodynamic oscillation has a frequency relevant to the intensity of pulse. This work grants us the prospect of optimizing the x-ray laser in two directions in an efficient way considering both the expansion vertical to the slab and the inhomogeneity of plasma along the line focus in practical experiments.

Reference

1 Matthews, D. L.: 'Demonstration of a Soft X-Ray Amplifier', *Phys. Rev. Lett.,* 54, 110-113, 1984

2 Lee, T. N.: 'Soft x-ray lasing in neonlike germanium and copper plasmas', *Phys. Rev. Lett.*, 59, 1185-1188, 1987

3 Boehly, T.: 'Demonstration of a narrow-divergence x-ray laser in neonlike titanium', *Phys. Rev. A*, 42, 6962-6965, 1990

4 Zhang, J.: 'Soft-x-ray lasing at 32.6 nm in Ne-like Ti ions driven by 40 J of energy from two 650-ps laser pulses', *Phys. Rev. A,* 53**,** 3640-3646, 1996

5 Nickles, P. V.: 'Short Pulse X-Ray Laser at 32.6 nm Based on Transient Gain in Ne-like Titanium', *Phys. Rev. Lett.*, 78**,** 2748-2751, 1997

6 Dunn, J.: 'Demonstration of X-Ray Amplification in Transient Gain Nickel-like Palladium Scheme', *Phys. Rev. Lett.*, 80, 2825-2828, 1998

7 Keenan, R.: 'High-Repetition-Rate Grazing-Incidence Pumped X-Ray Laser Operating at 18.9 nm', *Phys. Rev. Lett.*, 94, 103901(4pages), 2005

8 Li, Y. L.: 'Study of Ne- and Ni-like x-ray lasers using the prepulse technique', *Phys. Plasmas*, 4, 479-489, 1996

9 Li, Y. J.: 'Hydrodynamic characteristics of transient Ni-like x-ray lasers', *Phys. Rev. E*, 63, 036410(6pages), 2001

10 De Groot, J. S.: 'Distributed absorption model for moderate to high laser powers', *Phys. Fluids B*, 4, 701-707, 1992

Excitation Rates for Transitions in Ne-Like Ni XIX

K M Aggarwal and F P Keenan

Astrophysics Research Centre, School of Mathematics and Physics, Queen's University Belfast, Belfast BT7 1NN, Northern Ireland, UK

Abstract. In this paper we report results for collision strengths and effective collision strengths for transitions among 89 levels of the ($1s^2$) $2s^22p^6$, $2s^22p^53\ell$, $2s2p^63\ell$, $2s^22p^54\ell$, and $2s2p^64\ell$ configurations of Ni XIX, for which the DARC and FAC codes have been adopted. Resonances have been included in the calculations from DARC and their importance in determining the excitation rates is highlighted.

1 Introduction

Ne-like ions, such as Ti XIII, Fe XVII, and Ni XIX, are often used in lasing plasmas, and many transitions, particularly within the $n = 3$ configurations, have been measured in laboratory plasmas. Additionally, emission lines of Ni XIX have been observed in the spectrum of the solar corona, and are useful for plasma diagnostics. Nickel is also an important impurity element in fusion reactors, and hence atomic data are required for diagnostics as well as modelling of plasmas. Since there is a paucity of experimental data, theoretical results are desirable. Therefore, in a recent paper [1] we reported calculations for energy levels, lifetimes, and radiative rates for four types of transitions, namely electric dipole (E1), electric quadrupole (E2), magnetic dipole (M1), and magnetic quadrupole (M2), among 89 levels of the ($1s^2$) $2s^22p^6$, $2s^22p^53\ell$, $2s2p^63\ell$, $2s^22p^54\ell$, and $2s2p^64\ell$ configurations. Here in this paper we focus on our results for collision strengths (Ω) and effective collision strengths (Υ) or equivalently the excitation rates.

Considering the importance of Ni XIX a few calculations have appeared in the recent past, such as by Gu *et al* [2] and Chen *et al* ([3], [4]). However, these calculations are limited to comparisons between the theoretical and experimental line intensity ratios, and do not report any atomic data. Therefore, the *aim* of the present work is to report a *complete* set of results for *all* 3916 transitions among the lowest 89 levels of Ni XIX, which can be confidently applied in plasma modelling.

For our calculations of wavefunctions the GRASP (general-purpose relativistic atomic structure package) code has been adopted, and for calculating collision strengths (Ω) and subsequently the effective collision strengths (Υ), the *Dirac atomic R-matrix code* (DARC) is employed. Additionally, in order to make an accuracy assessment of our results, and particularly to estimate the contribution of resonances, which significantly enhance the values of Υ, especially for the forbidden transitions, we have also performed calculations from the *Flexible Atomic Code* (FAC) of Gu [5], available from the website `http://kipac-tree.stanford.edu/fac`. This is a fully relativistic code and provides a variety of atomic parameters. Furthermore, the code yields results comparable to those obtained from other atomic structure codes such as CIV3 and GRASP - see, for example, Aggarwal *et al* [6] and references therein, and scattering codes such as *R*-matrix - see, for example, Aggarwal *et al* [7] and references therein. Furthermore, FAC is based on the well known and widely used *distorted-wave* (DW) method and yields results for collision strengths (Ω) comparable with those from the *R*-matrix method, particularly for the allowed transitions, as demonstrated in several of our earlier papers on a variety of ions, including many of iron, i.e. Fe IX - Fe XXVI.

2 Collision strengths and effective collision strengths

For the scattering calculations, the *R*-matrix radius has been adopted to be 3.64 au, and 25 continuum orbitals have been included for each channel angular momentum for the expansion of the wavefunction. This allows us to compute Ω up to an energy of 250 Ryd, and subsequently the excitation rates up to $T_e = 10^7$ K. The maximum number of channels for a partial wave is 401, and the corresponding size of the Hamiltonian matrix is 10086. In order to obtain convergence of Ω for all transitions and at all energies, we have included all partial waves with angular momentum $J \leq 39.5$, although a higher range would have been preferable, particularly for the convergence of allowed transitions. However, to account for higher neglected partial waves, we have included a top-up, based on the Coulomb-Bethe approximation for allowed transitions and geometric series for forbidden transitions. Furthermore, resonances have been resolved in a fine energy mesh at over 21000 energies in the thresholds region.

For consistency we have included the same 89 levels in our calculations from FAC, as from DARC. The values of Ω are calculated up to an energy of $\sim$ 800 Ryd, more than sufficient to calculate values of Υ up to $T_e = 10^7$ K. However, as stated earlier, resonances are not included in this calculation. Nevertheless, these calculations from FAC are highly useful in assessing the importance of resonances with respect to the results obtained from DARC.

In Table 1 we list our values of Υ from DARC and FAC for all *resonance* transitions at three temperatures of $10^{5.2}$, 10^6, and 10^7 K. For level indices see Table 1 of [1]. In Fig. 1 we compare the Υ values for three resonance transitions, namely 1-2 ($2s^22p^6\ ^1S_0$ - $2s^22p^53s\ ^3P^o_2$), 1-3 ($2s^22p^6\ ^1S_0$ - $2s^22p^53s\ ^1P^o_1$), and 1-5 ($2s^22p^6\ ^1S_0$ - $2s^22p^53s\ ^3P^o_1$), over the entire temperature range. The 1-2 is forbidden, 1-3 is allowed, and 1-5 is an intercombination transition, but resonances have significantly enhanced the values of Υ, by over an order of magnitude, particularly towards the lower end of the temperature range. Even at the highest temperature of 10^7 K, our Υ values from DARC are higher than from FAC by up to a factor of five. In Fig. 2 we compare the Υ values for three other transitions among the excited levels, namely 2-6 ($2s^22p^53s\ ^3P^o_2$ - $2s^22p^53p\ ^3S_1$), 2-7 ($2s^22p^53s\ ^3P^o_2$ - $2s^22p^53p\ ^3D_2$), and 2-8 ($2s^22p^53s\ ^3P^o_2$ - $2s^22p^53p\ ^3D_3$). All these three transitions are *allowed* and have comparatively higher Υ values. However, as for transitions in Fig. 1, these transitions also show a considerable enhancement in the Υ values of up to a factor of three. Furthermore, as expected, differences between the two calculations decrease with increasing temperature. Nevertheless, the temperature of maximum abundance in ionisation equilibrium in solar plasma for Ni XIX is $10^{6.55}$; the laboratory measurements of line ratios are in the range of $\sim 10^7$ K, at which we have noted significant enhancement in the Υ values for many transitions. Therefore, the resonance included values of Υ from our DARC calculations may significantly affect the diagnostic and modelling of plasmas. We discuss this further below.

Of particular recent interest are the six resonance lines of Ni XIX, namely 3C ($2p^6\ ^1S_0$ - $2p^53d\ ^1P^o_1$: 1-27), 3D ($2p^6\ ^1S_0$ - $2p^53d\ ^3D^o_1$: 1-23), 3E ($2p^6\ ^1S_0$ - $2p^53d\ ^3P^o_1$: 1-17), 3F ($2p^6\ ^1S_0$ - $2p^53s\ ^3P^o_1$: 1-5), 3G ($2\ p^6$ $^1S_0$ - $2p^53s\ ^1P^o_1$: 1-3), and 3H ($2p^6\ ^1S_0$ - $2p^53s\ ^3P^o_2$: 1-2), at respective wavelengths of 12.434 Å, 12.658 Å, 12.810 Å, 13.779 Å, 14.043 Å, and 14.077 Å. These prominent strong lines have been observed in the x-ray

Table 1. Effective collision strengths (Υ) from DARC and FAC for resonance transitions in NI XIX at three temperatures of $10^{5.2}$, 10^6, and 10^7 K. $a{\pm}b \equiv a{\times}10^{\pm b}$.

Transition		DARC			FAC		
i	j	$10^{5.2}$	10^6	10^7	$10^{5.2}$	10^6	10^7 K
1	2	3.926-2	2.179-2	4.168-3	1.372E-03	1.295-3	8.495-4
1	3	5.553-2	2.822-2	7.612-3	1.545E-03	1.717-3	3.370-3
1	4	7.811-3	4.088-3	7.889-4	2.782E-04	2.624-4	1.716-4
1	5	5.573-2	2.797-2	6.349-3	1.351E-03	1.467-3	2.648-3
1	6	3.368-2	1.450-2	3.883-3	3.655E-03	3.436-3	2.211-3
1	7	2.266-2	1.279-2	4.645-3	3.279E-03	3.183-3	3.043-3
1	8	2.627-2	1.471-2	4.267-3	4.054E-03	3.777-3	2.302-3
1	9	2.845-2	1.195-2	2.517-3	1.486E-03	1.385-3	8.426-4
1	10	2.183-2	1.179-2	4.041-3	2.622E-03	2.567-3	2.677-3
1	11	2.527-2	1.178-2	5.058-3	3.239E-03	3.255-3	3.351-3
1	12	1.848-2	9.011-3	2.199-3	1.605E-03	1.495-3	9.102-4
1	13	1.965-2	9.811-3	2.398-3	1.692E-03	1.581-3	9.769-4
1	14	2.478-2	1.234-2	4.401-3	3.114E-03	3.038-3	3.082-3
1	15	4.041-2	3.954-2	3.634-2	3.398E-02	3.445-2	3.710-2
1	16	2.750-3	2.398-3	1.229-3	1.800E-03	1.667-3	9.809-4
1	17	7.574-3	7.187-3	3.960-3	5.261E-03	4.912-3	3.170-3
1	18	9.189-3	9.657-3	4.801-3	6.687E-03	6.185-3	3.606-3
1	19	8.837-3	9.600-3	4.464-3	6.119E-03	5.638-3	3.208-3
1	20	7.606-3	7.784-3	4.094-3	4.216E-03	3.981-3	3.249-3
1	21	5.090-3	6.057-3	2.388-3	2.418E-03	2.215-3	1.212-3
1	22	5.553-3	6.667-3	3.358-3	2.891E-03	2.753-3	2.591-3
1	23	2.512-2	3.019-2	3.760-2	2.524E-02	2.670-2	3.893-2
1	24	5.367-3	5.579-3	2.337-3	2.862E-03	2.632-3	1.478-3
1	25	8.244-3	7.943-3	3.265-3	3.907E-03	3.601-3	2.053-3
1	26	7.487-3	7.468-3	3.796-3	3.529E-03	3.358-3	3.049-3
1	27	6.262-2	7.507-2	1.029-1	6.830E-02	7.268-2	1.086-1
1	28	7.311-3	4.632-3	1.121-3	9.442E-04	8.776-4	5.279-4
1	29	1.563-2	1.509-2	1.434-2	1.278E-02	1.302-2	1.444-2
1	30	1.263-3	7.724-4	2.435-4	2.286E-04	2.151-4	1.388-4

Table 1. Effective collision strengths (ϒ) from DARC and FAC for resonance transitions in NI XIX at three temperatures of $10^{5.2}$, 10^6, and 10^7 K. $a{\pm}b \equiv a{\times}10^{\pm b}$.

Transition		DARC			FAC		
i	*j*	$10^{5.2}$	10^6	10^7	$10^{5.2}$	10^6	10^7 K
1	31	4.935-3	2.955-3	1.419-3	8.719E-04	8.809-4	1.111-3
1	32	6.326-3	3.736-3	1.179-3	1.098E-03	1.034-3	6.727-4
1	33	7.292-3	5.195-3	4.645-3	1.781E-03	2.045-3	4.678-3
1	34	1.960-3	2.083-3	1.075-3	1.651E-03	1.537-3	9.289-4
1	35	3.817-3	3.959-3	1.920-3	2.777E-03	2.590-3	1.602-3
1	36	4.884-3	5.003-3	2.531-3	3.823E-03	3.558-3	2.149-3
1	37	1.077-2	1.282-2	1.703-2	1.222E-02	1.305-2	1.880-2
1	38	6.035-3	2.230-3	5.656-4	4.293E-04	4.050-4	2.649-4
1	39	8.565-3	3.504-3	9.175-4	2.657E-04	2.820-4	4.708-4
1	40	3.272-3	1.666-3	6.976-4	9.054E-04	8.518-4	5.491-4
1	41	4.144-3	1.993-3	8.660-4	9.066E-04	8.705-4	7.002-4
1	42	4.143-3	2.283-3	9.830-4	1.352E-03	1.261-3	7.730-4
1	43	2.926-3	1.435-3	4.703-4	4.968E-04	4.638-4	2.845-4
1	44	2.705-3	1.442-3	6.835-4	7.071E-04	6.812-4	5.650-4
1	45	9.647-4	3.856-4	1.059-4	8.822E-05	8.313-5	5.407-5
1	46	8.839-3	3.137-3	7.647-4	2.824E-04	2.876-4	3.824-4
1	47	6.731-3	5.191-3	4.664-3	4.318E-03	4.379-3	4.735-3
1	48	2.125-3	1.107-3	4.354-4	5.608E-04	5.234-4	3.211-4
1	49	7.344-4	6.372-4	3.624-4	5.712E-04	5.306-4	3.167-4
1	50	2.095-3	1.807-3	1.094-3	1.578E-03	1.477-3	9.626-4
1	51	2.783-3	2.534-3	1.346-3	2.030E-03	1.877-3	1.085-3
1	52	2.461-3	2.200-3	1.203-3	1.884E-03	1.745-3	1.023-3
1	53	2.039-3	1.723-3	9.649-4	1.206E-03	1.135-3	8.013-4
1	54	1.801-3	1.045-3	4.541-4	6.073E-04	5.688-4	3.564-4
1	55	2.627-3	1.514-3	7.868-4	8.702E-04	8.356-4	6.736-4
1	56	1.650-3	1.368-3	6.182-4	8.446E-04	7.756-4	4.284-4
1	57	1.510-3	1.299-3	6.991-4	8.835E-04	8.282-4	5.839-4
1	58	8.003-3	7.081-3	6.754-3	6.290E-03	6.384-3	6.937-3
1	59	1.256-2	1.375-2	1.762-2	1.194E-02	1.262-2	1.810-2
1	60	3.541-4	3.075-4	1.465-4	2.902E-04	2.646-4	1.398-4

Table 1. Effective collision strengths (ϒ) from DARC and FAC for resonance transitions in NI XIX at three temperatures of $10^{5.2}$, 10^6, and 10^7 K. $a{\pm}b \equiv a{\times}10^{\pm b}$.

Transition		DARC			FAC		
i	j	$10^{5.2}$	10^6	10^7	$10^{5.2}$	10^6	10^7 K
1	61	5.535-4	4.754-4	2.686-4	3.313E-04	3.131-4	2.404-4
1	62	6.501-4	5.766-4	2.567-4	4.858E-04	4.411-4	2.279-4
1	63	7.894-4	7.238-4	5.912-4	5.789E-04	5.674-4	5.670-4
1	64	5.313-4	4.518-4	2.020-4	4.006E-04	3.640-4	1.878-4
1	65	9.964-4	9.743-4	1.188-3	6.731E-04	7.341-4	1.199-3
1	66	3.841-4	3.143-4	1.295-4	2.285E-04	2.063-4	1.022-4
1	67	3.947-4	3.358-4	1.566-4	2.540E-04	2.330-4	1.401-4
1	68	1.827-3	1.402-3	6.993-4	1.031E-03	9.516-4	5.450-4
1	69	2.268-3	1.889-3	1.006-3	1.580E-03	1.463-3	8.531-4
1	70	1.979-3	1.562-3	8.659-4	1.084E-03	1.020-3	7.263-4
1	71	1.198-2	1.343-2	1.741-2	1.117E-02	1.180-2	1.690-2
1	72	3.383-4	3.069-4	1.388-4	2.636E-04	2.391-4	1.221-4
1	73	4.407-4	3.878-4	2.066-4	2.958E-04	2.755-4	1.880-4
1	74	6.665-4	6.895-4	7.851-4	5.245E-04	5.514-4	7.874-4
1	75	5.027-4	4.617-4	2.168-4	4.281E-04	3.895-4	2.031-4
1	76	1.123-3	5.967-4	2.427-4	3.689E-04	3.431-4	2.071-4
1	77	3.097-3	3.015-3	3.013-3	2.541E-03	2.595-3	2.918-3
1	78	2.171-4	1.388-4	6.771-5	8.978E-05	8.428-5	5.367-5
1	79	8.434-4	5.680-4	4.751-4	3.560E-04	3.595-4	4.464-4
1	80	1.190-3	6.977-4	3.325-4	4.284E-04	4.025-4	2.578-4
1	81	1.333-3	1.089-3	1.506-3	6.376E-04	7.178-4	1.504-3
1	82	5.307-4	5.097-4	3.156-4	4.840E-04	4.528-4	2.821-4
1	83	8.899-4	8.592-4	5.421-4	8.152E-04	7.648-4	4.897-4
1	84	1.267-3	1.197-3	7.356-4	1.123E-03	1.051-3	6.541-4
1	85	2.415-3	2.710-3	3.607-3	2.438E-03	2.596-3	3.724-3
1	86	2.214-4	2.186-4	1.204-4	2.372E-04	2.175-4	1.193-4
1	87	3.163-4	3.172-4	1.999-4	3.352E-04	3.121-4	2.003-4
1	88	3.966-4	3.922-4	2.156-4	4.254E-04	3.900-4	2.139-4
1	89	4.135-4	4.765-4	6.498-4	4.048E-04	4.393-4	6.672-4

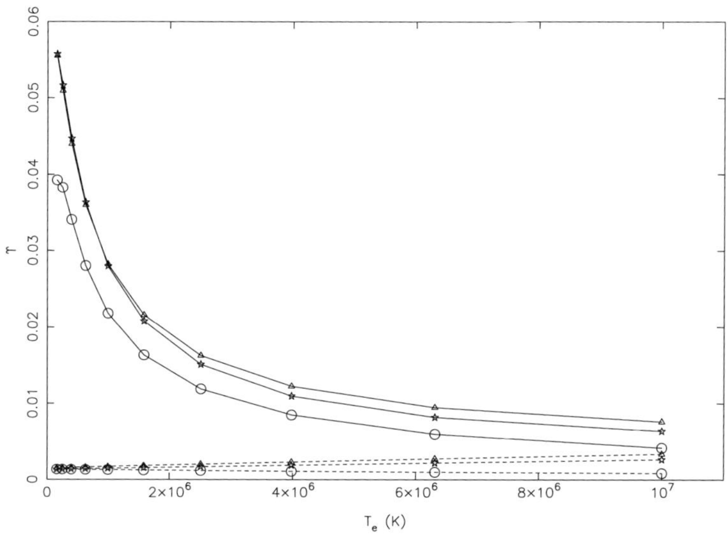

Figure 1. Comparison of effective collision strengths from DARC (continuous curves) and FAC (broken curves) for the 1-2 (circles: $2s^22p^6\ ^1S_0$ - $2s^22p^53s\ ^3P^o_2$), 1-3 (triangles: $2s^22p^6\ ^1S_0$ - $2s^22p^53s\ ^1P^o_1$), and 1-5 (stars: $2s^22p^6\ ^1S_0$ - $2s^22p^53s\ ^3P^o_1$) *resonance* transitions of Ni XIX.

spectra of solar and stellar coronae, active galactic nuclei, x-ray binaries, and supernovae from the *Chandra* and XMM-*Newton* satellites. Furthermore, line intensity ratios have also been measured on EBIT (electron beam ion trap) machines at the Lawrence Livermore National Laboratory (LLNL) by Brown *et al* [8] and Gu *et al* ([2], [9]), and at the National Institute of Standards and Technology (NIST) by Chen *et al* [4]. However, the laboratory measurements, solar observations, and theoretical results do not agree with one another. Since 3C and 3D lines are allowed, and hence resonances insignificantly affect their excitation rates, the theoretical and experimental ratios 3C/3D agree within $\sim 10\%$. Moreover, the energy/temperature variation of this ratio, in both theory and experiments, is within 10%. Similarly, the theoretical and experimental 3C/3E ratios vary within a factor of two from the lowest to the highest energy/temperature, but differences between theory and measurements are up to a factor of two, particularly at lower energies/temperatures. Furthermore, the EBIT measurements from spectrometer and calorimeter significantly differ, particularly at the high energy of 1.44 KeV [2]. Therefore, discrepancy between theory and measurements remains

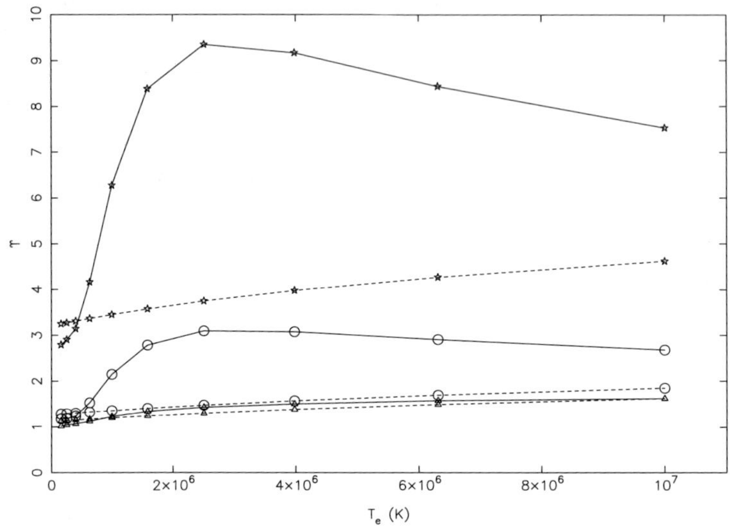

Figure 2. Comparison of effective collision strengths from DARC (continuous curves) and FAC (broken curves) for the 2-6 (circles: $2s^22p^53s\ ^3P^o_2$ - $2s^22p^53p\ ^3S_1$), 2-7 (triangles: $2s^22p^53s\ ^3P^o_2$ - $2s^22p^53p\ ^3D_2$), and 2-8 (stars: $2s^22p^53s\ ^3P^o_2$ - $2s^22p^53p\ ^3D_3$) transitions among excited levels of Ni XIX.

for the 3C/3E line ratio. However, the maximum discrepancy between theory and measurements is for the 3C/(3F+3G+3H) line ratios for which both spectrometer and calorimeter give consistent results throughout an energy range of 1.0 - 1.44 KeV. Moreover, the experimental ratios are (almost) energy independent whereas theoretical ones show a large variation of up to a factor of two. This is mainly because all 3F, 3G, and 3H lines are significantly affected by the presence of numerous closed-channel (Feshbach) resonances, as seen already in Fig. 1. A detailed comparison of theoretical and experimental line ratios is discussed in a separate paper by Aggarwal and Keenan [10]. Nevertheless, significant discrepancy between theory and measurements remains. An analysis of observations based on the available atomic data is even more difficult because many lines are blended. For this reason we will discuss the comparison between theory and observations in a separate paper.

To conclude, in this work we have performed two independent calculations from DARC and FAC codes to determine values of Ω and Υ for transitions in Ni XIX. The importance of resonances in determining the Υ values has been demonstrated, and discrepancy between theory and measurements for some prominent line ratios is discussed. However, results presented in this short paper are limited to only a few transitions. A complete set of results for Υ values for *all* transitions over a wide range of temperatures is reported in a separate paper [10]. Those results along with our reported A-values [1] can be confidently applied for the modelling of plasmas.

Acknowledgments

This work has been financed by the EPSRC and STFC of UK and FPK is grateful to AWE Aldermaston for the award of a William Penney Fellowship.

References

[1] Aggarwal K M and Keenan F P 2006 *Astron. Astrophys.* **460** 959

[2] Gu M F, Beiersdorfer P, Brown G V, Chen H, Boyce K R, Kelley R L, Kilbourne C A, Porter F S and Kahn S M 2004 *Astrophys. J.* **607** L143

[3] Chen G X, Kirby K and Brickhouse N S 2006 *Phys. Rev.* **A 73** 052708

[4] Chen G X, Kirby K, Silver E, Brickhouse N S, Gillaspy J D, Tan J N, Pomeroy J M and Laming J M 2006 *Phys. Rev. Lett.* **97** 143201

[5] Gu M F 2003 *Astrophys. J.* **582** 1241

[6] Aggarwal K M, Tayal V, Gupta G P and Keenan F P 2007 *At. Data Nucl. Data Tables* **93** 615

[7] Aggarwal K M, Hamada K, Igarashi A, Jonauskas V, Keenan F P and Nakazaki S 2008 *Astron. Astrophys.* **484** 879

[8] Brown G V, Beiersdorfer P and Widmann K 2001 *Phys. Rev.* **A 63** 032719

[9] Gu M F, Beiersdorfer P, Brown G V, Chen H, Thorn D B and Kahn S M 2007 *Astrophys. J.* **657** 1172

[10] Aggarwal K M and Keenan F P 2008 *Astron. Astrophys.* **488** 365

Conversion Efficiency Calculations for Soft X-Rays Emitted from Tin Plasma for Lithography Applications

P. Demir[1,2], P. Demir[3], E. Kacar[1,2], S. K. Bilikmen[3] and A. Demir[1,2]

[1] Faculty of Arts and Science, Department of Physics, Kocaeli University, Kocaeli, Turkey
[2] Laser Technologies Research and Application Centre, Kocaeli University, Kocaeli, Turkey
[3] Faculty of Arts and Science, Department of Physics, Middle East Technical University, Ankara, Turkey

Abstract. Laser-produced plasmas can be used as extreme ultraviolet (EUV) sources for lithography, operating in the 13.5 nm range. Soft X-rays emitted from laser-produced tin plasma were simulated using the EHYBRID code. Required atomic data for tin ions were calculated using the Cowan code. In the EHYBRID simulations tin slab target is assumed to be irradiated by 1064 nm wavelength Nd:YAG laser. Simulations were performed for different driving laser parameters.

1 Introduction

Extreme ultraviolet lithography (EUVL), using extreme ultraviolet (EUV) light with a wavelength in the range of 10 to 14 nm, is one of the most promising technologies of semiconductor production. Optimization and improvement of extreme ultraviolet light sources is important for the development of EUV lithography [1,2]. Lithography systems are being designed with Mo/Si multilayer mirrors. Soft X-ray wavelengths of 13.5 nm in 2% bandwidth have approximately 71% reflectivity for Mo/Si multilayer coated optics. Therefore this wavelength regime is preferred for the next generation of EUV technologies [3,4].

There are different methods to generate EUV radiation. For generating EUV light, plasma is regarded as the most promising source of EUV radiation. In the case of laser produced plasmas, the emitting plasma is produced by focusing of the pulsed laser on the target material [5,6]. Laser produced Sn or Xe plasmas and discharge produced Sn or Xe plasmas have been systematically studied and are under development [2,3].

One of the important issues in developing EUV lithography is optimization of 13.5 nm EUV light source to achieve the highest possible conversion efficiency. Industrial requirement of conversion efficiency for EUV lithography is 3% conversion into a 2% bandwidth at 13.5 nm. Studies have been shown that, in comparison with Xe plasmas, Sn plasmas provide better conversion efficiency with wavelength of 13.5 nm [2,4]. Computer modeling has been used to optimize EUV sources. For calculating the EUV radiation in the 2% bandwidth at 13.5 nm from tin or xenon plasma, transitions from outer shells $4d^n - 4d^{n-1}$ 5p must be considered in combination with the transitions $4d^n - 4d^{n-1}$ 4f and the transitions from the inner shells, such as $4p^6\,4d^n - 4p^5\,4d^{n+1}$ [7,8].

In this study, emission spectra of tin plasma from Sn XI and Sn XII ions in the range of 12 to 17 nm wavelength and conversion efficiency of EUV radiation in the 2% bandwidth at 13.5 nm from these ions in the tin plasma have been simulated. Calculations were carried out using EHYBRID code [9]. Required atomic data were obtained using Cowan code [10]. The transitions from $4d^n - 4d^{n-1}$ 4f, $4d^n - 4d^{n-1}$ 5p and $4p^6\,4d^n - 4p5\,4d^{n+1}$ were considered for Sn XI and Sn XII ions in the calculations. In the simulation, tin slab target is assumed to be irradiated using a single pulse with 1064 nm wavelength Nd:YAG laser. Duration and intensity of driving laser pulse were varied for optimizing conversion efficiency of EUV radiation in the 2% bandwidth at 13.5 nm into 2π steradians.

2 Simulation Results

In the simulation, a single pulse with 1064 nm wavelength Nd:YAG laser is assumed to irradiate tin slab target with 10 mm length, 100 μm width and 20 μm thickness. Intensity of the driving laser pulse was varied from $1\text{x}10^{11}$ W/cm^2 to $5\text{x}10^{11}$ W/cm^2 for 6 ns pulse duration. Figure – 1 (a) and (b) show variations of conversion efficiencies for Sn XI and Sn XII, respectively, within a 2% bandwidth around 13.5 nm into 2π steradians versus driving laser pulse intensity.

Durations of the driving laser pulse were varied from 6 ns to 21 ns for $4\text{x}10^{11}$ W/cm^2 pulse intensity. Figure – 2 (a) and (b) show variations of conversion efficiencies for Sn XI and Sn XII, respectively, within a 2% bandwidth around 13.5 nm into 2π steradians versus driving laser pulse duration.

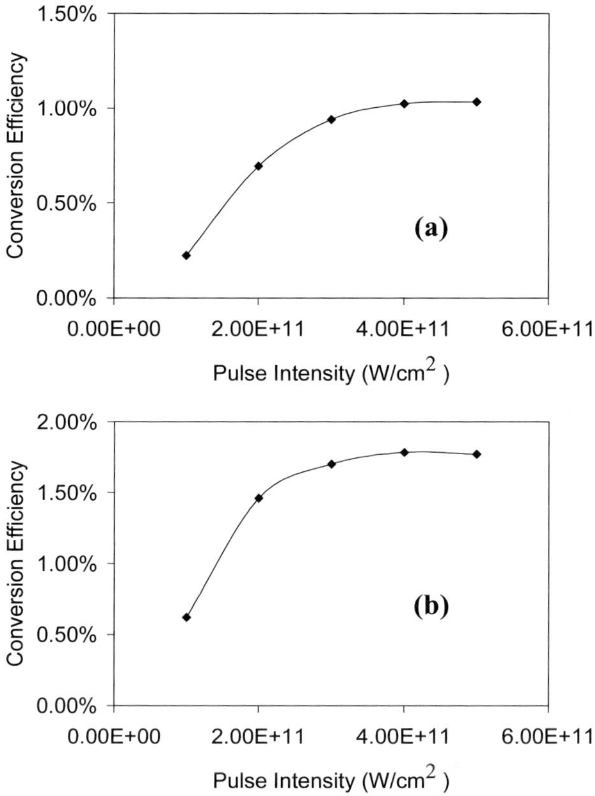

Fig. 1 Variations of conversion efficiencies for (a) Sn XI and (b) Sn XII within a 2% bandwidth around 13.5 nm into 2π steradians versus driving laser pulse intensity for 6 ns pulse duration.

Simulated emission spectra of tin arising from $4d^4 - 4d^3\ 4f$, $4d^4 - 4d^3\ 5p$, $4p^6\ 4d^4 - 4p^5\ 4d^5$ transitions for Sn XI and $4d^3 - 4d^2\ 4f$, $4d^3 - 4d^2\ 5p$, $4p^6\ 4d^3 - 4p^5\ 4d^4$ transitions for Sn XII have been shown in Figure – 3 (a) and (b), respectively, in the range of 12 to 17 nm wavelength. These figures are obtained for 6 ns pulse duration and $4x10^{11}$ W/cm^2 pulse intensity. Figure – 3 (a) and (b) show that contribution of 4d – 5p transition is less than contributions of $4p^5$ – 4d and 4d – 4f transitions.

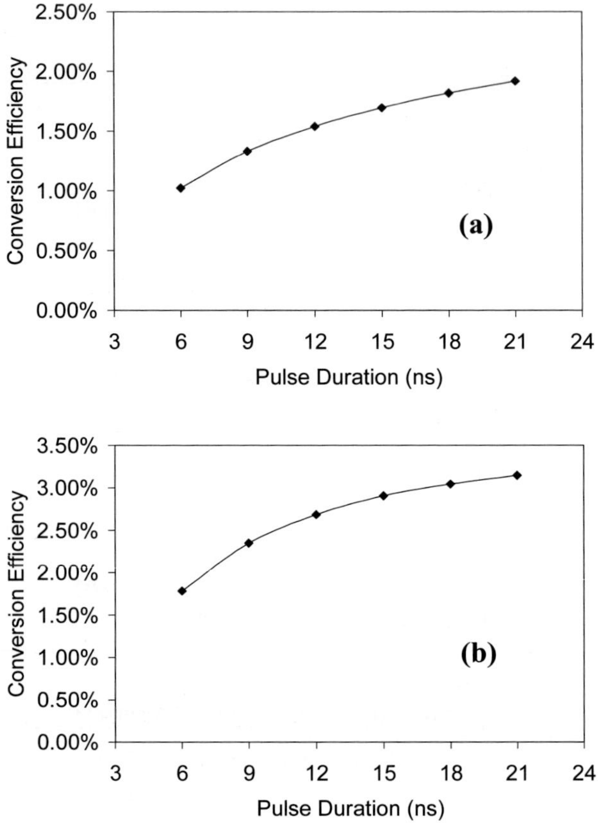

Fig. 2 Variations of conversion efficiencies for (a) Sn XI and (b) Sn XII within a 2% bandwidth around 13.5 nm into 2π steradians versus driving laser pulse duration.

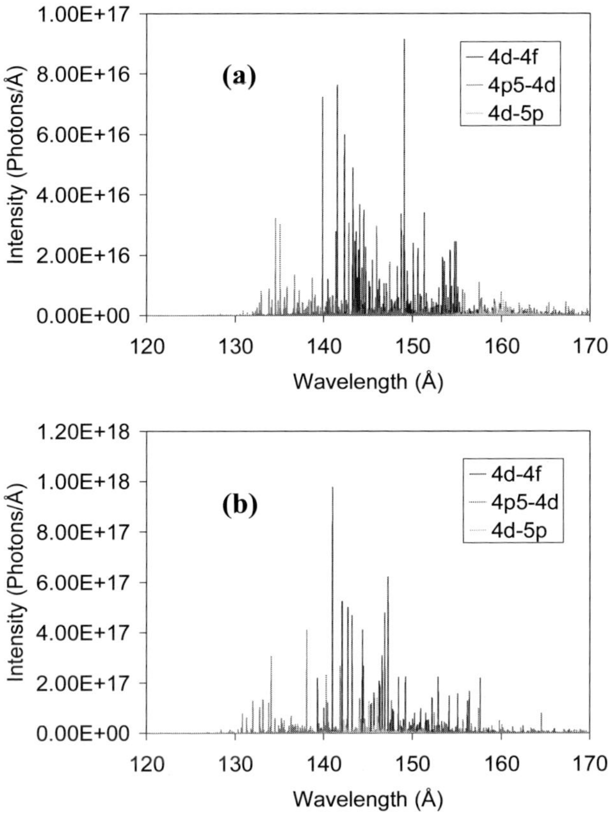

Fig. 3 Simulated emission spectra of tin arising from $4p^6\,4d^n - 4p^5\,4d^{n+1}+4d^{n-1}(4f+5p)$ transitions for (a) Sn XI and (b) Sn XII ions in the range of 12 to 17 nm wavelength.

Simulated emission spectra of tin plasma for Sn XI and Sn XII ions in the range of 13 to 14 nm wavelength are shown in Figure – 4 (a) and (b), respectively, for 6 ns pulse duration and $4x10^{11}$ W/cm^2 pulse intensity.

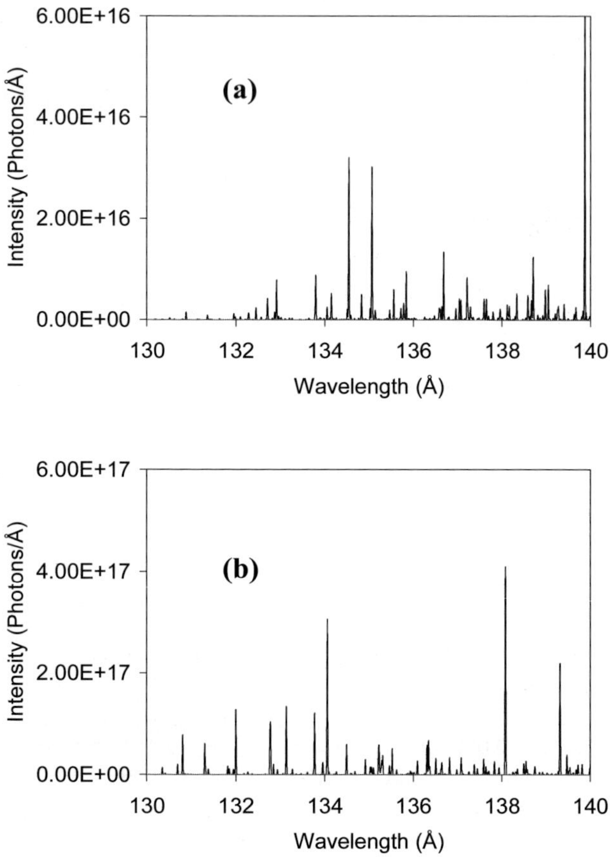

Fig. 4. Simulated emission spectra of tin plasma for (a) Sn XI and (b) Sn XII ions in the range of 13 to 14 nm wavelength for 6 ns pulse duration and $4x10^{11}$ W/cm^2 pulse intensity.

3 Conclusions

In this study, laser produced tin plasma is simulated as a EUV source to optimize conversion efficiency of EUV radiation in the 2% bandwidth at 13.5 nm into 2π steradians from SnXI and Sn XII ions in the tin plasma. In the simulation, tin slab target with 10 mm length, 100 μm width and 20 μm thickness is assumed to be irradiated using a single pulse with 1064 nm wavelength Nd:YAG laser. Effects of driving laser pulse duration and intensity variations on conversion efficiency of EUV radiation were obtained. Figures – 1 and 2 show that conversion efficiency of EUV radiation depends upon duration and intensity of driving laser pulse. Emission at 13.5 nm wavelength may be increased and thus high conversion efficiency may be achieved by further optimization of the plasma formation conditions.

4 Acknowledgements

The authors would like to thank Kocaeli University for supporting this work (Project No. 2007 – 34).

References

1. Gwyn, C. W. et al.: 'Extreme ultraviolet lithography', *J. Vac. Sci. Technol. B*, 16(6), 3142-3149, 1998
2. Zhang, C. H. et al.: 'Characterization of Solid Tin Target for Gas Discharges Produced EUV Plasmas', *6th International Siberian Workshop and Tutorial Edm, Tutorials, July 1-5, Erlagol*, 8-10, 2005
3. Yamaura, M. et al.: 'Characterization of extreme ultraviolet emission using the fourth harmonic of a Nd :YAG laser', *App. Phys. Lett.*, 86, 181107-181110, 2005
4. Cummings. A. et al.: 'Conversion efficiency of a laser-produced Sn plasma at 13.5 nm, simulated with a one-dimensional hydrodynamic model and treated as a multi-component blackbody', *J. Phys. D: Appl. Phys.*, 38, 604-616, 2005
5. Banine, V., Moors, R.: 'Plasma sources for EUV lithography exposure tools', *J. Phys. D: Appl. Phys.*, 37, 3207-3212, 2004
6. Stamm, U.: 'Extreme ultraviolet light sources for use in semiconductor lithography - state of the art and future development', *J. Phys. D: Appl. Phys.*, 37, 3244-3253, 2004
7. Morozov, V. et al.: 'Calculation of tin Atomic Data and Plasma Properties', *Energy Technology Division, Argonne National Laboratory, ANL-ET-04/24*, 2004
8. White, J. et al.: 'Simplified modeling of 13.5 nm unresolved transition array emission of a Sn plasma and comparison with experiment', *J. Appl. Phys.*, 98, 113301-113313, 2005
9. Pert, G. J.: 'The hybrid model and its application for studying free expansion', *Fluid Mech.*, 131, 401-426, 1983
10. Cowan, R. D.: *J. Opt. Soc. Am.*, 58, 808- , 1968.

Theoretical Investigation of Photo-pumping X-Ray Lasers Using Kα Linc from Solid Target

T. Kawachi[1] and Y. Kato[2]

[1]Quantum Beam Science Directorate, Japan Atomic Energy Agency (JAEA).
[2] The Graduate School for the Creation of New Photonics Industries

Abstract. New photo-pumping x-ray laser scheme is theoretically investigated, in which the wavelength matching between the aluminum Kα line (λ = 0.833816 nm) and the $2p^6$-$(2p_{1/2}, 4d_{3/2})_1$ transition of the neon-like zinc ions (λ = 0.83400 nm) is used. Population kinetics code of the neon-like zinc ions in plasma under the irradiation of the aluminum Kα line shows that substantial amplification gain can be expected in the transition of $(2p_{1/2}, 3p_{1/2})_0$ - $(2p_{1/2}, 4d_{3/2})_1$ at a wavelength of 3.5 nm. We also propose the experimental set-up for this scheme, which implies that this scheme is feasible under the present technology of ultra-short pulse laser and optics.

1 Introduction

In the photo-pumping x-ray laser scheme, spectral line emission from particular ions is absorbed by different element ions to create the population inversion in the latter. The success of this scheme as an x-ray laser depends upon exact spectral matching between the emission line and the absorption line. The widths of the spectral lines of ions in plasma, that are due mainly to the Doppler broadening and the Stark broadening, are typically ~ $\Delta\lambda/\lambda < 0.01\%$. Therefore high resolution spectroscopic studies are required to find an appropriate pair of the "emitter" and "absorber" ions. Indeed, in the early 1990s, the atomic physics group at the Lawrence Livermore National Laboratory (LLNL) has determined the accurate wavelengths of many spectral lines of highly charged ions by use of the electron beam ion trap (EBIT) [1, 2]. The data obtained by the EBIT group, in collaboration with the x-ray laser specialists at LLNL, was used to find the candidate pairs of ions for photo-pumping x-ray lasers.

However, the resonance photo-pumping scheme has a technical difficulty [3]: the emitter and the absorber ions should be located as close as possible so that the pumping emission reaches to the absorber ions efficiently, whereas the electron temperature should be high for the emitter to increase the emissivity of the ions, and lower temperature is favorable for the absorber ions to avoid the "thermal" population. Therefore the simultaneous satisfaction of these two conditions is key issue to work this scheme.

In this paper, we investigate the possibility of the use of Kα emission from solid target as the pumping source. This scheme has two advantages compared with the cases of ion-ion pair. Firstly, recent experimental investigation realizes efficient generation of sub-ps duration Kα pulse: the energy conversion efficiency from the pumping laser to Kα emission reaches ~ 10^{-4} per steradian [4]. Secondly, the Kα line emits in localized area on the solid surface, therefore the distance from the pumping source to the absorber can be kept small.

We focus the wavelength matching of the aluminum Kα line (0.833816 nm) and the $2p^6$-$(2p_{1/2}, 4d_{3/2})_1$ resonance line of the neon-like zinc ions (0.83400 nm) [5] and investigate the feasibility of this scheme using population kinetics code of the neon-like zinc ions under the Kα line irradiation.

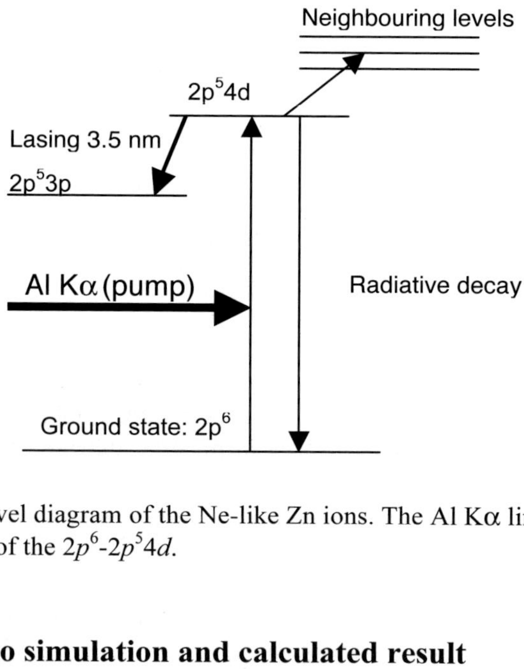

Fig. 1 Energy level diagram of the Ne-like Zn ions. The Al Kα line can be absorbed by the transition of the $2p^6$-$2p^5 4d$.

2 Monte-Carlo simulation and calculated result

Figure 1 shows the simplified energy level diagram of the neon-like zinc ions. In the following context, we denote the $(2p_{1/2}, 3p_{1/2})_0$ and $(2p_{1/2}, 4d_{3/2})_1$ as $2p^5 3p$ and $2p^5 4d$, respectively. We consider the ground state $2p^6$, the $2p^5 3p$ level, the $2p^5 4d$ level and the neighbouring levels of the $2p^5 4d$ levels. The Kα line emission is absorbed by the transition of $2p^6$-$2p^5 4d$ resonance line, and the following collisional-radiative processes transfer the populations into other levels. Atomic processes included in the calculation are the spontaneous transition, electron impact excitation and deexcitaion, electron impact ionization and the radiation trapping effect of the $2p^6$-$2p^5 4d$ resonance line. The spontaneous transition probability, and the electron impact excitation cross section were calculated by use of HULLAC code, and the electron impact ionization cross section from the ground state and the excited levels are derived from the scaled hydrogenic approximation. The rate coefficients of the excitation and ionization are calculated under the assumption of Maxwellian electron velocity distribution. The electron impact deexcitation rate coefficients are estimated from the principle of the detailed balance.

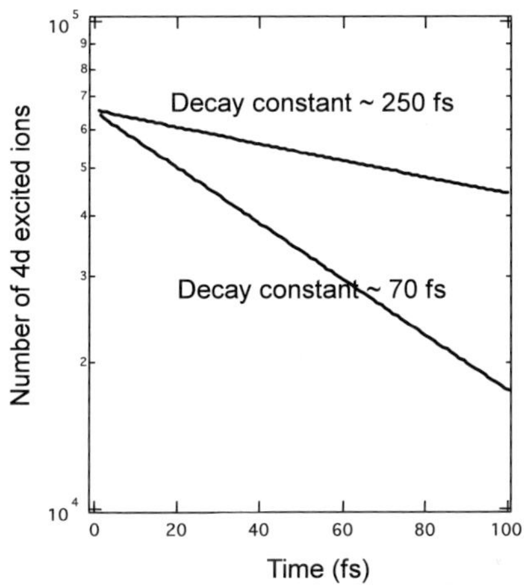

Fig. 2 The temporal evolution of the $2p^5 4d$ level under the condition that 3×10^5 photons comes into one sliced area in 1 fs. Thick solid line is with the radiation trapping effect, and the thin solid line is the case without radiation trapping. Due to the radiation trapping effect, the decay constant becomes around 4 times longer.

The size of the zinc plasma is defined as 20 μm × 20 μm × 5 mm in the *x*-, *y*- and *z*-directions, respectively. We consider a sliced area in the plasma with the size of 20 μm × 20 μm in the *x*-*y* plane and the thickness of 100 nm. This sliced area is divided into (*x, y*) mesh with the size of 100 nm × 100 nm. The plasma parameters, *i.e.*, the electron density, n_e, the electron temperature, T_e, the neon-like ion density, $n_{ne\text{-}like}$, and the ion temperature, T_i, are assumed to be 100 eV, 2 × 10^{21} cm^{-3}, 50 eV, 10^{20} cm^{-3}, respectively. The Kα photons come into the plasma from the *y*-direction. The linewidth ($\Delta\lambda/\lambda$) of the Kα line is assumed to be ~ 10^{-3}, and the linewidth of the $2p^6$-$2p^5 4d$ resonance line is determined by Doppler broadening.

The calculation procedure is as follows: Firstly the frequency of the Kα photon and the *x*-position are determined, randomly. Secondly the absorption length of the Kα photon in the plasma is calculated to determine the *y*-position where the Kα photon is absorbed. This routine is repeated *N* times, where *N* is the number of Kα photons coming into the calculation area in the time step. Thirdly, for each (*x, y*) position, we determine the absorbed photon is lost by collisional-radiative process to other excited levels or not. If the collisional-radiative processes take place, we count the increase of the other level populations. If not, it is determined that the resonance emission event occurs or not. In the case of the resonance emission process, the frequency of the resonance emission is determined randomly within the spectral linewidth, and the direction of the emission and the absorption length are calculated. From these information, we judge the photon is absorbed in new (*x, y*)-position or escapes from the plasma. The procedure 1~3 is iterated with 1 fs time step, and we calculated the temporal evolution of the population of each excited level.

Figure 2 shows a calculated result under the condition that the 3×10^5 photons comes into one sliced area in 1 fs. The lifetime of the $2p^54d$ level population is presented. From the number of the $2p^54d$ excited ions in the early time region ($t \sim 0$ fs), it is found that 20% of the Kα photons can be absorbed in the plasma. Under the absence of the radiation trapping effect, the lifetime of the $2p^54d$ level is ~ 70 fs, which is in accordance with the spontaneous transition probability of this level, whereas with the radiation trapping effect, the effective lifetime extends up to 250 fs.

Figure 3 shows the temporal evolution of the $2p^53p$ and $2p^54d$ level populations divided by their statistical weights (left-hand side ordinate) under the Kα pump with the duration of 500 fs (FWHM) Gaussian. The amplification gain of the $2p^53p$-$2p^54d$ line at a wavelength of 3.5 nm is also attached (right-hand side ordinate), where the line-shape of the lasing line is assumed to be Gaussian due to Doppler broadening.

In Fig. 3, the $n(4d)/g(4d)$ increases as time proceeds and decreases for $t >$ 600 fs. This is due to the decrease in the pumping rate together with the collisioinal depopulation whose time constant is ~ 300 fs. Substantial amplification gain is obtained in the time region of t = 300 – 800 fs. This gain duration is much longer than that of other x-ray laser scheme using inner-shell ionized atoms [6, 7], in which the gain generation process competes with Auger process with the typical time constant of around 10 fs. This quite short duration induces the practical difficulty of the inner-shell scheme so far. The present long duration of the gain (~ 500 fs) implies that Kα pumping scheme is feasible without any complicated traveling wave technique.

The $n(3p)/g(3p)$ increases gradually and at around t = 1000 fs, the $n(3p)/g(3p)$ becomes larger than $n(4d)/g(4d)$. This result may be slightly overestimate. Because in the present calculation, the collisional depopulation process from the $2p^53p$ to the neighbouring levels, *e.g.*, $2p^53s$, $2p^53d$ is not taken into account. Including these effect, the $n(3p)/g(3p)$ may become smaller, which leads to the enhancement of the amplification gain

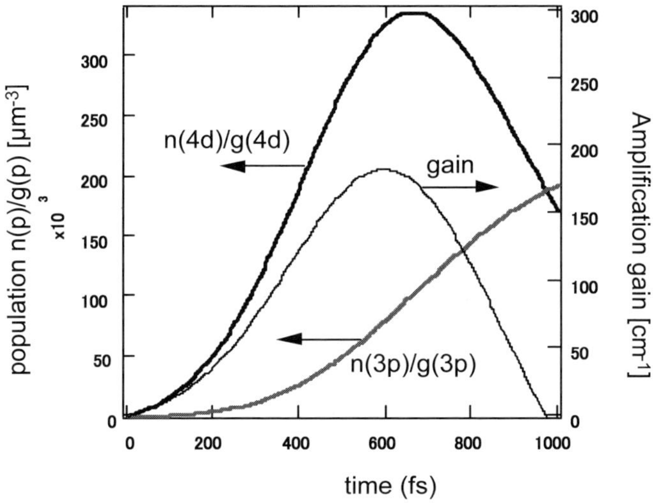

Fig. 3 The temporal evolution of the excited level populations of the $2p^5 3p$ (solid gray line), $2p^5 4d$ (solid thick line) levels divided by the statistical weights and the amplification gain (solid thin line).

3 Target design and pumping geometry

In Fig. 4, we propose the experimental set-up for this photo-pumping scheme. Target is aluminum foil with the thickness of 4 μm, and 0.2 μm-thick zinc is deposited on it. The zinc side is irradiated by double pulses of pumping laser with both the intensities of 10^{13} W/cm^2, separated by 100 ps, to prepare the neon-like zinc ions in plasma. After a certain time delay, the opposite side (aluminum side) is irradiated by ultra-short pulse with 300fs duration and 10^{15}-10^{16} W/cm^2 intensity to generate intense aluminum Kα. The energy conversion efficiency from the pump laser to Kα line is at least around 10^{-5}/Sr [4]. Since the distance from the Kα source to the absorber is around 10 μm, the effective solid angle of the Kα source to neon-like zinc plasma is 5 steradian, therefore we can deliver more than 200 μJ of Kα photons into the zinc plasma where the transmittance of Kα photons through the 4 μm-thick aluminum foil of 70 % is taken into account. This value is much larger than the assumption in the calculation in section 2.

It is noted that under the intensity of 10^{15}-10^{16} W/cm^2, several KeV electrons are generated on the surface of aluminum side, however the energy of these electrons are dumped due to 4 μm-thick aluminum foil, resulting in the suprathermal electron energy is transferred into bremsstrahlung. Since the aluminum foil works as a spectral window for these continuum emission, the heating of the plasma due to the continuum radiation becomes negligible.

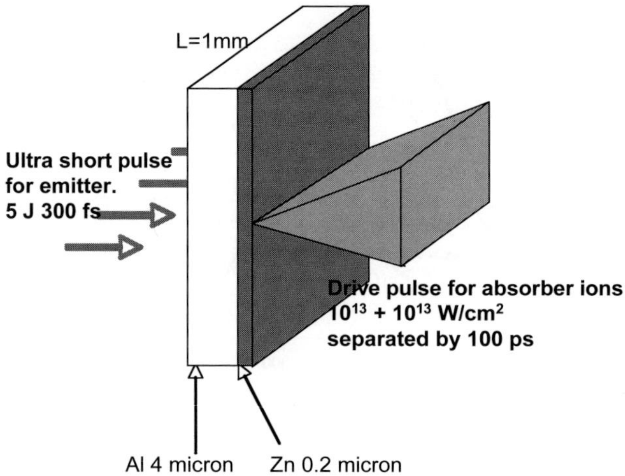

Fig. 4 Experimental set-up for the photo-pumping scheme.

4 Summary

We have investigated new photo-pumping scheme by use of aluminum Kα line and the $2p^6$-$2p^5 4d$ resonance line of neonlike zinc ions. The result of the population kinetics code shows that substantial amplification gain in the transition of $2p^5 3p$-$2p^5 4d$ line at a wavelength of 3.5 nm can be generated in this scheme. This scheme is one of the feasible "water-window" x-ray lasers using a laboratory size pumping source.

References

1. Nilsen J, Biersdorfer P, *et al.*, *Phys. Rev. A* 50 2143, 1994
2. Elliott S R, Biersdorfer P, *et al.*, *Phys. Rev. A* 52: 2689, 1995
3. Kato Y and Kawachi T, 2008, *Progress in Ultrafast Intense Laser Science IV*, (Berlin: Springer) 2008 (in press).
4. Rousse A, *et al.*, *Phys. Rev E* 50, 2200, 1994
5. Boiko V, Faenov A Ya, *et al.*, J. Quant. Spectrosc.Radiat. Transfer 19, 11, 1978
6. Kim D, Son S H, Kim J H, Toth C and Barty C P J, *Phys. Rev. A* **63**, 023806, 2001
7. Moon S J and Eder D C, *Phys. Rev A* **57**, 1391, 1998

Part 6 – High Harmonic Generation (HHG)

Coherent Water-Window X-Ray Generation by Phase-Matched High Harmonics in Neutral Media

Eiji J. Takahashi and Katsumi Midorikawa

Extreme Photonics Research Group, RIKEN Advanced Science Institute,
2-1 Hirosawa, Wako, Saitama 351-0198, Japan

Abstract. Life science and biology have advanced with the progress of microscope technology. A strong desire of life scientists is to observe live cells with high spatial resolution. The development of soft x-ray microscopy using a special wavelength called the water window started in the 1970s in order to fulfill this desire. However, the photon fluxes realized so far are still insufficient for this application. Here, we demonstrate the generation of a coherent water window x-ray by extending the plateau region of high-order harmonics under a neutral-medium condition. The observed maximum harmonic photon energy attained are 300 eV and 450 eV in Ne and He, respectively. Our proposed generation scheme, combining a 1.6 μm laser driver and a neutral Ne gas medium, is efficient and scalable in output yields of the water window x-ray. This powerful concept is expected to boost the development of soft x-ray microscopes based on high-order harmonics.

1 Introduction

The realization of a powerful coherent x-ray source has been the dream of researchers of laser physics and photonics. In particular, the spectral range between the K-absorption edges of carbon (284 eV) and oxygen (543 eV), which is called the *water window*, is attractive for high-contrast biological imaging. This spectral range has been one of the most important grails in the research and development of coherent x-ray sources. An intense ultrafast water-window x-ray pulse would allow us to capture images of live cells by instantaneously halting their motion, preserving structural information that is lost in the sample's preparation process for electron microscopy.

So far, the most practical source of coherent x-rays has been synchrotron radiation (SR) produced by an accelerator. Images of live cells, however, cannot be captured without freezing the sample even using the world's largest SR facility due to its low instantaneous power. On the other hand, an alternative approach of high-order harmonic generation (HHG) using a tabletop laser system, has been investigated over the last ten years for generating coherent water window x-rays [1]. HHG provides highly coherent ultrafast x-ray pulses and even enables the creation of a single burst or a train of attosecond pulses.

Although the generation of a large amount of photon flux with below 100 eV region by phase-matched HHG has been reported[2], the output photon flux at about 300 eV is still less than approximately 10^3 photons/shot with a conversion efficiency (CE) of less than 10^{-10} [1]. This is mainly due to phase mismatch induced by *plasma* free electrons, because a high intensity greater than the ionization threshold is required for 0.8 μm Ti:sapphire laser driving.

On the other hand, an alternative possible approach to extending the harmonic photon energy is to use a longer-wavelength laser for the driving field. From a theoretical point of view, the maximum photon energy of HHG is approximately given by E_{cutoff}[eV] = $I_p+3.17U_P$, where I_P is the binding energy of the electrons and U_P [eV] = $9.38x10^{-14}$ I [W/cm^2] (λ_0 [μm])^2 is the ponderomotive energy. This formula indicates that the high-order harmonic (HH) photon energy extends significantly into the water window region despite the neutral-medium condition[3] when the 1.6 μm laser is employed for Ne gas. Moreover, since neutral media enable us to adopt a *right* phase-matching (PM) technique without quasi-PM[4], we can apply an energy-scaling procedure that is well established using a loosely focused beam [2]. Here, we show the efficient generation of a coherent x-ray HH in the water window region using an IR driving laser and *neutral* media [3]. Our proposed procedure for generating the water window x-ray is efficient and scalable in output yield [5].

2 High-energy IR pulses by optical parametric amplifier

High-energy IR source[6] is accomplished in two-stage OPA with type II BBO crystal. The seed pulses are formed in OPA-1, which is designed with a two-pass configuration of white-light continuum (LIGHT CONVERSION Inc. TOPAS-C). Pump pulse for OPA is an TW-Ti:sapphire laser system (40 fs, 200 mJ, 800 nm, 10 Hz). To optimize the OPA process at the OPA-1 stage, spatial filtering for the pump pulse is performed. When the pumping energy for OPA-1 is ~2 mJ, the output energy of 1.4 μm is measured to be ~200 μJ with 40 fs pulse width. We use an appropriate mirror pair at the output of OPA-1 to collimate the seed pulses (1.4 μm) and to match its spatial volume with the pump pulses at the OPA-2 stage. Type II collinear phase matching BBO (θ = 27°, φ = 90°, t = 4 mm) is used in the OPA-2. Using hard aperture (φ = 15 mm) for the pump pulses at the OPA-2, we select the homogeneous intensity region of the Ti:sapphire laser beam (φ = 50 mm) in order to improve conversion efficiency and spatial quality of OPA scheme. Pump laser intensity at the OPA-2 is adjusted to ~0.25 TW/cm^2 using a telescope.

Figure 1 (a) depicts a typical output energy of OPA-2 versus the wavelength at the 20 mJ pumping energy. Note that the wavelength of the seed pulse is fixed to be 1.4 μm. The amplified pulse energy exceeds 1 mJ in almost the whole tuning range (1.2 ~ 2.1 μm) . When the pump energy is up to 25 mJ, output energy exceeding 6 mJ with 40 fs pulse width is achieved at the signal wavelength near 1.4 μm. Total output energy (signal + idler) of 10 mJ is recorded with ~ 40 % conversion efficiency. At an ultrafast OPA scheme, there are the highest energy ever produced from 1.2 to 2.1 μm region. A beam quality is very important for practical use of these IR beam. The typical spatial intensity distribution of the amplified 1.4 μm is depicted in Fig. 1 (b). The spatial profile of the amplified OPA-2 output adopts the top-hat shape of the pump pulses due to gain saturation.

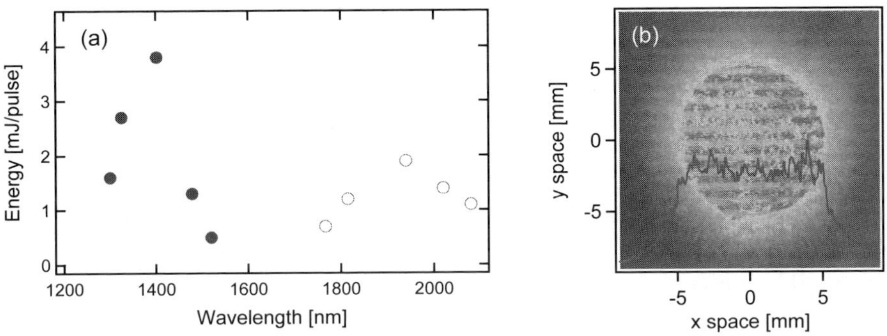

Fig. 1 (a) : Energy of the parametric pulses versus wavelength. Blue and red points correspond to signal and idler, respectively. (b) : Spatial profile at $\lambda = 1.4$ μm.

The output IR pulses are focused using a f=250 mm lens and are delivered into the target chamber. The target gas is supplied by a 2-mm diameter synchronized supersonic gas jet. The generated harmonics illuminate a slit of the spectrometer, which is placed 0.5 m from the gas jet position. A flat-field grating relays the image of the slit to a microchannel plate, where it is spectrally resolved along the horizontal axis. Here, we use two types of concave grating to measure the broadband wavelength region from 50 to 500 eV. One is specifically designed to cover the 50 – 200 eV photon energy range and has a nominal groove number of 1200 grooves/mm; the other is designed to cover the 200 – 500 eV photon energy range and has a nominal groove number of 2400 grooves/mm [7]. Note that a toroidal mirror is not set in front of the spectrometer, thus we can directly obtain a spatial profile information of HH beam.

3 Water window soft x-ray from neutral harmonic media

To attain efficient HHG, the absorption feature of a harmonic medium as well as the PM is important. In a broad plateau region of nearly constant harmonic intensity, the HH spectrum shape reflects the absorption feature of the medium under the absorption limited condition (ALC)[2]. For efficient HHG in the water window region, He and Ne gases are advantageous compared with Ar due to their lower absorption. However, since He has exceptionally small effective nonlinearity, Ne is expected to be the most practical gas for efficiently generating HH in the water window region.

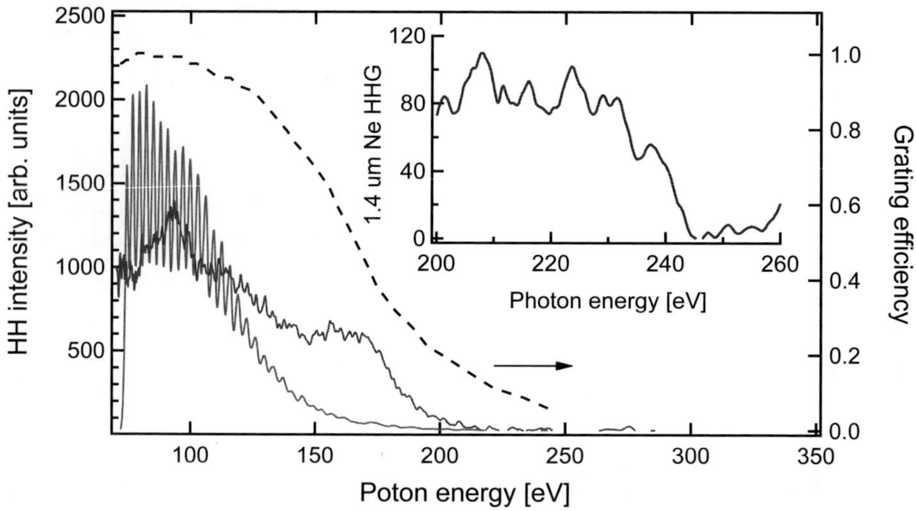

Fig. 2 Measured harmonic spectrum taken with a 1200 grooves/mm grating. Red and blue profile depict the harmonic from He with $\lambda_0 = 0.8$ μm pump and Ne with $\lambda_0 = 1.4$ μm pump, respectively. Inset shows the harmonic from Ne with $\lambda_0 = 1.4$ μm pump by using a 2400 grooves/mm grating.

First of all, we have performed harmonic experiment driven by 0.8 μm and 1.4 μm, in order to highlight the peculiarities of driving wavelength dependence of harmonic photon energy. The laser focused intensity is adjusted to generate HH from the neutral condition ($\eta < 1$ %) for the both two gases. Red profile in Fig.2 shows the measured He harmonic distribution driven by 0.8 μm with 35 fs pulse duration, which is taken with a 1200 grooves/mm grating. Pump energy, an estimated laser intensity, and a backing pressure of gas jet are 10 mJ, 6 x 10^{14} W/cm^2, and 10 atm, respectively. The well resolved HH spectrum from He driven by 0.8 μm shows dramatically dropping HH yield up to 100 eV. To increase the maximal photon energy of He HHG with 0.8 μm, we adjusted the focusing point and the backing pressure of gas jet. However, as far as the target gas condition is neutral, harmonic cut-off energy can not be extended more than the calculated prediction value. Blue profile in Fig. 2 shows the measured spectrum of HHG in Ne driven by 1.4 μm with a bucking pressure of 10 atm. The pump energy of 1.4 μm pulse was set to be 2-mJ with an attenuated pump pulse having a beam diameter of 10 mm. A focusing intensity is estimated to be 3.5 x 10^{14} W/cm^2 at the interaction region. It is obvious that the maximal Ne HH photon energy driven by 1.4 μm extends compared with the case of 0.8 μm driving field with He. Following the grating efficiency of the spectrometer, the intensity of Ne HH spectrum gradually decreases as photon energy increases. To measure the higher photon energy region, we replaced the grating from 1200 grooves/mm to 2400 grooves/mm. Inset shows the Ne HH spectrum driven by 1.4 μm in the higher photon energy region. The maximal harmonic photon energy is measured to be 240 eV with a linear plot of vertical axis. Unfortunately, the HH photon energy driven by 1.4 μm can't reach the water window region.

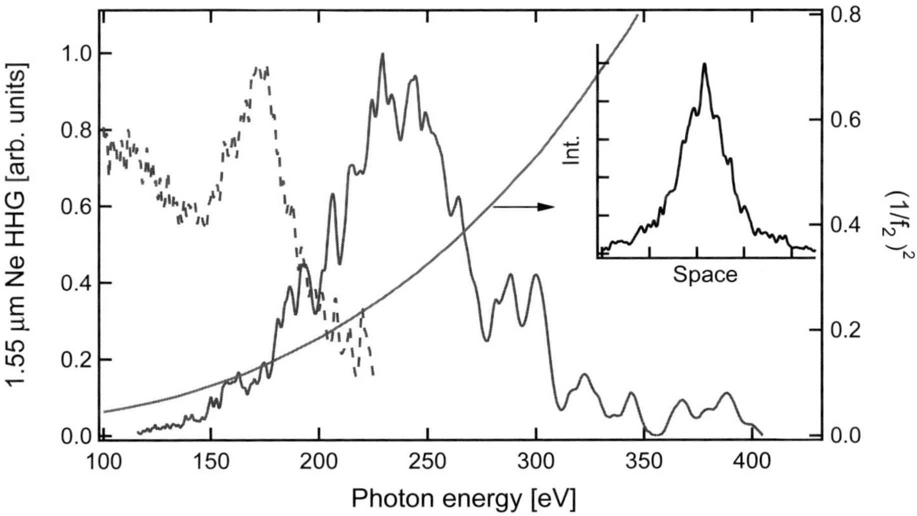

Fig. 3 Measured harmonic spectra from neutral Ne and evolution of the square of the reciprocal of imaginary component f_2. The inset shows the far-field spatial profile of the water window x-ray.

Our developed high-energy IR laser system adopted a collinear OPA scheme in order to amplify efficiently OPA signals. Since the wavelength of 1.6 μm corresponds to a degeneration point, it is difficult to spatially separate a signal and idler pulse. Therefore, we use 1.55 μm for driving field in this experiment. Figure 3 shows the measured Ne HH spectra driven by a 1.55 μm laser pulse at a focusing intensity of 3.5 x 10^{14} W/cm^2. The pump energy was set at 2.2 mJ. The measured harmonic spectrum gradually increases up to 250 eV. To evaluate the intensity distribution of the lower harmonics, we changed the spectrometer's grating to 1200 grooves/mm. The HH intensity (dotted blue profile) with the 1200 grooves/mm grating exhibits a broad plateau up to 150 eV. In contrast, the HH intensity rapidly increases above 150 eV, which agrees with the spectrum obtained from the grating with 2400 mm/grooves, then it starts to decrease at 170 eV because of the low diffraction efficiency of the 1200 grooves/mm grating above 170 eV. The HH yield is optimized by adjusting the focusing point and by varying the backing pressure of the gas jet. The HH intensity above the carbon K-edge is found to be maximum at a backing pressure of approximately 20 atm with a Gaussian spatial profile. The observed pressure dependence and spatial profile of the HH output indicate that macroscopic PM is satisfied at 20 atm. The red line shows the evolution of $(1/f_2)^2$ as a function of HH photon energy. $(1/f_2)^2$ indicates the effect of medium's absorption, where f_2 is the imaginary part of the atomic scattering factor. The measured HH spectrum is in sharp contrast to the conventional plateau of HH and clearly shows that the ALC is achieved by PM. The inset of Fig.3 shows the measured far-field spatial profile of the water window x-ray (280 ~ 310 eV). The beam divergence was measured and had a 7 mrad (FWHM). This good beam quality also indicates that the PM is substantially satisfied along the propagation axis.

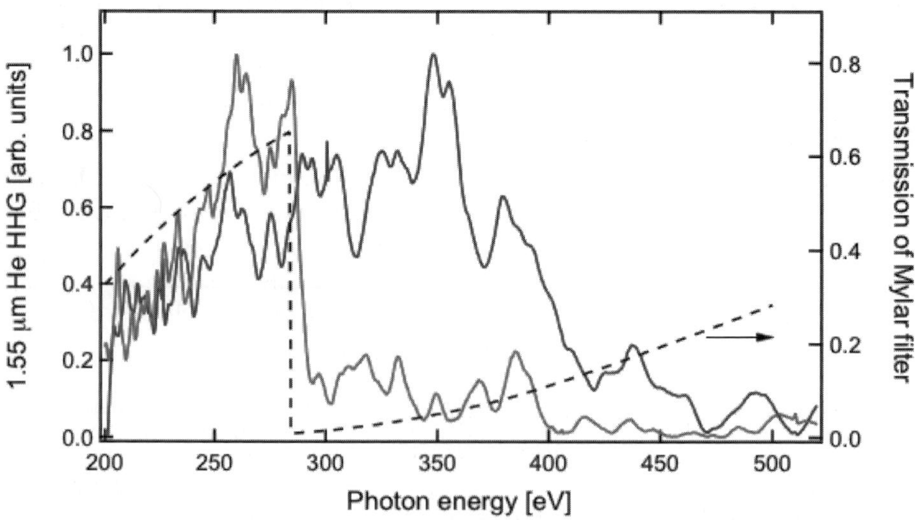

Fig. 4 Measured harmonic spectra from neutral He. The red and blue lines respectively correspond to the spectra with and without a 1-μm-thick Mylar filter ($C_{10}H_8O_4$).

We further explored the generation of HH under a neutral-medium condition by changing the nonlinear medium from Ne to He with the aim of obtaining a higher photon energy. Figure 4 shows the measured He HH spectra driven by a 1.55 μm pulse with a focusing intensity of 5.5×10^{14} W/cm^2, which is obtained with a 2400 grooves/mm grating. The pump energy, beam diameter and backing pressure are 4.5 mJ, 15 mm and 40 atm, respectively. We can clearly see the carbon K-edge on the spectral image by inserting a Mylar filter. Although the HH intensity above the carbon K-edge gradually increases with the He gas pressure, we cannot determine the optimum gas pressure because of the limitation of the backing pressure of the gas jet. Here, we emphasize that the yield of He HH at the carbon K-edge is two order of magnitudes lower than that of Ne HH. The maximum HH photon energy attained is 450 eV, which is the highest photon energy ever reported under a neutral-medium condition.

Taking our previous experimental results into account[2], we have evaluated the HH output yield, In a previous experiment, the CE of the Ne harmonic yield at 100 eV driven by a 0.8 μm laser was measured to be 2×10^{-7} [2]. When we apply the scaling formula $\lambda^{-5}{}_0 \times(1/ f_2)^2$, where the first term indicates the driving wavelength dependence of the HH yield [8] and the second term is the effect of the medium's absorption, we estimate the CE to be 6×10^{-8} at 250 eV with a driving laser wavelength of 1.55 μm. Since the evaluated photon yield for a Ne HH yield of approximately 2.5×10^6 photons/shot gives a CE of 5×10^{-8} at 250 eV in the present experiment, the HH output yield for neutral Ne can be accurately predicted by the above scaling formula. To obtain a sufficient harmonic yield around the water window region, the pump pulse wavelength should be extended to 1.65 μm because the maximum intensity of the Ne HH shifts to 300 eV. Consequently, a CE of 6.5×10^{-8} is estimated for a 300 eV HH driven with a 1.65 μm laser using the scaling formula. Using the estimated CE, we can obtain a single-shot fluence of 50 mJ/cm^2 over a 5 x 5 μm^2 area by upgrading the IR laser system to a pulse energy of 200 mJ and extending the focusing length to 5 m. This x-ray fluence is sufficient for obtaining a single-shot cell image by contact microscopy [9] with sub-100 nm spatial and sub-10 fs temporal resolutions. However, if we consider the HH loss due to focusing optics, we should further increase the output yield. The use of driving laser wave-front correction with a deformable mirror can improve the output yield by a factor of 2 – 3[10]. Although our estimation is based on a single harmonic order, a bundle of several harmonic orders can be used for microscopy, which greatly increases the effective output fluence.

References

[1] C. Spielmann *et al.*, *Science* 278, 661 (1997) ; Z. Chang *et al.*, *Phys. Rev. Lett.* 79, 2967 (1997)

[2] E. Takahashi *et al.*, Phys. Rev. A 66, 021802 (2002) : E. J. Takahashi *et al, IEEE J. Selct. Top. Quantum Electron.* 10, 1315 (2004)

[3] In this manuscript, a neutral condition means the effects of ionization is neglect during the duration of laser pulse.

[4] A. Paul *et al.*, *Nature (London)* 421, 51 (2003) ; J. Seres *et al.*, *Nature Phys.* 3, 878 (2007)

[5] E. J. Takahashi *et al.*, *Phys. Rev. Lett.* 101 253901 (2008)

[6] E. J. Takahashi *et al.*, *Appl. Phys. Lett.* 93, 041111 (2008)

[7] N. Nakano *et al.*, Appl. Opt. 23, 2386 (1984)

[8] J. Tate *et al.*, *Phys. Rev. Lett.* 98, 013901 (2007)

[9] T. Tomie *et al.*, Science 252, 691 (1991)

[10] E. J. Takahashi *et al.*, Appl. Phys. Lett. 84, 4 (2004)

Relativistically Oscillating Mirrors – an Ultrabright Attosecond Source

M. Zepf[1], B .Dromey[1], M. Geissler[1], R. Hörlein[2], Y. Nomura[2], G.D. Tsakiris[2], S. Rykovanov[2]

[1] Department of Physics and Astronomy, Queen's University, Belfast, UK
[2] Max-Planck-Institut for Quantumoptics, Garching, Germany

Abstract. The interaction of relativistically intense ($I\lambda^2$>>1.3 10^{18}Wcm^{-2}μm^2) laser pulses with a near step-like plasma density profile results in relativistic oscillations of the reflection point. This process results in efficient conversion of the incident laser to high harmonic spectrum. Recent experimental results show that the beam quality is near diffraction limited and consequently that very high focused intensities can be achieved opening up the possibility of ultra-intense X-ray interactions for the first time.

1 Introduction

High quality X-ray sources are a topic of diverse and intense international efforts ranging from large new synchrotron facilities such as the Diamond Light Source in the UK over Free Electron Lasers (FEL) being constructed in at SLAC in the US and DESY in Europe. In addition to these large facilities based sources, intense efforts are being made to use lasers as a route to intense XUV and X-ray sources. Laser based sources span X-ray lasers which have produced very impressive results, particularly in terms of pulse energy (see other articles in this volume and references therein), table-top FEL concepts[1] and sources that rely on the conversion of the incident laser light into very high order harmonics. The particular attraction of harmonic sources is that they can maintain the full spatial coherence of the laser and have pulse durations that can extend to the attosecond regime – which is typically achieved by letting a moderately intense femtosecond laser pulse interact with a gaseous target[2].

A far more recent development is the detailed theoretical and experimental investigation of harmonics emitted when a relativistically intense laser pulse creates a relativistically oscillating plasma surface (ROM)[3],[4],[5],[6],[7]. Unlike harmonics from gaseous targets, this process requires relativistically intense laser pulses, ideally significantly above the relativistic threshold of 1.3 10^{18}Wcm^{-2}μm^2. At such intensities the conversion process becomes efficient with the harmonic spectrum displaying a slow decay to higher orders[6] with a slope of $E_n/E_L \sim n^{2.6}$ (where E_n, E_L are n-th harmonic and laser energy respectively) and can extend to keV photon energies. Tantalizingly, ROM also offers the prospect of attosecond and zeptosecond duration pulses[4,5] and ultra-high intensity[3] X-ray pulses. The latter requires excellent, near-diffraction limited beam quality. Early results using ps lasers and a contrast ratio that led to an initial density gradient of $L>\lambda$ showed that the emission was essentially isotropic[8] – suggesting limited source brightness. This result was interpreted as resulting from critical surface turbulence due to hole-boring[9]. Consequently, for sharper density ramps and shorter pulses one might anticipate a marked improvement. Here we present our latest results showing that ROM can indeed be near diffraction limited in the case of high contrast, femtosecond pulses.

2 Background

ROM[10] essentially results from an oscillatory extension to Einstein's prediction for the frequency up-shift of light reflected off a perfect mirror moving at relativistic velocities – the relativistic Doppler effect[11]. In this theory a pulse of duration Δt at frequency ω_0 is shifted to a frequency of $\omega' = 4\gamma^2\omega_0$ (with the Lorentz factor $\gamma >> 1$). Since the number of cycles in an electromagnetic pulse is a Lorentz-invariant the pulse is also compressed by a commensurate amount and the resultant pulse duration is $\Delta t' = \Delta t/4\gamma^2$ [11]. From this one can see that in principle it should be possible to achieve substantial frequency upshifts and extremely short pulses by reflecting a high power laser pulse of a mirror with suitable properties. The key difficulty is, of course, to create a reflective structure moving at a velocity close to the speed of light.

One approach to achieving such a mirror is to illuminate an initially solid target with an ultra-short, intense laser pulse to create a near discontinuous plasma-vacuum boundary. The electric field of the laser can efficiently couple to the plasma surface, causing the electrons to oscillate in phase with the laser, thus constituting a relativistic mirror oscillating at the laser frequency ω_0. Since the plasma density is higher than the critical density for such a scenario, the plasma acts as an efficient reflector of the incident laser light. In a seminal paper by Paul Gibbon[10] it was shown that in such circumstances the incident laser-light is upshifted very efficiently to higher frequencies. The underlying process in the case of a relativistically oscillating mirror is in many ways similar to the process of the relativistic Doppler upshift described by Einstein[11]. The main difference is that instead of a constant value of γ describing the motion of the mirror surface, one now has Lorentz-factor that is a function of time $\gamma(t)$.

As the position of this mirror surface is a temporal function of the incident optical laser cycle, the phase of the reflected light wave is modulated such that it is no longer sinusoidal. As can be understood from Fourier theory, such a waveform must contain many high order harmonics of the fundamental frequency.

The most recent theoretical development in the field by Gordienko et al. and Baeva et al.[3,12], elegantly based on similarity theory, identifies the sharp spikes in the temporal variation of the Lorenz factor γ as the key to the production of the highest harmonics. That this should be the case can be easily understood when one considers the temporal variation of $\gamma(t)$. Even assuming a very smooth variation of the actual surface velocity with time (e.g. $v(t) \sim \sin(\omega t)$) the corresponding variation of $\gamma(t)$ is sharply spiked. From Einstein's theory of relativistic Doppler upshift one would therefore expect the upshifting process to be restricted to a timescale of the order of the temporal width of each 'γ– spike' – substantially shorter than an optical half cycle – and the maximum upshift to take place when the Lorentz factor reaches its maximum γ_{max}. The physical origin of this substantially larger frequency upshift also derives directly from the γ–spikes. Since the emission of high harmonic orders only takes place for large values of γ a sharp temporal localization of the emitted harmonics results. The temporal duration of the γ–spikes reduces for increasing intensity as $T_{Spike} \sim T_0/\gamma_{max}$ (with $T_0 = 2\pi/\omega_0$)[12]. The pulses of duration T_{spike} are upshifted and compressed by the factor of $4\gamma_{max}^2$ – familiar from the relativistic mirror. As a result the harmonics are emitted in short temporal bursts with $T_{burst} \sim T_{Spike}/\gamma_{max}^2 \sim T_0/\gamma_{max}^3$ and hence, from Fourier Theory, must contain significant spectral components up to frequencies of $O \sim \omega_0 \gamma_{max}^3$. In effect, the surprising new result of the theory of relativistic spikes is that the high energy cut-off and the ultimate slope of the spectrum is governed by the temporal compression and truncation of the electromagnetic pulse rather than the maximum upshift expected from a relativistic mirror moving at constant γ.

3 Beam Quality Considerations

The ultimate limits of focused intensity of ROM as a source of coherent X-rays is critically affected by the beam quality (transverse coherence), which ideally should be as close as possible to diffraction-limited divergence. Theoretical investigation suggests that the harmonic radiation can display excellent (near diffraction-limited) performance[3,13]. The somewhat surprising consequence of this is that the focused intensity of the emitted beam can, in principle, *exceed* the intensity of the driving laser and hence could allow electromagnetic radiation to be focused to the extreme intensities. Such a startling prediction merits experimental investigation and consideration of any deleterious effects that might prevent the harmonics from reaching their ultimate focused intensity.

Under perfect conditions (spatially constant intensity, plane wave conversion) the angular divergence of the nth harmonic (θ_n) should therefore simply be the diffraction limited divergence for the given wavelength, or, expressed in terms of the harmonic order n,

$$\theta_n \sim \frac{\theta_{Laser}}{n} \tag{1}$$

where θ_{Laser} is the divergence of the incident laser beam. In practice θ_{Laser} would be a fairly large angle (~20°), set by the focusing optics used to achieve high intensities.

Of course, in a real experimental situation both the laser quality and surface shape may deviate from this idealised case. There are four distinct areas which can provide an obstacle to reaching the desired beam quality and high focused X-ray intensities, which we will discuss briefly in the following paragraphs.

1) *Roughness* (surface shape variations on scale $<< \lambda_{Laser}$)

Typically, roughness of the target surface will primarily lead to an increased scattering of reflected radiation into wide angles. For surface roughness, ϕ_{rms}, significantly smaller than the wavelength of the reflected light, the angular distribution of the reflected light remains unaffected and all the energy is reflected into the specular diffraction limited angle. As is well know increasing surface roughness leads to a diffuse reflection component or halo. As ϕ_{rms} approaches the wavelength of the reflected light, the distribution becomes increasingly isotropic. This is of particular concern for the highest harmonics with wavelengths in the nm – Å range, where even the surface roughness of polished targets may be too high to allow the harmonics to remain highly collimated. Furthermore the growth of ripples in the critical density surface has been previously reported in the literature[9] and found to affect the angular distribution substantially for ps duration pulses[8].

2) *Surface curvature and aberrations* (variations on scale $\geq \lambda_{Laser}$):

As in all optical systems, the beam quality and divergence is controlled by the shape of the optical surfaces involved in beam transport. However, for the ROM the interaction is complicated by the fact that the driver and mirror surface are fundamentally linked. Even for an initially flat oscillating plasma, the intense laser pressure, $P \sim I/c$ ($>10^9$ bar for our parameters) results in the reflecting surface being initially pushed inwards due to the ponderomotive pressure[9]. The impact on the reflected harmonic radiation is that it is effectively reflected from a curved mirror and so will tend to exhibit a curved wavefront, passing through focus at some distance after the target. Since wavefront curvature of all harmonics is identical in this case all orders should display a constant divergence (unless diffraction is significant). Clearly, if the surface has the appropriate shape, the increased divergence does not necessarily lead to a reduced focusability of the harmonic beam and may lead to a very high intensity in the primary harmonic focus close to the target.

3) *Intrinsic Phase Effects*

However, even assuming ideal control of laser beam quality, a perfect initial surface and no movement of the ion background in response to the laser pressure (e.g. for few cycle pulses), an der Brügge et al.[14] report that the harmonic radiation is affected by a target density and (in the high intensity limit) a laser intensity dependent phase term. They derive an analytical expression for the intrinsic harmonic phase, given as

$$\varphi(a_0,N)=2.7\left(\frac{N}{a_0 N_{crit}}\right)^{-1}-0.32 \tag{2}$$

where N is peak density and $N_{\rm crit}$ the critical density for ω_{Laser}. This implies that for a Gaussian focal intensity distribution the harmonic radiation will focus at a distance

$$f_{SF}=\left(\frac{N}{2.7a_0 N_{crit}}\right)Z_R \tag{3}$$

before diverging again (where Z_R is the Rayleigh range of the driving laser).

The physical origin of this intrinsic phase can be understood in a relatively straightforward way. As with any oscillator, the relative phase of the oscillation ($\phi_{relative}$) and driving force is a function, F, of the ratio of the driving frequency (ω_{Laser}) to the resonance frequency (ω_{res}), such that $\phi_{relative}$=F($\omega_{Laser}/\omega_{res}$) and is therefore also a function of N/N_{crit}. More generally, for a strongly driven oscillator in an anharmonic potential this oscillation may become a function of the oscillation amplitude and hence the strength of the laser field. The dominant intensity dependent effect here is the cycle averaged offset of electron surface (from which the laser is reflected) with regards to the ion background. This effect is mediated by ponderomotive pressure[9] which pushes the plasma surface inwards and ultimately, in the case of a many cycle pulse, results in the motion of the ion background discussed above.

4) Laser phase-errors:

Since the phase of the surface motion at any point of the plasma surface is locked to the local phase of the driving laser field[14], laser phase variations will directly affect the phase front of the emitted harmonic radiation. Even small fluctuations in the local laser phase $\Delta\phi_{Laser}$ from an ideal flat (or spherical) wavefront will lead to substantial aberrations in the phase of the emitted harmonic radiation. The phase error for the nth harmonic preserves the absolute wavefront error and hence the relative phase error $\Delta\phi_n$, measured in terms of the harmonic wavelength λ_n, is $\Delta\phi_n = n(\Delta\phi_{Laser})$. This implies that the generation of high quality keV beams ($n>1000$) requires excellent quality of the high intensity laser driving the interaction. In the focus of near diffraction-limited laser beams, such as in our experiment, this may not present a significant limitation, as the harmonics are typically only generated over a fraction of the central diffraction limited peak[15] and hence in a region where the phase is highly flat. Consequently we will not be considering this effect further in this article.

4 Experimental Results

To investigate the dependence of angular distribution on laser parameters and surface roughness, on the Astra laser ($\sim 2 \times 10^{19}$ Wcm^{-2}, $\lambda_{Laser} \sim 800$nm, 50fs). For the Astra experiment, a plasma mirror[31] was inserted into the laser beam-line to increase the pulse contrast from $>10^7$:1 to $>10^9$:1 at <500fs.

Our experiment provides the first demonstration of diffraction limited harmonic emission from relativistically oscillating plasma surfaces generated during intense interactions on fused silica targets. Figure 1 shows the divergence of the harmonics in the range from 20 to 50 nm wavelength. As can be seen the harmonics from 40 to 35 nm wavelength are near diffraction limited in divergence. Note that the ROM harmonics display constant divergence. As explained in the considerations above, this is consistent with emission from a curved surface – with the divergence being set by surface curvature. Detailed comparison with PIC simulations show that the observed beam divergence is in very good agreement with that expected from the surface curvature due to hole-boring. As expected, the short pulse duration has suppressed the growth of large scale turbulence which caused the large divergence in the earlier experiments.

In order to investigate the effect of surface roughness on the emitted radiation, targets with surface roughness ranging from a few Å to 200nm were investigated. The striking observation is that the harmonics are insensitive to surface roughness on the scale of the *harmonic wavelength* – contrary to the commonly accepted principle that a surface that displays roughness on the order of the wavelength will lead to diffuse reflection.

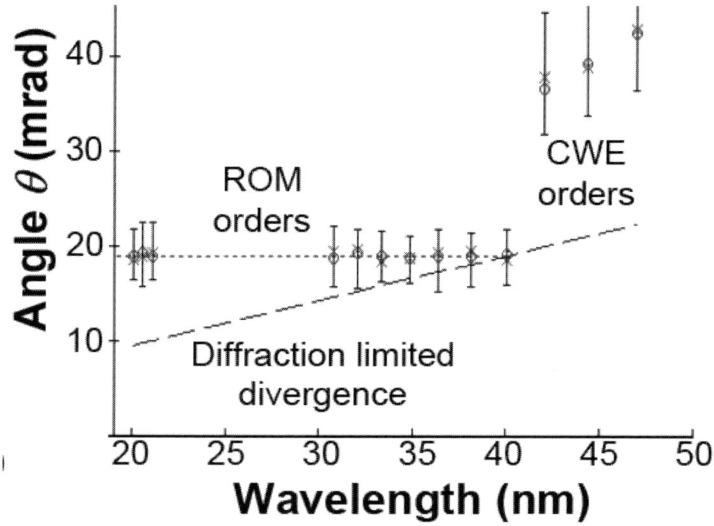

Fig. 1 Divergence of harmonics radiation. The dashed line shows the diffraction limited divergence for the given wavelength and source size. The ROM harmonics between 35-40nm can be seen to be diffraction limited with constant divergence to shorter wavelengths – consistent with a critical surface curved due to hole-boring. At longer wavelength orders due to a separate physical process (CWE emission[16]) with larger divergence are also visible.

To understand this highly surprising result PIC simulations[17] were performed for various values of surface roughness. In this study electron trajectories are observed to extend many times the original surface roughness for $\phi_{rms} << \lambda_{Laser}$ both transversely and longitudinally, which averages over the small-scale roughness and smoothes it considerably. This interpretation gains credence from the experimental observation that only roughness on the scale of the excursion amplitude substantially affects the harmonic generation process. When the surface roughness was increased to be comparable with λ_{Laser} (ϕ_{rms}=164nm) no ROM orders were observed.

We have calculated that for our experimental parameters the intrinsic phase effects discussed above would result in a divergence (from Eqn. 2) of ~3 mrad and are therefore only a small contribution to the overall observed divergence. For our experiments, this small correction on the observed divergence lies within the uncertainty of the measurements. The results presented here suggest that exceedingly good focusability and very low divergence can indeed be produced using ultra-short pulses to minimise the surface denting.

Conclusions

In conclusion harmonics emitted from relativistically oscillating plasma surfaces have bee shown to have near diffraction-limited beam quality – under appropriate experimental conditions of high contrast and fs pulse duration. This demonstrates that very high focused intensities are indeed possible opening up the realm of ultraintense X-ray interactions.

References

1 Grüner F et al., *Applied Physics B*, **86**, 431-435 (2007)
2 P. Corkum, F. Krausz, *Nature Physics*, **3**, 381 (2007).
3 Gordienko S., et al., *Phys. Rev. Lett.*, **94**, 103903 (2005)
4 Gordienko S. et al., *Phys. Rev. Lett.*, **93**, 115002 (2004)
5 Tsakiris GD et al., *New Journal of Physics,* **8**, 19 (2006)
6 Dromey B. et al., *Nature Phys*. **2,** 456 (2006)
7 Dromey B. et al., , *Phys. Rev. Lett.* **99**, 085001 (2007)
8 Norreys P. et al., *Phys. Rev. Lett.,* **76**, 1832-1835, (1996)
9 Wilks S. C. et al., *Phys. Rev. Lett.*, **69**, 1383 (1992)
10 P. Gibbon, *Phys. Rev. Lett.,* **76**, 50 (1996); S.V. Bulanov, et al., *Phys. Plasmas*, **1**, 745 (1994); R. Lichters, et al., *Phys. Plasmas*, **3**, 3425 (1996)
11 A. Einstein, *Ann. Phys. (Leipzig)*, 17, 891 (1905)
12 T. Baeva, S. Gordienko, A. Pukhov, *Phys. Rev. E*, **74**, 046404 (2006)
13 Geissler M. et al, *New Journal of Physics* **9,** 218 (2007)
14 An der Brugge D., Pukhov A., *Phys. Plasmas* **14**, 093104 (2007)
15 R. Lichters and J. Meyer-ter-Vehn, Multiphoton Processes 1996, Institute of physics Conf. Series 154, p221 (1997)
16 Quere F. *et al.*, *Phys. Rev. Lett.,* **96,** 125004 (2006)
17 S. Rykovanov et al. in preparation.

Spectral Characteristics of Strong High-Harmonics Generated in a Two-Color Laser Field

C. H. Nam[1], I J. Kim[1,2], G. H. Lee[1], S. B. Park[1], T. K. Kim[1], and C. M. Kim[2]

[1] Dept. of physics and Coherent X-ray Research Center, KAIST, Daejeon 305-701, Korea
[2] Femtoscience Lab, Advanced Photonics Research Institute, GIST, Gwangju 500-712, Korea

Abstract. High harmonics generated from gaseous atoms driven by a two-color laser field possess attractive properties of remarkable efficiency and unique spectral structure. With a two-color laser field synthesized by mixing fundamental and second-harmonic fields high harmonic generation processes can be manipulated. The spectral characteristics of high harmonics from He and Ne were investigated by simultaneously controlling the relative phase and polarization angle of the two-frequency field. The harmonic generation was optimized by obtaining guided laser propagation in a long gas jet. The strongest orders were the 38th (21.6 nm) with He and the 30th (27.3 nm) with Ne. In the case of He absorption-limited generation of high harmonics at 21 nm was achieved.

1 Introduction

High-order harmonics (HOH) generated from highly nonlinear interaction between intense femtosecond laser and atoms have unique properties of superb spatial coherence in the soft x-ray region and extremely short pulse duration on the attosecond time scale. These properties are quite beneficial to ultrafast spectroscopy and coherent XUV optics [1]. However, such applications generally require strong and efficient generation of HOH. For highly efficient generation of HOH, several methods have been applied. By adopting a long gas cell with long focusing optics [2-4] or a long gas jet in a guided propagation mode [5], the harmonic generation volume can be maximized, reaching the condition of absorption-limited harmonic generation. Recently, the use of a two-color laser field, consisting of fundamental and second harmonic (SH) fields, for high-order harmonic generation (HHG) has led to dramatic enhancement of harmonic generation efficiency [6]. As the two-color laser field contains much more control parameters, compared to a single laser field, an elaborate optimization of experimental parameters is required for strong and efficient HHG in the two-color laser field.

In this paper, we investigated the spectral characteristics of strong high harmonics from He and Ne by simultaneously controlling the relative phase and polarization between the fundamental and SH fields. Strong high harmonics with energy of ~μJ and conversion efficiency of $\sim 10^{-4}$ have been realized at low-order harmonics in the 50 nm or longer wavelength region [2, 7]. Previous studies also showed that as the harmonic order increases, the conversion efficiency rapidly decreases, e.g. 2×10^{-5} at 30 nm [3] and 5×10^{-7} at 13.5 nm [4]. Our earlier results on HHG in a two-color laser field, employing a target medium of a circular gas jet, well exceeded these values [6]. Efficient harmonic generation in a two-color laser field was realized in a simple setup. By simply inserting a SH crystal in the beam path of harmonic generation, exceptionally strong harmonics were generated. Strong harmonic generation was possible due to the formation of a quasi-linear field, the selection of short quantum path component which has denser electron wave packet, and high ionization rate, but not too high as in the case of a parallelly polarized two-color laser field [6,8]. In the recent work, we have significantly increased the harmonic conversion efficiency further by adopting a long gas jet medium and by improving the intensity of the SH field. The strongest orders were the 38th (21.6 nm) with He and the 30th (27.3 nm) with Ne. Among them, the achieved conversion efficiency over 10^{-4} at 21 nm from He is the highest value ever reported in the spectral region below 50 nm [9]. This study reveals the unique properties of high harmonics generated in the two-color field, which cannot be seen from usual one-color-field-driven high harmonics.

2 Experiment

The HHG in the two-color laser field can be optimized by controlling a range of experimental parameters of target and laser. In particular, the most sensitive parameters are laser intensities, relative phase, and polarization of the two fields as well as target density, length, and position. The experimental layout of the two-color HHG is shown in Fig. 1. Femtosecond laser pulses with an energy of 2.8 mJ and pulse duration of 30 fs, after being focused by a spherical mirror (f= 60 cm), were incident on a He or Ne gas jet. For SH generation, a type-I beta-barium borate (BBO) crystal was placed between the focusing mirror and the gas jet. With BBO crystals with thickness of 100 μm and 200 μm the SH conversion efficiencies of 20 % and 28 % were obtained, respectively. Gas jet with a circular nozzle of 0.5-mm diameter and with slit nozzles of 0.5-mm width and length L= 3, 6, and 9 mm were used for target optimization. Generated harmonics were detected using a flat-field extreme ultraviolet (XUV) spectrometer equipped with a back-illuminated x-ray charge coupled device (CCD; Princeton Instruments), and two aluminum filters of 1-μm thickness were installed to block scattered laser light and to prevent the saturation of CCD. The optimization of two-color HHG was performed by selecting the target length and position while controlling the relative phase and polarization between the two fields and increasing the SH conversion.

One critical parameter to determine the electric field configuration of the synthesized two-color laser field is the relative phase between the two fields. The relative phase greatly affects the electron trajectory in HHG and thus the harmonic yield. For the control of the relative phase between the fundamental and SH fields, a fused silica plate with a thickness of 150 μm was inserted in the beam path after the SH crystal, as shown in Fig.1.

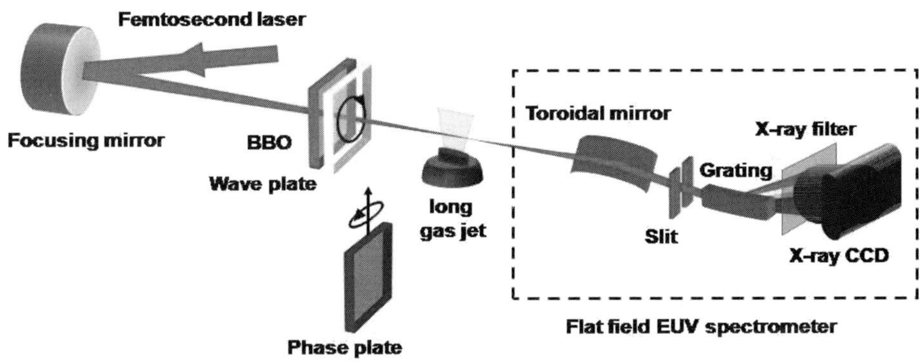

Fig. 1 Experimental setup for high-harmonic generation driven by a two-color laser field.

3 Analysis and Discussion

The relative phase of the two-color field is directly related to the electric field configuration in time, i.e. Lissajous diagram. It determines the nature of the quantum paths contributing to HHG. We observed the intensity modulation of harmonics as a function of relative phase using He and Ne gases. As shown in Fig. 2 we compared the strongest two harmonic orders such as 38th (21.6 nm) for He and 30th (27.3 nm) for Ne using 6 mm long gas jet and 200 μ m thick BBO crystals. The intensity modulation of this harmonic has π periodicity with respect to the change of relative phase irrespective of gas species. But the peak position of intensity modulation is different between two gases. These differences are due to the effect of phase shifting by material dispersion which is obtained by the incident laser pulse propagating the gas medium. The absolute value of relative phase is hard to determine, but the amount of the induced phase shift may be estimated from the material dispersion.

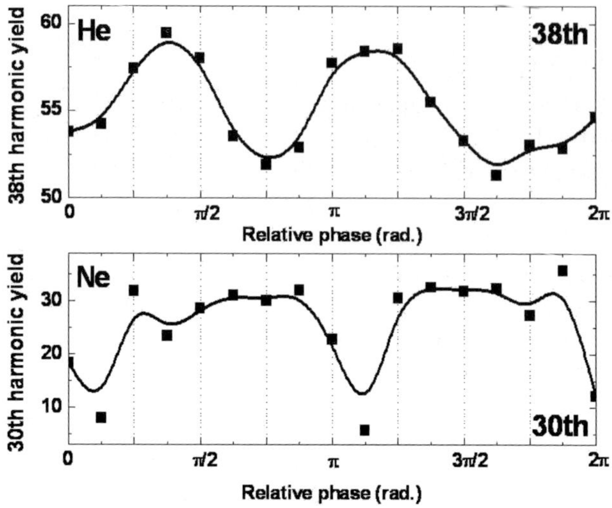

Fig. 2 Intensity modulation of 38th and 30th order harmonics from He and Ne gases obtained by varying the relative phase between the fundamental and SH field.

HHG in the two-color field depends also on the laser polarization angle between the fundamental and SH fields. We could decide the relative phase value which has the strongest harmonic intensity. From this condition, the HOH from orthogonal polarization was compared with HOH from parallel polarization. In orthogonal polarization, the harmonics at 2(2n+1)th orders are stronger than their neighboring harmonics while the harmonics at 2(2n)th and (2n+1)th of parallel polarization are stronger than that of orthogonal polarization. This inclination is similar to the previous result [6] even though the experimental condition such as the gas jet length is different.

In addition we continuously controlled the polarization angle between two-fields. The intensity variation of high-harmonics observed from He as a function of the polarization angle θ is shown in Fig. 3. From Fig.3 (a), we can find that there is spectral difference between 2(2n+1)th harmonics and the other harmonics such as 2(2n)th and (2n+1)th. The intensity of 2(2n+1)th harmonics is stronger in the vicinity of large polarization angle and weaker in the small polarization angle than those of 2(2n)th and (2n+1)th harmonics. The intensity modulation also appears with the control of the polarization. As shown in Fig. 3(b), 2(2n+1)th harmonics, such as 42nd harmonic, have clear spectral structure with 3 peaks during the change of polarization from orthogonal to parallel.

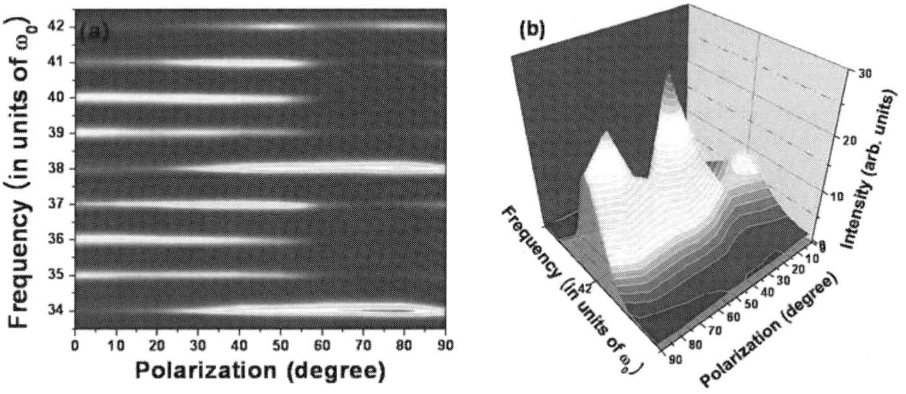

Fig. 3 Intensity modulation of HOH generated from He by controlling the polarization. (a) 2-dimensional graph from 34th to 42nd and (b) 3-dimensional graph for 42nd harmonic. These harmonic spectra were obtained from the relative phase value with the strongest harmonic yield.

By employing a long gas jet, much stronger HHG was achieved in the two-color laser field. First, the HHG was optimized for the target position. In the case of a single color HHG with a long gas jet the harmonic yield was found to be very sensitive to the target position as the plasma defocusing strongly affects the laser beam propagation even well within the Rayleigh length [5]. This was also true in the case of the two-color field. In Fig. 4 (a) the 38th harmonic yield is shown as a function of the target position for the gas jets with a 0.5-mm circular nozzle and a slit nozzle of 3, 6 and 9-mm length. The gas jet position of z = 0 corresponds to the position of the gas jet center at the laser beam focus, and the 'minus' sign in the target position means the gas jet placed before the laser focus. Here the BBO crystal of 200-μm thickness was used. The optimization at z=-10 mm is due to the guiding effect, i.e. a converging laser beam compensates for the plasma defocusing of the propagating laser field in the ionizing gas medium [5,10]. At the optimized target position, nearly uniform plasma column was observed along the entire length of the gas jet. The laser intensities, estimated from the cutoff orders, were 4×10^{14} W/cm^2 and 9×10^{15} W/cm^2 for the fundamental and SH, respectively.

The optimization of the medium length under the guided laser propagation was then pursued. The harmonic yield increases until the gas jet length of 6 mm, but it decreases for the 9-mm jet. Though an increase of the 38th harmonic at 21.6 nm is the largest one, other orders show similar trend. As the effective medium length becomes longer, the harmonic yield also increases, up to the point where the reabsorption of harmonics by the medium becomes significant. As the absorption length of 100-torr He is about 2 mm, the optimized medium length of 6 mm corresponds to three times the absorption length, which agrees with the calculation by E. Constant *et al.* [11] carried out for the optimum harmonic generation in an absorbing gas medium. Consequently, the harmonic yield of the long gas jet was obtained in the absorption-limited regime.

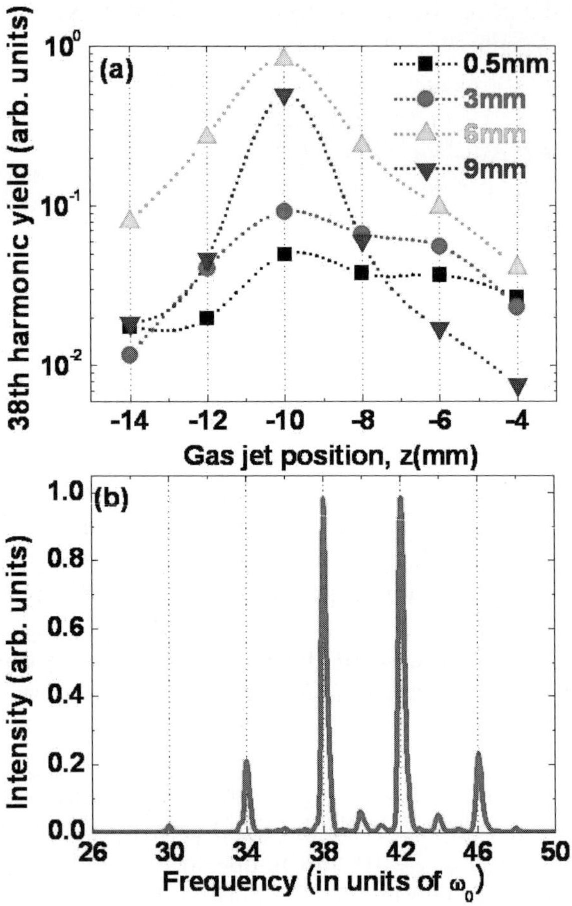

Fig. 4 (a) Optimization of the 38th-order harmonic energy with respect to gas jet for different gas jets. (b) Optimized harmonic spectrum obtained with the 6 mm gas jet.

Finally, by adjusting the medium length and position along with laser parameters such as the relative phase between the two fields and laser intensity, we have optimized HHG in the two-color laser field. The optimized harmonic spectrum in Fig. 4(b) shows that the harmonics at 2(2n+1)th orders are dominant, being much stronger than those of 2(2n)th and (2n+1)th orders. In particular, both 38th and 42nd harmonics, obtained with a single laser shot of 2.8 mJ, saturated the x-ray CCD even with two 1-μm-thick Al filters installed in front of the x-ray CCD. The 38th harmonic at 21.6 nm as well as the 42nd harmonic at 19.5 nm reached an energy of 0.6 μJ per shot, giving a conversion efficiency of 2×10^{-4}.

4 Conclusion

In conclusion, the spectral characteristics of strong high harmonics from He and Ne were investigated by controlling the relative phase and polarization angle of two-color laser fields. In particular, we achieved sub-μJ harmonics from He in the 21 nm spectral region, by optimizing the medium length and position as well as the relative phase and polarization between the two fields and SH conversion. With further increase of laser energy, the observed dramatic enhancement of HHG in the two-color laser scheme may be extended to even shorter wavelength region.

References

1. T. Sekikawa, A. Kosuge, T. Kanai, and S. Watanabe, Nature **432**, 605 (2004)
2. E. Takahashi, Y. Nabekawa, and K. Midorikawa, Opt. Lett. **27**, 1920 (2002)
3. E. Takahashi, Y. Nabekawa, T. Otsuka, M. Obara, and K. Midorikawa, Phys. Rev. A **66**, 021802 (2002)
4. E. Takahashi, Y. Nabekawa and K. Midorikawa, Appl. Phys. Lett. **84**, 4 (2004)
5. H. T. Kim, I J. Kim, D. G. Lee, K-H. Hong, Y. S. Lee, V. Tosa, and C. H. Nam, Phys. Rev. A **69**, 031805 (R) (2004)
6. I J. Kim, C. M. Kim, H. T. Kim, G. H. Lee, Y. S. Lee, J. Y. Park, D. J. Cho, and C. H. Nam, Phys. Rev. Lett. **94**, 243901 (2005)
7. J.-F. Hergott, M. Kovacev, H. Merdji, C. Hubert, Y. Mairesse, E. Jean, P. Breger, P. Agostini, B. Carré, and P. Salières, Phys. Rev. A **66**, 021801 (2002)
8/ C. M. Kim, I J. Kim and C. H. Nam, Phys. Rev. A **72**, 033817 (2005)
9. I J. Kim , G. H. Lee , S. B. Park , Y. S. Lee , T. K. Kim , C. H. Nam, T. Mocek and K. Jakubczak , Appl. Phys. Lett. **92**, 021125 (2008)
10. D. G. Lee, H. T. Kim, K. H. Hong, C.H. Nam, I. W. Choi, A. Bartnik, and H. Fiedorowicz, Appl. Phys. Lett. **81**, 3726 (2002)
11. E. Constant, D. Garzella, P. Breger, E. Mevel, C. Dorrer, C Le Blanc, F. Salin, and P. Agostini, Phys. Rev. Lett. **82,** 1668 (1999)

Diffraction Limited Harmonic Emission from Laser Produced Plasmas

B. Dromey[1]*, D. Adams[1], R. Hoerlein[2], Y. Nomura[2], D. Neely[3], G. Tsakiris[2], M. Zepf[1]

[1] Department of Physics and Astronomy, Queens University Belfast, BT7 1NN
[2] Max-Planck-Institut für Quantenoptik, Hans-Kopfermann-Straße 1,D-85748 Garching bei München, Germany
[3] Central Laser Facility, STFC Rutherford Appleton Laboratory, Didcot, OX11 0QX, UK
* b.dromey@qub.ac.uk

Abstract. High order harmonic generation (HOHG) from intense laser – solid density interactions has emerged as a promising route to the generation of attosecond pulses extending to keV energies. With efficiency of the nth harmonic, η_n, scaling as $\sim n^{-2.5}$-n^{-3} in the relativistic limit up to a maximum order $n_{max} \sim 8^{1/2}\gamma^3$ (where γ is the maximum relativistic Lorentz factor of the oscillating plasma surface), the potential for a bright solution to attosecond science is a distinct possibility. Some of the recent work performed in the field verifies for the first time that the exceptional coherence properties of the driving laser can be transferred directly to the high harmonic orders produced when plasma surfaces are driven to relativistic velocities under the action of intense laser pulses.

1 Introduction

The recent demonstration of attosecond phase locking[1] and diffraction limited performance of harmonics[2] generated during the interaction of intense lasers with solid density targets has benchmarked the unique source of high order harmonic generation (HOHG) from relativistic plasma surfaces as a next generation intense attosecond light source. Such a source can complement the range of ultrafast XUV science that is routinely performed at large scale free electron laser and synchrotron facilities worldwide.

Here we present the first demonstration of $n_{max} \propto \gamma^3$ cut off scaling[3] for ultra short pulses (<50fs), giving clear indication that bright soft x-ray pulses can be generated using even modestly sized ‘table-top’ generation lasers. The angular distribution of the HOHG signal is also investigated. For ultrashort pulses there is distinct behaviour observed between the recently described coherent wake emission[4] (CWE) process for harmonic generation (which cuts off at the plasma frequency) and relativistic oscillating mirror (ROM) harmonics[3], which extend to much higher orders[5,6]. Near diffraction limited performance for ROM HOHG with clear evidence of emission from a curved plasma surface is displayed. CWE is observed to be emitted into cone angles approximately twice that expected for diffraction limited performance. The relation of the angular distribution of the emitted harmonic radiation to target roughness (ϕ_{rms}) on a sub laser wavelength is investigated confirming that in the absence of significant laser prepulse there is considerable smoothing of the solid target surface.

2 Experiment

High order harmonic x-ray generation has the potential to open up the world of physical processes on an attosecond timescale[1-5]. The key to this is converting high-power optical laser pulses into broad, phase-locked harmonic spectra extending to multi-keV photon energies – which can be achieved, with unprecedented efficiency and brightness, by reflection off relativistically oscillating plasmas[2,3,5,6]. Of particular note is the implication this has for the production of high brightness attosecond pulses[3]. For a fixed fractional bandwidth at a given central frequency, $n_{cf}\omega_{Laser}$, for the attosecond pulse, the energy in the pulse scales as $\eta_{att}{\sim}n_{cf}^{-1.5}$ (Eqn. 1)[7], where n_{cf} is the harmonic order of the carrier frequency and ω_{Laser} the laser frequency.

When an intense laser pulse interacts with a near discontinuous plasma-vacuum boundary the electric field of the laser can efficiently couple to the plasma surface, causing the electrons to oscillate in phase[1-3], effectively constituting a relativistic mirror oscillating at the laser frequency ω_{laser}. As the position of this mirror surface is a temporal function of the incident optical laser cycle, the phase of the reflected light wave is modulated such that it is no longer sinusoidal and as can be understood from Fourier theory, contains many high order harmonics of the fundamental frequency.

The most recent theoretical development in the field, based on similarity theory [4], identifies the sharp spikes in the temporal variation of the Lorenz factor γ as the key to the production of the highest harmonics. From this theory of 'γ– spikes' the conversion efficiency in the relativistic limit for the nth harmonic is predicted to scale as $\eta(n){\sim}n^{-Prel}$ (Eqn. 1), with Prel = 8/3 [4]. Another important result of this theory is the prediction of the highest harmonic where Eqn. 1 still applies, up to an order $n_{max} \sim 8^{1/2}\gamma_{max}^{3}$, beyond which the conversion efficiency decreases more rapidly or 'rolls over' (where $\gamma_{max}=(1+3.6\times10^{-19}I\lambda^2)^{1/2}$ corresponds to the maximum surface velocity, I (Wcm^{-2}) is the peak intensity and λ (μm) the wavelength of the laser).

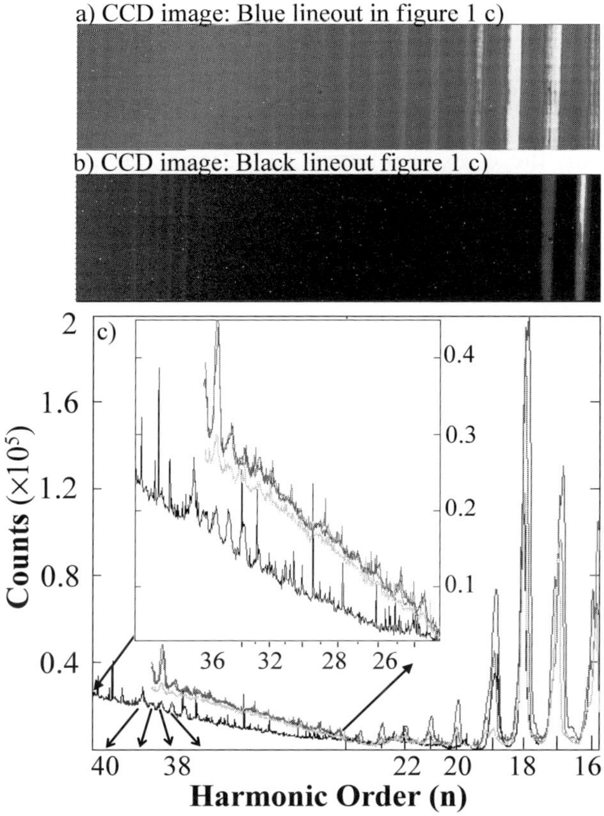

Fig. 1 Harmonic generation from sub nm – a), red trace, and b),blue trace – and 20nm – c), green trace, and d), black trace, – root mean square (rms) roughness targets. CWE harmonic radiation extending to ~19th order and ROM harmonics extending to ~40th order are routinely observed for intensities of $1.5\pm0.5\times10^{19}$Wcm^{-2}. Figures 2 e) and f) show the distinct nature of the two generating processes. For this data the harmonic radiation emitted along the specular axis was apertured using a narrow slit while the radiation ~2.5cm off the specular axis was collected using a glancing angle gold focusing optic and redirected onto the ccd detector. The signal is rising towards higher harmonic orders due to increased spectrometer response to shorter wavelengths.

Figure 1 shows the generation of harmonic orders using intensities (~10^{19}Wcm^{-2}) and orders (up to ~40th) using the Astra laser (~1J, 40fs) at the Central laser facility for sub nm, (a) and 20nm, (b), root mean square (rms) roughness targets. This is the first experimental observation of maximum harmonic order scaling to intensity dependent $n_{max}(\propto \gamma^3)$ predicted by Baeva et al[3] for ~50fs pulses. and indicates that a bright source of harmonic radiation from solid targets extending to soft x-ray wavelengths may indeed be possible using current generation table top lasers with modest intensity (~10^{19}Wcm^{-2}). For this experiment the contrast of the incident pulse was improved using plasma mirrors (either single or double as required)[8] to maintain a sharp plasma vacuum boundary suitable for efficient harmonic generation. The soft x-ray spectra were obtained using an XUV flatfield spectrometer and Andor CCD detector[2]. For these experimental conditions the plasma frequency corresponds to the ~19th harmonic order and consequently represents the cut off in the CWE harmonic spectrum[4].

3 Angular distribution

The angular distribution of the harmonics obtained from the results presented in Figure 1 indicate diffraction limited divergence of the ROM harmonics in the range 20th -40th (orders above the plasma frequency) for intensities of ~10^{19}Wcm^{-2} and 50fs pulse length.

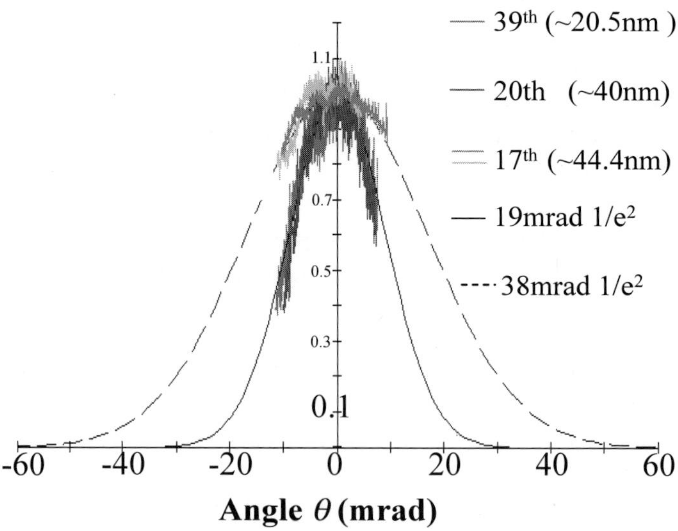

Fig. 2 Angular distribution of harmonics generated on targets with ϕ_{rms}<20 for intensities of 1.5±0.5×10^{19}Wcm^{-2} . Significantly the 39th order (red trace) is still beamed from a surface with $\lambda_{harmonic}$ roughness, and has identical angular distribution to that of the 20th harmonic. This is clear indication of emission from a curved surface. The CWE emission (17th order, green and pink traces) are beamed into approximately two times diffraction limited cones.

The angular divergence of the cut off orders is observed to be ~ $2 \times \alpha/n$, where n is harmonic order and α is laser cone angle. This is clear indication of emission from a dented plasma surface[2]. These spectra were only observed for targets with ϕ_{rms} on the order of 0-20nm. As can be seen from Figure 2, CWE harmonic radiation (<20th order) is emitted into much larger cone angles (~2-3 times that expected from a near diffraction limited source).

4 Surface smoothing

As highlighted in Figure 2 the significant result is the observation of beamed harmonic emission all the way to $\lambda_{\text{harmonic}}$ (the harmonic wavelength) at the cut off order for relativistic oscillating mirrors which is on the order of ϕ_{rms} i.e. 20nm radiation reflected from a surface with a root mean square roughness of 20nm. From Fourier theory it is clear that under normal conditions such reflection should be scattered into large angles. The observation presented here is clear evidence of extreme surface smoothing due to large scale transverse electron motion in the intense laser field.

To investigate this phenomenon we conduct numerical studies using the 1D3P PIC code PICWIG[9]. It permits, in planar geometry, simulation of the interaction of intense laser pulses with a pre-ionized non-collisional plasma. For these simulations the typical plasma density, N_e, was chosen to be $400Nc$ (and parameters corresponding approximately to the Astra experiment).

With respect to the surface smoothing effect mentioned above, the numerical integration of transverse momenta allowed the retrieval of the 3D trajectories of the particles. Figure 3 shows that the electrons undergo substantial excursions both along the target normal (longitudinal trajectories, x) – which leads to the harmonic generation in the first place – and also along the target surface (transverse trajectories, y). The excursion amplitude in both dimensions approaches an appreciable fraction of the laser wavelength in scale for relativistic intensities. This is substantially larger than the spatial and depth scale of the initial surface roughness.

Experimentally, for targets with ϕ_{rms}~150nm harmonic beaming was observed to be completely destroyed. This is direct evidence that as the surface roughness approaches an appreciable fraction of the driving laser wavelength surface smoothing cannot be achieved.

5 Conclusion

In conclusion it is clear that the exceptional spatial coherence properties of the laser can be transferred directly to the XUV via relativistically oscillating plasma interactions. Excitingly this is revealed to be permitted via extreme surface smoothing due to large scale transverse electron motion in the intense laser field. Uniform angular divergence of ROM harmonics indicates emission from a curved plasma surface. In summary this work improves the outlook for intense ultrafast XUV science possible using the relativistically oscillating plasma medium.

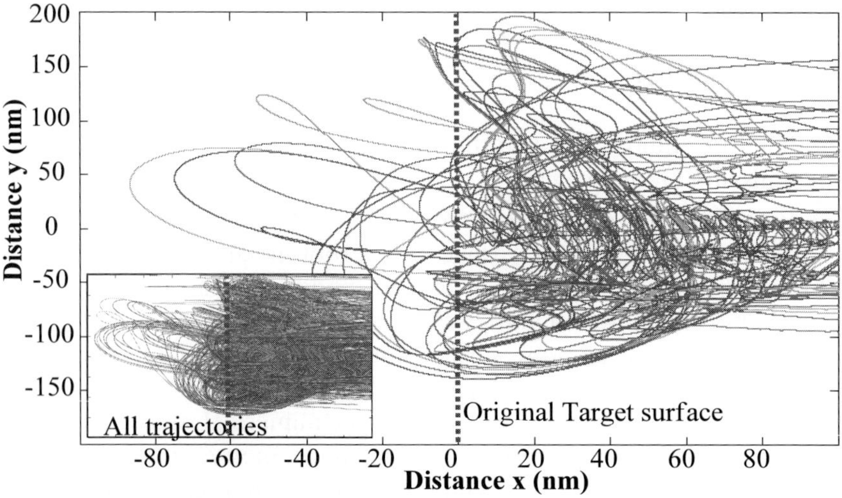

Fig. 3 The typical spatial extent of electron trajectories with respect to the original target surface, as obtained from PIC simulation, for Astra laser parameters – oblique incidence(30°) and a0=3. Other simulation parameters were chosen to reduce complexity, such as a two cycle pulse, but the principle remains the same. The inset shows all of the trajectories included in the simulation while the main figure shows every 20th trajectory plotted for clarity. It is clear that electrons traverse paths in both the transverse (y) and longitudinal (x) direction with respect to the driving laser that are far greater than the surface roughness that could affect harmonics generated in the Astra experiment.

References

1. Nomura, Y., *et al., Nature Phys.*, Advanced online publication 14 December 2008; doi:10.1038/nphys1155
2. Dromey, B., *et al.*, *Nature Phys.*, Advanced online publication 11 January 2009; doi:10.1038/nphys1158
3. Baeva, T., Gordienko, S., Pukhov, A., Phys. Rev. E, **74**, 046404 (2006), Pukhov, A., *Nature Phys*, **2,** 439 (2006), Gordienko, S., *et al.*, *Phys. Rev. Lett.*, **93**, 115002 (2004)
4. Quere, F., *et al.*, *Phys. Rev. Lett.* **96,** 125004 (2006)
5. Dromey, B., *et al.*, *Nature Phys*. **2,** 456 (2006)
6. Dromey, B., *et al.*, *Phys. Rev. Lett.* **99**, 085001 (2007)
7. Tsakiris, G.D., *et al.*, *New Journal of Physics* **8**, 19 (2006)
8. Dromey, B., *et al.*, *Rev. Sci. Instrum.*, **75**, 645-648 (2004)
9. Rykovanov, S.G., et al. New J. of Phys. **10**, 025025 (2008)

Part 7 – XUV Optics and Applications of X-Ray Lasers

X-Ray Lasers as Probes of Plasma Parameters

G J Tallents, N Booth, M H Edwards, L M R Gartside, H Huang, A K Rossall, E Wagenaars, D S Whittaker and Z Zhai

Department of Physics, University of York, York YO10 5DD, U.K.

Abstract. General issues relevant to the probing of plasmas with backlighters are first discussed. It is shown that soft x-ray wavelengths (> 10 nm) can generally probe thicker plasmas than harder radiation (< 1 nm) if the plasma is hot (> 100 eV). Demonstration experiments are discussed where the transmission of x-ray lasers through other laser-heated solid targets have been used (i) to measure the rate of laser ablation of solid targets and (ii) to probe the opacity of a plasma material. It is shown that narrow bandwidth radiation, such as from an x-ray laser, offers the ability to probe much more optically thick plasmas than backlighters composed of moderate to wide bandwidth spectral lines.

1 Introduction

Soft x-ray lasers have been developed to produce high repetition (~ 10 Hz) rate operation at wavelengths 10 – 40 nm from tabletop capillary [1] or laser-pumped [2] sources. Efficiencies are such that, for example, grazing-incidence pumping techniques can produce lasers at 13.9 nm in Ni-like Ag with less than one joule infra-red laser pumping energy [3]. These lasers have a much narrower spectral bandwidth and are cheaper and much more widely available than free-electron 4th generation sources [4]. However, the free-electron sources are brighter, can produce shorter pulses and promise laser output at much shorter wavelengths (down to 0.15 nm operation is planned [5]). Currently, plasma-based soft x-ray lasers are also not usually spatially uniform, though much greater x-ray laser uniformity should be achieved by seeding the amplification process with laser harmonic radiation [6].

This paper discusses the potential of plasma-based soft x-ray lasers for probing other laser-plasmas in order to elucidate properties of the laser-plasmas. Soft x-ray lasers have been used interferometrically to probe plasmas [7]. Some of the probing issues with interferometry are examined in section 2. A comparison of probing with soft x-rays (13.9 nm) and harder x-rays (1 nm) is presented in section 3. More recently, soft x-ray laser probing has been shown capable of elucidating laser ablation rates with solid targets [8]. Target materials such as iron which undergo a rapid opacity drop upon heating and expansion from the solid values have soft x-ray transmission dominated by the opacity of the cold solid material and so a measure of x-ray laser transmission can record the thickness of solid target remaining and hence the rate of ablation. The measurement of laser ablation from x-ray laser transmission is examined in section 4.

The opacity of material at x-ray and extreme ultra-violet (EUV) photon energies is difficult to measure. At high energy densities where materials are in the plasma state, there have been few direct measurements, though those that have been made have had important consequences. For example, opacity measurements of hot-iron plasmas, made with laser-plasma sources, led to revisions in opacity tables. These revised opacity values have impacted predictions of the behaviour of Cepheid variable stars (used as astronomical 'standard candles' [9]) and lead to revised values of the Hubble constant. However, opacity calculations using codes are usually all that is available, despite sometimes orders-of-magnitude variation in opacity σ between code predictions. As measurements of x-ray or EUV transmission T through material vary as $T = \exp\left(-\int \sigma\rho\, dx\right)$ where the integration is over the path through the sample opacity σ and density ρ, even an approximate measurement of T through a plasma can give a useful opacity measurement and eliminate some code treatments. Examples of opacity codes, include the Cassandra code operated at AWE [10], the OPAL code operated at LLNL [11], the TOPS code operated at Los Alamos [12], the French CEA Odalisc data [13] and the opacity project calculations [14] used to understand the opacity of the sun and other astrophysical objects. The use of x-ray lasers to measure plasma opacity is examined in section 5.

2 Soft x-ray plasma probing

In order to probe a plasma with impinging radiation, it is necessary to employ radiation of sufficiently short wavelength that the critical electron density n_c ($= 10^{27}/\lambda^2$ cm^{-3}) associated with the wavelength λ (in nm) is greater than the electron density n_e of the plasma. For a plasma of density ρ (in gcm^{-3}), the electron density (in cm^{-3}) is given by

$$n_e = 6 \times 10^{23} \rho \frac{Z^*}{M} \tag{1}$$

where Z^* is the degree of ionisation of the plasma and M is the mass number of the plasma ions. Wavelengths λ (in nm) such that

$$\lambda < \lambda_0 = 40\sqrt{\frac{M}{\rho Z^*}} \tag{2}$$

can thus be used to probe a plasma of density ρ (in gcm^{-3}) without collective absorption effects associated with electron densities close to the critical density affecting the measurement. For example, for fully ionised material $Z^*/M \approx 0.5$, the maximum density plasma that can be probed with 13.9 nm radiation is 17 gcm^{-3}. These densities able to be probed by x-ray lasers are several orders-of-magnitude higher than the maximum densities able to be probed by visible lasers. For example, 566 nm light can only probe a fully ionised plasma of maximum density 0.01 gcm^{-3}.

Interferometry measurements have often sought to probe laser-plasmas transversely and elucidate the profile of the electron density variation as a function of distance from the target [15]. In practise, refraction effects will not allow probing up to the critical densities if the beam path L through the plasma is long. With an electron density n_e varying with scalelength Δx, probing radiation is refracted to an angle of refraction θ (in radians) given approximately by

$$\theta \approx \frac{L}{\Delta x}\frac{n_e}{2n_c} = \frac{L}{\Delta x}3\times 10^{-4}\rho\frac{Z^*}{M}\lambda^2 . \tag{3}$$

Typically, probing a laser plasma transversely implies $L/\Delta x \approx 20$, so probing up to a density much less than an order-of-magnitude from the critical density is often regarded as the maximum feasible. Following this argument, the maximum density that 13.9 nm wavelength radiation can transversely probe in a full-ionised plasma is only 0.1 gcm^{-3} (with, say, $\theta < 0.6$ radian able to be collected with an imaging optic), significantly below the solid density of most materials. To probe high (up to solid) density plasma with soft x-ray lasers, it is necessary to probe longitudinally through the plasma in the direction of the density gradient.

The plasma refractive index μ is usually assumed to be dominated by the effects of electrons and so is usually assumed to be given by

$$\mu = \sqrt{1-\frac{n_e}{n_c}} = \sqrt{1-6\times 10^{-4}\rho\frac{Z^*}{M}\lambda^2} . \tag{4}$$

We have assumed equation (4) in deriving equation (3). However, close to strong, spectrally narrow radiation absorption features, the refractive index arising from plasma ions is also significant because of the Kramers-Kronig relationship [18] arising from the inter-relationship between imaginary refractive index (absorption) and the real refractive index. The deviation of the real refractive index from unity due to the Kramers-Kronig relationship is proportional to λK, where K is the plasma absorption coefficient for the probing radiaton. Typically, $K \propto (n_e)^2$, so higher plasma densities again have much more significant variations of μ . For Doppler broadened absorption lines, the maximum (and minimum) refractive index from the Kramers-Kronig relationship is approximately $1 \pm 0.05\ \lambda K$, so the effect becomes negligible if $K <\ <1/\lambda$. For our 13.9 nm x-ray laser wavelength, this implies absorption coefficient values below say 10^5 cm^{-1}. Solid materials have opacity $\sigma = K/\rho \sim 10^4 - 10^6$ cm^{2}g^{-1}, implying that plasmas at densities $\rho = 1 - 10$ gcm^{-3} have refractive indices that can be affected by the Kramers-Kronig effect.

The effect of the Kramers-Kronig refractive indices on plasma interferometry has been demonstrated by Filevich et al [16, 17]. It is worthwhile checking if the refractive index variation across a line profile associated with the effect could change the intensity of radiation used to probe a plasma close to an absorption line and not just affect the phase of the transmitted light. For a Gaussian profile, the refractive index varies with frequency ν across an absorption line of peak absorption coefficient K and FWHM frequency bandwidth $\Delta\nu$ as

$$\mu = 1 - \frac{\lambda K}{2\pi^{3/2}} F\left(2\sqrt{\ln 2}\,\frac{\upsilon}{\Delta\upsilon} \right) \tag{5}$$

where from [18] the function F varies as shown in Figure 1.

As a worst-case scenario of the effect of the refractive index variation across an absorbing line profile, we assume that the probing radiation has a spectral profile matching the Gaussian absorbing line profile. The relative value R of the transmitted radiation electric field amplitude allowing for the Kramers-Kronig effect to the amplitude ignoring the effect is then given by

$$R(KL) = \frac{\int_{-\infty}^{+\infty} \exp\left(-x^2/2\right) \exp[\frac{-KL}{2}\exp(-x^2)] \cos[\frac{\lambda}{L}\frac{1}{2\pi^{3/2}} KLF(x)]dx}{\int_{-\infty}^{+\infty} \exp\left(-x^2/2\right) \exp[\frac{-KL}{2}\exp(-x^2)]dx} \tag{6}$$

An example calculation of R is given in Figure 2 assuming $L/\lambda = 5$. From the form of equation (6) and Figure 2, the Kramers-Kronig effect becomes more apparent for low values of L/λ and high values of KL. However, high values of KL then imply very large absorption coefficient K. For example, a maximum value of $K = 10^6$ cm^{-1} for a plasma is feasible, which for $\lambda = 13.9$ nm implies $KL \cong 7$ for Figure 2. We see that there is little effect on the transmitted radiation flux at such typical values (see Figure 2).

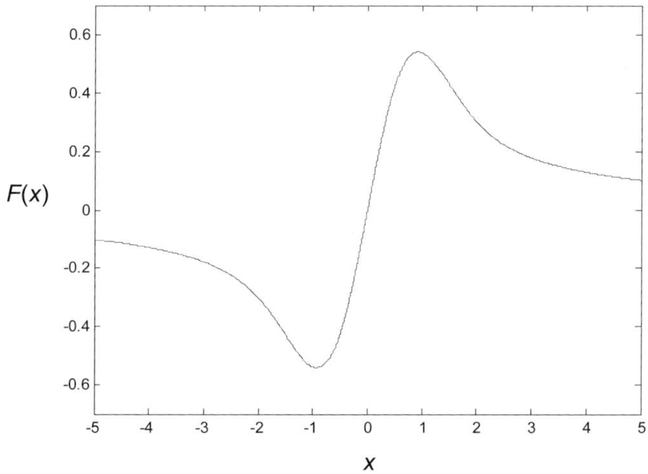

Fig. 1 Function F introduced in equation (5) giving the variation of refractive index due to the Kramers-Kronig effect for a Gaussian absorption line as a function of frequency (in half width half maximum frequency units).

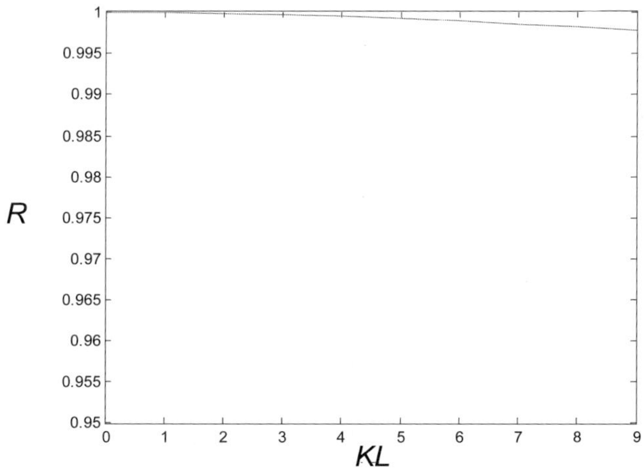

Fig. 2 The value of the electric field transmitted a length L through a plasma of peak absorption coefficient K assuming the Kramer-Kronig effect relative to the electric field neglecting the Kramers-Kronig effect. A value $L/\lambda = 5$ is assumed.

3 Plasma thicknesses able to be probed by x-ray radiation

Huang and Tallents [19] considered the issue of the maximum plasma thickness where transmitted x-ray radiation is more intense than self-emission from the plasma. Laser radiation can readily 'outshine' plasma self emission when used for probing because the detection optics can be optimised to select only a narrow solid angle corresponding to the laser divergence and a narrow spectral range around the narrow-band laser radiation. With short pulse duration x-ray lasers, it is also feasible to time resolve the detection in order to improve the signal (transmitted laser light) to noise (plasma self-emission). It is not widely appreciated that soft x-rays (for example, 13.9 nm wavelength corresponding to 89 eV photon energy) can more readily 'outshine' plasma self-emission than harder x-rays (for example, 0.1 nm or 1243 eV photon energy) when probing a hot (> 100 eV) plasma (see Figure 3).

4 Measurements of laser ablation

Most material opacities drop rapidly with increasing temperature and decreasing density. Increased temperatures results in increased ionisation so that energy gaps between bound quantum states and the ionisation energies increase. It becomes increasingly likely that a particular photon energy is insufficient to produce a bound-bound or bound-free transition. Similarly, free-free absorption for a single frequency drops with increasing temperature T_e with a cross-section $\propto T_e^{-3/2}$. Decreasing density results in reduced opacity at a typical rate $\propto (n_e)^2$. Ablation of a solid target results in the material rapidly increasing in temperature and as expansion occurs, rapidly dropping in density. However, there is evidence that 'warm dense matter' (temperatures < 10 eV say) have opacities that are close to those of the solid material [20]. We have demonstrated [8] that by measuring soft x-ray laser transmission through a target that is being irradiated by an infra-red laser, the rate of ablation of the target can be measured. The overall target transmission is determined almost entirely by the opacity of solid material that has not heated sufficiently to significantly expand. Once the material has heated and started to expand, the material becomes transparent. Measuring the x-ray laser transmission through the target thus gives a measure of the cold material thickness as this is the only absorbing component. From this measurement, the rate of laser ablation of the target is deduced. Further details have been presented [8] and results from a more recent experiment are presented in this volume [21].

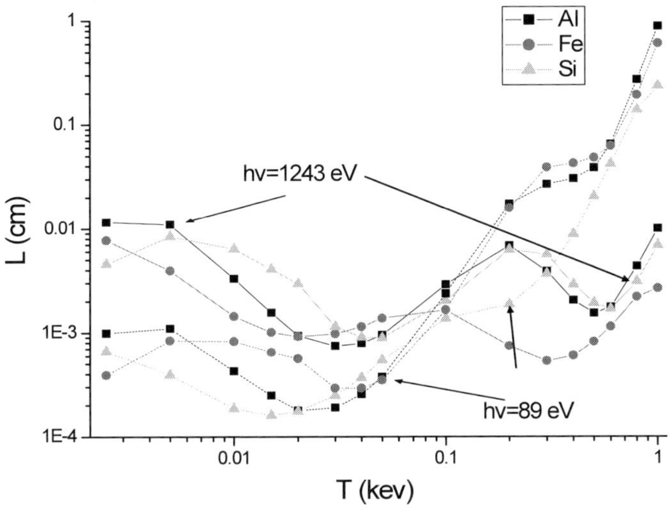

Fig. 3 The maximum plasma thickness where the transmitted radiation is more intense than the plasma self-emission. A uniform plasma with absorption given by the TOPS code [12] and LTE plasma emission is assumed. Other assumptions are that the plasma emission last 100 x longer than the probing radiation, the input radiation has irradiance 3 x 10^{10} Wcm^{-2}, and the detection solid angle at 2.5 x 10^{-5} sr records all the transmitted probing radiation flux. Results are shown for two photon energies for aluminium, silicon and iron plasmas (as labelled) of solid density.

5 Measurements of plasma opacity

The opacity of plasma can be probed using soft x-ray lasers to measure the transmission through laser-heated targets [22]. X-ray lasers are spectrally narrow ($\nu/\Delta\nu \sim 10^4$) because their linewidths are typically determined by Doppler broadening from plasma with cold ions (< 100 eV) and further narrowed by gain narrowing. The advantage of measuring the transmission with a spectrally narrow probe like an x-ray laser is illustrated in Figure 4. A 'saturation-like' effect on the transmission as a function of optical depth occurs when probing line radiation is spectrally comparable or broader than the absorbing line being probed. The instrument spectral resolution is usually much greater than the line widths of either x-ray laser or probed plasma (typically ($\nu/\Delta\nu \sim 100$), so the spectrally integrated transmission as shown in Figure 4 is observed. The effects illustrated in Figure 4 also occur when probing plasma with backlighter radiation composed of unresolved transition arrays [23].

Measuring plasma opacity using the intensity of transmitted radiation can be undertaken by tamping the material to be measured so that ablation occurs in a low opacity material (e.g. CH) and thermal conduction heats a buried layer of a high opacity material (e.g. Fe) [22]. It is desirable to employ a low opacity tamp so that changes in the overall target transmission are dominated by the changes in opacity of the buried layer. The effects of opacity changes in the tamp as it ablates then do not affect significantly the overall target transmission. An x-ray laser pulse duration can be ≈ 3 ps, so that using a short duration (< 1 ps) heating laser pulse ensures that a uniform plasma in the buried layer is achieved (see Figure 5) with a probe duration sufficiently short that temporal changes during probing are small. A short x-ray laser probing pulse also enables high density plasma to be probed before significant expansion can occur (see Figure 5a).

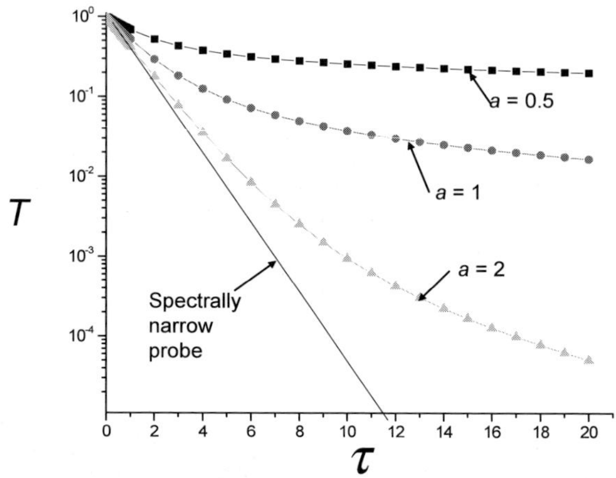

Fig. 4 The transmitted fraction T of Gaussian line radiation as a function of optical depth τ at line centre assuming absorption in a Gaussian absorbing line of width a (as labelled) relative to the probing linewidth. The greater range of T able to be recorded is apparent if a spectrally narrow probe radiation (high a) is employed. A probing linewidth that is infinitesimally narrow is also shown. The probing and absorbing lines are assumed to peak at the same frequency.

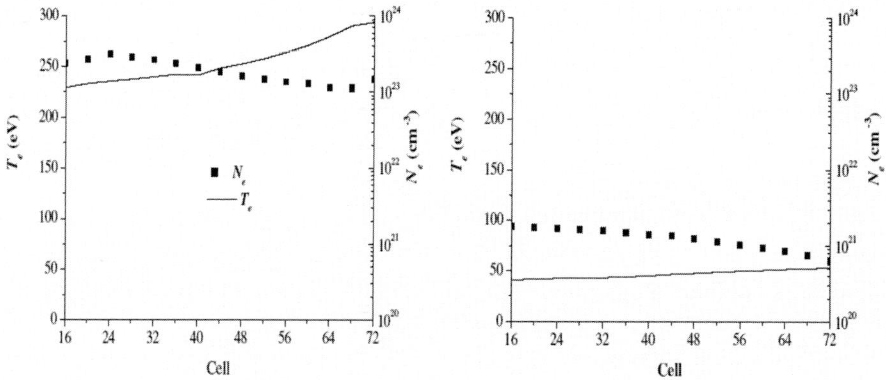

(a)

(b)

Fig. 5 Simulated densities and temperatures in an initially 50 nm iron layer buried below 20 nm plastic (CH) irradiated by an infra-red laser pulse of 3 ps duration and peak irradiance 4×10^{17} Wcm^{-2}. (a) Time 4 ps from the peak of irradiation (b) Time 100 ps after the peak of irradiation.

6 Conclusion

The use of soft x-ray laser to probe laser-plasma parameters has been examined. It has been shown that transverse probing such as interferometry to elucidate, for example, density profiles in laser-plasmas is limited due to refraction effects to densities < 0.1 gcm^{-3} typically. It has been shown that soft x-ray lasers (e.g. 13.9 nm) can more readily probe high temperature (> 100 eV) plasmas than harder x-rays. These results suggest a role for longitudinal probing of laser plasmas (in the direction of density gradients) in order to elucidate the plasma properties and behaviour at high density (up to solid density). Examples where soft x-ray lasers are used to probe the rate of ablation of laser irradiated solid targets and to measure the opacity of laser heated solid materials have been discussed.

Acknowledgements

Funding from the U.K. EPSRC and STFC research councils and from the EU Laserlab is gratefully acknowledged.

References

1 J J Rocca 1999 Rev. Sci. Instrum. **70**, 3799.

2 G J Tallents 2003 J. Phys. D**36**, R259.

3 Y Wang et al 2005 Phys. Rev. A**72**, 053807; R Keenan et al 2005 Phys Rev. Lett. **94**, 103901; J Tummler et al 2005 Phys. Rev. E**72**, 037401.

4 V Ayvazyan et al 2006 Eur. J. D**37**, 297.

5 http://www-ssrl.slac.stanford.edu/lcls/

6 P Zeitoun et al 2004 Nature **431**, 426; Y Wang et al 2006 Phys. Rev. Lett. **97**, 123901.

7 R F Smith et al 2002 Phys. Rev. Lett. **89**, 065004; J Grava et al 2008 Phys. Rev. A**78**, 016403.

8 M H Edwards et al 2007 Phys. Rev. Lett. **99**, 195002.

9 The first Cepheid variable star was observed in York by John Goodricke. Their periodic oscillation frequency $\propto$ luminosity and so enables their use as 'standard candles' for the measurement of astronomical distance. However, the frequency of oscillation also depends on the plasma opacity.

10 B J B Crowley and J W Harris 2001 JQSRT **71**, 257.

11 http://physci.llnl.gov/Research/OPAL/

12 http://www.t4.lanl.gov/cgi-bin/opacity/tops.pl

13 http://irfu.cea.fr/Projets/Odalisc/index.htm

14 C A Iglesias and F J Rogers 1996 Astrophys. J. **464**, 943

15 G J Tallents 1984 J Phys. D**17**, 721; L A Gizzi et al 1994 Phys. Rev. E**49**, 5628.

16 J Filevich et al 2005 Phys. Rev. Lett. **26**, 293.

17 J Filevich et al 2006 Phys. Rev. E**74**, 016404.

18 L Casperson and A Yariv 1971 Phys. Rev. Lett. **26**, 293.

19 H Huang and G J Tallents 2008 JQSRT **109**, 2272.

20 J Wark 2008 private communication.

21 N Booth et al 2008 these proceedings.

22 M H Edwards et al 2006 Phys. Rev. Lett. **97**, 035001

23 C A Iglesias 2006 JQSRT **99**, 295.

Advances in Nanoscale Resolution Soft X-Ray Laser Microscopy

C. S. Menoni[1], F. Brizuela[1], C. Brewer[1], D. Martz[1], P. Wachulak[1], S. Fernandez Jimenez[1], M. C. Marconi[1], J. J. Rocca[1], W. Chao[2], E. H. Anderson[2], D. T. Attwood[2], A. V. Vinogradov[3], I. A. Artioukov[3], Y. P. Pershyn[4], and V. V. Kondratenko[4]

[1] NSF ERC for Extreme Ultraviolet Science and Technology, and Department of Electrical and Computer Engineering, Colorado State University, Fort Collins, Colorado, USA
[2] NSF ERC for Extreme Ultraviolet Science and Technology, Center for X-ray Optics, Lawrence Berkeley National Laboratory, and University of California Berkeley, California, USA
[3] P.N. Lebedev Physical Institute, Moscow, Russia[4]Metal and Semiconductor Physics Department, National Technical University "KhPI", Kharkov, Ukraine

Abstract. We review our most recent results on table-top, nanometer-scale resolution, microscopy using compact soft x-ray lasers developed at Colorado State University. We have realized the first demonstration of wavelength-resolution microscopy in the soft x-ray spectral range. Images of carbon nanotubes, 50 nm in diameter, were obtained with a single ~1 ns duration laser pulse from a desk-top size capillary discharge 46.9 nm laser. We fully characterized the new microscope by measuring the modulation transfer function of the instrument for zone plate objectives with three different numerical apertures, demonstrating that 54 nm half-period structures can be resolved. The combination of near-wavelength spatial resolution with high temporal resolution imaging opens myriad opportunities for imaging nanoscale structures.

1 Introduction

Advances in nanotechnology and nanoscience are driving the need for practical microscopes with the ability to capture high quality full-field images with nanometer-scale spatial resolution and short exposure times. Conventional visible light microscopy, the most convenient method to image small objects, is limited in resolution to ~ 200 nm. This limitation can be overcome by exploiting the shorter wavelength nature of soft x-rays (SXR) since the resolution of a microscope, R= káλ/NA, is linearly proportional to the wavelength of the illumination, λ, and inversely proportional to the numerical aperture of the objective, NA. The constant k, of the order of 0.5, is illumination and test dependent. The use of SXR light illumination for microscopy has been successfully applied at third generation synchrotrons, where images with half-period spatial resolution as high as 15 nm have been obtained using zone plate lenses [1].

The desire for compact and more broadly accessible, full-field microscopes has motivated the use of newly developed compact EUV and SXR sources, such as high harmonic-light sources [2, 3], plasma sources [4, 5], and SXR lasers [6-10]. Most of these microscopes have achieved a spatial resolution several times the wavelength, and good quality images requiring exposure times from several seconds to minutes, thus limiting their use in studies of rapidly evolving phenomena. The challenging task of acquiring full-field images in single shot, has been limited to a few examples. Images with a single laser pulse and a spatial resolution of 50 nm were reported using a 4.48 nm wavelength SXR laser pumped by the large fusion-class NOVA-laser, but experiments were limited to only a few shots a day [11]. More recently, using a 13.9 nm wavelength plasma-based laser source, images with a reported spatial resolution of 200 nm were acquired with a single laser pulse [7].

In this paper, we report on the development of a full-field SXR microscope based on CSU's 46.9 nm capillary discharge SXR laser capable of single shot flash imaging of nanostructures with near-wavelength spatial resolution [12]. The combination of these attributes opens myriad opportunities for imaging, such as the ability to directly investigate dynamic processes of nanoscale structures.

2 Microscope Setup

The SXR microscope is based on a table-top capillary discharge laser that emits at 46.9 nm wavelength [13]. The laser is created via a highly ionized plasma column, generated by fast electrical discharge excitation of an argon-filled capillary. It emits pulses with ~10 μJ of energy ($2.4\text{x}10^{12}$ photons/ pulse) and high monochromaticity ($\Delta\lambda/\lambda \sim 5\text{x}10^{-5}$), at a repetition rate of up to 12 Hz. The degree of spatial coherence can be selected based on the length of the plasma column that constitutes its gain medium. The capillary length, 22 cm, selected for these experiments produces optimized, partially coherent pulses for illumination that limit speckle effects that can degrade image quality, while still providing sufficient photon flux for single shot imaging.

Fig. 1 a) Picture of the EUV microscope. The laser unit on the left is connected through standard vacuum fittings to the microscope chamber on the right. The entire system fits on a small optical table.

The microscope can operate in transmission or reflection mode. In either configuration, the output of the laser impinges on a multilayer coated Schwarzschild condenser that focuses the light on the sample. The light that is either reflected off or transmitted through the sample is collected by a free-standing objective zone plate lens that projects a magnified image onto a charge-coupled device (CCD) detector. The Schwarzschild condenser consists of two Sc/Si multilayer coated mirrors with a combined throughput of ~13% at 46.9 nm wavelength. Three zone plates lenses of numerical aperture NA=0.12, 0.20, and 0.32 were used as objectives. The freestanding zone plates were fabricated by electron beam lithography onto 100 nm thick Si_3N_4 using a design in which the different zones are connected with pseudo-random supportive spokes to allow for ample throughput of the λ = 46.9 nm light. As shown in Fig. 1, the entire system, including the source is very compact occupying an area of 0.4 m x 2.5 m on a small optical table.

3 Results

To determine the resolving power of the microscope, single-shot images of freestanding transmission gratings with half-periods ranging from 300 nm to 54 nm were obtained with each of the three objective zone plates. The gratings with a duty cycle nominally equal to one were made by electron beam lithography using a process similar to that used for the zone plates. The resolving ability of the microscope was evaluated by experimentally building modulation transfer functions (MTFs) for each zone plate objective. The MTFs for the three zone plates are shown in Fig. 2. To determine the modulation transfer for each half-period grating, multiple intensity lineouts were sampled within an image and averaged to obtain the values and error bars shown. The spatial resolution of a microscope equals the half-period for which the MTF equals 0.265, as derived from the Rayleigh criterion for two point sources [14]. For the NA=0.12, 0.20, and 0.32 zone plates, a diffraction-limited spatial resolution of 120 nm, 80 nm, and 54 nm, respectively, was measured. With a spatial resolution of 54 nm, this microscope is the first in the EUV region with the ability to resolve features of dimensions closely approaching the wavelength of illumination.

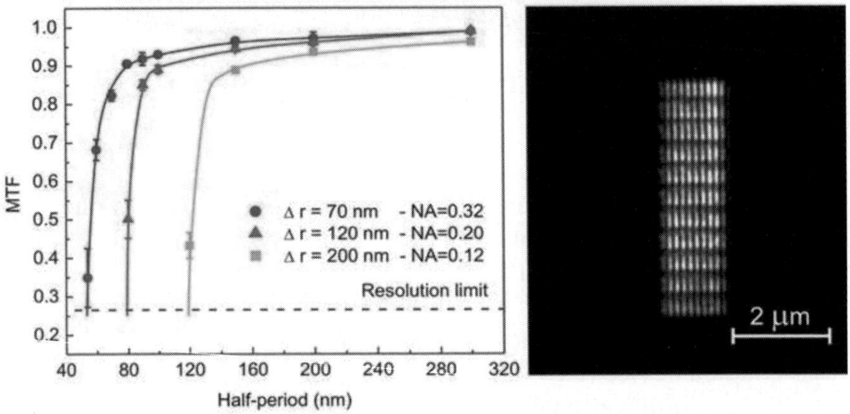

Fig. 2 Measured Modulation Transfer Functions (MTFs) for three objective zone plates. On the right, a single-shot SXR image of a 70 nm half period grating obtained with the 0.32 NA zone plate. With this objective, wavelength spatial resolution was achieved.

To illustrate the practical use of the microscope, we selected an object of dimensions similar to the measured resolution limit. We imaged carbon nanotubes with a diameter of ~50 nm, that were deposited onto a ~100 nm thick Si membrane with ~30% transmission at a wavelength of 46.9 nm. Fig. 3.a is a SXR image of a single nanotube obtained with one laser shot. For comparison, Fig. 3.b shows the same tube as imaged by a scanning electron microscope (SEM).

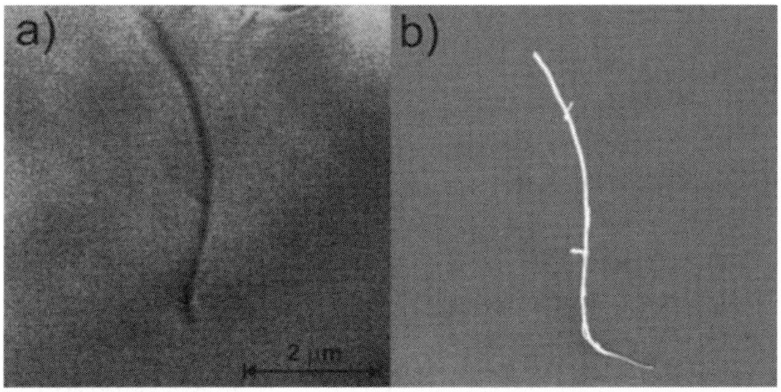

Fig. 3 a) Single shot SXR image of a ~50 nm wide carbon nanotube on a Si membrane. The image was obtained with a 0.32 NA zone plate. b) Same carbon nanotube as imaged by SEM.

The microscope can also render high resolution images in reflection mode. As an example, in Fig. 4., an a SXR image of a sample consisting of a GaN nanowire bridging two Ti contacts grown on a Si wafer is shown [15]. In the center of the SXR image one of these nanowires can be seen across the gap between the contacts. The image was obtained with the 0.20 NA objective and an exposure time of 5 seconds corresponding to 15 laser shots. In this case, the condenser was scanned during the acquisition of the image to improve the illumination of the sample.

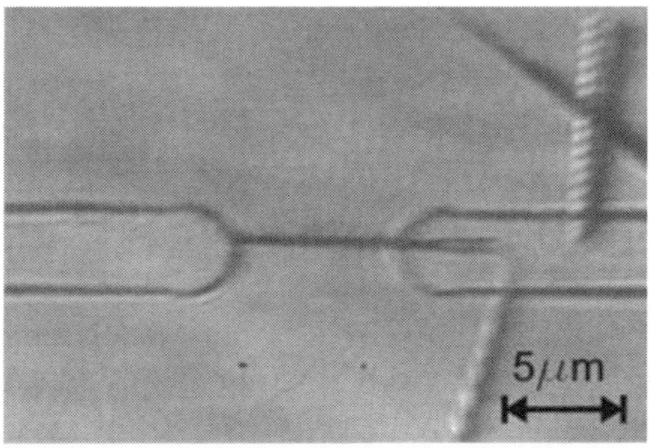

Fig. 4 Reflection mode image of a GaAs nanowire between two Ti contacts on a Si wafer. The SXR image was acquired with a 0.20 NA objective zone plate. In the image, other nanowires that did not successfully bridge the contacts can be seen.

4 Conclusions

We have developed a versatile high resolution full-field microscope that can operate in transmission and reflection mode based on a table-top 46.9 nm wavelength laser. The best performance of the microscope, diffraction limit spatial resolution of 50 nm and single shot flash exposure was demonstrated in transmission mode. These capabilities will enable the study of dynamics on nanosecond time scales. Based on these results, we anticipate that using recently developed high brightness tabletop lasers at shorter wavelengths it will be possible to develop compact full-field microscopes with a spatial resolution approaching 15 nm and picosecond temporal resolution for a wide range of nanoscience and nanotechnology applications.

We would like to acknowledge Dr. Bertness and Dr. Sanford from NIST for providing the GaAs nanowire sample. This work was supported by the Engineering Research Centers Program of the National Science Foundation under NSF Award Number EEC-0310717, and by the Cooperative Gran program of the U.S. Civilian Research and Development Foundation under Award Number 2845-MO-06.

References

1. B. D. H. Weilun Chao, J. Alexander Liddle, Erik H. Anderson & David T. Attwood, "Soft X-ray microscopy at a spatial resolution better than 15 nm," Nature **435**, 1210-1213 (2005).
2. M. Wieland, C. Spielmann, U. Kleineberg, T. Westerwalbesloh, U. Heinzmann, and T. Wilhein, "Toward time-resolved soft X-ray microscopy using pulsed fs-high-harmonic radiation," Ultramicroscopy **102**, 93-100 (2005).
3. R. Sandberg, L. , A. Paul, D. A. Raymondson, S. Hadrich, D. M. Gaudiosi, J. Holtsnider, R. a. I. Tobey, O. Cohen, M. M. Murnane, H. C. Kapteyn, C. Song, J. Miao, Y. Liu, and F. Salmassi, "Lensless diffractive imaging using tabletop coherent high-harmonic soft-x-ray beams " Physical Review Letters **99**, 098103 (2007).
4. M. Hoshino and S. Aoki, "Laser plasma soft x-ray microscope with wolter mirrors for observation of biological specimens in air," Jpn. J. App. Phys. **45**, 989-994 (2006).
5. P. A. C. Takman, H. Stollberg, G. A. Johansson, A. Holmberg, M. Lindblom, and H. M. Hertz, "High-resolution compact X-ray microscopy," Journal of Microscopy **226**, 175 (2007).
6. D. S. DiCicco, D. Kim, R. Rosser, and S. Suckewer, "First stage in the developement of a soft x-ray reflection imaging microscope in the Scharzschild configuration using a soft-x-ray laser at 18.2 nm.," Optics Letters **17**, 157-159 (1992).
7. M. Kishimoto, M. Tanaka, R. Tai, K. Sukegawa, M. Kado, N. Hasegawa, H. Tang, T. Kawachi, P. Lu, K. Nagashima, H. Daido, Y. Kato, K. Nagai, and H. Takenaka, "Developement of soft X-ray microscopy system using X-ray laser in JAERI Kansai " J. Phys. IV France **104**, 141-143 (2003).

8. G. Vaschenko, F. Brizuela, C. Brewer, M. Grisham, H. Mancini, C. S. Menoni, M. C. Marconi, J. J. Rocca, W. Chao, J. A. Liddle, E. H. Anderson, D. T. Attwood, A. V. Vinogradov, I. A. Artioukov, Y. P. Pershyn, and V. V. Kondratenko, "Nanoimaging with a compact extreme-ultraviolet laser," Optics Letters **30**, 2095-2097 (2005).
9. F. Brizuela, G. Vaschenko, C. Brewer, M. Grisham, C. S. Menoni, M. C. Marconi, J. J. Rocca, W. Chao, J. A. Liddle, E. H. Anderson, D. T. Attwood, A. V. Vinogradov, I. A. Artioukov, Y. P. Pershyn, and V. V. Kondratenko, "Reflection mode imaging with nanoscale resolution using a compact extreme ultraviolet laser," Optics Express **13**, 3983-3988 (2005).
10. G. Vaschenko, C. Brewer, E. Brizuela, Y. Wang, M. A. Larotonda, B. M. Luther, M. C. Marconi, J. J. Rocca, and C. S. Menoni, "Sub-38 nm resolution tabletop microscopy with 13 nm wavelength laser light," Optics Letters **31**, 1214-1216 (2006).
11. L. B. Da Silva, J. E. Trebes, R. Balhorn, S. Mrowka, E. Anderson, D. T. Attwood, T. W. J. Barbee, J. Brase, M. Corzett, J. Gray, J. A. Koch, C. Lee, D. Kern, R. A. London, B. J. MacGowan, D. L. Matthews, and G. Stone, "Xray laser microscopy of rat sperm nuclei," Science **258**, 269-271 (1992).
12. C. A. Brewer, F. Brizuela, P. Wachulak, D. H. Martz, W. Chao, E. H. Anderson, D. T. Attwood, A. V. Vinogradov, I. A. Artyukov, A. G. Ponomareko, V. V. Kondratenko, M. C. Marconi, J. J. Rocca, and C. S. Menoni, "Single-shot extreme ultraviolet laser imaging of nanostructures with wavelength resolution," Optics Letters **33**, 3 (2008).
13. S. Heinbuch, M. Grisham, D. Martz, and J. J. Rocca, "Demonstration of a desk-top size high repetition rate soft x-ray laser," Optics Express **13**, 40504055 (2005).
14. J. M. Heck, D. T. Attwood, W. Meyer-Ilse, and E. Anderson, "Resolution determination in X-ray microscopy: an analysis of the effects of partial coherence and illumination spectrum.," Journal of X-Ray Science and Technology **8**, 95-104 (1998).
15. S. M. Tanner, J. M. Gray, C. T. Rogers, K. A. Bertness, and N. A. Sanford, "High-Q GaN nanowire resonators and oscillators," Applied Physics Letters **91**(2007).

[illegible] "[illegible] with a compact [illegible] token," [illegible]

[illegible]

[illegible] "High-Q [illegible] nanowire [illegible] and oscillators," Applied Physics Letters [illegible]

Experimental Diagnosis of Plasma Jets by Using X-Ray Laser

Sun Jin-ren[1], Wang Chen[1], Fang Zhi-heng[1], Wang Wei[1], Xiong Jun[1], Fu Si-zu[1], Gu Yuan[1], Wang Shi-ji[1], Zheng Wu-di[2], Ye Wen-Hua[2], Qiao Xiu-Mei[2], Zhang Guo-ping[2]

[1] Shanghai Institute of Laser Plasma, Shanghai, 201800, P. R. China
[2] Beijing Institute of Applied Physics and Computational Mathematics, Beijing, 100088, P. R. China

Abstract. The supersonic jets and the interaction of strong shock waves are ubiquitous features of the nonlinear hydrodynamics of inertial-confinement fusion, astrophysics, and related fields of high energy-density science, so it is very important of laboratory studies on it. In this paper, a laboratory experimental study is reported. The jets were created by a ns laser pulse irradiating a specific target, and the shadow of jets were measured by an x-ray laser probe with wavelength of 13.9nm. The legible images of jets were recorded, and the experimental results and theory simulation were qualitative agreement.

1 Introduction

Jets are general nonlinear phenomena in the universe as well as in the nature, and attached importance to many fields from astrophysics to hydrodynamics[1-8]. Since the jets induced by various instabilities can exert great influence on the compression efficiency of target capsules in inertial confinement fusion (ICF) experiments, the jet phenomenon is one of the key problems in the ignition of ICF.

Most experiments on metallic jets of lower Mach numbers were included in the studies of shaped charges, different from which the jets produced with high power lasers in laboratories are of high Mach numbers and provide a proper means for profound understanding the interactions between jets and shockwaves, material properties and so on.

The diagnoses on the laser produced jets in laboratories are usually conducted by means of the shadow graphy with hard X-rays of keV photon energy. However this method results lower special resolutions. Nowadays the sophisticated soft X-ray laser is characterized by good directionality, ultrashort pulse duration, high brightness and quasi-monochromaticity, and has played a definite role in diagnoses of plasma[9-13]. Therefore, the soft X-ray laser can be also employed to measure the process of laser accelerated jets, and takes advantages of high special resolutions. Coordinating well the theoretical design into the experiments, good results can be expected.

The exploratory experiment of laser accelerated jets diagnosed with the soft laser beam has been fulfilled and described in this paper including the comparison between measured and calculated results on the laser facility SG-II at Shanghai, China.

2 Experiment

The experiment has been conducted on specially designed target structure irradiated with a ns laser beam of No.9 branch of the SG-II facility at the National laboratory of High Power Laser Physics, China, and by using the probe based on the soft X-ray laser beam driven by two 100ps Nd laser beams of the facility. The experimental scheme is shown in Fig. 1, where the probe beam of X-ray laser passed through the tested target, then the multi-layer optics, and focused the enlarged image of the target on the focal plane of the CCD. The amplification is about 10, and just suitable to the pixel size in CCD($24\times24\mu m^2$), so that the special resolution is 2.4μm.

The C_8H_8 slab target of thickness 160μm was made a blind hole on its pear surface, which was 140μm in depth and 60μm in diameter. The laser beam from the No9 branch was characterized by pulse duration 2.4ns, wave length 0.53μm, energy 1500J, and irradiated uniformally the target surface with a spot of size $400\times400\mu m^2$.

A strong shock wave was generated in the ablated target slab as irradiated with the driving laser beam, which in turn induced a rarefaction wave backward propagating into the slab when the shock wave reflect on the rear surface. The high speed jet from the hole could be produced as the incident shock wave propagated along the blind hole wall, then interactions of the rarefaction wave and the tail of shock wave would induced strong deformation and ejection an the rear surface. The shadow graphy using soft X-ray laser provided several lateral images of the target at the assigned moments, by which the jet and eject phenomenon could be diagnosed.

As the probe beam the nickel-like silver soft x-ray laser at wave length 13.98nm was generated from a planar target consisting of two joint silver-coated glass plates irradiated by two linearly focal laser beams, which were of pre-main pulsed, both from the eight branches of SG-II facility and with parameters as wave length 1.053um, pulse duration 90ps, energy 100J, the intensity ratio and time interval of precursor-mail pulses being 5% and 3ns, respectively.

A cylindrical system of lens array and non-spherical main lens was employed to focus the two laser beams on the planar glass target on whose surface a uniformal strip spot of size 27mm×80μm formed. The length of each glass plates was about 25mm to avoid the cold plasma generation at ends of the target. The distance between two glass plate targets was 260μm, and the angle included between the two plates was 2 *mrad*. The resulting nickel-like silver X-ray laser was of pulse duration~30ps and beam divergence angle~$3\times7 mrad^2$.

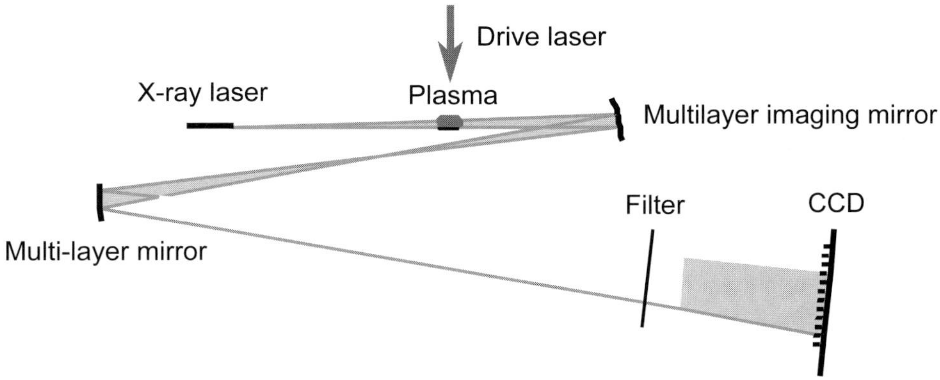

Fig. 1 The schematic diagram of experiment setup

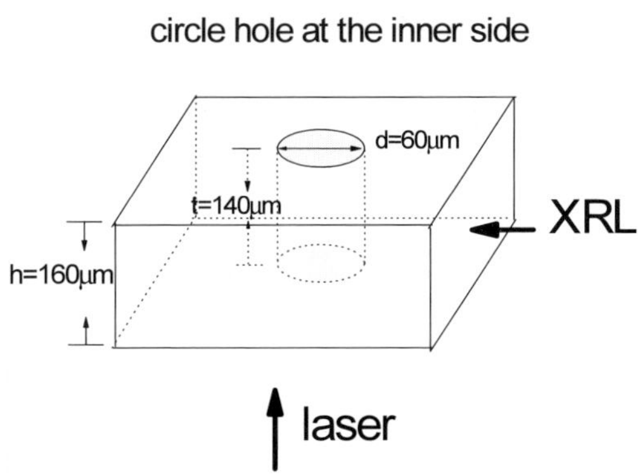

Fig. 2 The schematic diagram of the target structure

3 Comparison with Numerical Modeling

The measured typical results of three shots of laser accelerated jets are shown in Fig 3 in which the diagnosis moments of the probe beam of soft x-ray laser were 1.00ns, 2.15ns and 2.65ns from the half amplitude position along the descent of driving laser pulse (see Fig. 4). In addition, the pulse shapes of driving laser beam for every shot are shown in Fig. 4.

It is early indicated in Fig. 3 that there is a shadow of 400um in length and 400um in width left to the rear surface of the target slab, which can be attributed to the absorption of probe X-ray laser beam when passing through the spurt jet and eject plasmas. The density distributions of jet and eject plasmas calculated with the code XRL 2D are shown in Fig. 5, where the same structure of C_8H_8 is assumed in the numerical modeling, the target material density is about 1.1 g/cm3. As mentioned above the time profiles of the driving laser pulses are shown in Fig. 4, and their special intensity distributions are taken to be flat shaped, the assumed laser absorptivity 0.65 and fluence limit factor f_e=0.03. The snap pictures in Fig. 5 and Fig. 6 correspond to the moments (a), (b) and (c) in Fig. 3, respectively.

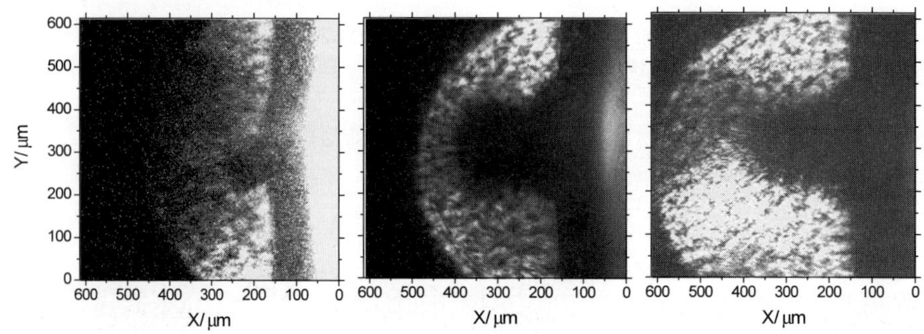

(a) 1.00ns (b) 2.15ns (c) 2.65ns

Fig. 3 The measured typical results of three shots of laser accelerated jets

The calculated shadow pictures to model that resulting from the absorption of probe X-ray laser beam are shown in Fig. 6. The absorption mechanism of CH plasmas to X-ray laser beams is considered to be both photoelectric absorption and inverse bremsstrahlung absorption, where the absorptivity due to inverse- bremsstrahlung is:

$$\alpha \approx 2.44\times10^{-37}\frac{\langle Z^2\rangle n_e n_i}{\sqrt{kT_e}(hv)^3}\left[1-\exp(\frac{-hv}{kT_e})\right]cm^{-1},$$

and that due to the photoelectric mechanism can be evaluated with the Kramer formula of absorption section for hydrogen-like atoms:

$$\sigma_{bf}(\nu) = \frac{64\pi^4}{3\sqrt{3}} \frac{e^{10} m_e Z^4 g_{bf}(\nu)}{h^6 c \nu^3 n^5},$$

where the Gaunt factor ~ $g_{bf}(\nu)$~1.33, hence the related absorptivity is:

$$\mu = 1.08\times10^{-19} N_n \frac{I_n^2}{n} \frac{1}{(h\nu)^3} \left(cm^{-1}\right).$$

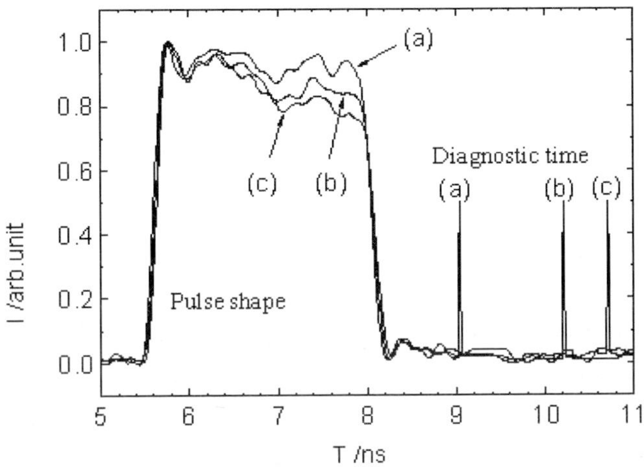

Fig. 4 The pulse shapes of driving laser beam for every shot

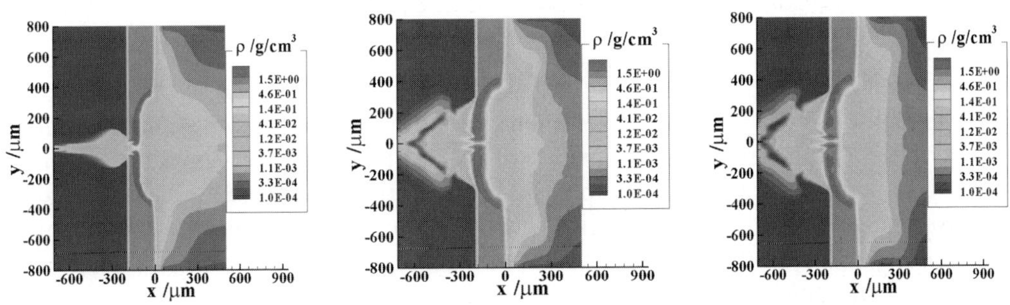

(a) 1.00ns (b) 2.15ns (c) 2.65ns

Fig. 5 The density distributions of jet and eject plasmas calculated with the code XRL 2D

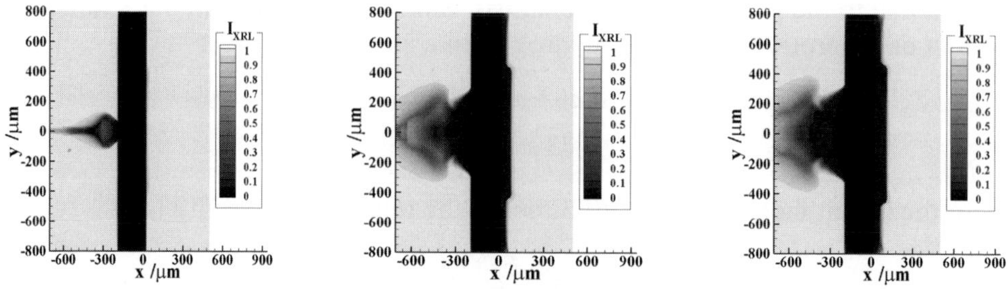

(a) 1.00ns (b) 2.15ns (c) 2.65ns

Fig. 6 The calculated shadow pictures to model that resulting from the absorption of probe X-ray laser beam

Consequently, the total absorptivity of CH plasma to nickel-like silver X-ray laser beams can be evaluated to be 3104cm2/g。 It is noted further that the spurt plasma is cylindrically shaped and a slight refraction of the passing X-ray laser beam can be neglected in the modeling. Compared Fig, 3 to Fig. 6, it is indicated that the shadow image forms and sizes of pictures (a) and (c) in both figures are well consistent with each other, however the shadow images in both pictures of (b) are close to each other in the size only, and markedly different in the forms. The difference can be attributed to the choices of modeling parameters, the simplification of driving laser in calculation, the error of target preperation and so on. A detailed investigation is necessary to make clear the main reasons responsible for the difference. The study describe in this paper is of great significance to further improve the physical model and parameters in the numerical simulation.

4 Conclusion

The ns laser pulse accelerated jet phenomenon in the specially designed target has been diagnosed with the probe soft X-ray laser of wave length 13.9nm. The experimental results are consistent qualitatively scheme is feasible and promising to explore new applications of X-ray lasers in plasma diagnoses related to ICF studies as well as to provide a proper means in checking the ICF related numerical simulations.

References

1. Whitham, G. B.: *Linear and nonlinear waves*, Academic Press, 1974
R. Ladenburg, C. C. Van Voorhis, J. Winckler, *Phys. Rev.,* **76,** 62, 1949
2. P. J. Finley, *J. Fluid Mech.,* **26, 337, 1966**

3. D. R. Farley, K. G. Estabrook, S. G. Glendinning *et al, Phys. Rev. Lett.,* 83, 1982, 1999
4. K. Shigemori, R. Kodama, D. R. Farley *et al, Phys. Rev. E,* **62,** 8838, 2000
5. D. D. Ryutov, R. P. Drake, J. Kane *et al, Astrophys. J.,* **518,** 821, 1999
6. D. D. Ryutov, B. A. Remington, H. F. Robey *et al, Phys. Plasmas,* **8,** 1804, 2001
7. J. M. Foster, B. H. Wilde, P. A. Rosen *et al, Phys. Plasmas,* **9,** 2251, 2002
8. B. E. Blue, H.F. Robey, S.G. Glendinning *et al, Phys. Plasmas,* **12,** 056313, 2003
9. L.B. DaSilva, T.W. Barbee, Jr., R. Cauble *et al, Phys.Rev.Lett.,* **74,** 3991, 1995
10. J.E. Trebes, S.B. Brown, E.M. Campbell *et al* 1987 *Science* **238** 517
11. L.B. DaSilva, J.E. Trebes, R. Balhorn *et al, Science,* **258,** 269, 1992
12. C. Wang, Y. Gu, S.Z. Fu *et al, Acta Phys.Sin.,* **51,** 847(in Chinese), 2002
13. C. Wang, W. Wang, J.R. Sun *et al, Acta Phys.Sin.,* **54,** 202(in Chinese), 2005
14. W.D. Zheng, G.P. Zhang, C. Wang *et al, Acta Phys.Sin.,* **56,** 3984(in Chinese), 2007

Soft X-Ray Holography with Wavelength Resolution

P.W. Wachulak, M.C. Marconi, R. Bartels, C.S. Menoni, J.J. Rocca

NSF Engineering Research Center for Extreme Ultraviolet Science and Technology and Electrical and Computer Engineering Department, Colorado State University, Fort Collins, Colorado

Abstract. We report the table-top acquisition of soft x-ray holographic images of nanostructures with a spatial resolution of 46±2 nm, nearly equal to the wavelength of the illumination. Holograms of carbon nanotubes were recorded using a compact table-top capillary discharge soft x-ray laser emitting at 46.9 nm in a high numerical aperture Gabor configuration.

1 Introduction

Imaging tools for nanoscicence involving sub-100nm scale objects have been dominated by atomic force microscopy (AFM), scanning tunneling microscopy (STM), and electron microscopy. These imaging techniques have contributed substantially to the development of nanoscience, providing a very powerful diagnostic tool capable of atomic resolution or as a subsidiary mechanism to arrange or modify surfaces also at the atomic scale. However, some important problems have resisted traditional nanoscale imaging techniques. For example when scanning a nanometer size object that can move more or less freely, the interaction with the tip will significantly perturb the specimen precluding the image acquisition. Electron microscopy often requires surface preparation, consisting of metallization to avoid surface charging the sample that has a direct influence in limiting the resolution. In both cases, if the sample is large (millimeter size) the image obtained with these conventional methods is only representative of a very small portion of the object.

We demonstrate wavelength-limited, sub-50 nm holographic imaging with a table-top soft X-ray (SXR) laser operating at 46.9 nm. The resolution achieved in this imaging was evaluated with standard resolution measurement methods and confirmed with a correlation analysis. The holography set up requires no optics or critical beam alignment; thus sensitive optical alignment is not required and no special sample preparation. In holography, image contrast requires absorption to provide scattering by the illuminating beam. The SXR laser wavelength employed in this experiment is advantageous in that carbon based materials typically exhibit very small absorption depths ~25 nm. The high absorption provokes that even small object volumes will produce high optical contrasts. The short absorption depth thus enables nearly full contrast for most objects, without applying forces to the imaged objects, no charge buildup, and without the need for complicated sample preparation. This simple and versatile technique allows to image millimeter size objects with sub-50 nm resolution.

Holography is a widely used imaging technique employed in several scientific and technological applications. Shortly after the introduction of in-line holography by Gabor [1], the feasibility of holographic imaging with a resolution superior to that obtained by optical microscopy was discussed by Baez [2] in 1952. It was not until 1987 that the high resolution x–ray imaging was realized by use of 2.5 – 3.2 nm soft x-ray radiation from the National Synchrotron Light Source (NSLS), where a spatial resolution of 40-nm was demonstrated. Fourier transform holography at the NSLS achieved a spatial resolution of 60-nm. Gabor holography with an early SXR laser pumped by two beams of the fusion-class NOVA laser at Lawrence Livermore National Laboratory demonstrated a spatial resolution of 5 μm [3]. Other experiments utilized synchrotron light to image biological samples, nano structures, and magnetic domains [4-6]. Recently, lensless diffractive imaging based on iterative phase retrieval following the proposal by Sayre [7] have demonstrated soft x-ray imaging with spatial resolution of 50 nm with $\lambda = 1.5$ nm source [8]. However the practical demonstration of SXR holography proved to be difficult in particular for the lack of sufficiently bright and coherent sources at short wavelengths. The widespread use has been further restricted because spatially and temporally coherent SXR laser sources have historically been restricted to large user facilities.

The emergence of table-top sources of laser-like coherent SXR radiation [9, 10] opens the feasibility of wide-spread applicability of table-top coherent imaging [11-13]. The first demonstration of coherent table-top imaging achieved 7 micron spatial resolution with a spatially-coherent high harmonic generation (HHG) source [14]; this resolution has been extended to 0.8 μm [15]. Time resolved holographic imaging was also implemented with HHG sources to study the ultrafast dynamics of surface deformation with a longitudinal resolution of < 100 nm and a lateral resolution of < 80 μm [16]. Lensless imaging with HHG sources has been recently demonstrated with 214 nm spatial resolution [17]. Recently, we have demonstrated sub-micron coherent in-line holography [18] and three-dimensional imaging with a single hologram with numerical optical sectioning [19].

2 Experiment

In this paper we present a demonstration of holographic microscopy with ultimate sub-50 nm resolution with a table-top 46.9 nm EUV laser. A sample composed of carbon nano-tubes placed on a thin semi-transparent silicon membrane was imaged using a table top SXR laser in a Gabor's in-line configuration. The hologram was recorded as a relief pattern in a high resolution photoresist, digitized using an AFM and numerically reconstructed with a Fresnel propagator code. Two independent techniques were used to assess the spatial resolution of the image. The 10-90% slope knife-edge resolution test giving ~46 nm and a correlation and convolution method yielded a resolution of 45.5 nm, which is approximately equal to the wavelength λ = 46.9 nm utilized in the recording step.

The spatial resolution of holography is determined by the numerical aperture (NA) of the recorded holographic interferogram. As holography is a coherent process, the hologram NA is set by the object-recording medium distance, the temporal and spatial coherence of the illuminating source, the resolution of the medium in which the hologram is recorded, and for numerically processed holograms, the NA of the digitization process. The design of the experimental apparatus was optimized so that each of the factors that limit the NA listed above did not degrade the imaging resolution.

The holograms in this work were recorded in the Gabor's (in-line) configuration. The test objects used were multiple wall carbon nano-tubes (CNT) with an outer diameter between 50 to 80 nm and a length of 10-20 μm. The sample was prepared by depositing a drop of water/CNT mix on the surface of a 100 nm thick Si membrane. The membrane was fabricated by sputtering a 100 nm thick layer of Si on the surface of a 2 × 0.6 mm^2 Si_3N_4 window previously defined in a Si wafer. The Si_3N_4 layer was subsequently removed by chemically assisted ion etching leaving only the 100 nm thick silicon membrane. The CNT remained attached to this membrane by van der Waals forces after the water evaporated. The Si membrane containing the nanotube sample was placed at few microns from a Si wafer spin-coated coated with a 120 nm thick layer of polymethyl methacrylate (PMMA) photoresist. The in-line hologram is recorded directly in the photoresist.

The CNT object is illuminated by a compact λ = 46.9 nm discharge-pumped capillary Ne-like Ar SXR laser. The laser and the vacuum chamber where the hologram was exposed are connected via a vacuum manifold that provides differential pumping of the chamber maintained at $\sim 10^{-5}$ Torr.

The hologram is written by exposing the PMMA photoresist to the in-line hologram interference pattern. PMMA exposure requires a photon flux in the range of $\sim 2\times10^7$ photons μm^{-2}. Sufficient doses for exposure were achieved in this experimental set up with approximately 150 laser shots. The exposure was kept low enough so that a linear response regime of the photoresist was maintained. In this way the holographic pattern was converted after the development to a height modulation in the surface of the photoresist proportional to the intensity. The temporal coherence of the EUV laser is approximately $\Delta\lambda/\lambda=10^{-4}$, and a spatial coherence radius at the source-object distance of 0.75-m of approximately 0.34-mm [20, 21]; the spatial resolution set by the laser coherence properties is ~28.6-nm [18].

The holograms recorded in the PMMA photoresist cannot be reconstructed in the conventional way by illuminating with a replica of the reference beam. Instead, we digitized the hologram scanning the surface of the photoresist with the AFM (Novascan 3D) operated in "tapping" mode. The holograms were digitized into 1024 × 1024 matrix with a pixel resolution of 9.7 nm and thus a spatial extent of ~9.9 microns. These sampling parameters limit the in-line holography imaging spatial resolution to 32.5 nm (NA ~0.88). Figure 1 (a) shows a digitized hologram.

The holograms were reconstructed by numerically recreating the illumination with a EUV readout wave. The amplitude and the phase distribution of the field in the image plane was obtained calculating the field emerging from the hologram illuminated by a plane reference wave and numerically back propagating the fields with a Fresnel propagator. The propagation distance to the image plane is estimated from the source-object and object-photoresist distances [22]; an optimal object-plane distance is determined through a correlation-based analysis of the hologram spatial resolution [19]. For the specific geometry employed in this experiment, the image-plane length is approximately the distance between the object and the recording medium. Using the image-plane distance of the hologram the Fresnel free space propagator is evaluated in the spatial frequency domain. The image is found by taking the two dimensional inverse fast Fourier transform (2D-FFT) of the spatial frequency product of the Fresnel propagator and the 2D-FFT of the hologram. The hologram processed with the Fresnel propagation code generated the reconstructed images shown in Fig. 1 (b).

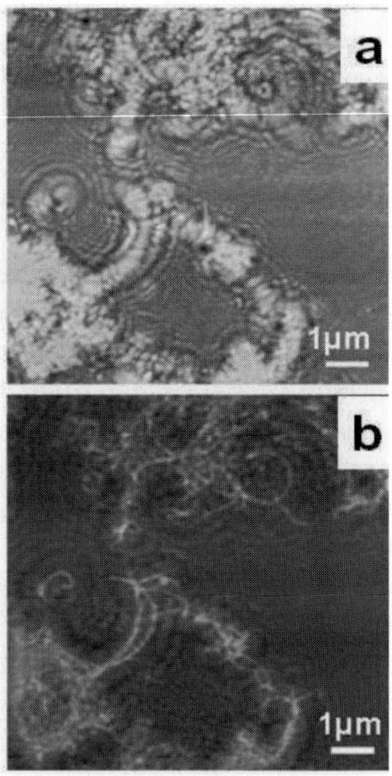

Fig.1. a) Hologram of CNT recorded in the surface of PMMA and digitized with an AFM. b) Numerical reconstruction of the hologram shown in (a) using a Fresnel propagator code

3 Image resolution assessment

The reconstructed in-line hologram presented shown in Fig. 1(b) shows the nanotubes. An initial assessment of the image resolution may be obtained by line-cuts through the image. This cut was realized in a region where we clearly reached a "plateau" in the maximum and minimum intensities to recreate the knife edge resolution test. Taking the 10%-90% rise of the intensity along an edge of the line-cut image yields an estimate of the spatial resolution approximately 46 nm. Other cuts realized in different points of the image indicated a resolution 45.8 ± 1.9 nm where the error bars were determined by the spread of the different measurement.

To obtain a more rigorous and global assessment of the imaging spatial resolution, a correlation analysis of the reconstructed in-line SXR holograms was performed with templates derived from the initial image [23]. This method is based on the correlation between the image and a template generated from a master binary image (pixels with values 0 and 1). From this master binary image a sub set of templates with decreasing resolution was obtained by applying a Gaussian filter of increasing width. Filtering the binary image causes a blurring that increases the size of the features in the image and reduces the resolution of the image by the factor inversely proportional to the filter width. In the spatial frequency domain this is equivalent to reduce the spectrum width by the same factor.

The algorithm executes by applying the following procedures. First the reconstructed image is filtered in the frequency domain to remove the low spatial frequency components. Second an intensity threshold is applied to the image to obtain a homogeneous background. The value of the applied threshold is arbitrarily adjusted to eliminate the fluctuations in the background and simultaneously to keep the object profile unchanged. We corroborate that the threshold value does not change the ultimate result of the algorithm because this step is only used for the construction of the reference template. Third, the image is "skeletonized" using the algorithm described in [24]. The skeletonizing algorithm produces a 1 pixel wide contour that follows the nanotube shape. Fourth, the skeleton is convolved with K different circular templates with different radii, ranging from 1 to 6 pixels, equivalent in this case to diameters from 19.4 nm to 116.4 nm. The convolution of the circular templates with the skeleton generates a set of K templates that resembled the image of the nanotube, each one with different thickness. All the templates generated in this manner are also binary images, with pixel values 0 or 1. Fifth, the K templates are degraded in resolution by applying L Gaussian filters with different FWHM widths and generate from each of the K binary templates, a set of L templates with resolutions ranging from 12 nm to 100 nm. The whole procedure produces a K x L set of templates of the nanotube image.

Finally the K x L templates were individually correlated with the original image and K x L correlation coefficients were obtained. Plotting the correlation coefficients in the $K \times L$ dimensional space corresponding to the K different template thickness and the L different resolutions gives a global maximum corresponding to a nanotube diameter equal to 70.5 nm and the spatial resolution equal to 45.5 nm. The values obtained from this analysis compare very well with the expected thickness of the nanotubes used as object (50-80 nm normal distribution) and also with the measured diameters obtained from the SEM measurements. The resolution also compares very well with the value obtained by the direct knife edge test 45.8 nm.

4 Summary and conclusions

We have demonstrated holographic imaging with spatial resolution comparable to the illumination wavelength. Combining the advantages of short wavelength SXR lasers with high resolution recording medium such as a photoresist enables holographic imaging maintaining sub-50 nm resolution in any point of the object that has millimeter scale extension. The method is very simple and can be easily extended to shorter wavelengths for imaging biological samples in the water-window range. The field of view is limited by the AFM scan size and consequently can be increased while the resolution in our case was limited by the wavelength of illumination. Extending this idea, future holographic imaging with sub-10nm resolution of biological samples using a suitable high resolution photoresist will be possible in a compact table-top system.

References

1. D. Gabor, "A new microscope pronciple". *Nature.* 161 777-778 1948.
2. A.V. Baez, "A study in diffraction microscopy with special reference to X-Rays". *Journal of the Optical Society of America.* 42 756-762 1952.
3. J.E. Trebes, S.B. Brown, E.M. Campbell, D.L. Matthews, D.G. Nilson, G.F. Stone, and D.A. Whelan, "Demonstration of X-ray holography with an X-ray laser". *Science.* 238 517-519 1987.
4. C. Jacobsen, M. Howells, J. Kirz, and S. Rothman, "X-Ray Holographic Microscopy Using Photoresists". *Journal Of The Optical Society Of America A-Optics Image Science And Vision.* 7 1847-1861 1990.
5. S. Lindaas, H. Howells, C. Jacobsen, and A. Kalinovsky, "X-ray holographic microscopy by means of photoresist recording and atomic-force microscope readout". *Journal of the Optical Society of America a-Optics Image Science and Vision.* 13 1788-1800 1996.

6. I. McNulty, J. Kirz, C. Jacobsen, E.H. Anderson, M.R. Howells, and D.P. Kern, “High Resolution Imaging by Fourier Transform X-ray Holography”. *Science*. 256 1009-1012 1992.
7. D. Sayre, H.N. Chapman, and J. Miao, “On the extendibility of X-ray crystallography to noncrystals”. *Acta Crystallographica Section A*. 54 232-239 1998.
8. S. Eisebitt, J. Luning, W.F. Schlotter, M. Lorgen, O. Hellwig, W. Eberhardt, and J. Stohr, “Lensless imaging of magnetic nanostructures by X-ray spectro-holography”. *Nature*. 432 885-888 2004.
9. B.R. Benware, C.D. Macchietto, C.H. Moreno, and J.J. Rocca, “Demonstration of a high average power tabletop soft X-ray laser”. *Physical Review Letters*. 81 5804-5807 1998.
10. Y. Wang, M.A. Larotonda, B.M. Luther, D. Alessi, M. Berrill, V.N. Shlyaptsev, and J.J. Rocca, “Demonstration of high-repetition-rate tabletop soft-x-ray lasers with saturated output at wavelengths down to 13.9 nm and gain down to 10.9 nm”. *Physical Review A*. 72 2005.
11. F. Brizuela, G. Vaschenko, C. Brewer, M. Grisham, C.S. Menoni, M.C. Marconi, J.J. Rocca, W. Chao, J.A. Liddle, E.H. Anderson, D.T. Attwood, A.V. Vinogradov, I.A. Artioukov, Y.P. Pershyn, and V.V. Kondratenko, “Reflection mode imaging with nanoscale resolution using a compact extreme ultraviolet laser”. *Optics Express*. 13 3983-3988 2005.
12. G. Vaschenko, C. Brewer, E. Brizuela, Y. Wang, M.A. Larotonda, B.M. Luther, M.C. Marconi, J.J. Rocca, and C.S. Menoni, “Sub-38 nm resolution tabletop microscopy with 13 nm wavelength laser light”. *Optics Letters*. 31 1214-1216 2006.
13. G. Vaschenko, E. Brizuela, C. Brewer, M. Grisham, H. Mancini, C.S. Menoni, M.C. Marconi, J.J. Rocca, W. Chao, J.A. Liddle, E.H. Anderson, D.T. Attwood, A.V. Vinogradov, I.A. Artioukov, Y.P. Pershyn, and V.V. Kondratenko, “Nanoimaging with a compact extreme-ultraviolet laser”. *Optics Letters*. 30 2095-2097 2005.
14. R.A. Bartels, A. Paul, H. Green, H.C. Kapteyn, M.M. Murnane, S. Backus, I.P. Christov, Y.W. Liu, D. Attwood, and C. Jacobsen, “Generation of spatially coherent light at extreme ultraviolet wavelengths”. *Science*. 297 376-378 2002.
15. A.S. Morlens, J. Gautier, G. Rey, P. Zeitoun, J.P. Caumes, M. Kos-Rosset, H. Merdji, S. Kazamias, K. Casson, and M. Fajardo, “Submicrometer digital in-line holographic microscopy at 32 nm with high-order harmonics”. *Optics Letters*. 31 3095-3097 2006.
16. R.I. Tobey, M.E. Siemens, O. Cohen, M.M. Murnane, H.C. Kapteyn, and K.A. Nelson, “Ultrafast extreme ultraviolet holography: dynamic monitoring of surface deformation”. *Optics Letters*. 32 286-288 2007.
17. R.L. Sandberg, A. Paul, D.A. Raymondson, S. Hadrich, D.M. Gaudiosi, J. Holtsnider, R.I. Tobey, O. Cohen, M.M. Murnane, and H.C. Kapteyn, “Lensless diffractive imaging using tabletop coherent high-harmonic soft-x-ray beams”. *Physical Review Letters*. 99 2007.
18. P. Wachulak, Bartels, R., Marconi, M.C., Menoni, C.S., Rocca, J.J., Lu, Y., Parkinson, B., “Sub 400 nm spatial resolution extreme ultraviolet holography with a table top laser”. *Optics Express*. 14 9636-9642 2006.

19. P. Wachulak, Marconi M.C., Bartels R., Menoni, C.S., Rocca, J.J, "Volume extreme ultraviolet holographic imaging with numerical optical sectioning". *Optics Express*. 15 10622-10628 2007.
20. Y. Liu, M. Seminario, F.G. Tomasel, C. Chang, J.J. Rocca, and D.T. Attwood, "Achievement of essentially full spatial coherence in a high-average-power soft-x-ray laser". *Physical Review A*. 6303 2001.
21. Y.W. Liu, M. Seminario, F.G. Tomasel, C. Chang, J.J. Rocca, and D.T. Attwood, "Spatial coherence measurement of a high average power table-top soft X-ray laser". *Journal De Physique Iv*. 11 123-126 2001.
22. U. Schnars and W.P.O. Juptner, "Digital recording and numerical reconstruction of holograms". *Measurement Science & Technology*. 13 R85-R101 2002.
23. P.W. Wachulak, Brewer, C.A., Brizuela, F., Chao, W., Anderson, E.H., Bartels, R.A., Menoni, C.S., Rocca, J.J., Marconi, M.C., "Simultaneous determination of feature size and resolution in soft x-ray microscopy images ". *Journal of the Optical Society of America B*. 25 B20-B26 2008.
24. T. Yatagai, M. Idesawa, Y. Yamaashi, and M. Suzuki, "Interactive Fringe Analysis system- Applications to Moire Contourogram and interferogram". *Optical Engineering*. 21 901-906 1982.

Ablation Measurements Using Ni-Like Ag X-Ray Laser Transmission

N. Booth[1], M.H. Edwards[1], Z. Zhai[1], G.J. Tallents[1], T. Dzelzainis[2], R. Ferrari[2], C.L.S. Lewis[2], G. Gregori[3], D. Neely[3]

[1] Department of Physics, University of York, Heslington, York, YO10 5DD, UK
[2] School of Mathematics and Physics, Queen's University of Belfast, Belfast, BT7 1NN, UK
[3] Central Laser Facility, STFC, Rutherford Appleton Laboratory, Chilton, Didcot, OX11 0QX, UK

Abstract. Results of a recent experiment to measure laser ablation rates from solid targets consisting of thin (50 nm) Fe at irradiances of ~ $2\mathrm{x}10^{13}$ Wcm^{-2} are presented. A Ni-like Ag X-ray laser at 13.9 nm was utilised to probe the targets and measurements of the transmission of the X-ray laser through the target were recorded by a flat-field spectrometer. The laser ablation rate is determined by the transmission of the X-ray laser through the remaining, unablated, iron layer. A self-regulating model of laser ablation is fitted to the experimental data to give a value of the absorption, A, of the heating laser in the target plasma of $A = 0.01$.

1 Introduction

Measurements of laser ablation rates are particularly important in laser fusion, where the outer surface of a target wall is ablated and results in the implosion of the target. Ablation rates are also important in laser deposition and cutting. Laser ablation rates are determined by the rate of energy absorption and transport within a laser-produced plasma. Absorption of the incident laser energy predominantly occurs by resonance absorption or by inverse bremsstrahlung.

Previous laser-ablation rate measurements have been performed in a number of ways. The ion flux from the target at a Faraday cup detector can be integrated over time and angle to the target to evaluate the full mass of the target ablated [1]. The emission from buried signature layers can be used to measure depth of heat penetration and hence the depth of ablation [2].

Earlier plasma probing experiments with X-ray lasers were undertaken using X-ray lasers as an interferometry tool in order to view expanding laser-produced plasma [3]. These plasma probing experiments also investigated transmission variations produced by the Rayleigh-Taylor growth of surface perturbations due to non-uniformities in the heating laser beams [4,5].

X-ray lasers are ideally suited to acting as plasma probes due to the fact that they have a high average intensity ($\approx 10^{10} - 10^{11}$ Wcm^{-2}), have a very narrow bandwidth (typically $\Delta\nu/\nu < 10^{-4}$) and have a low divergence (around 5 – 10 mrad). These properties ensure that the X-ray laser is brighter at a detector than self–emission from the sample plasma [6]. Edwards et al [7,8] showed that X-ray lasers can be used to measure plasma opacity and the rate of laser ablation of solid targets. In this paper, we present results where laser ablation of solid targets is measured from X-ray laser transmission of thin iron targets using a flat-field spectrometer as detector.

2 Experiment

The experiment was performed on the VULCAN target area west facility at the Central Laser Facility at the Rutherford Appleton Laboratory. A solid silver slab target was irradiated using two overlapping pulses to collisionally excite a 4d – 4p Ni-like Ag X-ray laser at 13.9 nm wavelength. The pre-pulse used to create the pre-plasma was frequency doubled to 2ω, had an energy of $\approx$ 10 J in 290 ps duration and was focussed to a line of dimension $\approx$ 75 μm x 5 mm on the target, at an irradiance of approximately $8\text{x}10^{12}$ Wcm^{-2}. The main CPA pulse contained approximately 35 J in 750 fs, resulting in an irradiance of $\approx 3\text{x}10^{15}$ Wcm^{-2} in the line focus. The CPA pulse remained at its original wavelength of $1\omega = 1.054$ μm.

The ablation target was heated and ablated using a third beam of the 1.054 μm VULCAN laser. The beam contained $\approx$ 30 J in 290 ps and was focussed to a spot $\approx$ 0.5 mm x 1.0 mm on target with an f/10 refracting lens at an incidence angle of 73° to the target normal. The average peak irradiance of the ablating pulse was 10^{13} Wcm^{-2}. The ablation targets were manufactured with a thickness of 50 nm of Fe supported on a 0.1 μm layer of CH and were mounted onto circular copper mounts. The transmission of the X-ray laser through the targets is dominated by the iron layer and is highly sensitive to changes in the thickness of the iron. This sensitivity enables accurate measurements of the changes in thickness of the iron layer as the iron material (closest to the laser) is ablated.

A flat-field spectrometer utilising an Andor CCD camera coupled as a detector to the back of the spectrometer was used to record the transmission of the X-ray laser through the iron targets. The CCD camera was filtered using various thicknesses and combinations of Al and CH foils in order to reduce the signal from the laser at the detector chip below saturation, and also to prevent background emission from the target from reaching the CCD detector. The X-ray laser light is transmitted through the ablation target at an angle of

45° to the ablation target normal and onto the flat-field spectrometer grating. Figure 1 shows a typical image from the flat-field spectrometer and figure 2 shows the corresponding lineout used to deduce the target transmission to the 13.9 nm radiation. The 13.9 nm lasing lines in both first and second order are clearly visible in the image. Around the position of the first order lasing line, a large amount of self emission from the heating of the ablation target is visible. Due to the problem of self emission, it is necessary to have a bright backlighter to probe the plasma. As can be seen in figure 1, the X-ray laser is still visible over the level of self emission. However, due to the fact that the transmission is more clearly visible through the target in second order, and there is a high level of self emission around the first order, all measurements of transmission given in section 3 are taken from the second order.

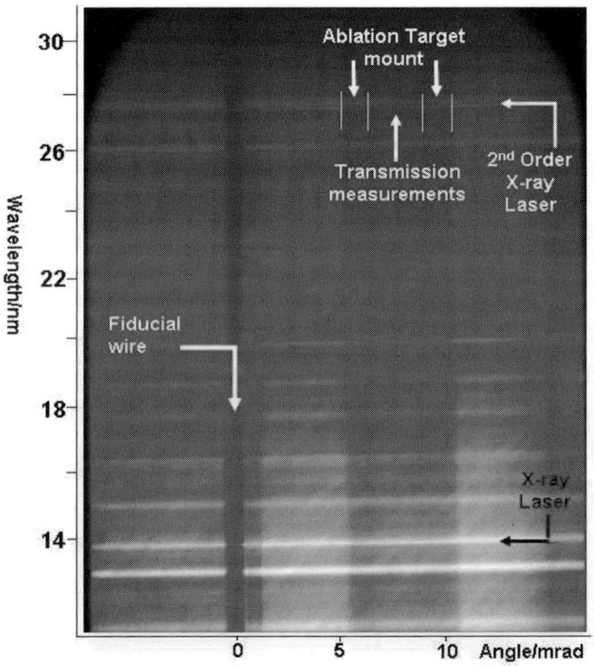

Fig. 1 Flat-field spectrometer image showing self emission from the ablation target plasma around the first order 13.9 nm lasing line. The ablation target transmission is apparent in the 6 – 9 mrad region in the x-axis.

The heating and subsequent ionisation of the ablation targets was monitored using a crystal spectrometer which consisted of a ≈ 100 μm vertical entrance slit, a flat KAP crystal and an Andor CCD camera detector. From the identification of the spectral lines using [9], the spectral emission is identified to be predominantly neon-like. This neon-like ionisation of the ablation target implies that ablated plasma is heated to approximately 500 eV.

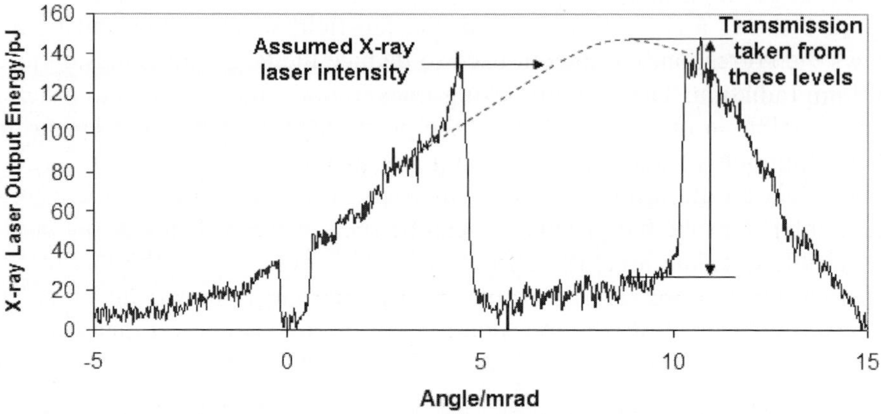

Fig. 2 Lineout of the second order X-ray laser transmission from the image shown in Figure 1.

A crossed-slit camera monitored the spatial variation of the x-ray emission from the ablation target (Figure 3). A lineout through the vertical direction of Figure 3 is shown in Figure 4. From the image in Figure 3 and the lineout in Figure 4, it is possible to see that the irradiation of the ablation target is approximately uniform, but with some evidence of 'hot spots'.

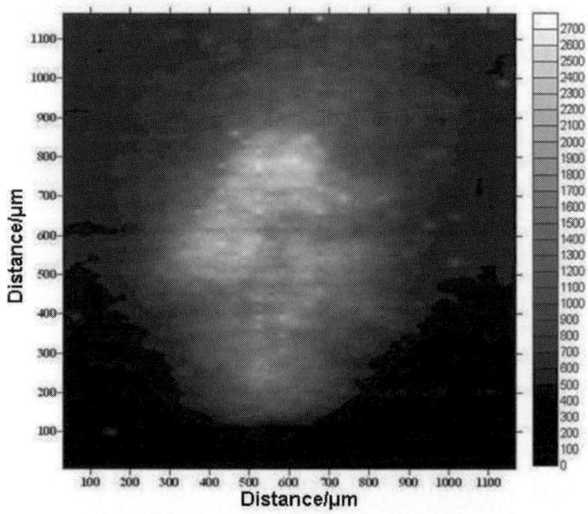

Fig. 3 Spatially resolved crossed-slit camera image of X-ray emission from the ablation target (photon energies > 2 keV). The camera was filtered with 8.4 μm of aluminium and 19 μm of mylar.

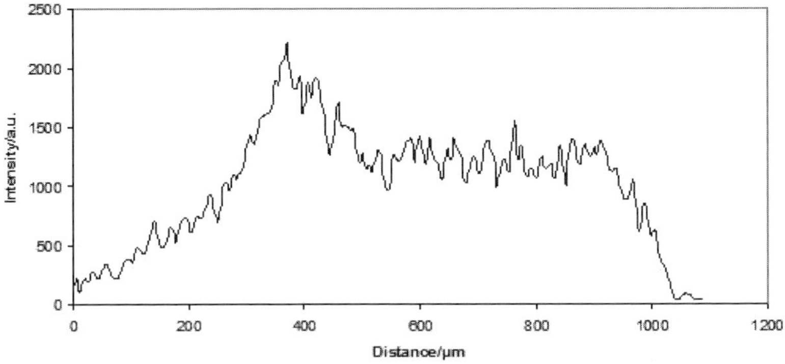

Fig. 4 Lineout in the vertical direction at a horizontal distance of 550 µm, of the crossed-slit camera image shown in Figure 3.

3 Laser Ablation Rate Measurements

Using the TOPS code developed at Los Alamos [10], it is possible to plot the transmission of the 13.9 nm Ni-like silver X-ray laser radiation through an iron target of 50 nm thickness at solid density assuming constant areal density for different densities as a function of target temperature (Figure 5). For ablated plasma with densities decreasing to below 0.1 gcm^{-3}, the iron is almost fully transparent at temperatures greater than 100 eV. From the crystal spectrometer images, Ni-like ionisation with a plasma temperature of 500 eV has been identified. At this temperature the transmission through iron plasma is close to 100% and so the transmission through the target is dominated by the remaining solid iron transmission.

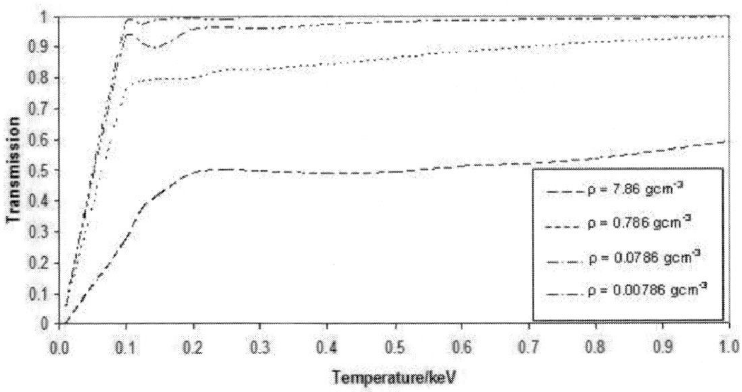

Fig. 5 Transmission of the 13.9 nm Ni-like silver X-ray laser through iron at varying densities as a function of temperature calculated using the TOPS code.

Using flat-field spectrometer images such as Figure 1, it is possible to measure the transmission of the 13.9 nm X-ray laser radiation through the ablation target. By changing the delay between the ablation laser pulse and the pulse producing the X-ray laser, the transmission as a function of time can be measured (Figure 6).

The self-regulating model of laser ablation [11] can be used to evaluate the material ablated by the laser and using the solid transmission values [12] for unablated material, the transmission through the target as a whole can be calculated. The self-regulating model applies where inverse bremsstrahlung is considered to be the heating source driving the expansion of the plasma plume. It is so called as the optical depth in the plasma for absorption of the heating beam must be approximately unity [13]. Previous measurements have shown that the self-regulating model applies for experiments with longer duration (≈ 500 ps) pulses and low irradiance of the heating beam on target [14]. The absorbed laser irradiance $I_a = AI$, where I is the irradiance on target of the heating beam and A represents the fraction of the irradiance absorbed by the target is employed in the self-regulating model.

The self-regulating model to calculate the mass ablation rate dm/dt ($\mathrm{gcm^{-2}s^{-1}}$) [11] is given by

$$\frac{dm}{dt} \approx 1.06x10^{-5} \frac{A_m^{7/8}}{Z^{9/8}} \frac{I_a^{1/2}}{\lambda^{1/2} t^{1/4}} \qquad (1)$$

where A_m is the mass number of the target material, Z is the average charge of the plasma, λ is the wavelength of the heating beam laser (in μm) and t is the time from the start of the laser pulse. From equation 1 it is possible to calculate the rate of mass ablation from the target as a function of time and to use this result to calculate the transmission of the X-ray laser through the target using the knowledge of the mass of target remaining from solid iron and CH tabulated transmissions [12]. Two curves for the model are plotted alongside the experimental data in Figure 6. The curve which most closely matches the data uses a value for the absorption of the ablating laser of $A = 0.01$ and the second curve is for a value of $A = 0.05$, which was found to be applicable in previous longer pulse duration experiments [9]. The expected value for absorption is typically $A = 0.3$ at wavelengths of ≈ 1 μm [15].

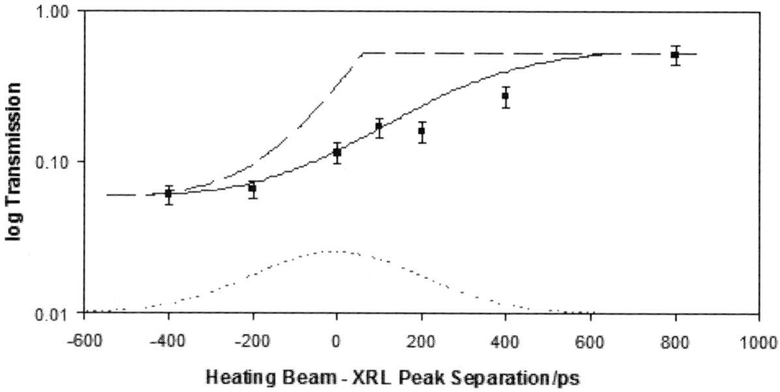

Fig. 6 Experimental (■) values for the transmission of the X-ray laser through the target as a function of time are compared with the self-regulating model for a value of A = 0.01 (—) and A = 0.05 (– – –). The experimental data points agree closely to an absorption of 1 %. The Gaussian curve (---) is added to display the heating beam intensity variation as a function of time.

Measurements of the ablation rate were also made with a large ablating beam (≈ 2 mm x 5 mm) focus (Figure 7). The irradiance on the targets is ≈ $1x10^{12}$ Wcm^{-2}, which is an order of magnitude lower than the previous measurements. The self-regulating model of transmission at this lower irradiance with $A = 0.01$ is superimposed in Figure 7. There is an approximate agreement of the experimental data points with the model.

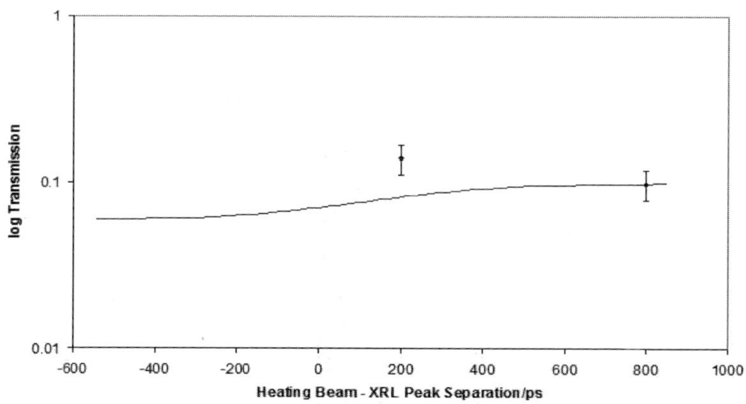

Fig. 7 Transmission measurements through the target measured experimentally (♦) for low heating laser irradiance of ≈ $1x10^{12}$ Wcm^{-2} and compared to the self-regulating model (—) for this irradiance at the target and for the same value of $A = 0.01$ as is used in Figure 4.

4 Conclusions

Results have been presented of the use of an X-ray laser as a probe to perform measurements of the rate of material ablation by a laser. It has been shown that a flat-field spectrometer may be considered a suitable diagnostic by which to measure the transmission of an X-ray laser through a sample target.

The absorption, *A*, of laser energy calculated by fitting the self-regulating model of laser ablation to the experimental transmission measurements was found to be low (with $A = 0.01$). The expected value for absorption at laser intensities of $\approx 2\times10^{13}$ Wcm^{-2} at an incident wavelength of ≈ 1 μm is typically $A = 0.3$. The absorption, *A*, measured with the technique represents the fraction of energy transferred to thermal plasma. Energy transferred to superthermal electrons or transported laterally does not contribute to ablation. As hot spots are visible in the crossed slit camera images (Figure 3), it is possible that these regions of higher irradiance may have produced hot electrons so further reducing the absorption to thermal plasma.

Acknowledgements

Funding from the United Kingdom Engineering and Physical Sciences Research Council and the Science and Technologies Facilities Council is gratefully acknowledged.

References

1. JP Anthes et al, *Phys. Rev. Letts.,* **41** 1300 (1978)
2. Y Al Hadithi et al, *Phys. Plas.*, **1** 1279 (1994)
3. RF Smith et al, *Phys. Rev. Letts.,* **89** 065004 (2002)
4. DH Kalantar et el, *Phys. Rev. Letts.,* **76** 3574 (1996)
5. E Wolfrum et al, *Phys. Plas.*, **5** 227 (1998)
6. MH Edwards et al, *Phys. Rev. Letts.,* **97** 035001 (2006)
7. MH Edwards et al, *Phys. Rev. Letts.,* **99** 195002 (2006)
8. H Huang et al, *JQSRT*, **109** 2272 (2008)
9. F Bijerk et al, *J. X-ray Sci. and Tech.,* **3** 133 (1992)
10. NH Magee et al, Astronomical Society of the Pacific Conference Series (Astrophysical Applications of Powerful New Databases, S. J. Adelman and W. L. Wiese eds.) 78 51 (1995)
11. H Puell, *Z Naturforsch A,* **24** 1807 (1970)
12. http://henke.lbl.gov/optical_constants/filter2.html
13. GJ Pert, *Phys. Rev. A,* **73** 033809 (2006)
14. MH Edwards et al, *Phys. Rev. Letts.,* **99** 195002 (2007)
15. DC Slater et al, *Phys. Rev. Letts.,* **46** 1199 (1981)

High Sensitive Characterization of Microdomain Structures in PZN-PT (91/09) by Means of Coherent Soft X-Ray Laser Speckle

K. Namikawa[1, 5, 6], R. Z. Tai [2, 5, 6], M. Matsushita[3, 6], K. Ohwada[4, 6], M. Kishimoto[5, 6]

[1]Department of Physics, Tokyo Gakugei University, Koganei-shi, Tokyo 184-8501, Japan
[2]Shanghai Synchrotron Facilities, Shanghai Institute of Applied Physics, Chinese Academy of Science, Shanghai 201204, China
[3]Functional Material Development Center, Research Laboratories, JFE Mineral Co., Ltd., Nishihama, Chuou-ku, Chiba, Chiba 260-0826, Japan
[4]Japan Atomic Energy Agency (Spring-8), Koto, Sayo-gun, Hyogo, 679-5148, Japan
[5]Quantum Beam Science Directorate, Japan Atomic Energy Agency, Kzugawa-shi, Kyoto 619-0215, Japan
[6]CREST, Japan Science and Technology Agency

Abstract. Coherent soft x-ray laser speckle has been examined to study the microdomain structures of relaxor ferroelectric 0.91PZN–0.09PT. Microdomains with nanometer size of about 400 nm were identified precisely. Temperature variation and electric field variation of the micro domain were characterized by a kind of foot print method with high sensitivity of a percent order.

1 Introduction

The relaxor ferroelectrics (1-x)Pb($Zn_{1/3}Nb_{2/3}$)O_3–xPbTiO_3 ((1-x)PZN–xPT) exhibit ultrahigh piezoelectric and electromechanical coupling constant, and has been used in ultrasonic devices, solid actuators etc.[1, 2] The high performance of these typical relaxor ferroelectrics have been considered to be the contribution of direct polarization rotation from [111] to [100] under an electric field, proposed by Hu and Cohen.[3] Experimental results show that the ultrahigh performance exists in a monoclinic phase near the morphotropic phase boundary (MPB).[4, 5] Noheda suggested that PZN–PT has a different polarization path which is distinguished from PbZrTiO_3 proposed by Cohen, and forms a different type of monoclinic distortion near MPB.[6] So investigating the response of domain structure among PZN–PT under an external electric field is vital for the understanding of relaxor's high performance. Recently, the method of piezo response force microscopy (PFM) has been invested to study the domain structures from submicron level to even nanometer scale among the relaxor ferroelectric (1-x)PZNxPT. [7, 8]

The relaxor materials exhibit relaxor characteristics most typically near the Curies temperature T_C, so it is meaningful to explore how the structure variations take place near T_C. An effect of external electric field applied on the relaxor material near T_C is believed to realize a polarization structure similar to that at lower temperatures, which contribute to the ultrahigh electromechanical character of material.[9] The variations of microdomain structure in relaxor ferroelectrics 0.91PZN-0.09PT versus the temperature variation and external electric field E were investigated by means of soft x-ray speckle method. Similar experiment has been conducted for the prototype ferroelectric $BaTiO_3$.[10]

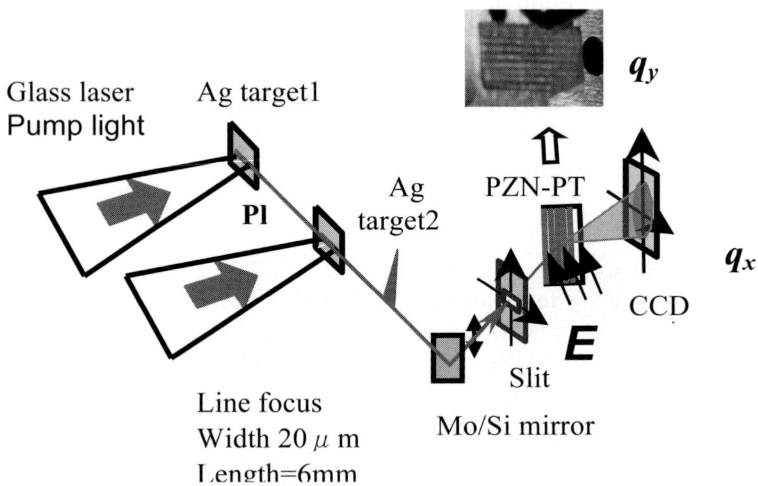

Fig. 1 Experimental setup. External electric field was applied along x axis.

2 Experiment and Results

The experimental setup is schematically shown in Fig. 1. X-ray laser is a type of amplified spontaneous emission (ASE) x-ray source. The x-ray laser media here is silver plasma, which is produced by two glass pump laser beam focusing on two silver targets in turns. The width of the pump laser is less than 20 μm, and the length is of about 6mm. The soft x-ray emitting from the silver target with a wavelength 13.9nm, bandwidth 10^{-4}, pulse duration 7ps, and 10^{12} photons with one pulse is reflected by a Mo/Si multilayer cylindrical mirror which is used to produce a vertically polarized convergent x-ray beam. A slit which located at 1.9m from the silver target was used to block the incoherent part of x-ray beam. The horizontal dimension of the slit was 80 μm, while the vertically dimension could be adjusted. The sample used in this experiment was a flux-grown relaxor ferroelectrics $0.91Pb(Mg_{1/3}Nb_{2/3})O_3$-$0.09PbTiO_3$ (0.91PZN-0.9PT) with dimensions of 5mm (h) × 12mm (v) and Curie temperature 175℃. The sample surface has been finished to optically flat at room temperature. Sub-millimetre size bright and dark bands have been observed in optically finished surface by naked eyes as shown in the insertion in Fig. 1. Polarization domain structures existing at room temperature have certain tilt angles among them. The tilt angle imbedded in the optically flat surface is a kind of foot print of the polarization domains existed at room temperature. The sample was located close to the slit with its band boundaries lay along the vertical direction (*y* axis). The temperature of the sample was controlled from room temperature up to 200℃ with precision of 0.5℃. The external electric field was applied perpendicular to the sample surface. The grazing angle was 10 degree. The speckle patterns recorded by a soft X-ray camera situated 0.5m distant from the slit.

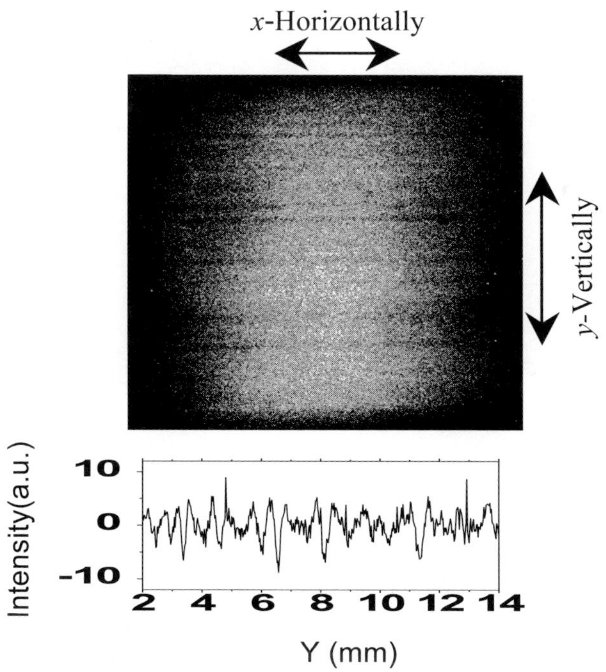

Fig. 2 Projected image of PZN-PT surface (the vertical sit width is 3mm), diffusive scattering pattern was observed. This pattern was measured at 150℃.

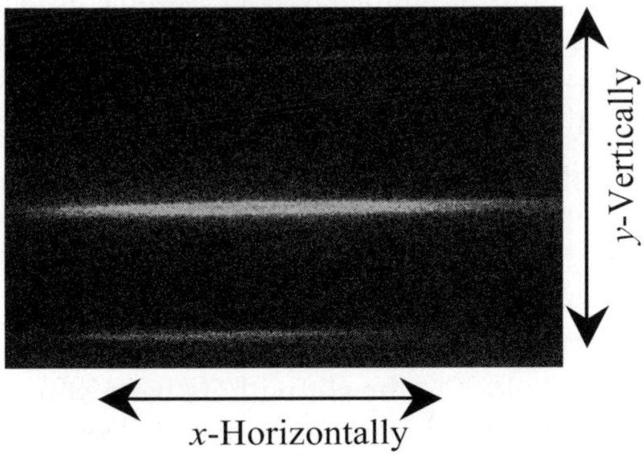

Fig. 3 Typical vertical scattering pattern when the vertical sit width was adjusted to be 80μm. The strong central line along x axis came from horizontally-focusing Mo/Si mirror, while the two symmetric and weak lines along x axis are the $\pm 1^{st}$ order diffraction lines.

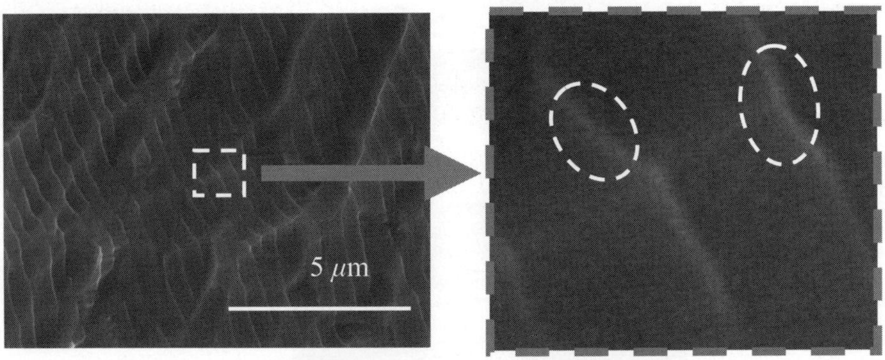

Fig. 4 SEM image of 0.91PZN-0.09PT sample surface.

A horizontally divergent projected reflection image of the 0.91PZN-0.9PT sample as illustrated in Fig. 2 are observed from room temperature up to 200 ℃, where the vertical slit width was adjusted to 3mm. A diffusive scattering pattern was observed, within which tremendous bright bands exist. The contrast of this image is attributed to the difference in reflectivity of the each band which is due to the performance of the optical polishing among them inherent from the crystallographic difference in mechanical resistivity for the polishing. The typical size of the bands estimated from the intensity profile inserted in Fig. 2 below is an order of 50μm. The scattering pattern when the vertical sit width was adjusted to 80μm is shown in Fig. 3. With the reduction of slit width, the diffusive scattering pattern shrinks to a pattern composed of three explicit and sharp diffraction lines. The central line with higher intensity came from the reflection of the horizontally-focusing Mo/Si mirror. Two scattering lines with relative weak intensity distribute symmetrically according to the central line, which are ascribed to the $\pm 1^{st}$ diffractions respectively. The scattering vector $\boldsymbol{q}$ is defined as $\boldsymbol{q} = k\,\xi\,/\,z\,\boldsymbol{e}_y$ in a reasonable approximation, where the ξ is the coordinate on CCD camera along the diffracting direction with a unit vector $\boldsymbol{e}_y$, z is the distance between the sample and CCD camera. The $\pm 1^{st}$ order diffraction peaks appear when $q = \pm 2\pi/d$, where d is the period of recorded microdomains. From the cross sectional intensity of Fig. 3, 400nm order domain structures were recognized, which are exist in 50μm order bands in Fig. 2. As a complementary way for soft x-ray speckle method, SEM is a traditional method in the observation of the surface morphology of relaxor ferroelectric.[11] Micro domain size of about 400nm achieved from the x-ray speckle pattern in Fig. 3 is consistent with the SEM image in Fig. 4. Clear micro domains with size of about 400nm were observed. The irregular domain boundaries between 400nm size domains are a chain of 50nm size micro domains as recognizable from the enlarged picture inserted.

Intensity profiles in Fig. 5 show the temperature dependence of the scattering pattern with vertical slit width 80μm. The central peak, which is considered to be the specular reflection, decreases in the tetragonal phase (150℃) and in cubic phase (200℃), while the intensity of the right hand side peaks (1st order) in tetragonal phase and cubic phase increases about 1/3 compared to that of in rhombohedral phase (24℃). The peak position of the diffraction line in rhombohedral phase shifts to the lower scattering angle by 3% comparing with the higher symmetric phases.

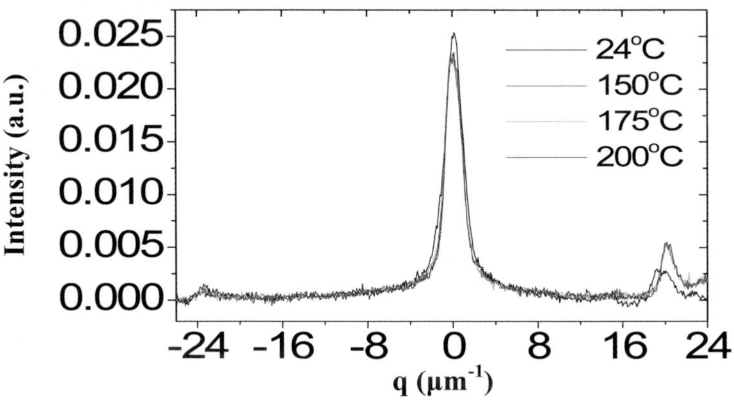

Fig. 5 Quantitative vertical intensity distribution at various temperatures

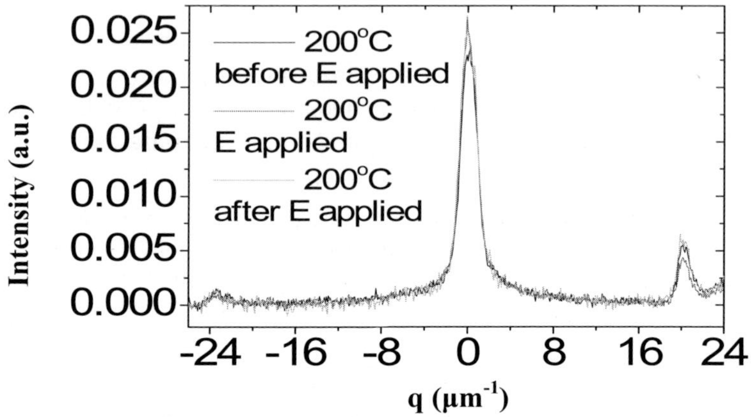

Fig. 6 Quantitative vertical intensity distributions before, when and after an external high DC field was applied at 200℃.

Intensity profiles in Fig. 6 show the electric field dependence of the scattering pattern with the vertical slit width 80μm. Intensity of the specular reflection increases when the external electric field ***E*** was applied perpendicular the c-plane in cubic phase (200℃), while the intensity of the 1st order diffraction line decreases about 25% comparing with that without the external electric field. The peak position of the 1st order diffraction line does not shift independent of the external electric field. X-ray diffraction data of the lattice structure change with the temperature were shown in Table 1. Table 1 exhibit the lattice constant along a, b, c axe and angle α, β, γ change slightly, in accordance with the corresponding structure phase at certain temperature. These variations affect the domain structures and their speckle patterns. However, we could not recognize any change in these constants even when the external electric field applied.

Table 1 X-ray observation results of the lattice structure

Temperature	Symmetry	*a, b, c* (0.1nm)	α, β, γ(°)
30 ℃	Rhombohedral	4.051, 4.076, 4.093	90.00, 90.02, 90.00
125 ℃	Tetragonal	4.063, 4.064, 4.094	90.01, 89.97, 90.00
200 ℃	Cubic	4.074, 4.073, 4.072	90.03, 89.96, 90.00

3 Discussion

Opposite behaviour in the temperature dependence of the intensity of the specular reflection and the 1st order diffraction line in Fig. 5, is due to appearance of the tilts in the polished surface between the 400nm size domain and the 50nm size domain: β changes from 90.02° to 89. 97° or 89.96°, and a difference in lattice constant appear between a-axis and b-axis when the structure changes from the rhombohedral phase to the tetragonal or cubic phases. Shift of the 1st order peak position is due to the shrink in size of the 400nm size domains in higher symmetric phases. The lattice constants in a, b and c axe tend to be equal in a higher symmetric phase, e.g., cubic phase, which has been confirmed by D. E. Cox in the phase diagram observation on the high piezoelectric perovskite system.[12] From the rhombohedral phase to the tetragonal phase, a shrink of lattice constant in *b*-axis occurs; while from the rhombohedral phase to the cubic phase, a shrink of lattice constant in *c*-axis occurs, just as Table 1 shows. The lattice constant shrinks lead to the slightly decreasing of microdomain size, which enlarges the distance between the 1st order diffraction line and specular reflection line. External electric field applied perpendicular to (100) face results in the orientation of the polarization of clusters (several nanometres) parallel to each of four [±1±11] direction close to applied ***E*** direction,[13, 14] or the similar degeneracy effect of microdomains.[15] This effect leads a symmetry change toward rhombohedral like phase. Thus the tilt between the 400nm size domains and the 50nm size domains decreases. As a result, intensity of the specular reflection increases and that of the 1st order diffraction peak decreases. However, the lattice constant does not change as confirmed by synchrotron radiation experiment, probably because the sample is still in paraelectric phase in average. As the mean width of two 400nm size domains does not change in this unknown phase, the peak position of the 1st order diffraction peak remains at the same scattering angle as that in the case without the electric field.

In summary, variations of micro domains structure in relaxor ferroelectric 0.91PZN-0.09PT were investigated by means of coherent soft x-ray speckle. The precise variation characteristics of micro domains with temperature and external electric field were obtained by soft x-ray speckle foot print observation. Direct observation of the foot print of the nanometer sized domain has an advantage for the hard x-ray diffraction method. This method has high sensitivity, typically a few percent, to the structural changes of micro domains comparing with the direct observation of SEM image.

References

1. J. Z. Xiao, M. Shao, and S. T. Yin, Journal of Crystal Growth **240**, 521 (2002).
2. M. Abplanalp, D. Barošová, P. Bridenbaugh, J. Erhart, J. Fousek, P. Günter, J. Nosek, and M. šulc, J. Appl. Phys. **91**, 3797 (2002).
3. H. X. Fu and R. E. Cohen, Nature **403**, 281 (2000).
4. M. Iwata, K. Katsuraya, I. Suzuki, M. Maeda, N. Yasuda, and Y. Ishibashi, Mater. Sci. Eng. B **120**, 88 (2005).
5. G. Xu, H. Luo, H. Xu, and Z. Yin, Phys. Rev. B **64**, 020102 (2001).
6. B. Noheda, D. E. Cox, G. Shirane, S. E. Park, L. E. Cross, and Z. Zhang, Phys. Rev. Lett. **86,** 3891 (2001).
7. S. Madeswaran, S. V. Rajasekaran, R. Jayavel, S. Ganesamoorthy, G. Behr, Mater. Sci. Eng. B **120**, 32 (2005).
8. M. Abplanalp, D. Barošová, P. Bridenbaugh, J. Erhart, J. Fousek, P. Günter, J. Nosek, M.šulc, Solid State Commu. **119**, 7 (2001).
9. ibid 3.
10. R. Z. Tai, K. Namikawa, M. Kishimoto, M. Tanaka, K. Sukegawa, N. Hasegawa, T. Kawachi, M. Kado, P. Lu, K. Nagashima, H. Daido, H. Maruyama, A. Sawada, M. Ando, and Y. Kato, Phys. Rev. Lett. **89,** 257602 (2002).
11. M. Abplanalp, D. Barošová, P. Bridenbaugh, J. Erhart, J. Fousek, P. Günter, J. Nosek, and M. šulc, J. Appl. Phys. **91**, 3797 (2002).
12. D. E. Cox, B. Noheda, and G. Shirane, Appl. Phys. Lett. **79**, 400 (2001).
13. Z. Guo, R.Z. Tai, H. Xu, C. Gao, H. Luo, G. Pan, C. Hu, Di Lin, R. Fan, R. Li, R. Yan, X.Z. Zhang, K. Namikawa, J. Appl. Phys. **101**, 053505 (2007).
14. Z. Guo, R. Z. Tai, H. Xu, Chen Gao, G. Pan, H. Luo, K. Namikawa, Appl. Phys. Lett. (in press).
15. S. E. Park and T. R. Shrout, J. Appl. Phys. **82**, 1804 (1997).

Warm Photoionized Plasmas Created by Soft X-Ray Laser Irradiation of Solid Targets

M. Berrill[1], F. Brizuela[1], B. Langdon[1], H. Bravo[1], C.S. Menoni[1] and J.J. Rocca[1]

[1] NSF ERC for Extreme Ultraviolet Science and Technology, Colorado State University, USA

Abstract. We report results of the study of warm plasmas generated by focusing 46.9 nm soft x-ray laser pulses of nanosecond duration onto Si, Cr and Ag targets. The absorption is dominated by single photon photoionization, resulting in plasmas that are significantly different from those created by visible lasers. Spectra from the soft x-ray laser-created plasmas agree with 1 ½ D simulations in showing that the Si plasmas are significantly colder and less ionized than the Cr and Ag plasmas, confirming that, in contrast to plasmas created by optical lasers, the plasma properties are largely dependent on the absorption coefficient of the target material.

1 Introduction

There is significant current interest in the study of the interaction of intense soft x-ray laser beams with materials [1-2]. The mechanism of soft x-ray laser energy deposition into the materials differs from that in optical laser-material interactions. The high photon energy of soft x-ray beams exceeds the ionization energy of all neutral atoms and low charge ions resulting in direct single-photon ionization of the plasma, an absorption mechanism unavailable to optical lasers. The fact that soft x-ray wavelengths are associated with critical densities that exceed solid density (eg. $n_c = 5 \times 10^{23}$ cm^{-3} for $\lambda = 46.9$ nm light) results in greatly decreased Bremsstralung absorption and the direct deposition of a large fraction of the laser pulse energy in the solid target, where the initial depth of the heated region is mainly determined by the absorption coefficient of the material. By using short soft x-ray laser pulses, it should be possible to create plasmas with very uniform conditions over relatively large volumes [1].

Table-top capillary soft x-ray laser beams generate intense pulses of ~ 1.2 ns duration [3] that when focused onto a solid target readily generates a plasma [4]. The recent demonstration of soft x-ray free electron lasers (FEL) [5] will allow the extension of the study to plasmas heated by femtosecond soft x-ray laser pulses. Motivated by the advent of FELs, theoretical studies have been conducted, that predicted the characteristics of plasmas created with intense monochromatic soft x-ray light will depend strongly on the target material and differ significantly from those created with visible lasers [2]. However, no experimental results of the study of the characteristics of such plasmas have been reported. A few groups have investigated the ablation of materials with focused soft x-ray laser beams [4,6-10], but these studies were focused on the effects on the ablated targets and not on the plasma phenomena.

We have conducted what to our knowledge is the first spectroscopic study of plasmas created by focused soft x-ray laser pulses on solid targets, and have compared the results to hydrodynamic/atomic physics model simulations [11]. Low temperature plasmas were generated by irradiating low Z (Si), and mid-Z (Cr and Ag) slab targets with focused laser pulses of 46.9 nm wavelength light (26.5 eV) from a Ne-like Ar capillary discharge soft x-ray laser. The pulses have a duration of ~1 ns duration. The measurements show, in accordance with hydrodynamic model calculations, that the much lower absorption coefficient for Si at this wavelength results in plasmas that are significantly colder and less ionized than Cr and Ag plasmas, in spite of the faster expansion of the Cr and Ag plasmas. In the nanosecond regime thermal heat conduction and expansion of the plasma into vacuum play a significant role by increasing the amount of ablated material and a significant amount of hydrodynamic cooling on this timescale. This is a different regime that differs and is complementary to the studies that will be conducted at FELs, which will employ pulses in the femtosecond time scale.

2 Experimental Setup

A schematic diagram of the experimental setup is shown in Figure 1. Si, Cr, and Ag targets were irradiated by focusing pulses of ~100 μJ energy and 1.2 ns duration created by a 46.9 nm Ne-like Ar capillary discharge soft x-ray laser [3,11]. The experiments were conducted in a vacuum chamber placed at 130 cm from the output of the laser. The laser beam was focused onto the target using a mirror coated with a Sc/Si multilayers [13], producing a focal spot of 10-15 μm in diameter. The targets consisted of thin slabs that intersected part of the incoming laser beam reducing the irradiation fluence on the target and making the focal spot slightly asymmetric. Shots with reduced irradiation energy were obtained by attenuating the beam with 0.2 μm thick, freestanding Al foils with measured transmissivity of ~17% at 46.9 nm. The targets were placed on a motorized translation stage that allowed the selection of the irradiated area and distance to the focusing mirror. The plasma light was collected by a 10 cm focal length fused silica lens to create a 1-to-1 image of the plasma on the entrance slit of a 0.3 m focal length visible/ultraviolet spectrometer. The light collection system was sufficiently efficient to produce line spectra from a single laser shot. However, most spectra were obtained accumulating several laser shots to improve the signal-to-noise ratio.

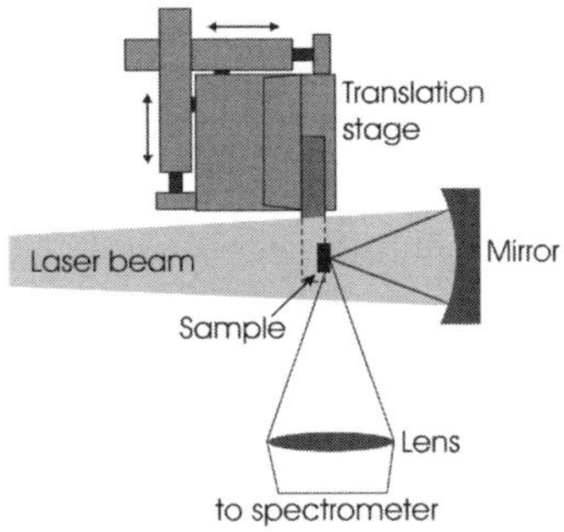

Fig. 1 Experimental setup used for the generation and spectroscopy of the soft x-ray laser-created plasmas.

3 Experiment and simulation results

Experiments were conducted irradiating the targets with three different soft x-ray laser pulse energies: 0.5 μJ, 3.1 μJ and 17 μJ; corresponding to peak intensities of $4\text{x}10^8$ W/cm^2, $2.5\text{x}10^9$ W/cm^2 and $1.4\text{x}10^{10}$ W/cm^2 respectively. The plasmas were simulated with a 1½ dimension hydrodynamic-atomic code [14]. The 1½ D code includes a collisional-radiative atomic model with multi-cell radiation transport which solves for the ground state and excited populations using a quasi-steady state solution. Both inverse Bremsstrahlung absorption as well as photoionization are included as energy deposition mechanisms for the soft x-ray laser. The standard treatment of inverse Bremsstrahlung breaks down for these short wavelengths and near solid density. We calculate inverse Bremsstrahlung absorption by computing the electron-ion collision frequency using the Lee-More model [15], which is valid for densities near solid density and includes a region of conductivity for the solid and liquid phases. We correct the coulomb logarithm to use the laser frequency and include additional corrections for solid density non-ideal degenerate plasma. Finally, we calculate the complex permittivity using the Drude model of electrical conduction. Values from the absorption model at room temperature and solid density were found to be in good agreement with those found in Palik [16]. Photoionization absorption is included by calculating its contribution to the complex permittivity using the ion distribution from the atomic model, with the photoionization cross section for each ion. The atomic data is obtained from the Flexible Atomic Code (FAC) [17] with photoionization cross sections from Reilman and Manson [18]. A post-processor was used to synthesize spectra based on multi-cell radiation transport and the computed populations and opacities. To improve the accuracy of the synthesized spectra the level energies and transition probabilities were calibrated using experimental data when available [19].

At the lowest irradiation energy investigated, 0.5 μJ, all three elements were found to be below the ablation threshold, in agreement with the simulations. At the intermediate energy of 3.1 μJ plasma radiation from neutral Cr and Ag atoms was observed (Fig. 2(b,c)). At this energy Si is very close to the ablation threshold, resulting in only very weak emission from two of the strongest Si I lines in the ultraviolet spectra (Fig. 2(a)). These observations agree with model simulations that predict that the Cr and Ag plasmas reach a peak temperature of 2.2 eV and 2.8 eV and a peak degree of ionization of Z=0.5 and 1.0 respectively, while Si, that is at the threshold of ablation, only reaches a peak temperature of 0.2 eV and a degree of ionization of $Z<0.01$. The different behavior of these materials is to be expected, as the absorption length of the 46.9 nm laser light in Si (~300 nm) greatly exceeds that of Cr and Ag (~18 nm and ~7.5 nm respectively). Consequently, in Si the soft x-ray light interacts with a much larger volume of material which results in a higher ablation threshold, and in a colder plasma.

At the highest irradiation energy (17 μJ) plasma radiation was observed for all three elements. Figure 2 shows the measured time integrated visible spectra for Si, Cr, and Ag plasmas for this irradiation condition (Fig. 2(d,e,f)). The Si spectra still only displays lines associated with neutral Si. Classified lines from singly charged Si (Si II) that fall within the spectral window of the measurement (eg: 207.27, 290.43, 290.57 nm) are not observed, indicating that this is a very low temperature plasma, in agreement with model calculations. In contrast, spectra of plasmas created by irradiation of Cr and Ag targets with the same soft x-ray laser intensity show lines of Cr II and Ag II. The peak temperature for Si is 1.1 eV while Cr reaches a significantly higher temperature of 4.5 eV. This again reflects the difference in the photoionization cross sections at 46.9 nm: 6.1×10^{-19} cm^2 and 9.5×10^{-18} cm^2 for Si and Cr respectively. This difference in the cross section results in more energy being absorbed by fewer atoms for Cr than for Si. The hotter Cr plasma temperature results in a computed peak degree of ionization of 1.6 for Cr, compared to a degree of ionization of only 0.25 for Si. Due to the higher temperature the Cr plasma also has an increased expansion velocity that results in significantly denser plasma away from the target surface.

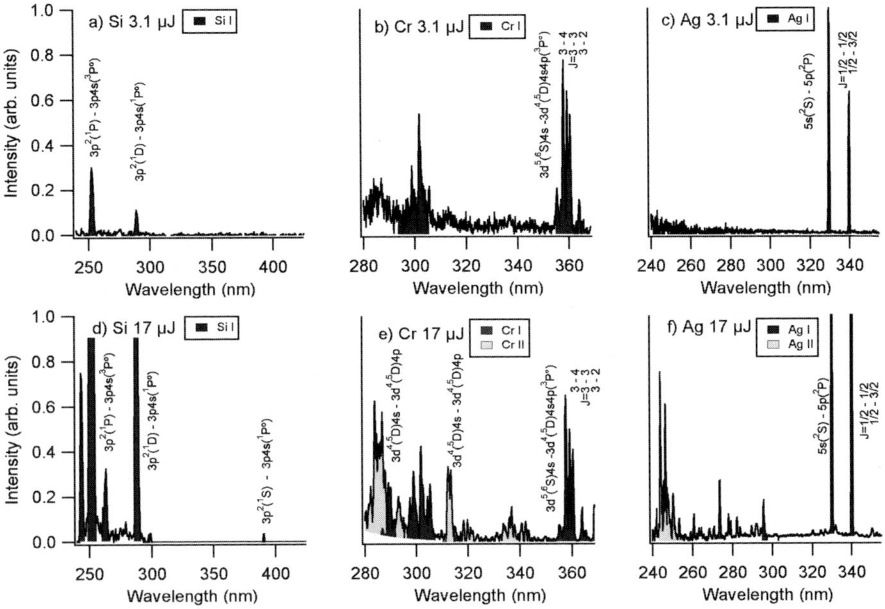

Fig. 2 Spectra of Si, Cr, and Ag plasmas created irradiating solid targets with 3.1 μJ and 17 μJ 46.9 nm soft x-ray laser pulses. At 3.1 μJ Si is only slightly above the ablation threshold (a), while for both Cr (b) and Ag (c) strong lines from the neutral atoms are present. At 17 μJ only neutral atom lines are present for Si (d), while the Cr (e) and Ag (f) spectra show lines from both the neutral atoms and singly charged ions.

Figure 3 shows simulated spectra for Si and Cr for the irradiation conditions corresponding to the experimental spectra of Fig. 2(d,e). The computed Si spectrum matches the experiments, showing only Si I lines. The synthesized Cr spectrum reproduces most of the features of the experimental data, but shows a slightly larger ratio between Cr II and Cr I lines and the presence of weak Cr III lines. This is in part due to the fact that the synthetic spectra are calculated for the plasma conditions on axis of the irradiated spot, where the plasma has the highest temperature. A 2D plasma model would be required to further improve the agreement between the computed and measured spectra. Some of the other differences may be due to an incomplete assignment of the energy levels and the bundling of lines in the calculations. Nevertheless, the computed and measured spectra agree, clearly showing the signatures of a significantly less ionized, colder Si plasma.

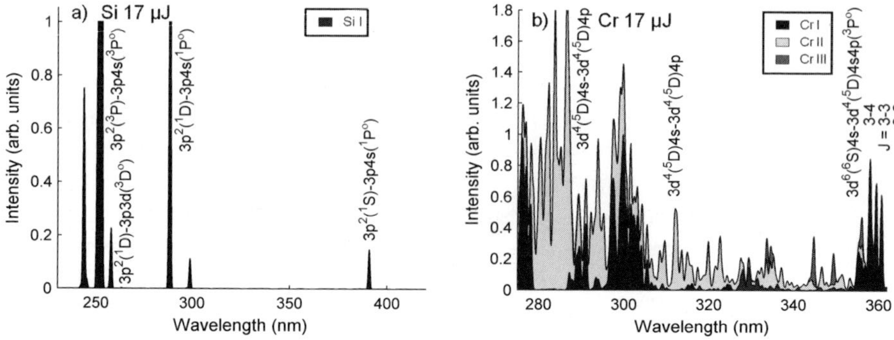

Fig. 3 Simulated spectra for Si and Cr plasmas created by 17 μJ 46.9 nm soft x-ray laser. The Si spectra is completely dominated be neutral atom lines (Si I) (a), while the hotter Cr spectra contains large numbers of both Cr I and Cr II lines (b).

Simulations were conducted for higher irradiation energies. The results show that as the intensity of the soft x-ray laser is increased, the behavior of the different elements tends to converge. Figure 4 shows the electron temperature and mean degree of ionization as a function of the soft x-ray laser intensity. When the laser energy is increased by a factor of 10, from $1.4\text{x}10^{10}$ W/cm^2 to $1.4\text{x}10^{11}$ W/cm^2, the temperature of the Si plasma is computed to increase from 1.1 eV to slightly above 3.5 eV, while that of the Cr and Ag plasmas increase from 4.5 eV and 4.5 eV to 9.0 eV and 6.5 eV respectively. Additionally, the mean degree of ionization increases to Z= 1.4 for Si and to 2.5 and 2.7 for both Cr and Ag. If the laser energy is further increased by a factor of 100, to 1.7 mJ, ($1.4\text{x}10^{12}$ W/cm^2) the electron temperature is computed to reach 7 eV for Si, 16 eV for Cr, and 10 eV for Ag, while the degree of ionization of the Si plasma, Z=2.5, is calculated to approach that of Cr and Ag, Z=3.0 and 3.2. The predicted convergence of the plasma parameters at higher irradiation flux is caused by the increase in the degree of ionization for Cr and Ag that causes the depletion of the neutral and singly charged ions that are responsible for most of the absorption of the 25 eV laser photons. The ratio of the Ag to Si temperature is also plotted. Cr exhibits a similar behavior as Ag, but requires higher intensities due to the larger ionization potentials of the neutral and first ion. At high irradiation intensity plasmas with a very uniform degree of ionization can be created. Figure 5 shows the simulated degree of ionization and electron density for Ag 1 ns after the peak of a $1.4\text{x}10^{10}$ W/cm^2 and $1.4\text{x}10^{12}$ W/cm^2 laser pulse. At the lower intensity there is a significant variation in the degree of ionization due to the different photo-absorption cross sections of the neutral and first few ions. At higher intensities the Ag plasma is able to reach a degree of ionization of Z=3 at which the absorption decreases allowing the hot region of the plasma to become transparent. As a result the laser heats a larger plasma volume resulting in a more uniform degree of ionization. From Figure 5, the degree of ionization is computed to vary by less than 10% over 90% of the entire plasma volume. However, the electron density within the plasma volume varies over 3 orders of magnitude due to expansion. The degree of ionization at which the plasma becomes transparent increases with decreasing laser wavelength. Therefore shorter wavelength soft x-ray lasers will creates hotter plasmas, and preserves the differences between elements until the plasma reaches a higher degree of ionization.

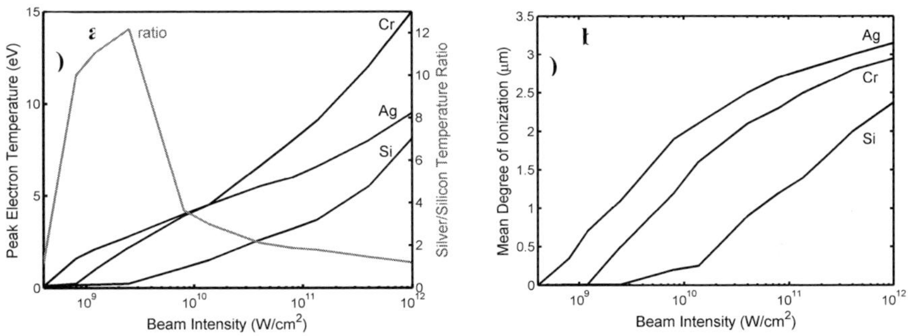

Fig. 4 Simulated electron temperature (a) and mean degree of ionization (b) for Si, Cr and Ag plasmas as a function of 46.9 nm soft x-ray laser irradiation intensity. The ratio of the Ag temperature to the Si temperature is also plotted (a).

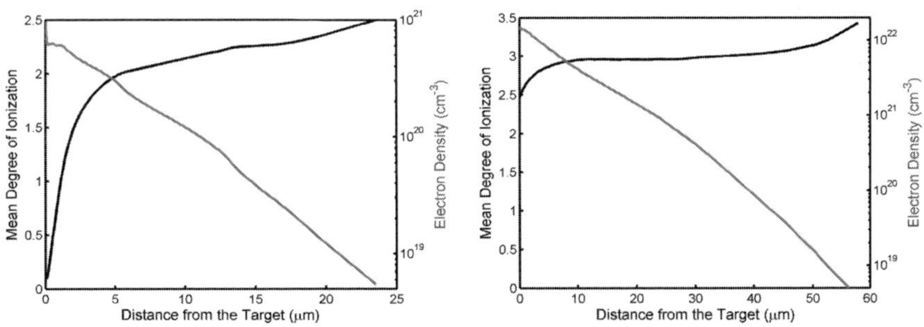

Fig. 5 Simulated mean degree of ionization and electron temperature of soft x-ray laser-created Ag plasma 1 ns after the peak of the irradiation with pulse intensities of 1.4×10^{10} W/cm^2 (left) and 1.4×10^{12} W/cm^2 (right).

4 Conclusions

We have conducted the first spectroscopy study of warm plasmas generated by soft x-ray laser irradiation and have compared the results to simulations. Plasmas were generated irradiating Si, Cr, and Ag targets with focused 46.9 nm soft x-ray laser pulses of nanosecond duration. The critical density corresponding to this wavelength (5×10^{23} cm^{-3}) exceeds solid density and absorption is dominated by single photon photoionization. Plasmas created by soft x-ray laser irradiation are shown to be strongly element-dependent, with characteristics largely determined by the relative position of the laser wavelength respect to absorption edges and resonances. Measured spectra agree with model simulations in showing that soft x-ray laser-created Si plasmas, a low absorption material at 46.9 eV, are significantly colder and less ionized than plasmas created from more highly absorbent materials such as Cr and Ag. This strong elemental dependence is computed to soften at higher nanosecond pulse irradiation intensities resulting in large plasma volumes with a very uniform degree of ionization. This is a result of the depletion of the low charge species, whose photoionization cross sections dominate the plasma absorption.

This work supported by NNSA under the Stewardship Science Academic Alliances program through U.S. DOE Grant #DE-FG52-06NA26152, using facilities from the NSF ERC Center for Extreme Ultraviolet Science and Technology, NSF Award Number EEC-0310717. M. Berrill acknowledges support from DOE CSGF under Grant No. DE-FG02-97ER25308.

References

1. R.W. Lee et al., "Finite temperature dense matter studies on next-generation light sources", JOSA B, **20**, 770-778, 2003.
2. M. Fajardo, et al., "Hydrodynamic simulation of XUV laser-produced plasmas", The European Physical Journal D, **29**, 69-75, 2004.
3. J. J. Rocca, et al., "Demonstration of a discharge pumped table-top soft-x-ray laser", Phys. Rev. Lett., **73**, 002192, 1994.
4. B. R. Benware, et al., "Focusing of a tabletop soft-x-ray laser beam and laser ablation", Optics Letters, **24**, Issue 23, pp. 1714-1716, 1999.
5. W. Ackermann, et al., "Operation of a free-electron laser from the extreme ultraviolet to the water window" Nature Photonics **1**, 336–342 (2007).
6. G. Vaschenko, et al., "Nanometer scale ablation with a table-top soft x-ray laser", Optics Letters, **31**, 3615-3617, 2006.
7. M. Grisham, et al., "Damage to extreme-ultraviolet Sc/Si multilayer mirrors exposed to intense 46.9-nm laser pulses," Optics Letters **29**, 620-622, 2004.
8. L. Juha, et al., "XUV-laser induced ablation of PMMA with nano-, pico-, and femtosecond pulses," Journal of Electron Spectroscopy and Related Phenomena **144**, 929-932, 2005.

9. Michal Bittner, et al., "Material ablation induced by focused 21.2-nm radiation from Ne-like Zn x-ray laser", Proceedings of SPIE **5777**, 965-969, 2005.
10. T. Mocek, et al., "Focusing a multimillijoule soft x-ray laser at 21 nm", Applied Physics Letters, **89**, 051501, 2006
11. M. Berrill, et al., "Warm photoionized plasmas created by soft-x-ray laser irradiation of solid targets", JOSA B, **25**, pp B32-B38, 2008.
12. B.R. Benware, et al., "Demonstration of a high average power tabletop soft x-ray laser", Phys. Rev. Lett., **81**, 5804, 1998.
13. Yu. A. Uspenskii, et al., "High-reflectivity multilayer mirrors for a vacuum-ultraviolet interval of 35-50nm",Optics Letters **23**, 771-773, 1998.
14. M. A. Berrill, *A Computer Model to Simulate Laser Created Plasmas Used for the Generation of Extreme Ultraviolet Light*, MS thesis, Colorado State Univ., 2006.
15. Y.T. Lee and R.M. More, "An electron conductivity model for dense plasmas", Physics of Fluids, **27**, pp. 1273-1285, 1984.
16. Edward D. Palik, *Handbook of Optical Constants of Solids*, Academic Press, 1998.
17. M. F. Hu, "Indirect X-Ray Line-Formation Processes in Iron L-shell Ions", The Astrophysical Journal, **582**, 1241-1250, 2003.
18. R.F. Reilman, S.T. Manson, "Photoabsorption Cross Sections for Positive Atomic Ions With $Z \leq 30$", The Astrophysical Journal Supplement Series, **40**, 815, 1979.
19. NIST Atomic Spectra Database, http://physics.nist.gov/PhysRefData/ASD

Development of Multilayer Optics in EUV, Soft X-Ray and X-Ray Range at IPOE

Zhanshan Wang*, Jingtao Zhu, Zhong Zhang, Xinbin Cheng, Jing Xu, Fengli Wang, Xiaoqiang Wang, Lingyan Chen

Institute of Precision Optical Engineering, Physics Department, Tongji University, Shanghai, 200092

Abstract. Multilayer coatings are key optical components in the EUV, soft X-ray and X-ray range. At Institute of Precision Optical Engineering (IPOE), the development of multilayer optics has been impelled for their wide applications in X-ray laser, plasma diagnostics, astronomical observation and synchrotron radiation. The paper presents our recent results of periodic multilayers of Mo/Si, Cr/C, Cr/Sc, La/B_4C, Mo/B_4C, Si/C, Si/SiC, Mg/SiC Mo/Y and Ru/Y. To improve the reflectivity of Ru/Y multilayer mirrors, Mo layers were inserted between Ru and Y layer. The Mo barrier layers suppress intermixing between Ru and Y, thereby increasing the reflectivity of Ru/Y multilayer. We also discuss the application of Mo/Si, Mo/Y, Mo/B_4C, La/B_4C non-periodic multilayers in EUV broadband polarization measurement.

1 Introduction

The research of multilayer structures having nanometer-scale thicknesses has been very active in recent years, numerous novel multilayer coatings (MLC) have been developed and used as optical key components in a variety of applications including plasma diagnostic[1], astronomical telescope[2], synchrotron and medical optics[3,4]. In this paper we outline ongoing research and development of EUV, soft X-ray and X-ray multilayer optics at IPOE.

The periodic multilayers have been designed as high reflectance normal incidence mirrors and polarized components working around at quasi-Brewster's angle. Multilayers with spacer materials of Si, SiC, C, B_4C, Y, Sc, in combination with the absorber materials Mo, Cr, La, Ru and Mg were investigated. By tailoring the period thickness and the thickness ratio for use at and below the resonance energy in normal incidence or near Brewster's angle, an excellent performance with respect to reflectance and polarisation was achieved.

To simplify experimental arrangements in EUV broadband polarization measurement, the concept of broadband polarized elements[5-7] using non-periodic MLC has been proposed. Using a pair of non-periodic reflective analyzers and phase retarder, a complete broadband polarization analysis system can be developed. Various broadband non-periodic polarized components operating at different wavelength ranges have been successfully realized in our laboratory, the following material combinations represent the absorber/spacer: Mo/Si (12.5-20 nm), Mo/Y (8-12 nm), Mo/B_4C (6.75-7.35 nm), La/ B_4C (6.72-8.32 nm).

All the MLC discussed in the following sections were fabricated by using a high vacuum direct current magnetron sputtering deposition system in Ar (99. 999 %) gas. The trial multilayers were characterized using small angle X-ray diffraction working at the copper Kα line (0. 154 nm) then the measured reflection curve is simulated as closely as possible to determine the individual layer thicknesses, so the deposition rate of each material can be calibrated. The synchrotron measurements of these samples were performed at the National Synchrotron Radiation Laboratory, China and at the beamline UE56/1-PGM-1 at ESSY II Berlin, Germany.

2 Periodic Multilayers

2.1 Mo/Si Multilayers

The Mo/Si multilayer system is perhaps the most well-known and best-studied owing largely to its important application in photo-lithography as well as in solar physics. The normal-incidence reflectance attainable with Mo/Si approaches 70% just below the Si L-edge. Figure 1 shows the measured reflectivities of Mo/Si multilayers working at incident angle of 5°. At the wavelengths of 13.5 nm, 19.5 nm and 29.5 nm, the reflectivities measured on BESSY and NSRF are 68.5%, 38.7% and 19.8%, respectively. Although the reflectivity decreases with the increasing of the wavelength, the Mo/Si multilayers provide considerable reflectivities in a broad wavelength range from 13 nm to 30 nm, which are widely used in EUV and soft X- ray regions.

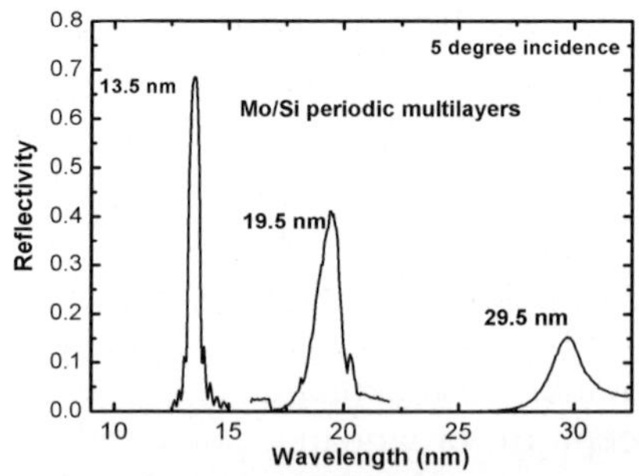

Fig. 1 The measured reflectance curves of Mo/Si multilayers with peak wavelengths at 13.5 nm, 19.5 nm and 29.5 nm.

2.2 Multilayers working at He-II 30.4nm

He-II (λ=30.4 nm) emission is very important for imaging the solar corona or for studying the structure and dynamics of cold plasma in Earth's plasmasphere. At this wavelength, the Mo/Si multilayers are less than optimal. Si-based EUV multilayers[8, 9] other than Mo/Si that have been investigated previously include Si/SiC, Si/C and Si/B_4C. We have recently compared experimentally the performance of all of these coatings, as well as a new multilayer system comprising Mg/SiC bilayers, and we present some of our results here. Shown in Fig. 2 are the results obtained for an array of Si/B_4C, Si/SiC, Mg/SiC and Si/C multilayers, designed for use at an incident angle of 10°. Clearly, the reflectivity of Mg/SiC system (43.80%) is higher than those of Si-based ones. In addition to its higher reflectance, Mg/SiC multilayer has a narrower bandwidth of 1.35 nm, which can provide considerably greater spectral selectivity relative to Si/SiC, Si/C and Si/B_4C.

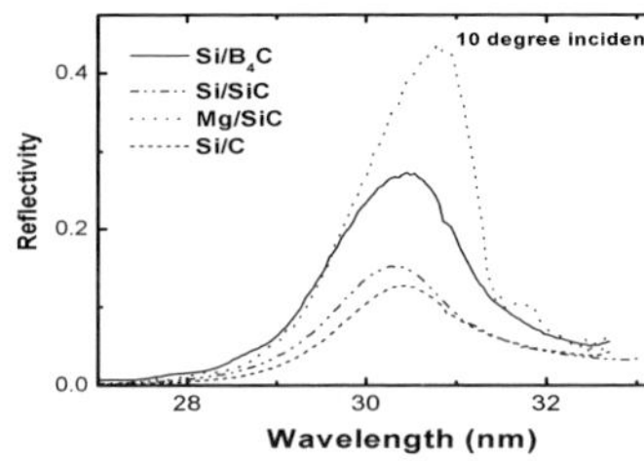

Fig. 2. The measured reflectance curves of Si/B4C, Si/SiC, Mg/SiC, Si/C MLC.

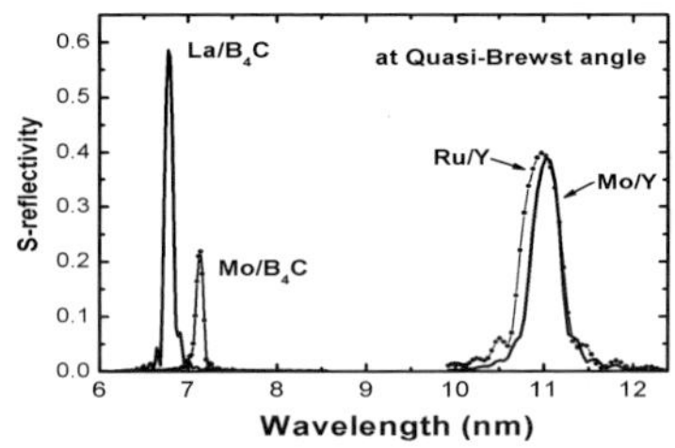

Fig. 3. The measured reflectance curves of Mo/Y, Ru/Y, La/B4C, Mo/B4C MLC.

2.3 Short-wavelength EUV Multilayers

At wavelengths shorter than the Si L-edge, a number of efficient new multilayer systems have recently emerged. We have recently compared the multilayer polarizing elements comprising Mo/Y, Ru/Y, La/B_4C and Mo/B_4C systems, and the results, obtained using synchrotron radiation at the beamline UE56/1-PGM-1 at ESSY II Berlin, Germany, are shown in Fig. 3. In this study, Ru/Y indeed provides the higher s-reflectance of 39.5% compared with 39% for Mo/Y system. Another new multilayer system showing high reflectance below 10 nm is the La/B_4C multilayer system[10], for which peak normal incidence reflectance of 58% has been achieved near the Si VII line at 6.77 nm, whereas the Mo/B_4C gives a low s-reflectance of 21.8% at 7.145 nm.

2.4 Mo diffusion barriers for Ru/Y multilayers

The s-reflectivity of an ideal Ru/Y multilayer is as high as 65.6%, because of the lower absorption of Ru around 11 nm. Ru/Y Interface imperfection including interfacial roughness and interfacial diffuseness significantly reduces the reflectivity. In this work, the interdiffusion processes between Ru and Y is limited by using Mo as a diffusion barrier. After theoretical and experimental investigations, the optimal thicknesses of Mo diffusion barriers at Ru-on-Y and Y-on-Ru interfaces are 0.5 nm respectively. Design parameters of Mo/Y, Ru/Y and Ru/Mo/Y/Mo multilayers with ideal interfaces for λ=11 nm are summarized in Table 2 and the theoretical reflection curves of these samples are given in Fig. 4a. Polarization performances of these multilayers were measured using the high precision 8-axis polarimeter on beamline UE56/1-PGM-1 at BESSY-II, the measured reflectance curves are shown in Fig. 4b. Although the ideal Ru/Mo/Y/Mo system provides the lower s-reflectance of 63.64% compared with 65.63% for Ru/Y multilayer, the measured peak s-reflectivity of Ru/Mo/Y/Mo system is 43.3 % which is higher than 39.5% for Ru/Y multilayer. This evidences that the Mo barrier layer suppresses intermixing of Ru and Y, thereby increasing the reflectivity of the multilayer. Further investigations are necessary to achieve the theoretical limit.

Table 1 Design parameters for Mo/Y, Ru/Y and Ru/Mo/Y/Mo MLC.

Sample	N	Design Layer thickness (nm)		θ (deg)	Rs %	P
Mo/Y	45	d_{Mo}=3.41	d_Y=4.53	46.7	54.87	0.9998
Ru/Y	30	d_{Ru}=3.25	d_Y=4.69	47.2	65.63	0.9996
Ru/Mo/Y/Mo	25	d_{Ru}=2.78	d_Y=4.19	47.2	63.64	0.9996
		d_{Mo}=0.5	d_{Mo}=0.5			

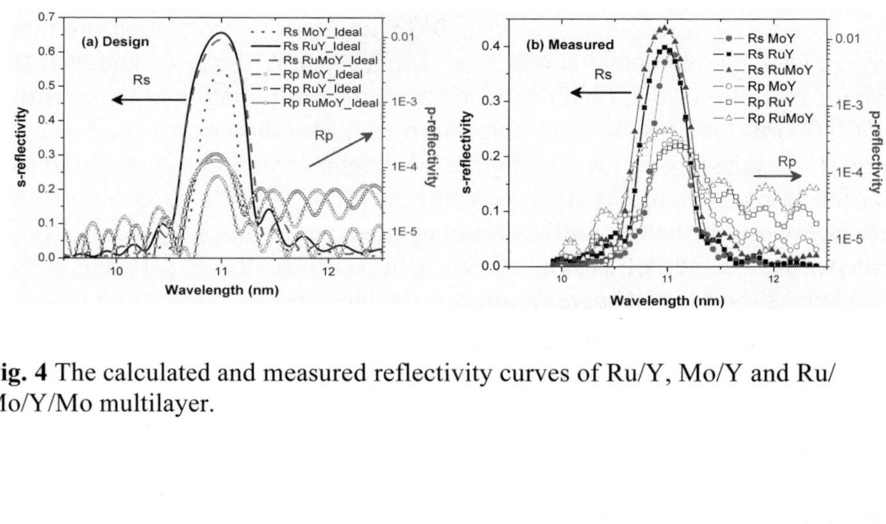

Fig. 4 The calculated and measured reflectivity curves of Ru/Y, Mo/Y and Ru/Mo/Y/Mo multilayer.

2.5 Ultra-Short-period Soft X-ray Multilayers

We discuss here the availability of ultra-short-period multilayer coatings operating near normal incidence in the soft X-ray region between the absorption edges of Oxygen (2.28 nm) and Carbon (4.36 nm). This spectral region provides a good contrast between organic materials and water[11], therefore it is called "water window". In the soft X-ray region, the absorption of candidate multilayer materials is significantly lower than in the EUV, suggesting, in principle, that efficient reflective multilayers can be constructed containing hundreds of periods. However, interface imperfections including roughness and diffuseness have greatly limited the achievable peak reflectance. Cr/Sc and Cr/C are promising material combinations for multilayer mirrors in the water window region. High reflectivities can be achieved if the interfaces between adjacent layers are extremely smooth. Hence, the deposition parameters of the magnetron sputtering process for these materials have to be optimized carefully. Figure 5 show the reflectance curves measured near normal incidence (5 deg) as a function of wavelength at the beamline UE56/1-PGM-1 at BESSY-II, Germany. The Cr/Sc and Cr/C multilayers both contain N=200 bilayers, with periods d=2.22 nm and 2.26 nm respectively. The highest reflectivity of Cr/Sc is 5.3%, and Cr/C multilayer provides the peak reflectance of 7.5%.

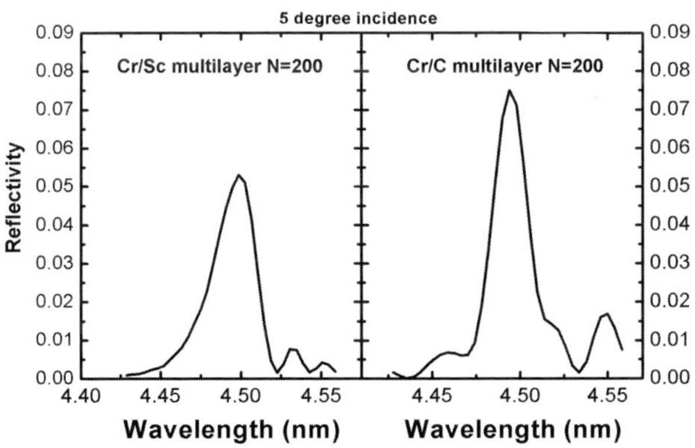

Fig. 5 Normal-incidence reflectance of periodic Cr/Sc and Cr/C multilayers, measured at BESSY-II.

3 Non-periodic Multilayers

To simplify experimental arrangements in EUV polarization measurement, non-periodical multilayers has been proposed to construct broadband reflective analyzers and broadband phase retarders. This kind of non-periodic multilayer can work in a broad spectral bandwidth at fixed Brewster angle, or exhibit wide angular range at fixed wavelength. So, these broadband multilayer polarization optical elements will greatly simplify experimental arrangements in EUV polarization measurement. This section will summarize the Mo/Si broadband polarization elements and their application in broadband polarization analysis. Figure 6 shows the measured results of Mo/Si non-periodic multilayer reflection analyzers with different bandwidths. The measured p-reflectivities were very low in all cases. Mo/Si non-periodic multilayer transmission phase retarder has also been designed and deposited on SiN membranes. The phase shift and transmission of phase retarder are shown in Fig. 7. The calculated results with the interface roughness of 1.0nm are in good agreement with experimental data.

Using the developed non-periodic transmission phase retarder and reflection analyzer, we have performed a complete broadband polarization analysis of synchrotron radiation from the BESSY-II UE56/1-PGM-1 beamline. The measured Stokes-Poincaré parameters of the radiation as a function of wavelength are shown in Fig. 8. The circularly polarized parameter P3 is close to unity while the linearly polarized values, with averages P1=0.007±0.026 and P2=0.053±0.005, oscillate around zero over the wavelength range. The fitted behaviors agree quite well with model predictions of the circularly polarized radiation. The wavelength dependence of polarization of synchrotron radiation can thus be characterized by a complete polarization analysis without changing the incidence angles of the broadband phase retarder and analyzer.

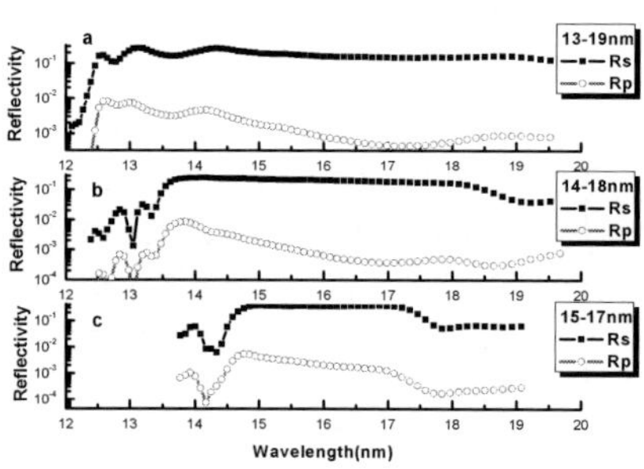

Fig. 6 The measured s- and p- reflectance curves of Mo/Si multilayer analyzers.

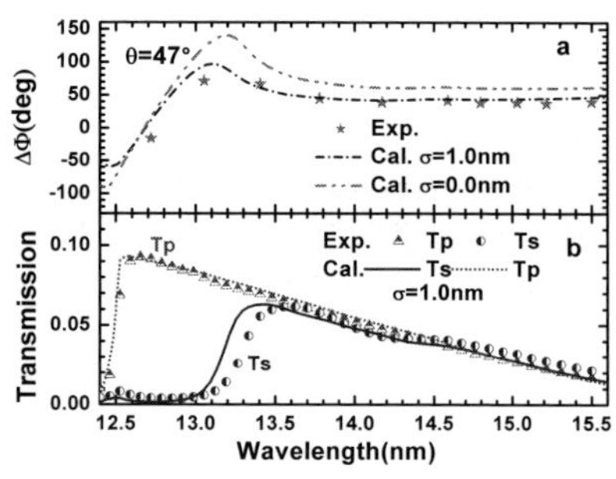

Fig. 7 (a) The phase shifts, (b) The Ts and Tp of Mo/Si phase retarder.

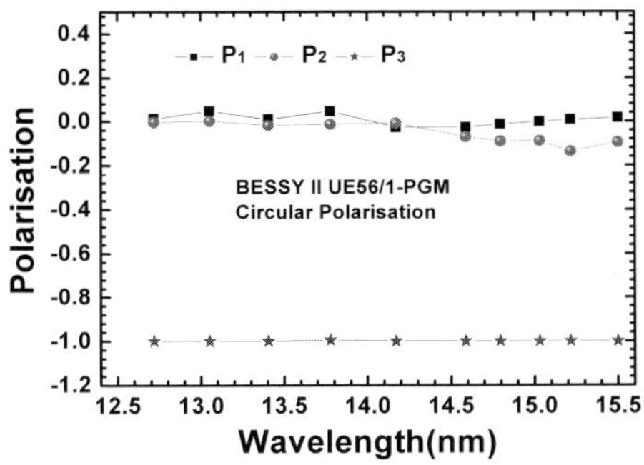

Fig. 8 Stokes-Poincaré parameters of the circular radiation from the BESSY-II UE56/1-PGM1 beamline, at fixed angles of the Mo/Si broadband phase retarder (45°) and analyzer (47°).

Various broadband non-periodic polarized components operating at different wavelength ranges have been designed and fabricated in our laboratory and the summary were listed in Table 2. All non-periodic MLC provided moderate s-polarized reflectivity and high polarization degrees over the wide reflection band, which is unattainable using periodic MLC. They will greatly simplify future polarization experiments in shorter EUV wavelength region.

4 Summary

We have attempted to provide an overview of EUV, soft X-ray and X-ray multilayers developed in IPOE. We also outline prospects for operation at both longer and shorter EUV wavelengths, and also the ultra-short-period multilayers that work near normal incidence in the soft X-ray region. Application of non-periodic multilayers in EUV broadband polarization measurement is also discussed.

Table 2 Measurement results of various non-periodic multilayer reflection analyzers working at different wavelength ranges.

Sample	Bilayer	Grazing angle (deg)	λ range (nm) measured	Mean R_s (%) measured	Mean P (%) measured
Mo/Y	100	46	8.5-10.1	5.5±1.4	98.79±0.32
Mo/Y	100	45	9.1-11.7	6.1±0.7	96.48±0.7
Mo/B_4C	80	45.6	6.75-7.35	6.23±0.61	99.93±0.003
La/B_4C	80	45.8	6.72-8.32	6.68±0.98	98.87±0.21

Acknowledgements

This work was supported by the National 863 Sustentation Fund under contract No. 2006AA12Z139 and the National Natural Science Foundation of China under contract No. 10435050, 10675092 and 10675091. I would like to thank all my colleagues, Postdoc, PHD and M.S. students for their valuable contribution to the progress of the laboratory. Continuous support from NSFC is also highly acknowledged. Many thanks to Hongjun Zhou (NSRL), Tonglin Huo (NSRL), Mingqi Cui (BSRF), Yidong Zhao (BSRF) for their kindly assistance during the synchrotron measurements. I thank Alan G. Michette, Slawka J. Pfauntsch, A. Keith Powell in King's College London for useful discussions and application of beam time in BESSY. Polarization measurements in EUV spectral region were done at the beamline UE56/1-PGM-1 at BESSY, special thanks to Franz Schäfers and Andreas Gaupp for their excellent cooperation. I gratefully acknowledge helpful discussions with Igor V Kozhevnikov (Shubnikov Institute of Crystallography), Mike MacDonald (CCLRC Daresbury Laboratory), Makoto Watanabe (Shangai Dianji University), Koujun, Yamashita (Nagoya University).

References

1. Champeaux, J.-Ph. Troussel, Ph. Villier, B, Vidal, V, Khachroum, T, Vidal, B and Krumerey, M.: 'Development and realization of non-periodic W/Si multilayer mirrors for 5–14 keV X-ray plasma diagnostic', *Nucl. Instrum. Methods Phys. Res., Sect. A* 581, 687-694, 2007
2. Windt, D. L. Donguy, S. Seely, J. F. Kjornrattanawanich, B. Gullikson, E. M. Walton, C. C. Golub, L. and DeLuca. E.: 'EUV multilayers for solar physics', *Proc. SPIE* 5168, 1-11, 2004
3. Ziegler, E.: 'Multilayer optics for synchrotron x-ray applications', *Proc. SPIE* 2253, 248-259, 1994
4. Kantsyrev, V. L, Bruch, R, Phaneuf, R and Publicover, N. G.: 'New Concepts for X-Ray, Soft X-Ray, and EUV Optical Instrumentation Including Applications in Spectroscopy, Plasma Diagnostics, and Biomedical Microscopy: A Status Report', *J. X-Ray Sci. Technol*, 7, 139-158, 1997
5. Z. S. Wang, H. C. Wang, J. T. Zhu, Y. Xu, Z. Zhang, L.Y. Chen, A G. Michette, A.K. Powell, S.J. Pfauntsch, F. Schäfers, A. Gaupp. M. MacDonald.: 'Extreme ultraviolet broadband Mo/Y multilayer analyzers', *Appl. Phys. Lett.* 89, 24120, 2006
6. Z. S. Wang, H. C. Wang, J. T. Zhu, Y. Xu, Z. Zhang, L.Y. Chen, A G. Michette, A.K. Powell, S.J. Pfauntsch, F. Schäfers, A. Gaupp. M. MacDonald.: 'Broadband Mo/Si multilayer transmission phase retarders for the extreme ultraviolet', *Appl. Phys. Lett.* 90, 031901, 2007

7. Z. S. Wang, H. C. Wang, J. T. Zhu, Z. Zhang, Y. Xu, L.Y. Chen, A G. Michette, A.K. Powell, S.J. Pfauntsch, F. Schäfers, A. Gaupp. M. Q. Cui, L. J. Sun, M. MacDonald.: 'Complete polarization analysis of extreme ultraviolet radiation with a broadband phase retarder and analyzer', *Appl. Phys. Lett.* 90. 081910, 2007
8. Slaughter, J. M. Medower, B. S. Watts, R. N. Tarrio, C. Lucatorto, T. B. and Falco, C. M.: 'Si/B_4C narrow-bandpass mirrors for the Extreme Ultraviolet', *Opt. Lett.*, 19, 1786-1788, 1994
9. Grigonis, M. and Knystautas, E. J.: 'C/Si multilayer mirrors for the 25–30 nm wavelength region', *Appl. Opt*. 36, 2839-2842, 1997
10. Ricardo, P. Wiesmann, R. G. Nowak, C. Michaelsen, C. Bormann, D.: 'Improved Analyzer Multilayers for Aluminiumand Boron Detection with X-Ray Fluorescence', *App. Opt*, 40, 2747-2754, 2001
11. Thieme, J. Schmahl, G. Rudolph, D. Umbach, E. (eds.): *in X-Ray Microscopy and Spectromicroscopy* Springer Verlag, 1998

Highly Efficient Surface Modification of Solids by Dual Action of XUV/Vis-NIR Laser Pulses

T. Mocek[1], K. Jakubczak[1], J. Polan[1], P. Homer[1], B. Rus[1], I.J. Kim[2], C.M. Kim[2], S.B. Park[2], T.K. Kim[2], G.H. Lee[2], C.H. Nam[2], J. Chalupský[3], V. Hájková[3], L. Juha[3]

[1] Department of X-ray Lasers, Institute of Physics, Na Slovance 2, Prague 8, Czech Republic
[2] Coherent X-ray Research Center, Korea Advanced Institute of Science and Technology, 373-1 Guseong-dong, Yuseong-gu, Daejeon 305-701, Korea
[3] Department of Laser Plasmas, Institute of Physics, Na Slovance 2, Prague 8, Czech Republic

Abstract. We demonstrate a novel experimental method for unusual structural surface modification of various solids (PMMA, amorphous carbon) achieved by simultaneous action of XUV (21.6 nm), obtained from high-order harmonic generation, and Vis-NIR (410/820 nm) laser pulses. Although the fluence of each individual pulse was far below the surface ablation threshold, very efficient and specific material expansion was observed after irradiation by a few shots of mixed XUV/Vis-NIR radiation.

1 Introduction

The study of irreversible surface responses of various organic and inorganic solids to extreme ultraviolet (XUV) and soft X-ray radiation is motivated mostly by nanotechnology interests [1-3]. The main objective is to realize short-wavelength ($\lambda < 100$ nm), direct nanostructuring of materials resistant to optical radiation. Several research groups have investigated the behavior of poly(methyl methacrylate) (PMMA) at short wavelengths under irradiation conditions both above [2-4] and also below [1, 5-7] its single-shot ablation threshold. Erosion of the organic polymer induced by high-energy photons was reported [1-7]. In addition to that, surface modification of organic polymer has been extensively investigated with conventional UV-Vis-IR lasers.

In this contribution, we introduce a very effective method for structural surface modification of solids achieved by dual action of XUV, obtained from high harmonic generation (HHG), and Vis-NIR ultrashort laser pulses. While the separate use of the Vis-NIR or XUV beam did not cause any change to the sample surface, the application of a mixed XUV/Vis-NIR field resulted in clearly visible, irreversible modifications after irradiation by a few shots only.

2 Experimental setup

Fig. 1 shows the experimental setup. For strong HHG a two-color laser field, consisting of fundamental and second harmonic (SH) of a femtosecond laser pulse, was applied to a gas jet of He [8]. Femtosecond laser pulses at 820 nm with an energy of 2.8 mJ and pulse duration of 32 fs were focused by a spherical mirror (f=600 mm) into a He gas jet. For SH generation, a 200-μm-thick beta-barium borate (BBO) crystal was placed between the focusing mirror and gas jet so that, after the BBO crystal, the laser field consisted of both the SH and the residual fundamental laser fields. For the optimum SH conversion the BBO crystal was placed ~40 cm from the focusing mirror and the energy conversion efficiency was about 27%. A gas jet with a slit nozzle of 0.5 mm width and length of 6 mm was used for efficient HHG in the two-color laser field [9]. The gas pressure in the interaction region was 150 Torr. Generated HHG were first characterized using a flat-field XUV spectrometer equipped with a back-illuminated X-ray charge coupled device. The strongest harmonic at the 38th order (21.6 nm) reached an energy of about 50 nJ, as shown in Fig 2.

After HHG source optimization, the spectrometer was replaced with a 1 inch diameter off-axis parabolic mirror (OAP, f= 125 mm at 13 degrees) with a Mo:Si multilayer coating (R = 30% at 21 nm) placed 245 mm from the HHG source. The sample targets (e.g. 500-nm or 5 μm thick layer of PMMA coated on silicon substrate, 46-nm or 890-nm thick layer of amorphous carbon on Si substrate) were positioned 125 mm from the OAP, perpendicularly to the incident beam. The measured reflectivity of the OAP in the optical region was 37%. The measured diameter (full-width-at-half-maximum) of the HHG beam incident on the OAP was 280 μm while the diameter of the fundamental and SH laser beams was ~4 mm. The morphology of irradiated target surface was first investigated by Nomarski differential interference contrast optical microscopy, and then with an atomic force microscope (AFM) operated in the tapping mode to preserve high resolution.

To estimate the spot size for both XUV and Vis-NIR beams at the target position, we have performed numerical simulations of the employed optical layout using the ZEMAX optical software [10]. The modelling predicted an XUV spot size of 140 μm, which is about 17 times smaller than that of Vis-NIR beams. Thus the estimated fluence on the sample surface was 0.1 mJ/cm^2 at 21.6 nm, 15 mJ/cm^2 at 820 nm, and 6 mJ/cm^2 at 410 nm per shot, respectively. As all these values lie far below the signle-shot ablation threshold for PMMA by infrared (2.6 J/cm^2 [11]) as well as by XUV (2 mJ/cm^2 [12]) radiation, no damage of target surface is expected.

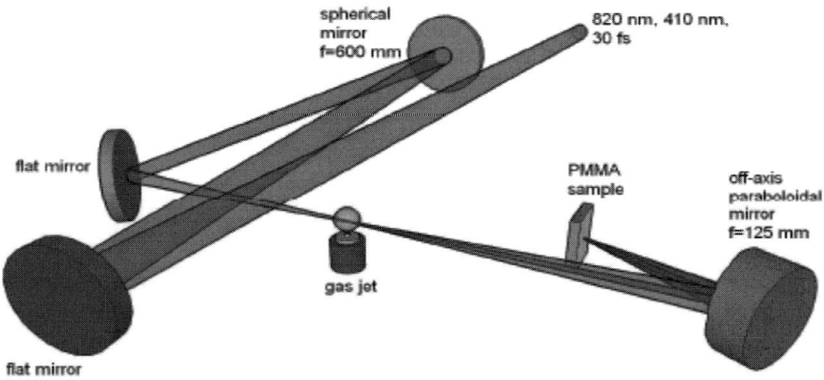

Fig. 1 Schematic of the experimental setup.

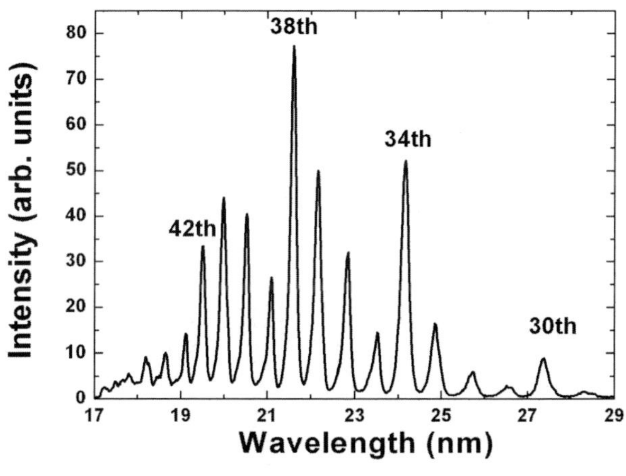

Fig. 2 XUV spectrum of HHG from the two-color laser field.

3 Results

The PMMA sample was irradiated under the following conditions: a) Vis-NIR beams only (no gas jet in operation), b) XUV beam only (0.4 μm Al filter placed in front of the OAP), and c) mixed XUV/Vis-NIR field (21.6 nm + 820 nm + SH). In cases a) and b) we have not observed any signs of surface damage after irradiation by ~3000 shots, although the accumulated dose was significantly higher than in case c). The target surface remained virtually unaffected, preserving its original quality and roughness. In striking contrast, the application of a mixed XUV/Vis-NIR field resulted in clearly visible, irreversible surface modification after irradiation with only a few shots.

Figure 3 shows images of the 500-nm thin PMMA exposed to the simultaneous action of XUV and Vis-NIR ultrashort pulses. The damage is characteristic of material expansion, quite different from ablation craters observed in experiments with pure XUV pulses [2-4,12-14]. A very interesting feature observed in AFM images was the formation of nanoscale spike(s) in the center of a uniform flat pedestal. The size of pedestal increased with the number of shots applied to the target while the spikes had a radius of the order of few microns down to a few tens of nanometers in length.

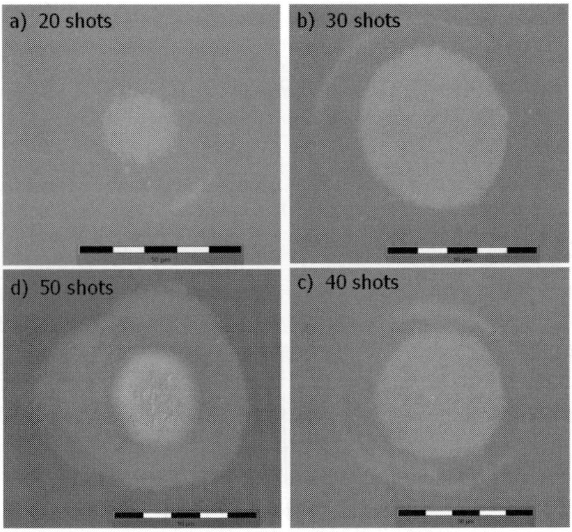

Fig. 3 Nomarski images of 500-nm thin PMMA surface simultaneously irradiated with XUV and Vis-NIR laser pulses.

The experiment was also performed with 5 μm thick PMMA exposed to higher XUV flux (200 nJ per pulse). In this case, we have observed a weak effect even in the absence of XUV beam (Fig. 4(b)) however AFM analysis did not reveal any surface modification, but only radiative coloration of the sample. In contrast, material expansion was clearly observed after irradiation by mixed XUV/Vis-NIR field (Fig. 4(a)).

Qualitatively similar results were achieved with a-C samples of various thickness, i.e. the application of mixed field has resulted in strong surface modification (Fig. 5(a)) while no effect could be observed in the absence of XUV photons. Additionaly, laser induced periodic surface structure (with period of about 460 nm) were identified in the central part of the impact by mixed field. Detailed analysis of a-C samples in underway.

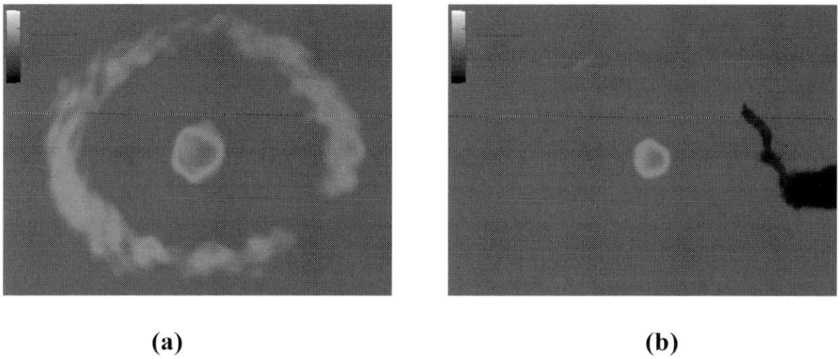

Fig. 4 Optical Nomarski images of 5-μm thick PMMA irradiated by single shot of (a) mixed XUV/Vis-NIR field, and (b) Vis-NIR

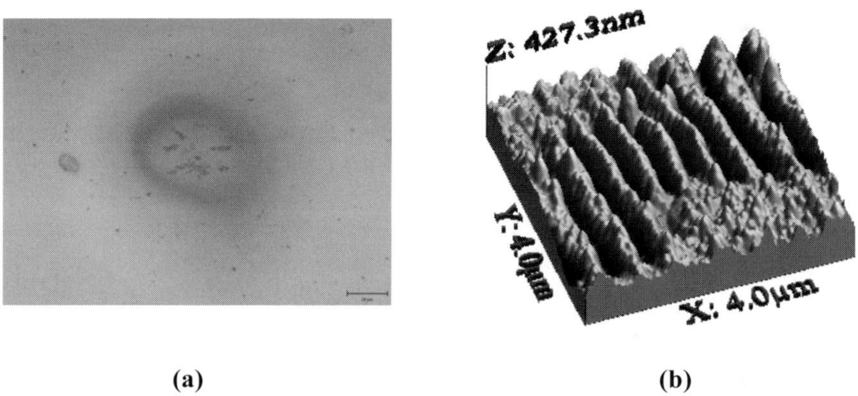

Fig. 5 (a) Nomarski image of 890-nm thin a-C sample irradiated by 10 shots of mixed XUV/Vis-NIR field, and (b) corresponding AFM image of the central part.

4 Discussion

As the separate application of Vis-NIR or XUV beams did not lead to any observable change on the irradiated surface, the expansion must be attributed to the dual action of 21.6-nm high-order harmonic and 820-nm laser radiation, contaminated with SH in the visible spectral region. The mechanism of this dual action can be described in terms of a simplified physical picture. It is well known that PMMA exhibits a very weak linear absorption at 820-nm. Therefore, the single-shot ablation threshold lies for the near-infrared beam at a level of 2.6 J/cm^2 even for femtosecond laser pulses [11]. However, if the beam contains 21.6-nm radiation, the high-energy XUV photons do indeed generate free charge carriers in the nearly transparent polymer dielectrics. The

Vis-NIR radiation can then interact strongly with the charge carriers formed through simultaneous irradiation by XUV so that its absorption dramatically increases in PMMA. The XUV radiation thus creates irreversible rise of Vis-NIR absorption in the material which is a well known effect of radiation coloration of transparent dielectrics which is dose dependent. The irreversible coloration might be responsible for the observed nonlinear dependence of growth rate of expanded structure on the number of accumulated shots.

The laser pulse energy deposited in the system due to radiation-enhanced absorption also heats the material. It has been shown with synchrotron radiation, as well as using the laser-plasma emission, that the efficiency of soft X-ray radiation-induced etching processes strongly depend on temperature [1, 15]. In this way the 820-nm radiation enhances the molecular fragmentation and re-arrangement processes following the initial ultrafast action of energetic XUV photons on ns-ms time-scales. The influence of a temperature jump on the photo-etching of PMMA has been suggested to explain the dual action of nanosecond pulses of focused X-rays and ultraviolet-to-optical radiation emitted by a laser-produced plasma [7] which causes fragmentation of PMMA and forms a liquid, oligomeric phase [16]. The viscous liquid subsequently solidifies, forming the central spike(s) through extrusion of the fluid core pressurized by surrounding, already solidified, material [17, 18].

5 Conclusion

To summarise, we have experimentally demonstrated that the dual action of XUV/Vis-NIR ultrashort pulses is capable of producing highly efficient structural surface modification of various solid materials. The observed phenomena might be utilized in applications such as surface nanopatterning with widely available, table-top radiation sources.

References

1. Zhang, Y.: 'Synchrotron radiation direct photo-etching of polymers', *Adv. Polym. Sci.*, **168**, 291-340, 2004
2. Juha, L., et al.: 'Short-wavelength ablation of molecular solids: pulse duration and wavelength effects', *J. Microlith. Microfab. Microsyst.*, **4**, 033007, 2005
3. Vaschenko, G., et al.: 'Nanometer-scale ablation with a table-top soft x-ray laser', *Opt. Lett.* **31**, 3615-3617, 2006
4. Krzywinski, J., et al.: 'Conductors, semiconductors, and insulators irradiated with short-wavelength free-electron laser', *J. Appl. Phys.*, **101**, 043107, 2007

5. Bartnik, A., et al.: 'Micromachining of organic polymers by X-ray photo-etching using a 10 Hz laser-plasma radiation source', *Microelectronic Eng.*, **78-79**, 452-456, 2005
6. De Grazia, M., et al.: 'Applications of intense ultra-short XUV pulses to solid state physics: time-resolved luminescence spectroscopy and radiation damage studies', *Proc. SPIE*, **6586**, 65860I-1, 2007
7. Barkusky, F., et al.: 'Direct photo-etching of poly(methyl methacrylate) using focused extreme ultraviolet radiation from a table-top laser-induced plasma source', *J. Appl. Phys.*, **101**, 124908, 2007
8. Kim, I J., et al.: 'Highly Efficient High-Harmonic Generation in an Orthogonally Polarized Two-Color Laser Field', *Phys. Rev. Lett.*, **94**, 243901, 2005
9. Kim, I J., et al.: 'Generation of submicrojoule high harmonics using a long gas jet in a two-color laser field', *Appl. Phys. Lett.*, **92**, 021125, 2008
10. ZEMAX Development Corporation, www.zemax.com
11. Baudach, S., et al.: 'Ultrashort pulse laser ablation of polycarbonate and polymethylmethacrylate', *Appl. Surf. Sci.*, **154-155**, 555-560, 2000
12. Chalupský, J., et al.: 'Characteristics of focused soft X-ray free-electron laser beam determined by ablation of organic molecular solids', *Opt. Express*, **15**, 6036-6043, 2007
13. Juha, L., et al.: 'Ablation of organic polymers by 46.9-nm-laser radiation', *Appl. Phys. Lett.*, **86**, 034109, 2005
14. Mocek, T., et al.: 'Focusing a multimillijoule soft x-ray laser at 21 nm', *Appl. Phys. Lett.,* **89**, 051501, 2006
15. Bartnik, A., et al.: 'Strong temperature effect on X-ray photo-etching of polytetrafluoroethylene using a 10 Hz laser-plasma radiation source based on a gas puff target', *Appl. Phys. B*, **82**, 529-532, 2006
16. Juha, L., et al.: 'Ablation of poly(methyl methacrylate) by a single pulse of soft X-rays emitted from Z-pinch and laser-produced plasmas', *Surf. Rev. Lett.*, **9**, 347-352, 2002
17. Libbrecht, K.G., et al.: 'An investigation of laboratory-grown ice spikes', *J. Glaciology*, **50**, 371-374, 2004
18. Mills, D., et al.: 'Solidification driven extrusion of spikes during laser melting of silicon pillars', Nanotechnology, **17**, 2741-2744, 2006

Strand Breaks in DNA Samples Induced with LASERIX

B. Zielbauer[1], J. Habib[1], S. Kazamias[1], O. Guilbaud[1], M. Pittman[1], D. Ros[1], M.-A. Hervé du Penhoat[2], A. Touati[2], C. Le Sech[3], E. Porcel[3], S. Lacombe[3]

[1] Université Paris Sud 11, CNRS / LIXAM / LASERIX, Orsay, France
[2] Institut de Minéralogie et de Physique des Milieux Condensés, Université Pierre et Marie Curie – Paris 6, Paris, France
[3] Université Paris Sud 11, CNRS / LCAM, Orsay, France

Abstract. Samples of plasmid DNA were irradiated with pulsed 18.9 nm radiation originating from a Mo X-ray laser pumped in GRIP configuration at the LASERIX facility [1]. Surface doses of up to 500 kGray were delivered with a repetition rate of 10 Hz and average pulse energy of 200 nJ. The occurrence of single and double strand breaks not present in controls having undergone the same treatment except for the XRL irradiation can be seen as a clear effect of the XRL irradiation. This confirms the role of direct effects in DNA damages as previously seen with low energy ions [2, 3] and electrons.

1 Motivation

Many experimental investigations have been carried out over the last decades to understand the effects of DNA irradiation. When being exposed to natural or artificially created radiation, the genetic code can be totally or partially damaged and induce the death of the cells or a mutation of the genetic code. The understanding of such processes is crucial for medical purposes. Radiotherapy for example is used in cancer treatment to destroy anomalous cells while the damage to the healthy cells should be as small as possible. The aim of extended bio-physical studies on DNA irradiation is thus to improve the efficiency and accuracy of such treatments and find the most appropriate irradiation parameters.

The irradiation can be achieved by sending either particles or photons at a given energy onto the sample. They both induce ionization of the medium which has direct and indirect effects on the DNA structure itself. Two degrees of damage are generally distinguished: Single strand breaks (SSB) can be auto-repaired by a living cell while the more dramatic double stand breaks (DSB) can induce the death of the cell.

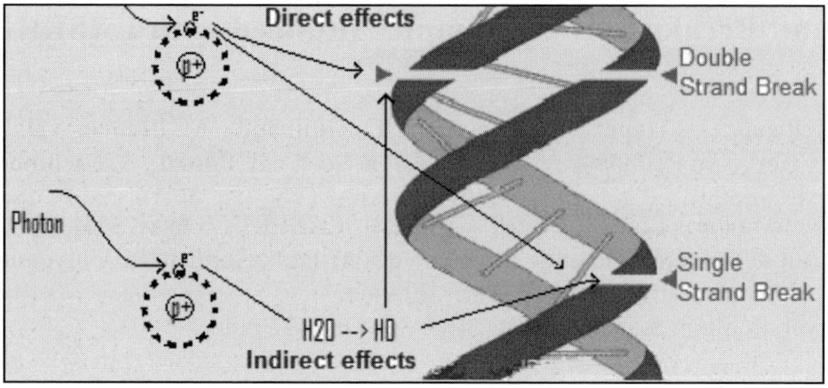

Fig. 1 Single and double strand break of DNA structure

These breaks can be produced via direct pathways such as direct ionization of binding electrons in the sugar or the phosphate groups or indirect pathways due to water radicals produced in the track. Recently, new processes of bond breaking have been observed when charged particles (electrons or ions) interact directly with DNA. For example, an important publication [3] has shown that, even with quite low energy, slow electrons can induce a large number of single and double strand breaks in DNA. These effects are due to resonant dissociative electron attachment on DNA. The abundance of secondary electrons with low energy (from 1 to 20 eV) thus plays a key role for the radiation induced damage of DNA. Such an electronic energy can be obtained by irradiating DNA with soft x-rays in the 50 to 100 eV photon energy range. More recently, the group of University Paris 11 has demonstrated that low energy ions can also induce severe DNA damage which is attributed to charge transfer and binary collisions [4].

Up to now, no extended study on low energy x-ray photon interaction with DNA has been carried out whereas many have been done at high energies [5, 6, 7]. This can in part be explained by a lack of easily accessible soft x-ray photon beam time, especially with elevated temporal resolution and coherence. Our team at LIXAM-LASERIX has been collaborating since a few years on such a subject with colleagues from biology. A preliminary study has been carried out in 2004 at the PALS x-ray laser in Prague, and gave first indication of single and double strand breaks of dehydrated DNA plasmids [8]. At that time, quality issues of the DNA samples and their inability to undergo long irradiation periods under vacuum induced a large fluctuation of SSB and DSB, even in absence of soft x-ray irradiation. Moreover, the number of shots that could be delivered on the samples was limited by the laser repetition rate, which was no more than 3 shots per hour. The 2004 experimental results thus consist of very few points with low signal to noise ratio. At the LASERIX installation, as we show here, we benefit from a more convenient mode of operation for extended and quantitative measurements.

2 Experiment

The LASERIX configuration that was used for the present study is the low energy/high repetition rate part of the full system. The 1.8 Joules of uncompressed infrared energy per pulse coming out from the last Ti:Sa amplifier stage are first equally split into two beams. The first one remains uncompressed (700 ps) and the second one enters a in-vacuum compressor capable of providing durations from 40 fs to 10 ps. The optimized duration in our configuration was 4 ps. These two beams are focussed into lines with 3 to 4 mm length and 100μm width. The first one (uncompressed beam) arrives on the 4 mm long Molybdenum target at normal incidence while the second one is incident after 400 ps at a grazing angle of 19° (GRIP configuration). The line foci optimization is done in IR with a high resolution imaging device (RILF). The XRL optimization in terms of pulse energy and stability at 10 Hz is done using an XUV near field imaging system.

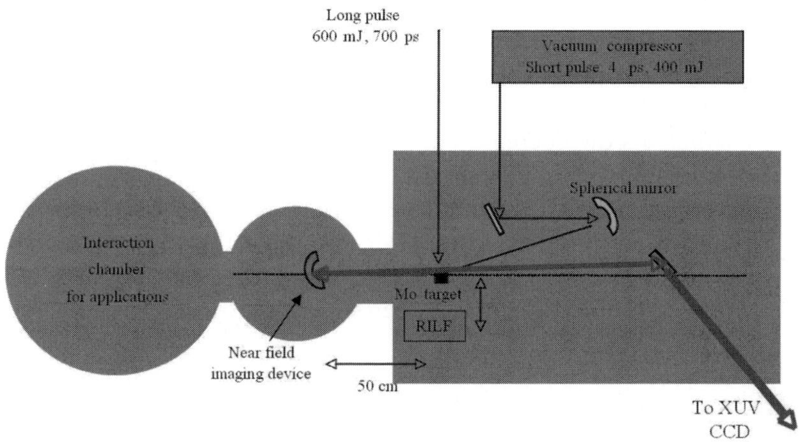

Fig. 2 Experimental set-up for the XRL generation

The dose that had to be delivered on the DNA samples to observe significant amount of DSB is so high that a large number of shots are to be sent: This is of the order of a few ten thousands for 200 nJ energy per shot in the soft x-ray beam. A dedicated investigation was carried out for the characterization of the energy stability both in time and space. We first did 10 Hz films of the soft x-ray near-field source. The results are presented in Figure 3 which shows the energy as a function of shot number for the same target position.

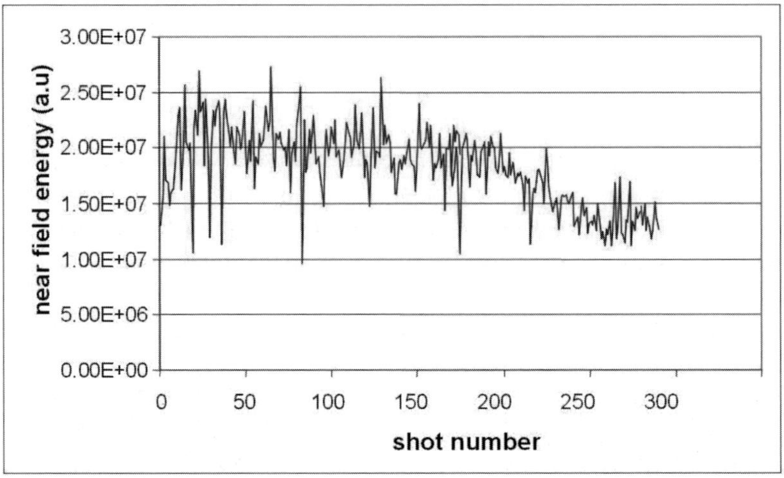

Fig. 3 Near-field energy as a function of shot number for the same target position.

Good energy stability is the result of a specific optimization that includes a study of the effect of pre-pulses in the uncompressed beam. In the best configuration, the same energy level can be maintained over approximately 200 shots, after which a slow decrease occurs. The shot to shot energy fluctuations are of the order of 20 % but we checked that from one 200 shot dose accumulation to another one, the integrated dose fluctuations are limited to less than 5 %. The latter consideration appears to be the most relevant for our biology users. We thus decided to irradiate DNA samples by shooting 200 times on the same Molybdenum target groove, then moving it by 200 microns in height and redo the same operation over the entire available target height of 50 mm. If the shot number that has to be delivered is greater than what is achievable with one target, a pause of twenty minutes is needed to replace it by a new one, including pump-down time. The DNA samples stay under vacuum continuously. An on-line measurement of the photon number delivered on target during the irradiation has been used to monitor the irradiation process. This device is composed of a copper grid with more than 80 % XUV transmission connected to a picoamperemeter. The photocurrent caused by the photoelectric effect of the soft x-ray photons on the copper surface has been cross-calibrated with a photodiode and we performed a statistic analysis to check the linearity of the process. A measurement system developed under LabView can then provide the integrated dose on sample and could stop the irradiation after a given dose has been delivered. The automation of such a process is in progress.

The spatial homogeneity of the dose deposition on the samples is obtained despite the intrinsic speckle structure of the soft x-ray laser beam. As this structure is purely random, the position of the maxima in intensity for each single-shot is changed from one shot to another and the structure contrast decreases as $\frac{1}{\sqrt{N}}$ where N is the number of shots. As a consequence, the large number of shots necessary for the DNA irradiation appears here to be an advantage if the repetition rate is large enough. We experimentally checked this effect and results are presented on Figure 4. These images prove that a few ten thousand shot irradiation will provide excellent homogeneous profile for photon deposition.

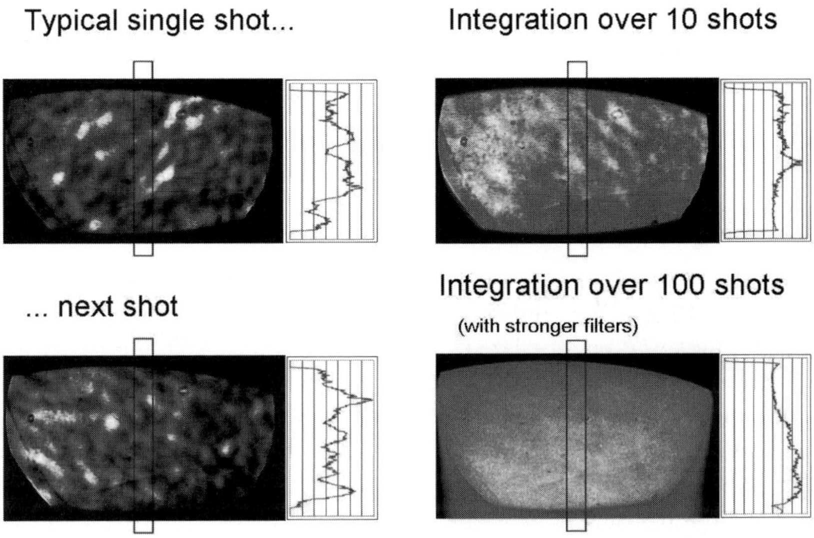

Fig. 4 Comparison of spatial homogeneity between different integration times.

Plasmid DNA (pBr322 Euromedex) was purified (Qiagen kit) and dissolved in pure water. The DNA was then deposited and dried on Mylar foils under controlled atmosphere to form a thin homogeneous sample of 4 mm^2 surface. Several samples were placed in the irradiation chamber on a multi-target holder wheel and aligned under vacuum in the XRL beam using a combination of visible and XUV cameras. The soft x-ray laser beam was filtered by a 200 nm Al filter and focussed by an f = 300 mm spherical multilayer mirror aligned slightly off-axis to match the sample surface. Surface doses of up to 500 kGray were delivered during a time of 90 minutes operation at a repetition rate of 10 Hz and average pulse energy of 200 nJ with Mo target surface changes every 20 s. After irradiation, the DNA was re-dissolved from the Mylar surface and analyzed via electrophoresis.

3 Results and conclusion

The results obtained up to date by irradiating DNA samples show reproducible and quantitative effects induced by the soft x-ray laser photons. Significant single and double strand break amounts were obtained. Figure 5 shows one example of a 90 000 shot irradiation. T1 and T2 denominate DNA samples not placed inside the interaction chamber, 1 is the control (witness sample) that remained under vacuum during all the irradiation time and 2 is the irradiated sample.

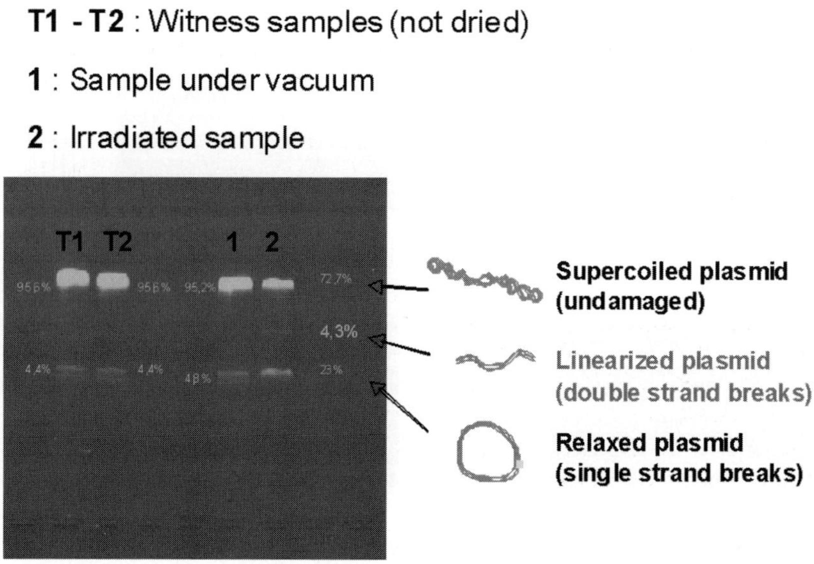

Fig. 5 Picture of the 1% agarose gel

As compared to the control (witness), sample 2 shows 23% SSB with 5 % noise (SNR>4) and 4.3% DSB with undetectable noise for the control under vacuum. This type of irradiation was repeated three times and produced very similar results. We thus have the proof of principle for this kind of experiment: the occurrence of double strand breaks not present in control samples having undergone the same treatment except for the XRL irradiation can be seen as a clear effect of the XRL irradiation. This confirms the presence of specific effects in DNA damages as previously seen with low energy ions [2, 3] and electrons. From the LASERIX team part, the experimental stabilization and characterization of soft x-ray irradiation is now sufficient to enter a phase of systematic and quantitative study of such effects. This will be done in the coming months, by varying the dose deposition, the soft x-ray wavelength and the spatial profile of DNA sample. It is evident that this will help understanding the direct effects leading to DNA breaks.

Acknowledgment

We gratefully thank Jerôme Guigand for the manufacturing of the sample wheel and his efficiency in general.

References

1. D. Ros et al., these proceedings (2008)
2. S. Lacombe, V.A. Esaulov and C. Le Sech. DNA strand breaks induced by low keV energy heavy ions. *Physics in Medecine and Biology 49 N65 (2004)*
3. B. Boudaïffa, P. Cloutier, D. Hunting, M. A. Huels and L. Sanche (2000) Resonant formation of DNA strand breaks by low-energy (3 to 20 eV) electrons *Science 287, 1658-1660 (2000)*
4. L. Sellami, S. Lacombe, D. Hunting, R. Wagner and M. A. Huels. Novel apparatus to measure Hyperthermal Heavy Ion Damage to DNA: strand breaks, base loss, and fragmentation. *Review of Scientific Instruments 78, 1, 085111 (2007)*
5. A. Eschenbrenner, M.-A. Hervé du Penhoat, A. Boissière, G. Eot-Houllier, F. Abel, M.-F. Politis, A. Touati, E. Sage, A. Chetioui., Strand breaks induced in plasmid DNA by ultra-soft X-rays: Influence of hydration and packing. *International Journal of Radiation Biology 83, 687-697 (2007).*
6. C. Le Sech, K. Takakura, C. Saint-Marc, H Frohlich, M. Charlier, N. Usami and K. Kobayashi (2000) – Strand break induction by photoabsorption in DNA-Bound molecules, *Radiation Research 153, 454-458*
7. H. C. Newmann, K. M. Price, M. Forlkard, et al., DNA strand break distribution in X-ray and alpha-particle irradiated V79: Evidence for non-random breakage. *International Journal of Radiation Biology 71, (1997) 347-363*
8. K. Cassou, D. Ros, S. Kazamias, A. Klisnick, G. Jamelot, O. Guilbaud, B. Rus, M. Koslova, J. Polan, AR. Präg, M. Stupka, G. Eot-Houllier, E. Sage, M. Begusova, V. Stisova, MA Du Penhouat, A. Touati, A. Chetioui, Etude des dommages induits dans l'ADN par irradiation laser UV à 21,2 nm, *Journal de Physique IV (2004)*

High Resolution X-Ray Laser Backlighting of Plasmas Using Spatial Filtering Technique

M. Kozlová[1], B.Rus, T. Mocek[1], J. Polan[1], P. Homer[1], D. Snopek[1], K. Jakubczak[1], M. Fajardo[2], A. Barszczak Sardinha[2]

[1] Institute of Physics v.v.i. / PALS Centre, AS CR, Czech Republic
[2] Centro de Física dos Plasmas, Instituto Superior Técnico, Lisbon, Portugal

Abstract. We developed and experimentally demonstrated a novel X-ray laser backlighting technique based on spatial filtering, which allows to dramatically reduce the contribution of plasma self-emission. This technique was used for probing of plasmas relevant to Ne-like and Ni-like lasers schemes. We report on measurements of 2D electron density profiles of line plasmas produced under conditions corresponding to X-ray laser amplifiers operating in TCE and QSS regimes. The probed plasmas were produced on 1-mm long targets by ~300-ps long laser pulses of 1.3 μm light, at intensities of up to $4x10^{13}$ Wcm^{-2}. The investigated plasmas were actively probed by Ne-like Zn laser, delivering multimillijoule output at 21.2 nm.

1 Experimental setup

The experiment was performed at the PALS Centre. The plasma produced by the auxiliary 150 mm diameter beam delivering 80 J at the fundamental wavelength (1.3μm) in 300-ps (FWHM) pulses, was backlighted by the neon-like zinc X-ray laser at 21.2 nm [1]. A doublet of spherical f=120cm and cylindrical f=400cm lens produced 4mm long line focus (with minimal lateral size ~150μm) overfilling 1mm slab targets. The plasmas investigated are produced either by a single pulse or by sequence of a weak prepulse followed by the main pulse. In the latter case, dogleg arrangement (see Fig 1) was employed: the incident beam is split by a 10/90% (T/R) beamsplitter into two beams that are recombined back by a 20/80% beamsplitter. The dogleg makes changing the delay between the prepulse and the main beam from 2 to 10 ns possible. Additional lens in the optical path of the prepulse is used to produce prepulse focus larger than the focus of the main pulse.

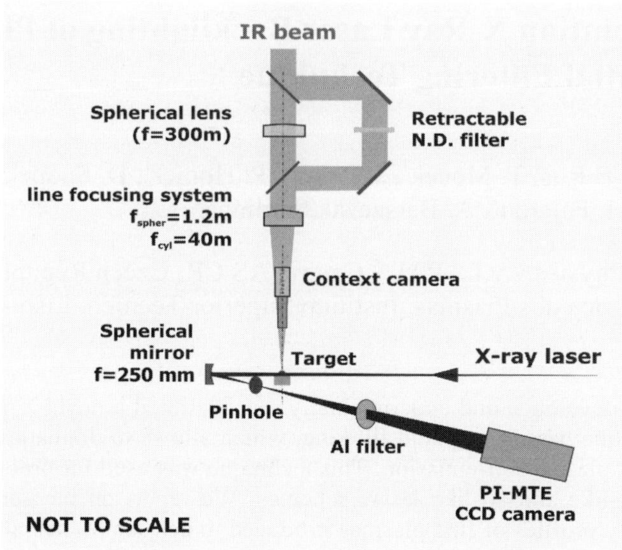

Fig. 1 Experimental scheme. Targets consisting of 1- or 3-mm wide slabs are irradiated either by a single pulse or by sequence prepulse / main pulse produced in the dogleg. The pulses are focused down to a ~150 μm × 4 mm line by f=120 cm spherical / f=400 cm cylindrical lens doublet; the prepulse is separately defocused by additional f=3000 cm lens. Upon probing the plasma, the X-ray laser at 21.2 nm is reflected by f=250 mm MoSi multilayer spherical mirror and passes through a pinhole located in the focal plane of the mirror.

The size of the studied plasmas was monitored via VIS emission by an optical telemicroscope (shown as "context camera" in Fig.1), viewing the target under a small angle with respect to the incident laser beam. In the soft X-ray spectrum near 21 nm, the output plane of the plasma was imaged to a back-side illuminated X-ray CCD camera PI-MTE (2048×2048 pixels with size 13.5 μm). The imaging was carried out by a spherical (f=250 mm) MoSi multilayer mirror with magnification of 8. A small pinhole (300 μm) in the focal plane of the X-ray laser acted effectively as a spatial filter improving strongly the contrast of the X-ray laser to plasma self-emission in the CCD data.

2 Spatial filtering for enhancing contrast in XRL plasma probing

The principle of spatial filtering can be shown on rays of a (near) parallel beam passing freely through a pinhole located in the focal plane of the lens/mirror, while majority of (near) isotropic emission from the probed object is stopped by the pinhole [2]. For illustration see Fig. 2 showing the paths of the optical trajectories for one point on the object.

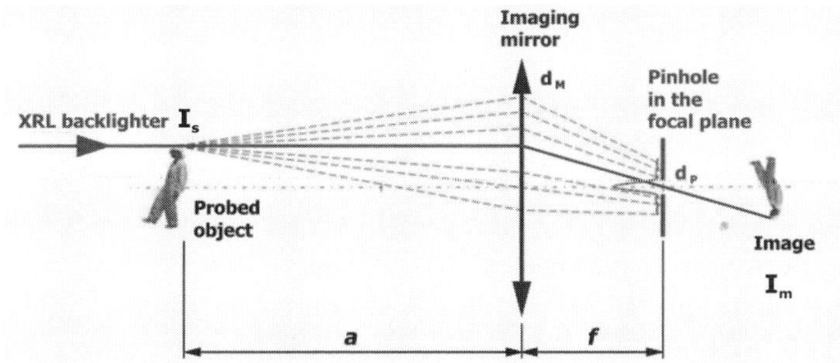

Fig. 2 Principle of spatial filtering. The object being itself an isotropic emitter with intensity I_s is backlighted by nearly parallel X-ray beam. The self-emission signal is collected by a mirror of size d_m and forms image with intensity I_m. When the pinhole is inserted into the focal plane of the mirror, the intensity in the image plane is reduced to I_p. The ratio between I_m and I_p gives the contrast enhancement.

To assess contrast enhancement created by the pinhole, we consider a point on the axis and assume that the source is isotropic, yielding total intensity I_s. The emitter is located at a distance a from the imaging mirror. When rays reflected from the mirror (focal length f) propagate unobstructed, the intensity I_m produced in the image plane is

$$I_m = I_s \frac{\Omega_m}{4\pi}, \tag{1}$$

where the spatial angle Ω_m is given by size of the mirror d_m and the distance a between the object and the mirror

$$\Omega_m = \frac{\pi d_m^2}{4a^2}. \tag{2}$$

When a part of the rays is stopped by the pinhole, the intensity in the image plane I_p will be similarly

$$I_p = I_s \frac{\Omega_p}{4\pi}, \tag{3}$$

and the spatial angle Ω_p directly depends on the pinhole size p projected to the mirror plane, and on the distance a between the object and the mirror

$$\Omega_p = \frac{\pi p^2}{4a^2}, \tag{4}$$

where p is given by Equation (5), d_p denoting the diameter of the pinhole

$$p = \frac{d_p a}{f} \tag{5}$$

Using equations (1) and (3) we find the searched contrast enhancement in the image plane:

$$\frac{I_m}{I_p} = \frac{\Omega_m}{\Omega_p} = \frac{d_m^2}{p^2} = \frac{d_m^2}{d_p^2}\frac{f^2}{a^2} \tag{6}$$

For the experimental parameters ($f = 250$mm, $a = 305$mm, $d_m = 25.4$mm, $d_p = 0.3$mm) the contrast enhancement amounts to about 5000. Figure 3, showing experimental data from two otherwise identical shots but obtained without and with pinhole inserted into the focal plane, illustrates the strength of the technique.

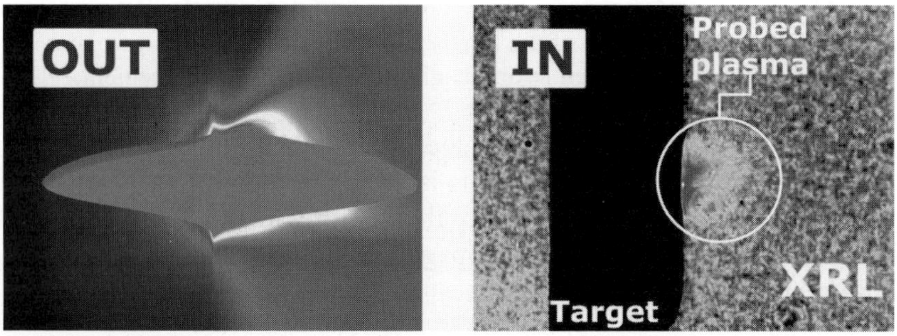

Fig. 3 Comparative CCD data obtained with same experimental parameters but differing by the presence of the pinhole. Left picture: shot taken without the pinhole, the plasma self-emission completely saturating the CCD. Right image: with the pinhole in the focal plane, contribution of the self-emission is dramatically reduced.

3 Results

The experiment was dedicated to study 1mm long line plasma generated from solid targets. The plasma was created by linearly focused IR (1.3 μm, 300 ps) pulses in two different configurations. The first configuration is using tight (in the lateral direction) focus of the main pulse or sequence of the main pulse preceded by loosely focused prepulse, and XRL probe delivered at 0ns or 16ns after the main pulse (see Fig. 4). The prepulse and the main pulse have nominal intensity $3\text{x}10^{10}$ Wcm^{-2} and $4\text{x}10^{13}$ Wcm^{-2}, respectively. The second configuration is similar to the previous one; however, the main pulse is laterally shaped to have "hollow" profile, with minimum intensity at the axis. The main pulse has in this case $2\text{x}10^{13}$ Wcm^{-2} at the peaks, and the prepulse produces $1\text{x}10^{10}$ Wcm^{-2} uniformly over the illuminated surface. The plasmas produced by the "hollow" focus are studied in the context of searching ways for generation of "slab" X-ray laser amplifiers [3].

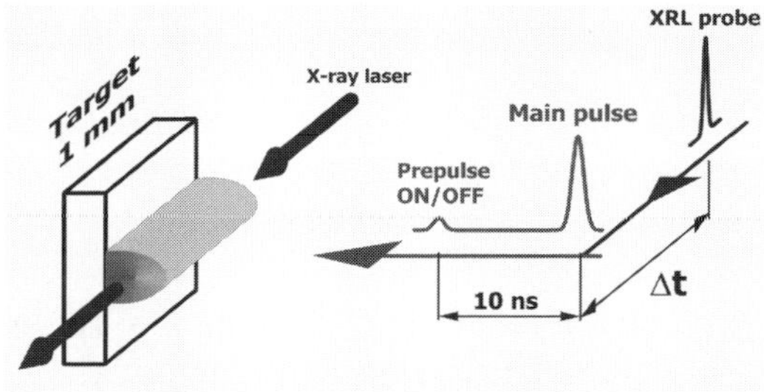

Fig. 4 The generic scheme of the experimental setup. The XRL backlighter is arriving at $\Delta t = 0$ns or 16ns after the main pulse. When the sequence is used, the prepulse is arriving 10 ns before the main pulse.

3.1 Zn line plasmas produced by laterally tight focus

Plasma hydrodynamics is an essential component involved in the performance of the zinc laser. Figures 5 and 6 show the morphology of Zn plasmas generated by the prepulse and the main pulse, applied both separately and in a sequence, and probed by the X-ray laser in $\Delta t = 0$ ns (Fig. 5) and $\Delta t = 16$ ns (Fig. 6) with respect to the main pulse. It is seen that the plasma produced by the isolated main pulse (Fig. 5a) significantly differs from the plasma produced by the sequence prepulse-main pulse (Fig. 5b). When the same plasma as in Fig. 5b is backlighted by maximum intensity (non-attenuated by filter) of the XRL probe available (Fig. 5c), the data indicate the presence of strongly absorbing plasma on the periphery of the surface irradiated by the laser. Signature of this -presumably cold- plasma is clearly seen in Fig. 6, where the XRL backlighter is introduced 16ns later. Although hydrodynamics of this plasma is presently not understood, Fig. 6 shows that it plays a role in forming morphology of sequentially pumped X-ray laser amplifiers, as it is generated even by low-intensity prepulse. This is seen in Fig. 6a, displaying Zn plasma generated by prepulse with intensity I=3x10^{10} Wcm^{-2} nominally employed for pumping the X-ray laser. Figs 6b and 6c show respectively plasmas produced by the main pulse alone (I=4x10^{13} Wcm^{-2}), and by the sequence of the prepulse and the main pulse, with intensity contrast 1.3x10^{3}.

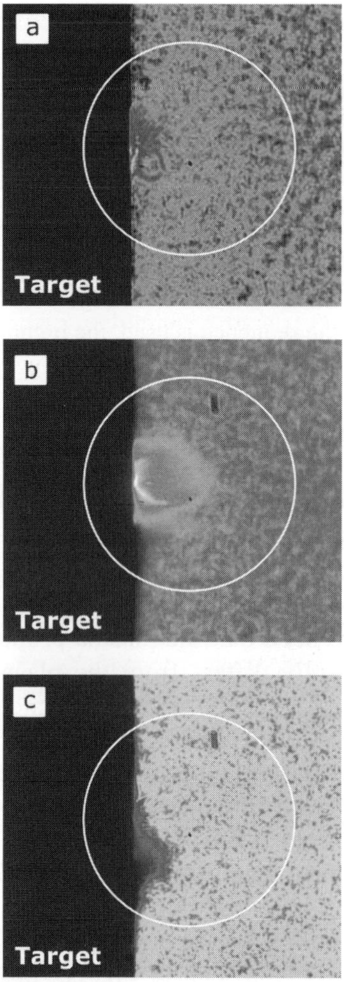

Fig. 5 Columns of 1mm Zn plasmas probed by XRL at Δt=0ns: (a) main pulse at $4x10^{13}$ Wcm^{-2}, (b) sequence of the prepulse and the main pulse with contrast $1.5x10^{3}$, (c) same sequence as in (b), with non-attenuated XRL probe. The scale is indicated by the circle with diameter of 1.3 mm.

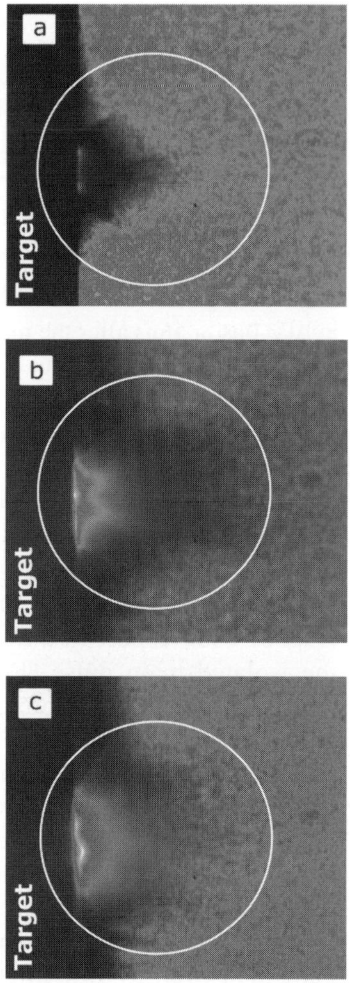

Fig. 6 Columns of 1mm Zn plasmas probed by XRL at Δt=16ns with respect to the arrival of the main pulse: (a) prepulse alone at $3x10^{10}$ Wcm^{-2}, (b) main pulse alone at $4x10^{13}$ Wcm^{-2}, (c) sequence of the prepulse and the main pulse. The scale is indicated by the circle with diameter of 1.3 mm.

3.2 Zn and Ta plasmas produced with hollow focus of the main beam

The 2D plasma density profile is an important factor in the development of amplifiers for seeded XRLs. To increase the amount of amplified signal, the most practicable way is to increase the lateral size of the active region. This is because the amplification length is limited by refraction of the injected XRL beam in the plasma, and because the active region size perpendicular to the target is given by the density and temperature profiles that arise due to plasma expansion (although the active region width can be somewhat increased by using long prepulse delays, e.g. 50 ns [3]). On the other hand, the lateral width of the plasma can be increased as free parameter. Broadening the plasma parallel with target surface increases the volume of plasma in which the signal is amplified; moreover since the ASE will be very divergent due to comparatively small geometrical aspect ratio, ASE contribution to the output signal will be small.

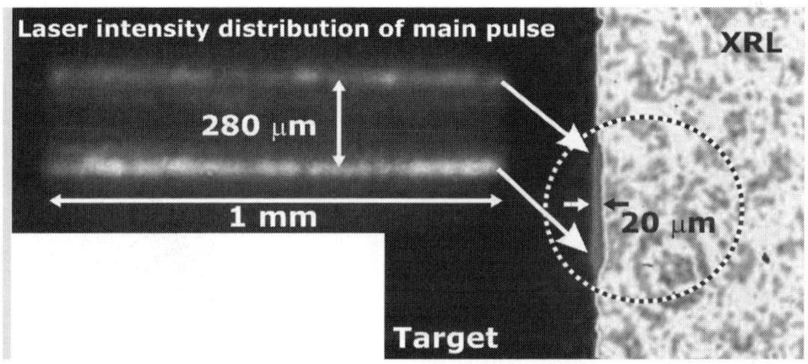

Fig. 7 The left side of the picture shows the intensity distribution of the main driving IR pulse (prepulse is not applied) on the 1mm long Zn target, creating the "hollow" profile with the distance 280μm between the peaks. The right side shows the geometry of the created plasma, probed by XRL at $\Delta t=0$ (at peak of the driving laser pulse). The absorption zone width is ~20μm.

In searching for generation of laterally uniform line plasmas with width significantly greater than 100 μm we used "hollow" line focus, with intensity dip on the axis and the distance 280μm between the peaks (see Fig. 7), with peak to valley intensity modulation of ~3. When probed by the zinc X-ray laser, the probed plasma appears as thin dark region (see Fig.7 right) from which the X-ray laser is "pushed out". To assess the electron density, we used ray tracing to follow individual rays propagating near the target surface. This tracing assumes exponential electron density fall-off, $n_e = n_{e0}\exp(-x/L)$, where n_{e0} is density to be inferred, and L has a value compatible with the extent of the detected plasma (20 μm). The result, shown in Fig. 8, indicates that the electron density in the region close to the surface is $\sim 9\times 10^{21}$ cm^{-3}.

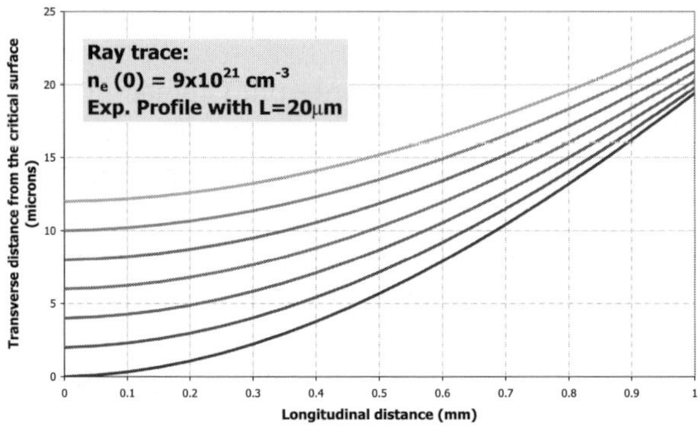

Fig. 8 Trajectories of the probing X-rays, down the axis of the 1mm long line plasma. The rays are injected parallel to the target and are refracted away, producing a dark region near the surface at the output plasma plane (at 1mm). The electron density $n_{e0} = 9\times10^{21}$ cm^{-3} at x=0 is adjusted in order for the shadow size to equal to the experimentally found value, and to be nearly insensitive to the scale length L.

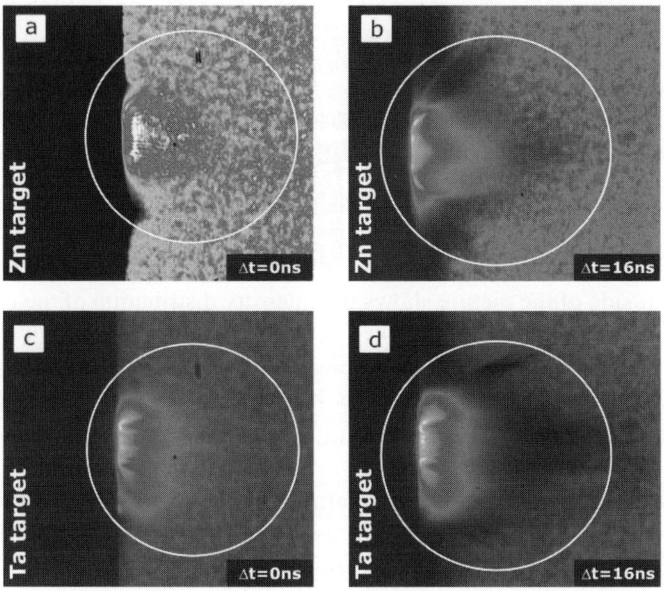

Fig. 9 Plasmas generated by the "hollow" focus from targets with different Z numbers: Zn plasmas probed by XRL at Δt = 0ns (a) and Δt = 16ns (b); Ta plasmas probed at Δt = 0ns (c) and Δt = 16ns (d). The circle with diameter 1.3 mm is dimension indicator.

We further investigated morphology of line plasmas produced by sequence of a prepulse and a main pulse, where the prepulse focus has near top-hat intensity distribution and the main pulse is delivered in the hollow focus with parameters displayed in Fig. 7. We investigated two targets with different Z numbers, Zn (Z=30) and Ta (Z=73), and probed the plasmas at two different times Δt=0 ns and Δt=16 ns. The example of results is shown in Figure 9.

Both in Zn and Ta the generated plasmas are laterally fairly homogeneous, despite hollow profile of the intensity of the main beam. This may suggest that hollow focus may be used for producing "slab" X-ray laser amplifiers. However, imprint of the hollow focus is seen in the plasma corona, as well as filamentation at Δt=0, and further investigation of this setup is necessary.

References

1. B. Rus, T. Mocek, A.R. Präg, M. Kozlová, G. Jamelot, A. Carillon, D. Ros, D. Joyeux, D. Phalippou, "*Multimillijoule, highly coherent x-ray laser at 21 nm operating in deep saturation through double-pass amplification*", PRA **66**, 063806 (2002)
2. M. Kozlová et al, *Development of Plasma X-ray Amplifiers Based on Solid Targets for the Injector-Amplifier Scheme*, ICXRL Proceedings, Berlin, 2007
3. J. Polan, T. Mocek, M. Kozlová, P. Homer, and B us, "*Spatial and temporal profiles of the 21.2-nm saturated X-ray laser output*", ICXRL Proceedings, Berlin, 2007

Development of Soft X-Ray Fourier Transform Holography with Fresnel Zone Plate

Masaharu Nishikino, Hiroshi Yamatani, Keisuke Nagashima, and Tetsuya Kawachi

X-Ray Laser Group, Advanced Photon Research Center
Quantum Beam Science Directorate, Japan Atomic Energy Agency
8-1 Umemidai, Kizugawa, Kyoto, 619-0215, JAPAN

Abstract. We report the soft x-ray Fourier transform holography using a high brilliance and highly coherent x-ray laser at 13.9nm. A Fresnel zone plate generates a point reference source and works as a beam splitter. The object is placed close to the first order focal spot and the transmitted zeroth order beam illuminates the object directly. A Fourier transform hologram is generated by the interference between the object wave at a large distance from the object and the spherical reference wave. The Fourier transform hologram of a grid pattern was successfully recorded with only one XRL pulse and an x-ray CCD camera and successfully numerically reconstructed. The spatial resolution is estimated to be about 1 μm.

1 Introduction

The holographic imaging was introduced by Gabor in 1948[1], and x-ray holography was suggested by Baez in 1952[2]. The optical holographic imaging has found various application fields and has developed by various lasers and recording materials. In soft x-ray region, the x-ray holography has been investigated by the development of bright coherent x-ray sources, such as synchrotron source[3] and soft x-ray laser[4]. Gabor[3] and Fourier transformation[5,6] holographic imaging techniques with sub-100nm resolution have been developed for soft and hard x-rays. One of these holographic imaging techniques is the Fourier-transform holography (FTH) which was already successfully investigated with a soft x-ray undulator source[6,7]. Recently, the lensless FTH[8,9] is well known an imaging technique for the high resolution x-ray microscopy. In order to generate a Fourier-transform hologram the object is illuminated by a plane object wave and a reference point source has to be situated close to the object. In x-ray region, a Fresnel zone plate (FZP) generate a reference point source by its first order diffracted beam, and works as a beam splitter. The transmitted zeroth order beam illuminates the object directly. A Fourier transform hologram is generated by the interference between the object wave diffracted from the object and the spherical reference wave in the far field. The wave in the far field. The amplitude and the phase of the object wave can be reconstructed by the inverse Fourier transform of hologram. A schematic view of the experimental set up is shown in Fig. 1. Here we describe experiments of the generation of submicron x-ray beam and the FTH using only one coherent soft XRL pulse.

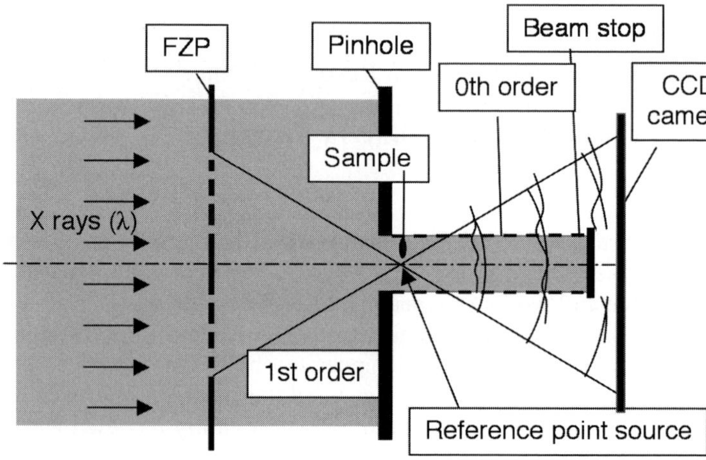

Fig. 1 Schematic view of FTH

2 Experiment

2.1 Generation of submicron soft x-ray beam with FZP

In the FTH, the imaging resolution is limited by the reference source size. A focused x-ray spot is generated by a highly coherent XRL with FZP. The experiment was performed at the JAEA Plasma X-Ray Laser (JAEA-XRL) beam line. A highly coherent XRL[10] at a wavelength of 13.9 nm was generated by the oscillator-amplifier configuration with two targets. The output energy of the XRL beam was about 0.1 μJ, corresponding to a flux of about 10^{10} photons/pulse. The XRL beam divergences along the vertical and horizontal directions were 0.2 and 0.5 mrad, respectively. The beam pointing stabilities along the horizontal and vertical directions were about ±0.12 and ±0.36 mrad, respectively. The pulse width of the XRL was about 7 ps.[11] Figure 2(a) shows a schematic view of the experimental setup used for the generation and characterization of the focused x-ray beam. The XRL beam was reflected with a molybdenum–silicon (Mo/Si) multilayered mirror with an incidence angle of 45° to cut the hard x-ray radiation from the plasma, and filtered with a zirconium (Zr) foil to reduce the scattered optical radiation.

The image and time-integrated energy of the first-order diffracted x-rays were obtained with an x-ray CCD camera. The distance between the XRL source and the FZP was 2620 mm, and the distance between the FZP and the CCD camera was 240 mm. A pinhole as an order-selecting aperture (OSA) with a diameter of 100 μm was placed 8.5 mm away from the FZP in order to select the first-order diffraction. The FZP contains a poly(methylmethacrylate) (PMMA) zone on a 0.75×0.75 mm^2 silicon nitride (SiN) membrane with a thickness of 50 nm. The diameter of the FZP was 0.5 mm, and the total zone number and outermost zone width were 450 and 280 nm, respectively. The focal length was 10 mm for a λ of 13.9 nm. The theoretical spot diameter was 680 nm (290 nm at FWHM) at the focal position. We have performed horizontal scans at various positions of the focused x-ray beam using the knife-edge plate. A knife-edge scan profile close to the best focus is plotted in Fig. 2(b). The line-spread function of the focused beam is derived from the numerical differential of the measured knife-edge scan profiles. Figure 2(c) shows the differentiated profile of the submicron x-ray beam in the horizontal direction. The beam diameter of the x-ray beam was 0.5 μm at FWHM. The measured focal spot diameter was 1.7 times greater than the theoretically spot diameter.[12]

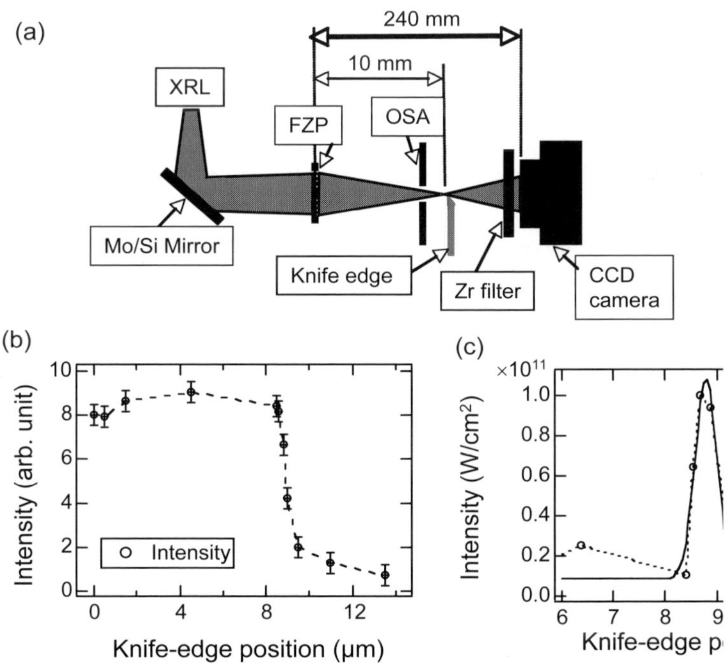

Fig. 2 Generation of submicron soft x-ray spot with FZP

2.2 Generation of Fourier Transform Holography

The experimental setup of FTH is very similar to the generation of submicron soft x-ray beam expect for the sample and the FZP. Figure 3(a) shows the experimental setup of the FTH. The FZP contains an Au zone with a thickness of 50 nm on a 0.75×0.75 mm^2 silicon nitride (SiN) membrane with a thickness of 100 nm. The diameter of the FZP was 0.434 mm, and the total zone number and outermost zone width were 1700 and 64 nm, respectively. The focal length was 2 mm for a λ of 13.9 nm. The theoretical spot diameter was 66 nm at the focal position. A test pattern was placed at the focal plane of the zone plate. A pinhole of 100 μm in diameter was placed 0.5 mm upstream from the test pattern. A Fourier transform hologram was recorded by an x-ray CCD camera, which was placed at 0.23 m from a test pattern. Figure 3(b) shows the tungsten test pattern. The test pattern has vertical 10 lines of each line width. Figure 3(c) is the hologram and Figure 3(d) is the whole reconstructed intensity image. Figure 3(e) shows a part of Figure 3(d) in liner scale. The vertical and horizontal 1 μm line-and-space pattern can be resolved.

3 Summary

We have generated a submicron x-ray spot using the FZP. The spot diameter and peak intensity of the submicron x-ray beam were 0.5 μm at FWHM and 1×10^{11} W/cm^2, respectively. We have developed the Fourier transform holography by using a FZP as one of the application of submicron x-ray beam. Fourier transform holograms of a grid test pattern were successfully recorded with one XRL pulse, and numerically reconstructed. The 1-μm line-and-space pattern could be resolved at 13.9 nm. It is necessary for the clearly reconstruction image to improve the numerically process of reconstruction and the x-ray photons.

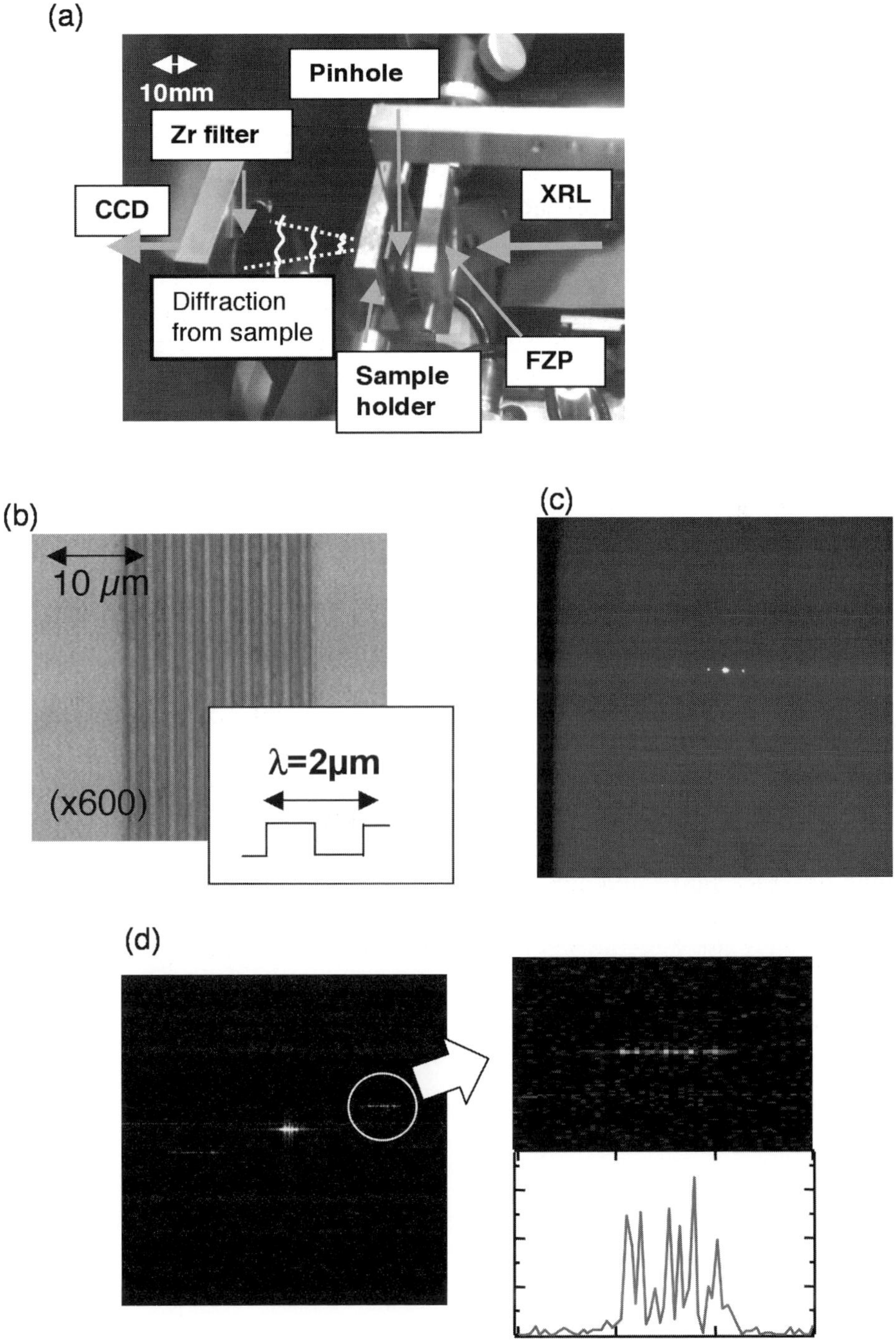

Fig. 3 Fourier transform hologram of the test pattern.

Acknowledgement

This work was partly supported by a Grant-in-Aid for Young Scientist (A) No. 17684030 and (B) No. 20710046 from the Japan Society for the Promotion of Science. The authors are grateful to Dr. T. Kimura, Dr. T. Tajima, and Professor Y. Kato of the Japan Atomic Energy Agency and the staff of the x-ray laser research group for their technical support.

References

1. D .Gabor, Nature **161**, 777 (1948).
2. A. V. Baez, J. Opt. Soc. Am. **42**, 756 (1952).
3. C. Jacobsen et al., J. Opt. Soc. Am. A **7** 1847 (1990).
4. J. E. Trebes et al., Science **238**, 517 (1987).
5. I. McNulty et al., Science **256** (1992).
6. W. Leithenberger and Anatoly Snigirev, J. Appl. Phys. **90**, 538 (2001).
7. N. Watanabe *et al.*, J. Phys. IV France **104**, 551 (2003).
8. S. Eisebitt et al., Nature **432**, 885 (2004).
9. W. F. Schlotter et al., Appl. Phys. Lett. **89**, 163112 (2006).
10. M. Nishikino *et al.*, Appl. Opt . **47**, 1129 (2008).
11. Y. Ochi *et al.*, Appl. Phys. B **78**, 961 (2004).
12. M. Nishikino *et al.*, Jpn. J. Appl. Phys. **47**, 983 (2008).

Lensless Imaging Using Table-Top Soft X-Ray Lasers and High Harmonics Sources Reaching 70 nm Resolution

R.L. Sandberg[1,2], P.W. Wachulak[1,3], D.A. Raymondson[1,2], A.Paul[1,2], A.E. Sakdinawat[1,4], B. Amirbekian[1,5] , E. Lee[1,5] , Y.Liu[1,4] , C. La-O-Vorakiat [1,2], C. Song [1,5] ,M.C. Marconi [1,3], C.S. Menoni[1,3], M.M. Murnane[1,2], J.J. Rocca[1,3], H.C. Kapteyn[1,2], and J. Miao[1,5]

[1] National Science Foundation Center for Extreme Ultraviolet Science and Technology, USA
[2] JILA and University of Colorado at Boulder, Boulder, CO, USA
[3] Colorado State University, Fort Collins, CO, USA
[4] CXRO, Lawrence Berkeley National laboratory, Berkeley, CA, USA
[5] Coherent Imaging Group, University of California at Los Angeles, Los Angeles, CA, USA

Abstract. A very promising high resolution imaging technique currently under rapid development is x-ray diffraction microscopy (or lensless imaging), in which a coherent x-ray diffraction pattern of a non-crystalline specimen is measured. In this technique, a sample is illuminated by coherent light, and the detailed scatter pattern is sampled at a rate higher than the Nyquist frequency. A computerized, iterative phase-retrival algorithm replaces the imaging optics, thus making it possible to reconstruct the image of the sample. Here we demonstrate soft x-ray diffraction microscopy with two complementary coherent soft x-ray sources- a capillary discharge soft x-ray laser at 47 nm [1] and a high harmonic generation source at 29 nm [2]. By using novel field curvature correction in the image reconstruction at high numerical aperture, we retrieve images with resolution as los was 70 nm using set-ups that fit on a 4'x8' optical table- including both the soft x-ray source and the lensless microscope.

1 Introduction

New and innovative imaging techniques have lead to a deeper understanding of the nanometer-scale world, making possible technological advances. One approach to obtaining nanometer-scale resolution in imaging is to use short-wavelength light in the soft x-ray (SXR) region. SXR light is highly desirable for nondestructive imaging applications requiring high spatial resolution and thick samples (> 0.5μm). Lensless imaging is a relatively new technique particularly suited to SXR imaging that enables high-resolution imaging of noncrystalline, aperiodic samples such as biological materials [3]. This technique provides a large depth of field, insensitivity to alignment, and (as we demonstrate here) near-wavelength-limited resolution without the need of complex optical systems. In past work, we demonstrated lensless diffraction imaging using a tabletop high harmonic (HHG) soft x-ray source at 29 nm, achieving resolutions ~200 nm [4,5]. In this work, we significantly enhance our diffractive imaging resolution by implementing a new high numerical aperture (up to NA=0.62) scheme and field curvature correction. We achieve a resolution of 94 nm with a 29 nm HHG source, and a resolution of 71 nm using a 47 nm capillary discharge SXR laser [5]. This high numerical aperture scheme, a first for diffractive imaging with x-rays, will make possible near-λ resolution even as shorter wavelength sources become available. This technique thus promises an important advance in full-field nanoscale imaging on a tabletop system.

2 Experimental Setup and Sample

The experimental setup for lensless diffractive microscopy is compact and straightforward (Fig.1). The sample is illuminated by coherent, narrowband, SXR light. In our microscope, a set of narrowband multilayer mirrors acts as both a monochromator and a condenser, to gently focus the beam onto a sample. Samples are positioned using a closed loop, x-y piezo stage, and are illuminated with either a HHG beam at 29 nm, or a SXR laser beam at 47 nm. The diffraction of the SXR light from the sample is then recorded on a large-area CCD camera (13.5 μm pixel size). Beam blocks of varying size are placed in the center of the diffraction pattern to collect the diffracted light at different spatial frequencies, and to maximize the dynamic range of the camera. Although the CCD camera only records an intensity pattern, if the diffraction pattern is sampled at a frequency greater than twice the Nyquist frequency, the image can be reconstructed using an iterative phase retrieval algorithm [6].

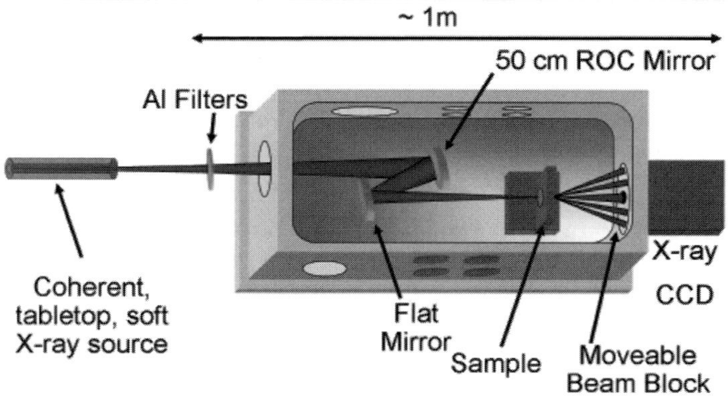

Fig. 1 Compact lensless diffractive microscope used with both a 29 nm HHG source and a 47 nm SXR laser source [5].

3 Results and Discussion

Figure 2 shows images obtained using a 29 nm HHG source [5]. Figure 2 (a) shows a scanning electron microscope (SEM) image of a "waving stick figure" transmission sample, which was fabricated using e-beam lithography on a composite sample consisting of 200 nm of gold on a 100 nm Si_3N_4 substrate. The curvature corrected, high numerical aperture (NA=0.31, sample-to-CCD distance is 42 mm) diffraction pattern is shown in Figure 2 (b). The full spatial coherence of the HHG source enables very high image contrast. However, the femtosecond duration of the HHG pulses means that the temporal coherence after the two multilayer mirrors is moderate ($\lambda/\Delta\lambda$~300). The reconstructed image is shown in Figure 2 (c); the lineout in Figure 2 (d) demonstrates a resolution of 94 nm.

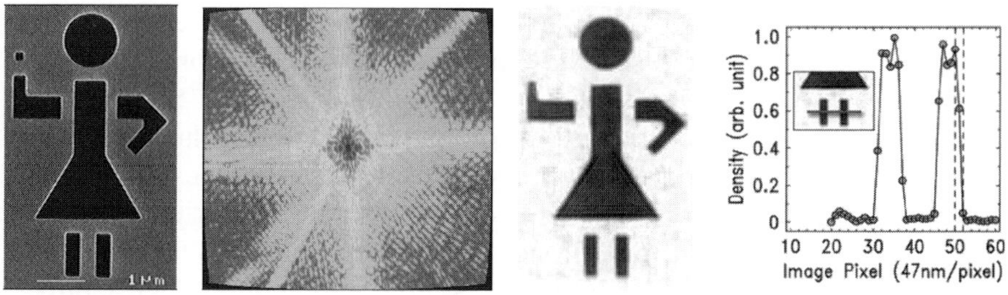

a b c d

Fig. 2 a) Electron microscope image of a waving stick figure transmission sample; b) curvature corrected diffraction pattern using a 29 nm HHG source c) reconstructed image, showing a 94 nm resolution, as shown in (d) [5].

Figure 3 shows the first results of lensless imaging using a table-top SXR laser, here using a numerical aperture of 0.62 (sample-to-CCD distance of 17 mm) [5]. This laser source has extremely high temporal coherence ($\lambda/\Delta\lambda > 5000$) but not full spatial coherence. Thus, a 1 mm diameter pinhole was placed about 1.5 m from the source— where the beam is about 1 cm in diameter— to ensure full spatial coherence. The waving stick figure sample is shown in Figure 3 (a) and was fabricated through e-beam lithography on a 100 nm thick Si_3N_4 window. Figure 3 (b) shows the curvature corrected diffraction pattern, while Figure 3 (c) shows the final reconstruction and 3 (d) a lineout, demonstrating a resolution of 72 nm or 1.5λ. In this case, the higher temporal coherence results in higher spatial resolution but with slight noise in the reconstruction due to lower spatial coherence.

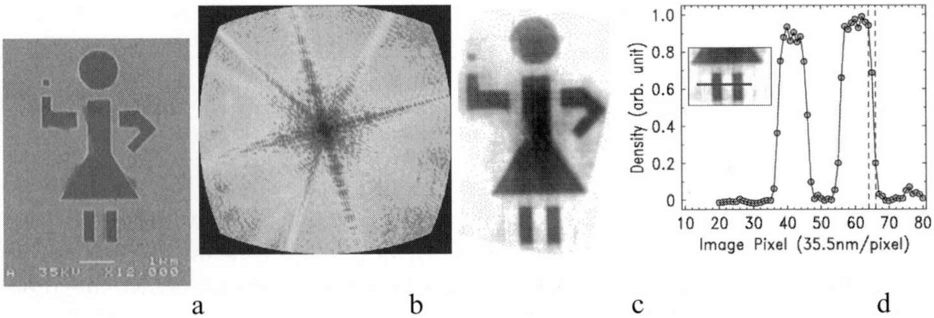

Fig. 3 a) Electron microscope of waving stick figure transmission sample; b) curvature corrected diffraction pattern using a 47 nm SXR laser source c) reconstruction of sample, showing a 71 nm resolution (d) [5].

4 Conclusions

In the future, shorter wavelength tabletop SXR sources will enable versatile soft x-ray diffraction microscopy with sub-10 nm resolution for dynamics imaging in biology, materials science, and nanotechnology. Straightforward extension of the current work will allow imaging at the technologically important 13 nm wavelength using both soft X-ray laser sources and high harmonic sources. Higher repetition-rate lasers, faster phase retrieval algorithms using parallel processors, and improved phase matching techniques will improve the flux of the high harmonic source and reduce data acquisition times, allowing extension of this work to materials and biological imaging throughout the soft-x-ray region of the spectrum. At the same time, the short-pulse nature of the high harmonic source will allow time-resolved imaging of dynamic processes with femtosecond timescales.

We thank Y. Liu and F. Salmassi at the Center for X-Ray Optics and A. Ponomareko and V. Kondratenko at the National Technical University Kharkov Polytechnical Institute for multilayer mirror coatings. This work was supported by the National Science Foundation (NSF) Engineering Research Center for Extreme Ultraviolet Science and Technology, the Department of Energy (DOE) National Nuclear Security Administration, and the JILA Instrument Shop and the Lehnert laboratories. R.L.S. was supported in part by an NSF Integrative Graduate Education and Research.

References

1. Benware B., Macchietto C. D., Moreno C. H., and Rocca J. J.: 'Demonstration of a High Average Power Tabletop Soft X-Ray Laser', *Phys. Rev. Lett.* 81, 5804, 1998
2. Rundquist, A., *et al.*: 'Phase Matched Generation of Coherent Soft X-Rays', *Science*, 280, 1412, 1998
3. Miao, J., *et al:* 'Extending the methodology of X-ray crystallography to allow imaging of micrometre-sized non-crystalline specimens', *Nature,* 400, 342-344 1999
4. Sandberg, R. L., *et al.*: 'Lensless diffractive imaging using tabletop coherent high-harmonic soft X-ray beams', *Phys. Rev. Lett.*, 99, 098103, 2007
5. Sandberg, R. L., *et al.*: 'High Numerical Aperture Table Top Soft X Ray Diffraction Microscopy with 70 nm Resolution', *Proc. Nat. Acad. Sci. of USA*, 105, 24, 2008
6. Miao, J. *et al.:* 'The oversampling phasing method', *Acta Crystallogr. D,* 56, 1312–1315, 2000

Gas Phase Study of The Reactivity of Optical Coating Materials with Hydrocarbons Using a Compact Soft X-Ray Laser

S. Heinbuch[1,2], F. Dong[1,3], E.R. Bernstein[1,3] and J.J. Rocca[1,2,4]

[1] NSF ERC for Extreme Ultraviolet Science and Technology, Colorado State University, USA
[2] Department of Electrical and Computer Engineering, Colorado State University
[3] Department of Chemistry, Colorado State University
[4] Department of Physics, Colorado State University

Abstract. The reactivity of prospective capping layer extreme ultraviolet (EUV) mirror materials with hydrocarbons, alcohols, and water is studied in the gas phase using mass spectroscopy of metal-oxide clusters. We report the results of chemistry studies for Si, Ti, Hf, Zr, and Ru metal oxide nanoclusters in which the reaction products were ionized with little or no fragmentation by 26.5 eV photons from a desktop size 46.9 nm Ne-like Ar laser. Hf and Zr oxides are found to be much less reactive than Si or Ti oxides in the presence soft x-ray light. Ru oxides show different chemistry and are inactive with water/alcohols, and active with saturated/unsaturated hydrocarbons.

1 Introduction

The implementation of EUV lithography as a manufacturing tool for the fabrication of the future generations of computer chips requires a solution to the problem of the degradation of the reflectivity of EUV mirrors and masks caused by carbon deposition in the presence of soft x-ray light. This motivates the study of the reaction of prospective metal oxide capping layer materials with hydrocarbons.

In general, two primary reactive threats contribute to a loss of EUV mirror reflectivity: (1) growth of a carbonaceous layer on the mirror surface caused by the soft x-ray induced dissociation of adsorbed hydrocarbons; and (2) oxidation of the mirror surface by the radiative dissociation of adsorbed water [1-4].

Silicon is a material that is widely used as the top layer in multilayer EUV optics that easily oxidizes on the surface leading to the growth of oxide layers upon soft x-ray radiation [2-4]. During the oxidation process, adsorbed water molecules are dissociated by secondary electrons from the incident soft x-ray radiation and the oxygen atoms react with the Si surface to create SiO_2 [7-9]. This is considered to be an irreversible process causing reflection loss. Similarly, carbon contamination occurs when soft x-ray induced photoelectrons crack hydrocarbon molecules that are adsorbed on the mirror surface causing reflection loss [4,10,11]. Carbon contamination is a reversible process in which adding O_2 to the system increases time between cleanings of the optics [2,12].

We have recently completed a series of studies that constitute the first demonstration of the use of soft x-ray lasers in photophysics and photochemistry [14]. These studies have primarily focused on finding active catalysts for the improvement of industrial processes. The soft x-ray laser was used to provide single photon ionization of neutral clusters for mass spectroscopy studies, with the significant advantage of reduced cluster fragmentation with respect to multi-photon or electron ionization sources. In the present study, we are searching for materials that are not active, in the presence of common vacuum system contaminants and soft x-ray light, in the process of oxidation of or carbon deposition on optical surfaces. For this purpose, Si/Ti/Hf/Zr/Ru oxide neutral clusters are generated in a conventional laser vaporization/supersonic expansion cluster source by laser ablation of the appropriate metal wafer/foil into a He carrier gas mixed with 5% O_2 at 80 psig. Neutral clusters pass through a reactor cell into which a reaction gas is input by a pulsed valve. Single photon ionization of the neutral clusters and reaction products takes place using a 46.9 nm (26.5 eV) laser [20] in the ionization region of a time of flight mass spectrometer (TOFMS). The capillary discharge soft x-ray laser emits pulses of about 1.5 ns duration with an energy/pulse of ~10 μJ at a repetition rate of up to 12 Hz [20] . A time of flight (linear/reflectron) mass spectrometer is employed for mass analysis.

2 Results

2.1 Silicon oxide nanocluster (Si_mO_n) reactions (reactive)

Figure 1 depicts mass spectra of the Si_mO_n cluster distribution with no reactant present (1a), and reaction products with water (1b), acetylene (1c), ethylene (1d), and ethane (1e). In Figure 1a, the cluster distribution, with no reactant, is dominated by oxygen deficient clusters such as $Si_2O_{2,3}$, and $Si_3O_{4,5}$. No oxygen rich clusters such as Si_2O_5, or Si_3O_7 are observed. These respective clusters are labeled oxygen rich/deficient because the most stable cluster configuration is of the nature $(SiO_2)_n$; for example, Si_2O_4, and Si_3O_6.

Shown in Fig. 1b, when water is added to the reactor products are generated through an association forming products SiO_2H_2O, $Si_2O_4H_2O$, and $Si_3O_6H_2O$. Generally, the most stable silicon clusters are more active with water than oxygen deficient clusters. The most stable silicon oxide clusters are also observed to take hydrogen atoms from water to form SiO_2H, $Si_2O_4H_{1,2}$, and $Si_3O_6H_{1,2}$.

When C_2H_2 gas is added to the reactor, many new product signals are observed in the mass spectra, as shown in Fig. 1c. The major products, $SiO_2C_2H_2$, $Si_2O_4C_2H_2$, $Si_3O_6C_2H_2$, etc., are generated from the association reactions. Additionally, products $SiOCH_2$, $Si_2O_4H_2$, and $Si_3O_6H_2$, are found. In general, the most stable silicon oxide clusters are more active with acetylene than oxygen deficient clusters.

As shown in Fig. 1d, the major products of the reaction $Si_mO_n + C_2H_4$ are assigned as $SiO_2C_2H_4$, $Si_2O_4C_2H_4$, and $Si_3O_6C_2H_4$ generated from an association reaction channel.

Figure 1e illustrates that all cluster signals decrease in roughly the same proportion when the saturated hydrocarbon C_2H_6 gas is added to the reactor cell. A similar result is also observed when inert gases are added to the reactor cell. No major reaction channel is observed.

2.2 Titanium oxide nanocluster (Ti_mO_n) reactions (reactive)

The Ti_mO_n cluster distribution is dominated by oxygen deficient clusters. Oxygen rich clusters such as TiO_3, or Ti_2O_5 are observed although the signals are weak. Also, the mass resolution of our experimental setup allows us to observe the isotopic structure of the Ti_mO_n clusters.

When water is added to the reactor, products are generated through an association channel to form $Ti_3O_4H_2O$, and products where Ti_mO_n clusters take H atoms from water to form Ti_3O_4H, and Ti_4O_8H are generated.

When acetylene gas is added to the reactor, the most stable titanium clusters are more active with acetylene than oxygen deficient clusters generating association products. Additionally, oxygen stable and rich clusters generate new products, TiO_2H, TiO_2C_2, $Ti_2O_4H_{1,2}$, and $Ti_2O_6H_2$.

When ethylene is added to the reactor, we observe association products generated to form products such as $TiO_2C_2H_{2,3,4}$, and $Ti_2O_4C_2H_4$. A few reactions involving oxygen deficient clusters are also observed. The most stable titanium clusters are also observed to take hydrogen atoms from ethylene and break apart ethylene to form products such as TiO_2CH_2, TiO_2CH_3, and $Ti_2O_4H_{1,2}$.

No major reaction channel is observed when the saturated hydrocarbon C_2H_6 gas is added to the reactor cell.

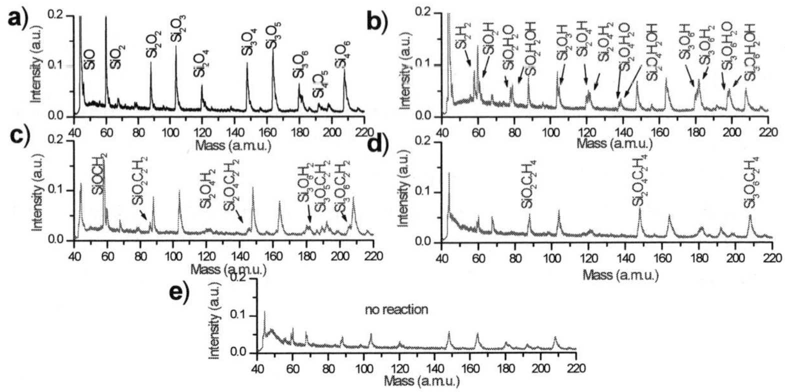

Fig. 1 (a) SimOn cluster distribution with no reactant in the reaction cell, (b) with water in the reaction cell, (c) with acetylene (C2H2) in the reaction cell, (d) with ethylene (C2H4) in the reaction cell, and (e) with ethane (C2H6) in the reaction cell.

2.3 Hafnium oxide nanocluster (HfmOn) reactions (unreactive)

The cluster distribution dominance is shared by both oxygen deficient, and oxygen stable clusters of the type $(Hf_nO_{2n-1})_n$ and $(HfO_2)_n$, respectively. Oxygen rich clusters, $(Hf_nO_{2n+1})_n$ are observed although the signals are smaller than the oxygen deficient and oxygen stable clusters.

When water is added to the reactor, products from the reaction Hf_mO_n + H_2O are generated through an association channel.

When acetylene gas is added to the reactor, very few new product signals, formed in the reaction of $Hf_mO_n + C_2H_2$, are observed in the mass spectra. Only the most stable hafnium clusters and oxygen rich hafnium clusters are active with acetylene generating association products $Hf_2O_4C_2H_2$, $Hf_2O_5C_2H_2$, and $Hf_4O_8C_2H_2$. No reactions occur involving oxygen deficient clusters.

No major reaction channel is observed when ethylene, or ethane are added to the reactor.

2.4 Zirconium oxide nanocluster (Zr_mO_n) reactions (unreactive)

The cluster distribution dominance is shared by both oxygen deficient, and oxygen stable clusters of the type $(Zr_nO_{2n-1})_n$ and $(ZrO_2)_n$, respectively. Oxygen rich clusters, $(Zr_nO_{2n+1})_n$, are observed although the signals are smaller than the oxygen deficient and oxygen stable clusters. We find that Zr_mO_n clusters are highly unreactive with any reactant that we introduce into the system.

2.5 Ruthenium oxide nanocluster (Ru_mO_n) reactions

The reactivity of the ruthenium oxide clusters is difficult to assess due to the abundance of isotopes present in the spectra. However, when water is added to the system, one observes and intensity switch at mass number 100 and a decrease at mass number 99 due to a reaction in pure Ru. This was the only observable reaction for Ru_mO_n clusters with water.

When acetylene gas is added to the reactor, new product signals, formed in the reaction of $Ru_mO_n + C_2H_2$, are observed in the mass spectra, involving oxygen deficient, stable, and rich ruthenium clusters to generate hydrogen abstraction products such as $^{102}RuOH$, and $^{104}Ru_2O_4H_2$. The reactions are not isotopically dependant.

When ethylene is added to the reactor, we observe the hydrogen abstraction products. Again, the oxygen deficient, stable, and rich ruthenium clusters are active with ethylene generating hydrogen abstraction products.

When the saturated hydrocarbon C_2H_6 gas is added to the reactor cell, once again one can observe the same hydrogen abstraction products. This is a different from Si, Ti, Hf, and Zr in that none of them react with saturated hydrocarbons.

3 Conclusions

Our spectra involving Si_mO_n and Ti_mO_n metal oxide nanoclusters show that they are highly reactive in the gas phase. This suggests that an oxidized Si/Ti surface in the condensed phase will be highly reactive. Silicon and titanium oxide capping layers in optical coatings may be contaminated easily by residual gasses in vacuum and should be avoided/protected in the presence of soft x-ray irradiation.

Conversely, our spectra involving zirconium and hafnium metal oxide nanoclusters show that they are unreactive in the gas phase. These data suggest that oxidized hafnium and zirconium surfaces in the condensed phase are much less reactive than Ti or Si oxide surfaces. Zirconium oxide is less reactive than hafnium oxide. Hf and Zr oxides should make good protective coatings for EUV optical surfaces and extend optical lifetimes upon soft x-ray irradiation.

Ruthenium oxide clusters have different and interesting chemistry. They show almost no reactivity with water and alcohols, where as they are fairly active with saturated and unsaturated hydrocarbons. These data suggest in a low carbon environment [1], Ru_mO_n protective capping layers may be of great importance.

Also, if a surface can be tailored to be oxygen poor, this condition could also help to protect an EUV optic surface from carbon contamination and help to increase lifetime of EUV optics.

Acknowledgments

This work is supported by the NSF ERC for Extreme Ultraviolet Science and Technology under NSF Award No. 0310717, Philip Morris, U.S.A., and the U.S. DOE BES program

References

1. J. Hollenshead, and L. Klebanoff, J. Vac. Sci. Technol. B **24(1)**, 118 (2006).
2. B. Mertens, M. Weiss, H. Meiling, R. Klein, E. Louis, R. Kurt, M. Wedowski, H. Trenkler, B. Wolschrijn, R. Jansen, A. Runstraat, R. Moors, K. Spee, S. Pl€oger, R. Kruijs, Microelec. Eng. **73-74**, 16 (2004).
3. H. Meiling, B. Mertens, F. Stietz, M. Wedowski, R. Klein, R. Kurt, E. Loius, and A. Yakshin, SPIE **4506**, 93 (2001).
4. K. Boller, R.-P. Haelbich, H. Hogrefe, W. Jark, and C. Kunz, Nucl. Inst. Meth. **208**, 273 (1983).
6. G. Kyriakou, D.J. Davis, R.B. Grant, D.J. Watson, A. Keen, M.S. Tikhov, and R.M. Lambert, J. Phys. Chem. C **111**, 4491 (2007).
7. T. Engel, Surf. Sci. Rep. 18, 91(1993).
8. G.Y. Mcdaniel, S.T. Fenstermaker, D.E. Walker Jr., W.V. Lampert, S.M. Mukhopadhyay, and P.H. Holloway, Surf. Sci. **445**, 159 (2000).
9. T. Miyake, S. Soeki, H. Kato, A. Namiki, H. Kamba, and T. Suzaki, Phys. Rev. B **42**, 11801 (1990).
10. R. Roseberg et. al., Nucl. Inst. and Meth. in Phys. Research A 291, 101 (1990).
11. T. Koide et. al., Nucl. Inst. and Meth. in Phys. Research A 246, 215 (1986).
12. M.E. Malinowski, P. Grunow, C. Steinhaus, W.M. Clift, and L.E. Klebanoff, SPIE **4343**, 347 (2001).
13. C. Tarrio, and S. Grantham, Rev. Sci. Instrum. **76**, 056101 (2005).
14. F. Dong, S. Heinbuch, J.J. Rocca, and E.R. Bernstein, J. Chem. Phys. **124**, 224319 (2006); S. Heinbuch, F. Dong, J.J. Rocca, and E.R. Bernstein, J. Chem. Phys. **125**, 154316 (2006); F. Dong, S. Heinbuch, J.J. Rocca, and E.R. Bernstein, J. Chem. Phys. **125**, 154317 (2006); F. Dong, S. Heinbuch, J.J. Rocca, and E.R. Bernstein, J. Chem. Phys. **125**, 164318 (2006); S. Heinbuch, F. Dong, J.J. Rocca, and E.R. Bernstein, J. Chem. Phys. **126**, 244301 (2007); F. Dong, S. Heinbuch, Y. Xie, J.J. Rocca, Z. Wang, K. Deng, S. He, and E.R. Bernstein, J. Am. Chem. Soc. **130**, 1932 (2008).
20. S. Heinbuch, M. Grisham, D. Martz, and J.J. Rocca, Opt. Exp. **13**, 4050-4055 (2005); S. Heinbuch, M. Grisham, D. Martz, F. Dong, E. R. Bernstein, and J. J. Rocca, SPIE 5919, 591907 (2005).
22. M. E. Geusic, M. D. Morse, S. C. O'Brien, R. E. Smalley, Rev. Sci. Instrum. **56**, 2123 (1985).
23. Y. Matsuda and E.R. Bernstein, J. Phys. Chem. A, **109 (2)**, 314 -319 (2005).
25. M. Ritala, M. Leskelä, L. Niinistö, T. Prohaska, G. Friedbacher and M. Grasserbauer, Thin Solid Films **250**, 72 (1994).

Gas Phase Studies of Catalytic Processes Involving V_mo_n Clusters and their Reaction with Alcohols, Alkenes, No_x, and Nh_3 Using a Desk-Top Size Soft X-Ray Laser

S. Heinbuch[1,2], F. Dong[1,3], E.R. Bernstein[1,3] and J.J. Rocca[1,2,4]

[1] NSF ERC for Extreme Ultraviolet Science and Technology, Colorado State University, USA
[2] Department of Electrical and Computer Engineering, Colorado State University
[3] Department of Chemistry, Colorado State University
[4] Department of Physics, Colorado State University

Abstract. Reactions of neutral vanadium oxide clusters with (1) methanol, ethanol, (2) alkenes, and (3) NO_x and NH_3, in a flow tube reactor, are investigated by time of flight mass spectroscopy experiments and density functional theory (DFT). Single photon ionization through using a compact soft x-ray (46.9 nm, 26.5 eV) laser is employed to detect both neutral cluster distributions and reaction products. Results and insights to catalytic processes are presented.

1 Introduction

Metal oxides are widely used in chemical reactions for industrial processes, and as of yet, there is not a proper understanding of these chemical reaction processes due to very complicated environments on condensed phase surfaces. Particularly, vanadium oxide is widely useful as a catalytic agent for many industrial processes. In our experiments, we explore three of these important processes.

(1) The oxidation of methanol is interesting because of its importance in catalytic processes[1, 2]; for example, oxidative reforming of methanol is an important catalytic process in fuel cells [3]. Selective oxidation of methanol on supported vanadium oxide catalysts has been considered as a simple probe reaction for a number of other selective oxidation reactions [4, 5]. The major product of methanol oxidation over V_2O_5 is found to be formaldehyde, with minor products of dimethyl ether, etc [6]. The mechanism for oxidation of methanol on supported vanadium oxide is suggested to be methanol oxidation by the catalyst, and not by gas phase molecular O_2 [7].

(2) The oxidation of alkenes over supported metal oxide catalysts is a very important catalytic process in the chemical industry. For example, catalytic

partial oxidation of propylene (CH_2=$CHCH_3$) produces acrolein (CH_2CHCHO), one of the more employed industrial chemical intermediates. [8-10].

(3) Nitrogen oxides NO_x produced by fuel combustion are among the main atmospheric pollutants responsible for acid rain and smog [11]. They are greenhouse gases produced both by industry and agriculture and are blamed for the formation of ozone and many respiratory problems [12]. Great effort has been made to decompose NO_x into harmless N_2 and O_2. Even though the decomposition is exothermic, it does not occur without an appropriate catalyst.

2 Experimental Procedures

The reactions of neutral vanadium oxide clusters with different reagents are investigated using a time of flight mass spectrometer (TOFMS) coupled with single photon ionization by a desk-top 26.5 eV soft x-ray laser. Since the experimental apparatus has been described in detail elsewhere, [13] only a general outline of the experimental scheme will be presented in this report. Briefly, neutral vanadium oxide clusters are generated in a conventional laser vaporization/supersonic expansion source. A 532 nm wavelength laser is employed to ablate a target of vanadium metal foil at 10-20 mJ/pulse. A mixture of 0.5% O_2 and He is used as a carrier/reaction gas at 80 psig for the generation of V_mO_n neutral species. The reactant gas, formed by flowing He through a reservoir containing methanol/ethanol, alkenes, NO, or NH_3 at a pressure of 15 psi, is pulsed into the reactor tube located ~ 20 mm downstream from the exit of the expansion channel. Vanadium oxide clusters generated from the ablation source react with reactant gases in the flow tube reactor [14]. Neutral clusters and reaction products pass into the ionization region of a time of flight mass spectrometer, where these neutral species are ionized by the focused soft x-ray laser [15].

The soft x-ray laser (26.5 eV/photon energy) emits pulses of about 1 ns duration with an energy/pulse of 10 μJ at a repetition rate of up to 12 Hz. A pair of gold coated mirrors, a torodial and a plane mirror, is placed in a grazing incidence Z-fold configuration just before the ionization region of the TOFMS to provide alignment and focusing capabilities for the laser with respect to the molecular beam at the ionization region. Soft x-ray laser light is not tightly focused in the ionization region to avoid multiphoton ionization and a space charge Coulomb effect due to He^+ ions. A large number of He^+ ions can be produced by 26.5 eV ionization of He in the molecular beam, and these could broaden the V_mO_n, etc. mass spectral features. The microchannel plate (MCP) ion detector voltage is gated to reduce the MCP gain when He^+ arrives at the mass detector, in order to prevent detector circuit overload and saturation.

3 Results

In the present experiments, a 26.5 eV laser is employed to ionize neutral clusters and their reaction products. The V_mO_n cluster distribution is displayed in the mass spectrum of Figure 1(a). Three categories of vanadium oxide clusters can be identified. Oxygen stable clusters VO_2, V_2O_5, V_3O_7, V_4V_{10}, V_5O_{12}, V_6O_{15}, etc. can be expressed as stoichiometries of the form $(VO_2)_{0,1}(V_2O_5)_n$. These clusters are the most stable clusters (highest intensities for V_mO_n within a given V_m cluster family) in the neutral V_mO_n cluster distribution based on both experiments and theoretical calculations. Oxygen rich clusters VO_3, V_2O_6, V_3O_8, V_5O_{13}, etc. that have one more oxygen atom compared to the most stable clusters. These oxygen rich clusters are found to associate one or two hydrogen atoms to make more stable structures. Oxygen deficient clusters VO, V_2O_3, $V_3O_{5,6}$, $V_4O_{8,9}$, etc. They have one or more oxygen atoms fewer than the most stable clusters.

The reactions of neutral vanadium oxide clusters are investigated employing 26.5 eV soft x-ray laser and 10.5 eV nm laser ionization coupled with TOFMS. In the experiments, nearly identical reaction products are detected using 26.5 eV and 10.5 eV laser ionizations. It is well known that no fragmentation occurs with 118 nm ionization because very little energy above the vertical ionization energy is left in the molecule for dissociation. We conclude that neutral vanadium oxide clusters and their reaction products are not fragmented or photodissociated by 26.5 eV photons.

3.1 $V_mO_n + CH_3OH/CD_3OH$ reactions

Three major reactions are identified for $V_mO_n + CH_3OH/CD_3OH$:

Association reactions

Association products $V_mO_nCH_3OH$ are observed for most of vanadium oxide clusters in the experiments, indicating that methanol molecules are easily adsorbed on neutral vanadium oxide clusters. In the condensed phase, the surface of a vanadium oxide catalyst will be very active toward methanol adsorption.

Hydrogen abstraction reactions

Oxygen deficient vanadium oxide clusters (VO, V_2O_3, V_3O_6, etc.) can abstract only one hydrogen atom from a CH_3/CD_3 unit of CH_3OH/CD_3OH to form

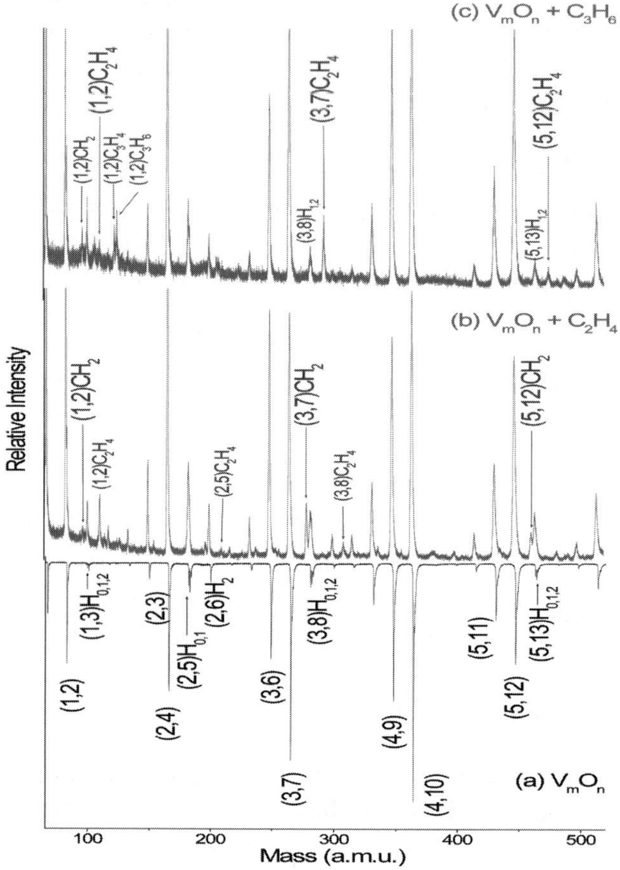

Fig. 1 (a) V_mO_n cluster distribution generated with 0.5% O2/He expansion gas. Reactant gases, (b) pure C2H4 (c) pure C3H6.

$V_xO_{y-1}D/H$ products. The most stable vanadium oxide clusters can abstract more than one hydrogen atom from CH_3 and/or OH to form $V_xO_yDH_{0,1,2}$ products. Oxygen rich vanadium oxide clusters can abstract more than one H atom from any kind of hydrogen source in a high vacuum system to form $V_xO_{y+1}H_{0,1,2}$ products.

Intermolecular dehydration reactions

Strong signals of $VO_2(CH_3)_2$, $VO_3(CH_3)_2$, $V_2O_5(CH_3)_2$, and $V_3O_8(CH_3)_2$ are observed in the experiments, indicating that the intermolecular dehydration reaction, $V_mO_n + 2CH_3OH \rightarrow V_mO_n(CH_3OH)_2 \rightarrow V_mO_{n+1}(CH_3)_2$, is one of the major reaction channels for the V_mO_n reactions with methanol. A concomitant product H_2O is derived from the OH radicals of two CH_3OH molecules.

3.2 V_mO_n + alkenes reactions

$V_mO_n + C_2H_4/C_3H_6$

Shown in Figure 1(b), when C_2H_4 ($H_2C{=}CH_2$) is added to the reactor, the products VO_2CH_2, $V_3O_7CH_2$, and $V_5O_{12}CH_2$, generated from $V_mO_n + C_2H_4$ reactions, are identified as the main products, as well as several association products are observed.

In Figure 1(c), if reactant propylene ($H_2C{=}CHCH_3$) is added into the reactor, a series of new signals is assigned to products $VO_2C_2H_4$, $V_3O_7C_2H_4$, and $V_5O_{12}C_2H_4$ for the reactions $V_mO_n + C_3H_6$.

$V_mO_n + C_4H_8/C_4H_6$

The complexes VOC_3H_6, $V_3O_7C_3H_6$, and $V_5O_{12}C_3H_6$, are identified as the major products, while several small signals at masses corresponding to VOC_4H_6, $VO_2C_2H_4$, and $VO_2C_4H_8$ are also observed. 1, 3-butadiene ($H_2C{=}CH{-}CH{=}CH_2$) is another alkene with a C=C double bond used as a reactant.

$V_mO_n + C_6H_6/C_2F_4$

Benzene has a delocalized π double bond system with no particular localized single or double bonds; the delocalization of electrons makes it more stable typically than alkenes. Products are observed for reactions $V_mO_n + C_6H_6$. These products are different with that generated from V_mO_n + $C_2H_4/C_3H_6/C_4H_6/C_4H_8$.

3.3 Preliminary results of SCR reactions of NO with NH_3 on V_m/Ta_mO_n catalysts

Vanadium oxide reactions

When NO is added to the reactor (top Fig. 3), no observable products are found in the new spectrum. The same inactivity is found if the NO reactant is exchanged for NO_2. Contrarily, when the reactant gas is NH_3, V_mO_n clusters become active. One can observe many reactions including association products.

In an attempt to model the SCR of NO using NH_3 on a vanadium oxide catalyst, NO and NH_3 are added to the reaction cell. There is no observable co-adsorption in the spectrum.

Tantalum oxide reactions

Ta_mO_n clusters react with NO, mainly forming association products with oxygen rich clusters. This result is much different than V_mO_n cluster reactions. When the reactant gas is NH_3, Ta_mO_n clusters remain active and behave very similarly to V_mO_n clusters.

In an attempt to model the SCR of NO using NH_3 on a tantalum oxide catalyst, NO and NH_3 are added to the reaction cell (Figure 4). Using Ta_mO_n clusters for the catalyst, one can observe reaction products involving both reactants.

Acknowledgments

This work is supported by the NSF ERC for Extreme Ultraviolet Science and Technology under NSF Award No. 0310717, Philip Morris, U.S.A., and the U.S. DOE BES program.

References

1. R. G. Herman, Q. Sun, C. Shi, K. Klier, C. Wang, H. Hu, I, E. Wachs, M. M. Bhasin, Catalysis Today, **37**, 1 (1997).
2. I. E. Wachs, Catalysis Today, **100**, 79 (2005).
3. Z. Jusys, R. J. Behm, J Phys. Chem. B **105,** 10874 (2001); K. Drew, G. Girishkumar, K. Vinodgopal, P. V. Kamat, J Phys. Chem. B **109**, 11851 (2005).
4. I. E. Wachs, G. Geo, M. V. Juskelis, B. M. Weckhuysen, Stud. Surf. Sci. Catal. **109**, 305 (1997).
5. B. M. Weckhuysen and D. E. Keller, Catalysis Today, 78, 25 (2003).
6. J. Sambeth, L. Gambaro, H. Thomas, Adsorp. Sci. Technol. **12**, 171 (1995); J. Sambeth, M. Centeno, A. Paul, L. Briand, H. Y. Thomas, J. Odriozola, J. Mol. Catal. A **161**, 89 (2000); L. Gambaro, J. Mol. Catal. A **214**, 287 (2000).
7. G. Froment, K. B. Bischoff, Chemical Reactor Analysis and Design, Wily, New York, 1979; S. Chio, I. E. Wachs, 223rd ACS National Meeting, Orlando, FL, United States, April 7-11, 2002, FUEL-509.
8. J. R. Valbert, J. G. Zajacek, D. I. Orenbuch, E*ncyclopedia of Chemical Processing and Design*, Marcel Dekker: New York, **1993**, p88.
9. C. Zhao, I. E. Washs, *Catalysis Today*, **2006**, *118*, 332.
10. T. A. Nijhuis, M. M akkee, J. A. Moulijn, B. M. Weckuysen, *Ind. Eng. Chem. Res.* **2006**, *45*, 3447.
11. V.I. Parvulescu, P. Grange, B. Delmon, Catal.Today **46**, 233 (1998).
12. P. Forzatti, Catal. Appl. A. Gen. **222**, 236 (2001).

13. F. Dong, S. Heinbuch, J. J. Rocca, E. R. Bernstein, J. Chem. Phys. **125**, 164318 (2006); S. Heinbuch, F. Dong, J. J. Rocca, E. R. Bernstein, J. Chem. Phys. **126**, 244301 (2007); F. Dong, S. Heinbuch, J. J. Rocca, E. R. Bernstein, J. Chem. Phys. **125**, 154317 (2006); S. Heinbuch, F. Dong, J. J. Rocca, E. R. Bernstein, J. Chem. Phys. **125**, 154316 (2006); F. Dong, S. Heinbuch, J. J. Rocca, E. R. Bernstein, J. Chem. Phys. **124**, 224319 (2006); Y. Matsuda, E. R. Bernstein, J. Phys. Chem. A **109**, 3803 (2005).
14. M. E. Geusic, M. D. Morse, S. C. O'Brien, R. E. Smalley, Rev. Sci. Instrum. **56**, 2123 (1985).
15. S. Heinbuch, M. Grisham, D. Martz, J. J. Rocca, Opt. Express, **13**, 4050 (2005).

Time-of-Flight Measurements of Ion and Electron from Xenon Clusters Irradiated with a Soft X-Ray Laser Pulse

S. Namba[1], N. Hasegawa[2], M. Nishikino[2], M. Kishimoto[2], T. Kawachi[2], M. Tanaka[2], Y. Ochi[2], K. Nagashima[2] and K. Takiyama[1]

[1] Graduate school of Engineering, Hiroshima University, Kagamiyama 1-4-1, Higashi-Hiroshima, Hiroshima, 739-8527 JAPAN
[2] Quantum Beam Science Directorate, Japan Atomic Energy Agency, Umemidai 8-1, Kizugawa, Kyoto, 619-0215 JAPAN

Abstract. The photoionization process in xenon clusters irradiated with a soft x-ray laser pulse (hν=89.2 eV, intensity: ~10^{10} W/cm^2) was examined using a time-of-flight mass spectrometer, where the laser photon energy was high enough to ionize Xe 4*d* inner electrons (threshold: 67.55 eV for $4d_{5/2}$ and 69.54 eV for $4d_{3/2}$). The dominant ion yield resulting from Auger decay of 4*d* hole was Xe^{3+} ion. This showed that the double Auger transition probability would be enhanced in the cluster environments. In order to clarify the decay dynamics of 4*d* inner holes, the electron energy spectrum was measured. It was found that the distinct photo- and Auger electron lines were not observed, while the energy distribution implies the production of a low temperature cluster plasma.

1 Introduction

The photoionization process involving an inner shell electron in free rare gas clusters has been extensively studied using synchrotron radiation (SR) [1]. As one of the applications for it, the site-specific electronic structures that correspond to an isolated atom and to the surface and bulk of an argon cluster can be determined using photoabsorption and photoelectron spectroscopy [2]. Since the atomic cluster comprises a relatively simple system, a deep understanding of the ionization and expansion behaviors is essential for the analysis of a more complicated structure, such as bio-molecules.

SR has proved useful for investigations in various fields of physics and chemistry, while high-intensity plasma-excited x-ray laser radiation is also expected to become a powerful diagnostic tool in these fields [3]. In addition, since the typical brightness of an x-ray laser pulse is approximately six orders of magnitude higher than that of radiation from a third-generation synchrotron [4], it is feasible to generate fascinating plasmas by the illumination of high-intensity x-ray laser pulses, such as extremely strongly coupled plasma [5] and plasmas induced by inner-shell ionization.

On the other hand, the interaction of the clusters with ultrashort, high-intensity laser pulses from IR to UV wavelength regime has attracted a great deal of interest for fundamental condensed matter physics and various applications [6]. For shorter wavelength region, Wabnitz *et al* investigated the interaction of xenon (Xe) clusters and a vacuum UV free electron laser (VUV-FEL, wavelength: 98 nm, pulse width: 100 fs, intensity: 2×10^{13} W/cm^2) [7]. However, little is known about the ionization process and expansion dynamics of the clusters irradiated with an intense x-ray laser pulse, in which the photoionization of inner-shell electrons, followed by Auger decays, plays an important role in the generation of highly-charged ions [8].

In this study, the ionization dynamics in Xe clusters irradiated with a soft x-ray laser (13.9 nm) was investigated using a time-of-flight mass spectrometer (TOF-MS). The photon energy of the x-ray laser (89.2 eV) is sufficiently high to photoionize Xe 4*d* inner electrons (threshold energy: ~70eV). It should be noted that the cross section for the inner-shell ionization of 4*d* electron is ~16 Mb, whereas for the outermost valence electron ~1.5 Mb at a photon energy of ~90 eV [9]. Subsequently, this inner-hole is filled by the valence electron with a decay time of less than or equal to a few femtoseconds, resulting in the productions of an Auger electron and a Xe^{2+} ion. On the other hand, Xe^{3+} ions can also be generated by double Auger (DA) decay of a 4*d* vacancy, although the transition probability for this process is small. In the present study, it was found that the ion mass spectra up to Xe^{4+} were observed, which was in contrast to the results of SR experiments. In order to clarify the decay dynamics of 4*d* holes, the electron energy distribution was measured. Although the photo- and Auger electron spectra were not observed, the energy distribution implied the production of low temperature cluster plasmas.

2 Experimental Setup

The experiment was carried out using the x-ray laser facility at JAEA, Kansai. The x-ray laser with a wavelength of 13.9 nm, a pulse duration ~7 ps and a resolution of $\lambda/\Delta\lambda < 1 \times 10^{-4}$ was generated by transient collisional excitation scheme for nickel-like silver [10]. The maximum laser intensity focused using a Mo/Si multilayer spherical mirror was $\sim2\times10^{10}$ W/cm^2 on target.

The Xe cluster targets were prepared by injecting the high stagnation pressure gas into vacuum using a supersonic conical nozzle with a throat diameter of 200 μm [11]. In order to extract the central part of the injected beam, a skimmer with a horn aperture of 3 mmϕ was also used. The average cluster size was estimated by Hagena scaling law and its range was estimated to be

1×10^2 to 3×10^5 atoms/cluster [12]. Moreover, a seeding technique (Xe 30%-He 70%) was also employed to promote the formation of larger clusters.

The Xe^{q+} ion stages and their energies were measured using a TOF-MS. A dual microchannel plate (MCP) located at the end of a 50 cm long field free flight tube was employed to detect ions. In order to prevent the gas break-down between the electrodes, the TOF electrode assembly was covered by a shroud vessel with second skimmer (1 mmϕ), as shown in Fig. 1. On the other hand, the electron energy spectrum was observed using the other TOF device and the flight tube made of μ-metal had a length of 170 mm (not shown). The cluster gas jets also included a large amount of free atoms, and thus the ions and electrons emanating from these uncondensed atoms would be preferably emitted into the polarization direction, while for cluster the distribution is likely to be isotropic. To obtain the contribution from the clusters, therefore, the observations were achieved in the direction perpendicular to the laser polarization and the propagation axes. To measure the incoming laser intensity, a soft x-ray CCD camera was also installed 50 cm from the intersection of the laser and cluster beam.

3 Results and Discussion

TOF spectra for the Xe clusters irradiated with the x-ray laser pulse are shown in Fig. 2 for Xe and Xe-He gas jets of backing pressures of 4.4 and 11.2 atm. As clearly seen, the Xe^{3+} ions became the dominant ionic fragment. The Xe^{4+} ions might be generated by the sequential two photon absorptions, that is, Xe $\rightarrow Xe^{2+}\rightarrow Xe^{4+}$. The radiative decay of the inner vacancy yielding Xe^{+} ions is negligibly small [13]. The Xe^{+} ion spectrum observed, therefore, was ascribed to a charge exchange process of the product ions with ambient gas, such as $Xe^{q+} + Xe \rightarrow Xe^{(q-1)+} + Xe^{+}$. The similar experiments with synchrotron radiations involving isolated atoms revealed that the dominant decay process following the inner vacancy is the normal Auger process resulting in the production of Xe^{2+} ions [13]. The ratio of the Xe^{2+} yield to the Xe^{3+} yield resulting from the decay of the $4d^{-1}$ state in the free atom was found to be ~4.3 (average value for the multiplet states of j=3/2 and 5/2) and was independent of the incidence photon energy in the 4*d* shape resonance region [13]. The shake-off process associated with the simultaneous emission of inner and valence electrons does not significantly contribute to the Xe^{3+} yield. The distribution of ionic stage, therefore, cannot be explained by the conventional interpretations.

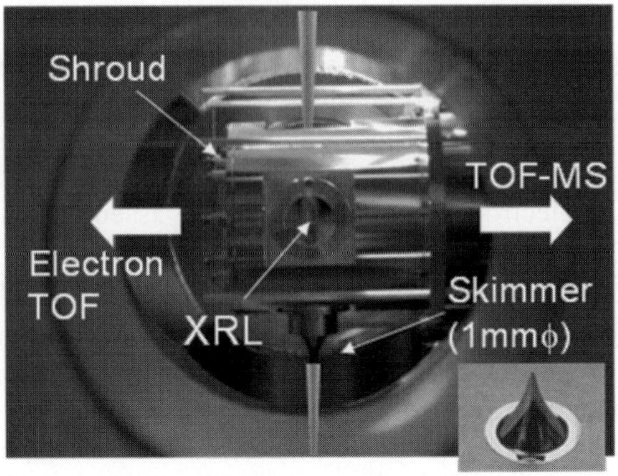

Fig. 1 Experimental setup for the interaction of Xe clusters and X-ray laser pulse. The Xe cluster was injected from the bottom, and the ions and electrons emitted were measured in the direction perpendicular to both of the laser polarization axis and the gas jet.

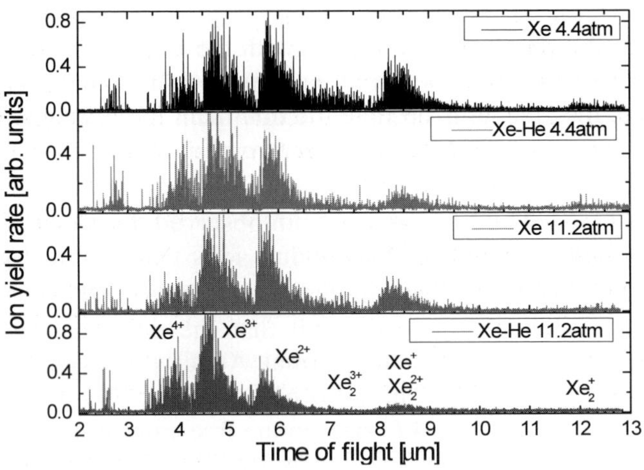

Fig. 2 TOF spectra observed for Xe and Xe-He admixture gas jets.

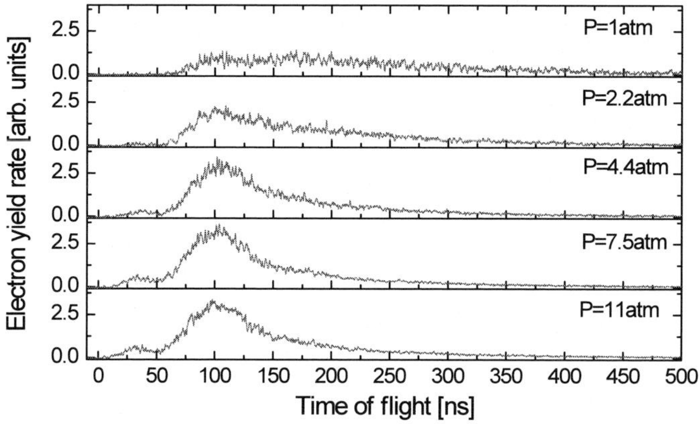

Fig. 3 Electron TOF spectra observed for various backing pressures of Xe gas jets.

In order to clarify the decay dynamics of 4*d* inner hole, the electron spectra were measured using the TOF type electron spectrometer. Figure 3 shows the TOF signals observed for various backing pressures of Xe gas jets. Figure 4 shows the corresponding electron energy spectrum, where the background components were subtracted. Since the accurate energies and its relative intensities for the photo- and Auger electron lines have been examined by SR experiments [14], the relevant expected energy peaks taking into account the instrumental resolution (Lorentzian profile) are also shown in the figure. The experimental curve for photoelectron peaks around 20 eV and $N_{4,5}OO$ Auger lines from 10 to 18 eV had a similar structure with the expected ones, while the higher energy components were not reproduced at all. On the other hand, the energy distribution obtained had a peak value around a few eV. The reason might be explained by that the energy relaxations due to electron-electron and electron–phonon interactions could be very effective within the cluster. Moreover, the simultaneous DA decay, in which the Auger electrons share continuously the excess energy, contributes to generate the low energy electrons [15]. The creation of the low temperature and solid density cluster plasma, therefore, is probably feasible by the irradiation of high-intensity x-ray laser pulses.

From the analysis of the electron energy distribution, we conclude that the most probable explanation for the enhancement in the Xe^{3+} yield could be a reduction in the ionization threshold within the strongly coupled cluster plasmas. Indeed, theoretical study shows that the plasma screening effect in such cluster plasma is likely to reduce the ionization potential significantly [16]. If the energies of the Xe^{3+} state are reduced within a cluster, the number of possible final Xe^{3+} states will increase and consequently the probability of 4*d* decaying to the Xe^{3+} states via the DA decay will also increase [8].

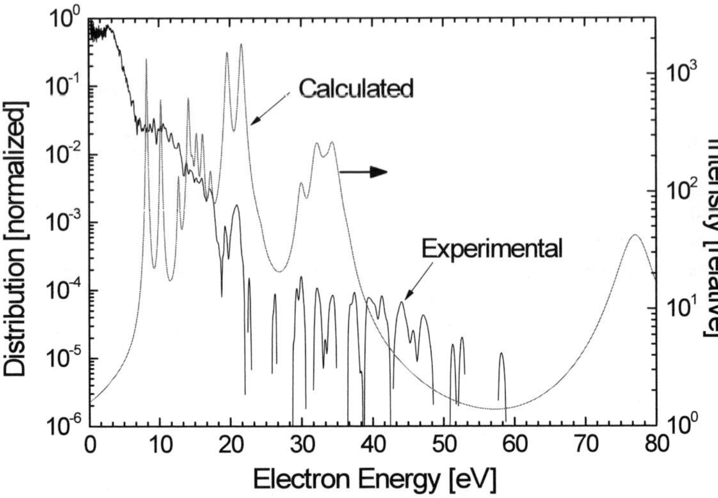

Fig. 4 Electron energy spectrum observed for Xe gas jets. The expected curve taking into account instrumental resolution is also shown in the figure.

4 Summary

The interaction of Xe clusters with an x-ray laser pulse with a wavelength of 13.9 nm was investigated using the ion and electron TOF spectrometers. In contrast to the results observed in SR experiments involving free atoms, an enhancement of Xe^{3+} yield was observed. The electron energy distribution showed the electrons with a few eV were efficiently generated due to the electron energy relaxation processes. The most probable explanation for the increase of Xe^{3+} yield could be a reduction in the ionization threshold within the cluster plasmas.

Acknowledgements

This work was partly supported by a Grand-in-Aid for Young Scientists (A) (No. 19684021) from the Japan Society for the Promotion of Science and the Matsuo Foundation.

References

1. Rühl, E.: 'Core level excitation, ionization, relaxation, and fragmentation of free clusters', *Int. J. Mass. Spectrom.* 229, 117-142, 2003
2. Björneholm, O. *et al* : 'Core Level Photoelectron and X-Ray Absorption Spectroscopy of Free Argon Clusters: Size-Dependent Energy Shifts and Determination of Surface Atom Coordination', *Phys. Rev. Lett.* 74, 3017-3020, 1995
3. Tai, R. Z. *et al* : ' Picosecond View of Microscopic-Scale Polarization Clusters in Paraelectric $BaTiO_3$', *Phys. Rev. Lett.* 93, 087601, 2004
4. Turcu, I.C.E. and Dance, J. B.: *X-rays from laser plasmas*, Chichester: Wiley 1998
5. Jungreuthmayer, C. *et al* : 'Intense VUV laser cluster interaction in the strong coupling regime', *J. Phys. B: At. Mol. Opt. Phys.* 38, 3029-3036, 2005
6. Krainov, V. P. and Smirnov, M. B. 'Cluster beams in the super-intense femtosecond laser pulse', *Phys. Rep.* 370, 237-331, 2002
7. Wabnitz, H. *et al* : 'Multiple ionization of atom clusters by intense soft X-rays from a free electron laser', *Nature* 420, 482-485, 2002
8. Namba, S. *et al* : 'Enhancement of Double Auger Decay Probability in Xenon Clusters Irradiated with a Soft X-Ray Laser Pulse', *Phys. Rev. Lett.* 99, 043004, 2007
9. Holland, D. M. et al : 'Multiple photoionization in the rare gases from threshold to 280 eV', *J. Phys. B: Atom. Mol. Phys.* 12, 2465-2484, 1979
10. Kawachi, T. *et al*: ' Gain saturation of nickel-like silver and tin x-ray lasers by use of a table top pumping laser system ', *Phys. Rev. A* 66, 033815, 2002
11. Namba, S. *et al* :' Efficient Electron Heating in Nitrogen Clusters Irradiated with Intense Femtosecond Laser Pulse ', *Phys. Rev. A* 73, 013205, 2006
12. Hagena, O. F. and Obert, W. : 'Cluster Formation in Expanding Supersonic Jets: Effect of Pressure, Temperature, Nozzle Size, and Test Gas', *J. Chem. Phys.* 56, 1793-1802, 1972
13. Luhmann, T. *et al* : 'Final ion-charge resolving electron spectroscopy for the investigation of atomic photoionization process: Xe in the region of the 4d-ef resonance', *Phys. Rev. A* 57, 282-291, 1998
14. Kivimaki, A. et al : 'Intensities of the xenon $N_{4,5}OO$ Auger electron spectrum revisited', *J. Electron Spectrosc. Relat. Phenom.* 101-103, 43-47, 1999.
15. Penent, F. et al: 'Multielectron Spectroscopy: The Xenon 4d Hole Double Auger Decay', *Phys. Rev. Lett.* 95, 083002, 2005
16. Gets, A. V. and Krainov, V. P. : 'The ionization potentials of atomic ions in laser-irradiated Ar, Kr and Xe clusters', *J. Phys. B: At. Mol. Opt. Phys.* 39, 1787-1795, 2006

Calibration of a High Resolution Soft X-Ray Spectrometer

J. Dunn, P. Beiersdorfer, G.V. Brown and E.W. Magee

Lawrence Livermore National Laboratory, Physical Sciences Directorate, Livermore, CA 94551, USA

Abstract. A high resolution grating spectrometer (HRGS) with 2400 line/mm variable line spacing grating for the 10 – 50 Å wavelength range has been designed for laser–produced plasma experiments at the Lawrence Livermore National Laboratory (LLNL). The spectrometer has a large radius of curvature, R=44.3 m, is operated at a 2° grazing angle and can record high signal-to-noise spectra when used with a low-noise, cooled, charge-coupled device detector. The instrument can be operated with a 10 – 25 μm wide slit to achieve the best spectral resolving power on laser plasma sources, approaching 2000, or in slitless mode with a small symmetrical emission source. Results will be presented for the spectral response of the spectrometer cross-calibrated at the LLNL Electron Beam Ion Trap facility using the broadband x-ray energy EBIT Calorimeter Spectrometer (ECS).

1 Introduction

A high resolution grating spectrometer (HRGS) has been designed to make measurements of spectral line shapes as well as precise wavelength positions of highly ionized lines emitted from laser-produced plasmas [1]. For this study, the main lines of interest are the K-shell oxygen and carbon series lines in the 10 – 50 Å wavelength band. An important part of this work is to be able to diagnose the emission conditions of these plasmas, in particular the electron temperature T_e and density n_e. Several spectroscopic methods have been described that allow the determination of these plasma parameters [2, 3]. These include Stark broadening of the higher n transitions (for n_e), density-sensitive dielectronic satellite line intensity ratios, x-ray continuum slope (for T_e), resonance line intensity ratios in the He-like and H-like ion series (for T_e) as well as satellite to resonance line ratios (for T_e). Knowledge of the instrument spectral sensitivity response curve across the waveband is required to determine accurately some of the above electron temperature parameters. Previous work calibrating a 2400 line/mm flat field variable spacing grating spectrometer was achieved by observing a laser-produced plasma source simultaneously with a transmission grating spectrometer of known efficiency [4]. In this work we report the calibration of the HRGS instrument using the

SuperEBIT electron beam ion trap x-ray source located at LLNL. The spectral sensitivity consisting of the grating reflectivity *R*(*E*) and detector quantum efficiency *Q*(*E*) are determined against the EBIT Calorimeter Spectrometer (ECS) [5]. The thin 200 nm aluminum foil filter used as a light tight filter in the laser plasma experiments is calibrated separately using the ECS and SuperEBIT [6].

In the next section the HRGS instrument, the SuperEBIT source, the ECS and the methods for calibrating the spectral sensitivity are described in detail. In section 3 the results of the measurements are presented and compared with other values in the literature.

2 Experimental Description

The high resolution grating spectrometer uses a large radius of curvature, *R*=44.3 m, 2400 line/mm variable-spaced grating as the wavelength dispersion element. The grating, with a large active area of 10 × 5 cm^2 (*L* × *W*), is inclined at an angle of incidence of approximately 2° to the source. The instrument is run on laser-produced plasma experiments with a 25 µm slit placed at 150 cm from the grating and defines the instrument spectral resolution. For the operation on SuperEBIT the slit was removed and the instrument was set up to focus onto the x-ray emission region of the ion trap. A low noise LN-cooled, back-thinned CCD 1300 × 1340 (20 × 20 μm^2 pixel) charge-coupled device records the spectrum and is located approximately 150 cm from the grating. This instrument is enclosed in a separate vacuum chamber and is bolted and aligned to the laser target or SuperEBIT x-ray source [1]. Single-shot spectra with lines resolved to about one part in 2000 are recorded from the laser-irradiated targets with energy on target of 0.5 J or more. For the SuperEBIT calibration runs the integration times were varied from 10 minutes to 60 minutes to record a spectrum with good line intensity. Figure 1 shows the layout of the HRGS instrument on SuperEBIT.

The LLNL EBIT-I was first built in 1986. SuperEBIT has been in operation for over 15 years [7]. It has been used to generate precise atomic data, detailed spectroscopy of highly charged ions as well as a calibration source for x-ray instrumentation. For this study neutral atoms using the gases CO_2, Kr, Ne, and SF_6, were injected into the trap and collisionally ionized by the electron beam. The beam electrons are confined and focused by a 3T magnetic field generated by superconducting Helmoltz coils. The ions are longitudinally confined by three voltages applied to the drift tube. Radial confinement is produced by the electron beam. This produces a trap dimension of 2 cm × 60 µm (*H* × *W*) where the electron density n_e is in the range $2 \times 10^{10} - 5 \times 10^{12}$ cm^{-3}. The ionization of the gas is adjusted by the electron beam voltage and the trap time defined by gating the drift voltages.

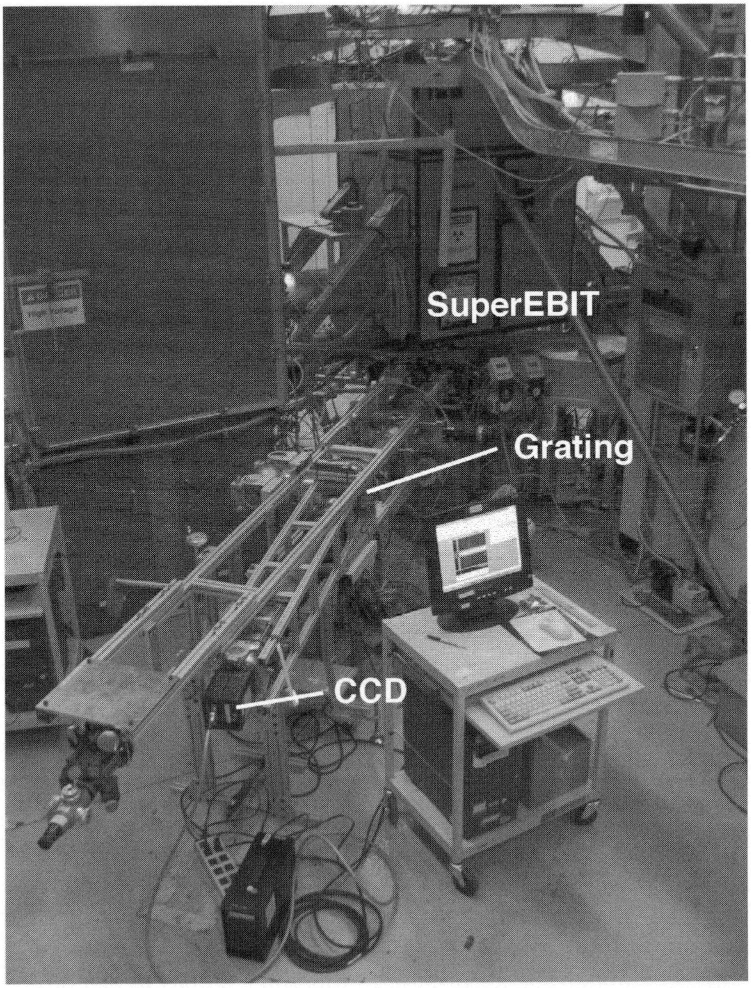

Fig. 1 High resolution grating spectrometer on the LLNL SuperEBIT. Grating and CCD detector are separated by 150 cm. Other instrumentation used in the experiment is located out of view on the opposite side of SuperEBIT.

A second instrument very similar to the HRGS designed for use on EBIT used the same 2400 line/mm R=44.3 m, 2400 line/mm grating and a LN-cooled CCD as described above [8]. It was set up like the HRGS and run in the same way where the instrument was focused on the EBIT x-ray source.

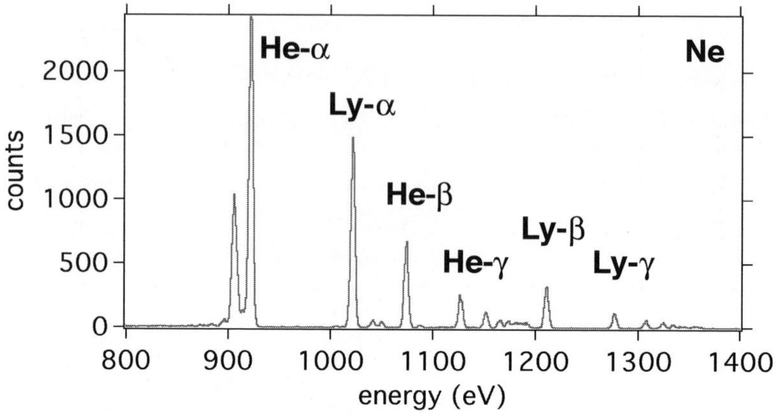

Fig. 2 Spectrum emitted from He-like and H-like Ne produced by SuperEBIT and measured with the ECS. Strongest transitions are labeled.

The main calibration spectrometer, the EBIT calorimeter spectrometer (ECS), is a solid-state device first developed at NASA's Goddard Space Flight Center in 1984 [5]. The present ECS consists of a 6 × 6 array of HgTe pixels cryogenically-cooled to 50 mK using an adiabatic de-magnetization refrigerator in a liquid $^3He/^4He$ bath. The array consists of 625 × 625 μm^2 × 8 μm thick pixels for mid-energy 0.1 – 10 keV photons interspersed with 625 × 500 μm^2 × 100 μm thick pixels for high energy 0.5 – 100 keV photon detection. HgTe is chosen as the detector material because of high x-ray absorption and low heat capacity. X-ray absorption is 100% for photon energy below 4 keV. The energy resolution when operated at the 50 mK temperature is $\Delta E \sim 5$ eV at 6 keV and $\Delta E \sim 25$ eV at 60 keV photon energy. Thermal isolation of the calorimeter is achieved with 4 thin foils of aluminized polyimide ($C_{22}H_{10}N_2O_5$) with a total thickness of 143.4 nm Al/218.2 nm polyimide. The absorption of this filter set has to be corrected in the calibration for low energy photons under study here. A typical K-shell spectrum for Ne is shown in Fig. 2 where the x-ray photons are detected by the mid-energy pixels in the ECS array. High $n = 7$ transitions can be clearly resolved in the He-like and H-like series where $E/\Delta E$ is 200 at 1 keV photon energy. For filter calibration, the filter was placed on a translation stage between the ECS and the x-ray source. Several runs of equal integration time were conducted with the filter in and out. The signal intensity and ionization balance of the x-ray source were monitored in all runs with the two grating spectrometers and the ECS data was corrected for small changes.

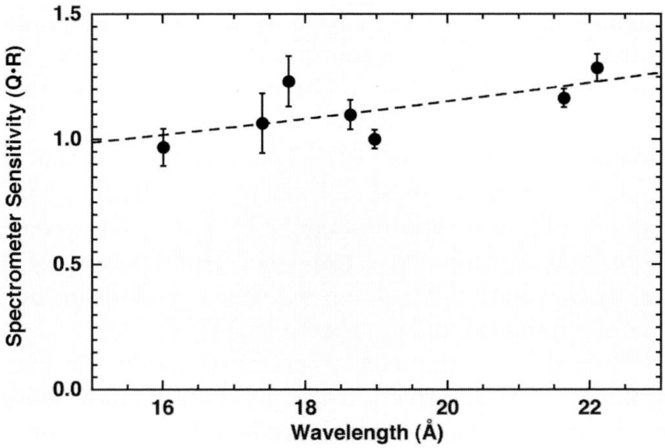

Fig. 3 Relative HRGS spectrometer sensitivity (detector efficiency and grating reflectivity) normalized against the oxygen Ly-α spectral line at 18.97 Å. Dashed line is second order polynomial fit to data points.

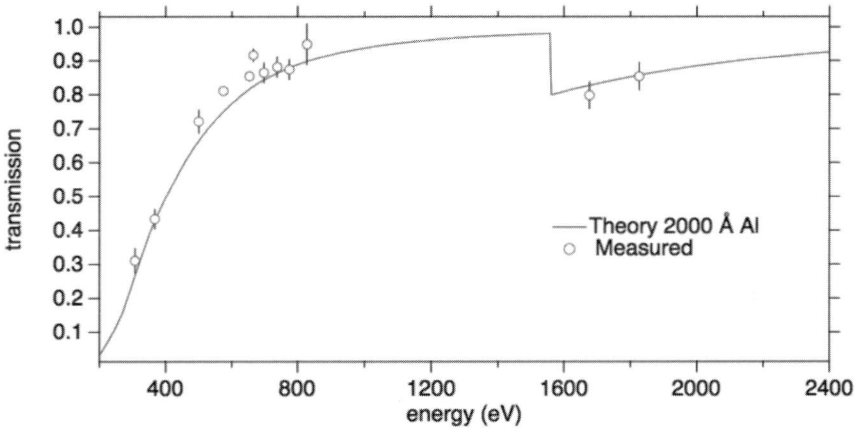

Fig. 4 Measured transmission of a 200 nm Al filter using the ECS. Solid line is theoretical data obtained from ref. 9.

3 Experimental Results

A preliminary calibration of the relative sensitivity $Q(E)\bullet R(E)$ is determined for the HRGS by measuring the change in the sensitivity in the 15 – 23 Å range normalized to the O H-like $1s - 2p$ line at 18.97 Å. The integrated signal in the spectral lines was measured for both the ECS and HRGS instruments running and integrating x-ray signal simultaneously. The filter response of the ECS was corrected for in the number of detected photons to determine the photon fluence incident on the HRGS instrument. The photon energy in each spectral line was taken into account in the analysis of the HRGS results. Figure 3 shows the results of the relative spectrometer sensitivity for the HRGS. The dashed line is a quadratic fit to the data points. Error bars arc determined mainly by the photon statistics recorded in each line with the largest errors coming from the weaker, higher *n* transitions. There is a general trend of sensitivity falling by approximately 25% with decreasing wavelength largely from the grating reflectivity response. The reported reflectivity for a smaller radius of curvature 2400 line/mm Hitachi grating in the same waveband shows a more rapid fall off in reflectivity [4].

Figure 4 shows the spectrometer filter transmission for the 200 nm Al measured by the ECS as described in the previous section. Good agreement between the measured values and the predicted transmission curve [9] is shown for the energy range investigated here. The results presented in Figs. 3 and 4 indicate that the high resolution grating instrument response can be accurately calibrated on EBIT. The HRGS is found to have useable reflectivity to below 12 Å. More details of the calibration will be reported at a later date, including using the ECS for absolute efficiency calibration.

The authors would like to thank Joel Clementson and Miriam Frankel for their assistance in running SuperEBIT. This work performed under the auspices of the U.S. Department of Energy by Lawrence Livermore National Laboratory under Contract DE-AC52-07NA27344.

References

1. J. Dunn, E.W. Magee, R. Shepherd, H. Chen, S.B. Hansen, S.J. Moon, G.V. Brown, M.-F. Gu, and P. Beiersdorfer, and M.A. Purvis, "High resolution soft x-ray spectroscopy of low Z K-shell emission from laser-produced plasmas", accepted for publication in *Rev. Sci. Instrum.* (2008).
2. H.R. Griem, "Plasma Spectroscopy", (McGraw-Hill, New York, 1964).
3. C. De Michelis and M. Mattioli, "Soft-x-ray spectroscopic diagnostics of laboratory plasmas", *Nucl. Fus.* **21**(6), 677 (1981).
4. A. Saemann and K. Eidmann, "Absolute calibration of a flat field spectrometer in the wavelength range 10–70 Å". *Rev. Sci. Instrum.* **69**, 1949 (1998).
5. F.S. Porter, B.R. Beck, P. Beiersdorfer, K.R. Boyce, G.V. Brown, H. Chen, J. Gygax, S.M. Kahn, R.L. Kelley, C.A. Kilbourne, E. Magee, and D.B. Thorn, "The XRS microcalorimeter spectrometer at the Livermore electron beam ion trap", *Can. J. Phys.* **86**, 231 (2008).
6. G.V. Brown, P. Beiersdorfer, J. Emig, M. Frankel, M.F. Gu, R.F. Heeter, E. Magee, D.B. Thorn, K. Widmann, R.L. Kelley, C.A. Kilbourne, and F.S. Porter, "Rapid, Absolute Calibration of X-ray Filters Employed By Laser-Produced Plasma Diagnostics", accepted for publication in *Rev. Sci. Instrum.* (2008).

7. P. Beiersdorfer, "A "brief" history of spectroscopy on EBIT", *Can. J. Phys.* **86**(1), 1 – 10 (2008); More references and work on EBIT can be found in this volume.
8. P. Beiersdorfer, E.W. Magee, E. Träbert, H. Chen, J.K. Lepson, M.-F. Gu, and M. Schmidt, "Flat-field grating spectrometer for high-resolution soft x-ray and extreme ultraviolet measurements on an electron beam ion trap", *Rev. Sci. Instrum.* **75**, 3723 (2004).
9. Theoretical filter transmission obtained from the Lawrence Berkeley National Laboratory website at http://www-cxro.lbl.gov/

XUV Probing as a Diagnostic of Rayleigh-Taylor Instability Growth

L M R Gartside[1], G J Tallents[1], J Pasley[1,2], J Gaffney[3] and S Rose[3]

[1] Department of Physics, University of York, York, YO10 5DD, UK
[2] STFC, Central Laser Facility, Rutherford Appleton Laboratory, Chilton, Didcot, OX11 0QX, UK
[3] Plasma Physics Group, Imperial College of Science, Technology and Medicine, London, SW7 2AZ, UK

Abstract. The application of extreme ultra-violet (XUV) lasers to the probing of Rayleigh Taylor instability (RTI) growth on planar targets has been investigated using the 2D hydrodynamic code POLLUX. Simulations of XUV probing have allowed characterisation of the expected effects of lateral movement of material on the transmission of plastic and iron targets. The effects on XUV transmission of seeding the Rayleigh-Taylor Instability with varying laser modulations has been studied.

1 Introduction

The Rayleigh-Taylor instability (RTI) arises when a more dense fluid is pushed by a less dense fluid. The RTI occurs when small spatial perturbations on the surface between the fluids grow exponentially in size. RTI growth can occur under gravity and, equivalently, from acceleration of a lighter fluid into a denser fluid [1]. Following Takabe et al [2] the growth rate of a laser driven instability, γ, in terms of fluid acceleration, g, and velocity, v, for surface perturbation wavenumber, k, is given by

$$\gamma = 0.9\sqrt{kg} - 3.1kv. \quad (1)$$

Equation (1) indicates that shorter wavelength perturbations (higher k) are dampened by material flow. RTI growth can potentially limit the densities achieved in laser-fusion [3].

Extreme ultra-violet (XUV) probing was applied to studies of the RTI in the mid to late 1990's [4-10]. Spatial perturbations in laser intensity were used to seed the RTI and the transmission of an XUV laser (~10 nm wavelength) through the foil targets used to determine the RTI growth. The use of XUV lasers for target probing allowed small variations in areal density to be measured, particularly during the imprinting of the instability. More recently XUV probing has been applied to iron targets, with iron opacity being of interest [11,12]. As the target material is ablated, the opacity rapidly drops. At such times it is the transmission of the remaining solid target that is being measured. Ideally the contributions of target ablation and changes in areal density can be quantified for different target designs [12]. It is shown that interpretation of XUV transmission data to produce areal density changes is highly dependent on an accurate knowledge of the plasma opacity of the target material.

2 POLLUX code and Ionised Material Package (IMP) opacity data

In order to explore RTI growth, profiles of density, temperature and XUV probe transmission have been obtained using the two dimensional code POLLUX. To calculate XUV transmissions, the POLLUX-deduced densities and temperatures are post-processed, using IMP [13] opacities. Examples of IMP calculated opacities as a function of density and temperature are shown in figure 1. IMP assumes a plasma in local thermodynamic equilibrium. The total opacity is obtained by summing bound-bound, bound-free, free-free and scattering contributions [12].

The hydrodynamic code, POLLUX, written by Pert [14] and later modified for studying the growth of seeded RTIs [15], was used to simulate laser ablation on CH and Fe targets. Utilising a cylindrical Eulerian geometry in 2D, POLLUX allows simulation of laser irradiation of targets in spot focus. Electron energy absorption by inverse bremsstrahlung is assumed with a dump of laser energy (20%) reaching the critical surface. Energy is transported beyond the critical region by thermal conduction with a longitudinal flux limiter of 0.06 [14-17].

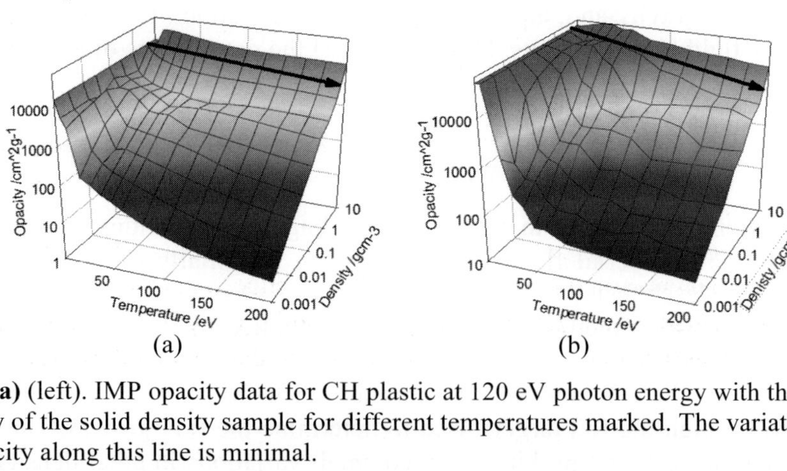

Fig. 1(a) (left). IMP opacity data for CH plastic at 120 eV photon energy with the opacity of the solid density sample for different temperatures marked. The variation of opacity along this line is minimal.

Fig. 1(b) (right). As (a) for iron at 120 eV photon energy. There is notable opacity variation along the line of solid density.

3 Model parameters

Simulations were carried out for small (10 μm) and large (100 μm) plastic and iron targets of 1μm thickness using irradiation focal diameters much greater than the target size. A 1μm thick target of solid state density was assumed embedded in an Eulerian mesh comprising low ambient density plasma (density = 0.001 gcm^{-3}). A peak intensity of $2x10^{14}$ Wcm^{-2} was chosen, consistent with original experiments [10]. A sinusoidal modulation pattern of the laser irradiance was imposed in the radial direction with modulation wavelength ranging from 200 μm to 1.25 μm in order to induce RTI. The incident irradiance was assumed to be temporally flat-top with a duration of 1 ns.

4 Simulated profiles

Calculated XUV transmissions through the RTI perturbations as shown in Figure 2 can be deduced (Figure 3). The XUV transmission varies corresponding to the imposed irradiance perturbation. Smaller perturbation wavelengths show some evidence for lateral thermal smoothing.

In order to investigate smoothing of the RTI growth imprinted by the laser irradiance variation, a value for rms areal density amplitude was deduced from the calculated transmission of the XUV probe. We follow the experimental procedure originally used to interpret the transmission of experimental XUV data, given by Wolfrum et al [10]. The density integrated through the target has differing maximum and minimum values as determined from the XUV transmission as follows.

$$(\rho r)_{max} - (\rho r)_{min} = \ln\left(\frac{I_{min}}{I_{max}}\right)\frac{1}{\kappa} = 2.86\Delta\rho r. \qquad (2)$$

Equation (2) relates the areal density, ρr, to the minima and maxima of the radiograph fringe pattern, scaled by the mass absorption coefficient, κ, here taken to be the solid value. The 2.86 factor arises in conversion to the rms value from the maximum value for the sinusoidal modulation pattern [10].

The ρr results deduced using equation (2) can be compared with density variations taken directly from the simulation code output (Fig. 4). At early times agreement is good. Both show the roll-over in $\Delta\rho r$ for smaller modulation wavelengths, where the imprint is smoothed due to lateral transport.

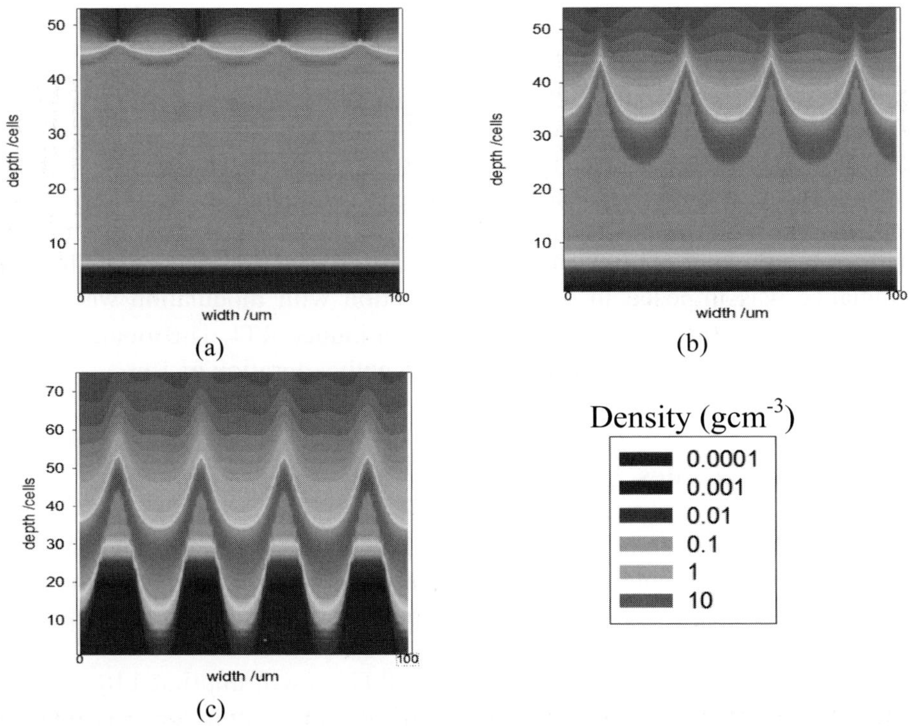

Fig. 2 POLLUX simulation mass density contours for an ablated planar CH target under modulated laser irradiance. The laser peak intensity is $2x10^{14}$ Wcm^{-2} with a modulation wavelength of 25 μm. Plots are shown for times a) 10 ps, b) 50 ps and c) 100 ps.

Accurate determination of the RTI growth rate is reliant on reliable opacity data. As the CH target is ablated, the opacity remains approximately constant (see the contour marked on Fig. 1 (a)) which means that equation (2) will give the areal density variation if κ is assumed constant. For the iron target (Fig. 1 (b)) the opacity changes more rapidly with ablation and so density measurements assuming constant κ will be less accurate. It is important to use a target material where the opacity does not change significantly on ablation for these RTI growth measurements.

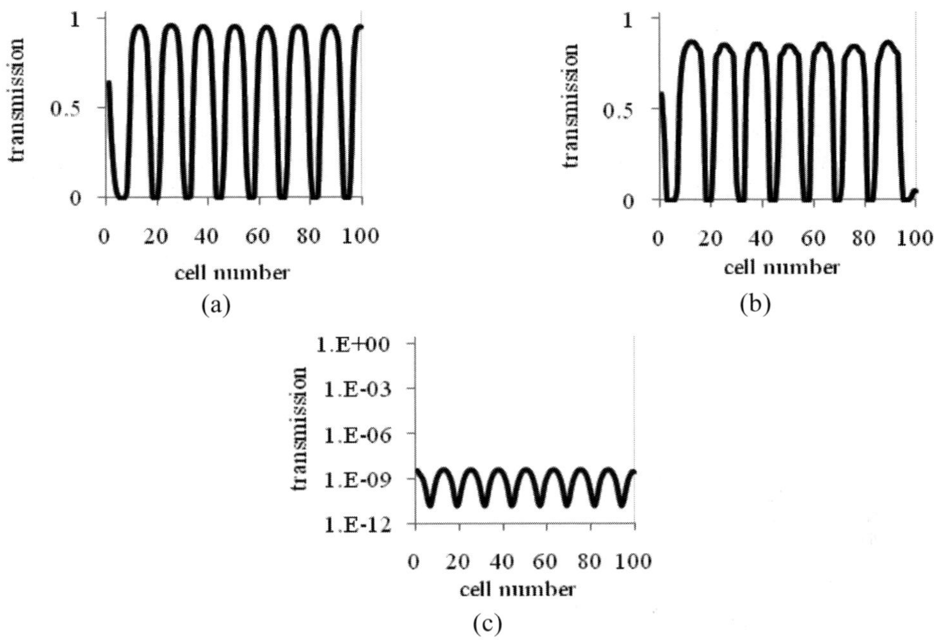

Fig. 3(a) (left) Transverse transmission of a 120 eV probe for a 100 μm CH target, at 220ps. with a laser modulation of wavelength 12.5 μm shown.

Fig. 3(b) (centre) As for (a) for a 10 μm CH target with 1.25 μm modulation wavelength.

Fig. 3(c) (right) As for (b) for an Fe target and 300 eV probe.

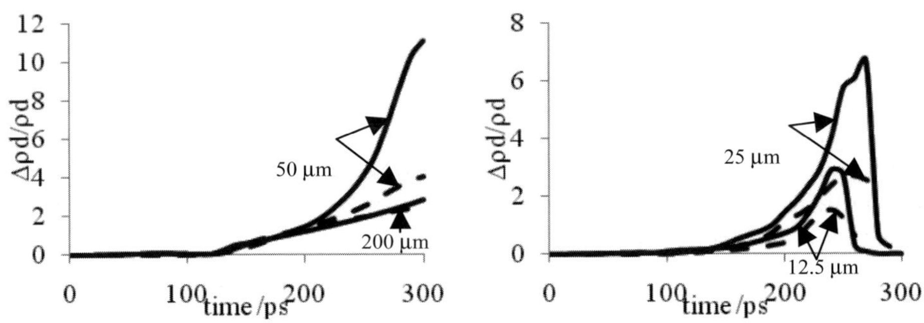

Fig. 4 (a) (left) Rms areal density amplitude for a 100 μm CH target, predicted by equation 2 and obtained directly from the simulation code are here shown as solid and broken curves respectively. 200 μm and 50 μm wavelength modulations are shown.
Fig. 4 (b) (right) As (a) with 25 μm and 12.5 μm wavelength modulations.

5 Conclusion

XUV probing can produce high visibility variations of transmission for plasmas irradiated by modulated laser intensity variations. The relative rms areal density amplitude indicates lateral transport smoothing for modulated irradiation. Consideration of the method by which RTI growth rates are calculated from experimental data has highlighted the importance of accurate opacity of the target material. Lateral transport is an important factor in determining the ablation pattern of planar targets undergoing RTI. Our simulations show that lateral smoothing occurs for modulation wavelengths less than ~ 50 μm.

Acknowledgements

EPSRC is gratefully acknowledged for funding support.

References

1. H J Kull, Phys. Rep. **206** 197 (1991)
2. H Takabe et al, Phys. Fluids **28** 2676 (1985)
3. J Nuckolls et a, Nature **239** 139 (1972)
4. D H Kalantar et al, Phys. Plasmas **4** 1985 (1997)
5. M H Key et al, JQSRT **54** 2221 (1995)
6. R Cauble et al, Phys. Re. Lett. **74** 3816 (1995)
7. R J Taylor et al, Phys. Rev. Lett. **76** 1643 (1996)
8. D H Kalantar et al, Rev. Sci. Instrum. **67** 781 (1996)
9. D H Kalantar et al, Rev. Sci. Instrum. **68** 802 (1996)
10. E Wolfrum et al, Phys. Plasmas **5** 227 (1997)
11. M H Edwards et al, Phys. Rev. Lett. **99** 195002 (2007)
12. M Edwards et al, Phys. Rev. Lett. **97** 035001 (2006)
13. S J Rose, J. Phys. B: At. Mol. Opt. Phys. **25** 1667 (1992)
14. G J Pert, J. Comp. Phys. **43** 111 (1981)
15. R Taylor, J Edwards and R Evans, POLLUX modifications 1993-1996
16. A Al-Khateeb et al, Appl. Phys. A **69** S479 (1999)
17. M J De, C Henshaw, G J Pert and D L Youngs, Plasma Phys. Control. Fus. **29** 405 (1987)

Line Focus Geometry for Grazing Incidence Pumped X-Ray Lasers

Z. Zhai, M.H.Edwards, N.Booth and G.Tallents

Department of Physics, University of York, York YO10 5DD, U.K.

Abstract. A new X-ray laser pumping concept Grazing Incidence Pumping (GRIP) based on collisional excitation of Ne-like or Ni-like schemes has recently been found to increase the laser pumping efficiency. Various optical arrangements to achieve grazing incidence irradiation in a line focus are studied and evaluated.

1 Introduction

The main goal pursued in the development of X-ray lasers is to reduce the pump requirements and provide an easily-accessible source for applications. The pump energy required has been reduced from 120 kJ [1] in 1984 to <1 J now and has resulted from several major breakthroughs. Introduction of multi-pulse pumping around 1994 reduced the pump energy to a few hundred joules [2]. Taking advantage of the chirped pulse amplification (CPA) technique developed for laser drivers ensured that plasma production was separated from the heating needed for inversion [3]. The final major advance has been the introduction of grazing incidence pumping where the main heating beam is brought in at near grazing incidence to the target surface to localize the energy deposition to the region of optimum gain. The advent of the grazing incidence pumping geometry enables saturation at wavelengths between 13 nm and 19 nm with total pump energy of about 1 J [4]. This paper investigates the various optical arrangements that can be used to produce the line focus especially for grazing incidence pumping of X-ray lasers.

2 Tilted off-axis mirrors

This technique of line focus production was originally proposed by Ross [5]. The line focus axis is defined by a point focus and the radius of curvature of the mirror. The grazing incidence angle of the pump laser onto the target is controlled by adjusting the point focus position relative to the line focus. It is possible to achieve a grazing incidence irradiation in a line focus with the geometry shown in Fig.1.(a). In Fig.1a, a spot focus C is imaged by a spherical mirror AB into a line focus A' B'.

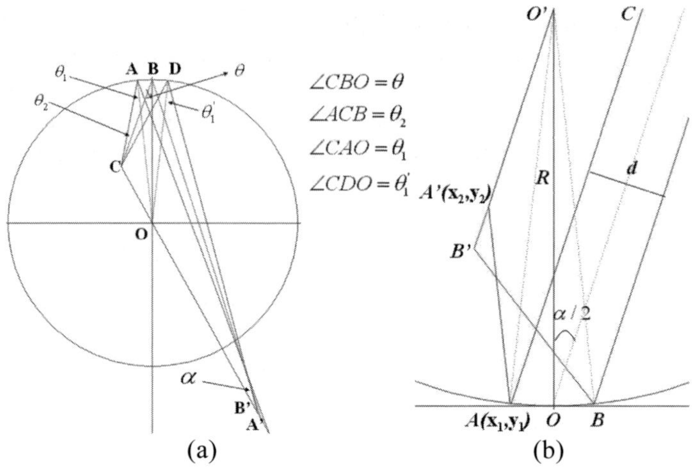

Fig. 1 (a) 2D Schematic of a line focus produced using a combination of off-axis spherical mirror and parabolic mirror producing the spot focus C. (b) 2D Schematic of a line focus produced with grazing incidence irradiation using a parabolic mirror. A pumping beam with diameter *d* is incident into the mirror surface with an intersection $A(x_1, y_1)$ and B and reflected into $A'(x_2, y_2)$ and B'. The focal plane is parallel to the pumping beam and passes through the central radius of curvature *R* at point O'.

Following Fig.1(a), we define α as the grazing incidence angle at the line focus of the centre ray, θ as the angle of incidence of the central ray onto the mirror and consequently also the tilted angle of the spherical mirror and the numerical aperture $\theta_2 = \angle ACB$. The angle θ_1 is represented by $\theta_1 = \angle CAO$. The line length *L* can be calculated using as follow:

$$L = R\left(\frac{\sin(-\theta_1 + 2\theta + \alpha + \theta_2)}{\sin(\theta_2 + 2\theta + \alpha)}\frac{\sin(2\theta_1')}{\sin(\theta_2 + 2\theta + \alpha - 2\theta_1)} - \frac{\sin(-\theta_1 + 2\theta + \alpha - \theta_2)}{\sin(-\theta_2 + 2\theta + \alpha)}\frac{\sin(2\theta_1)}{\sin(-\theta_2 + 2\theta + \alpha - 2\theta_1)}\right) \tag{1}$$

$$\text{where} \begin{cases} \theta_1 = \arcsin\left(\frac{r}{R}\sin(\theta_2 + 2\theta + \alpha)\right) \\ \theta_1' = \arcsin\left(\frac{r}{R}\sin(-\theta_2 + 2\theta + \alpha)\right). \end{cases}$$

The length *L* of the line focus is dependent on the radius of curvature *R* of the spherical mirror, the angle of incident angle θ of the laser onto the spherical mirror, the *f*-number of the optics ($=1/2\theta_2$) and the grazing incidence angle α. In previous GRIP experiment [4], the optimum X-ray laser output occurred for grazing incidence angles ranged from 14° to 25° depending on the experimental parameters. Fig. 2 shows the variation in length of the line focus as a function of incidence angle θ for different f-number optics obtained using the above equations. We assume that the radius of curvature of the spherical mirror is 203.2 mm and the achieved grazing incidence angle is 20°. The length of line focus is particularly sensitive to the mirror incident angle.

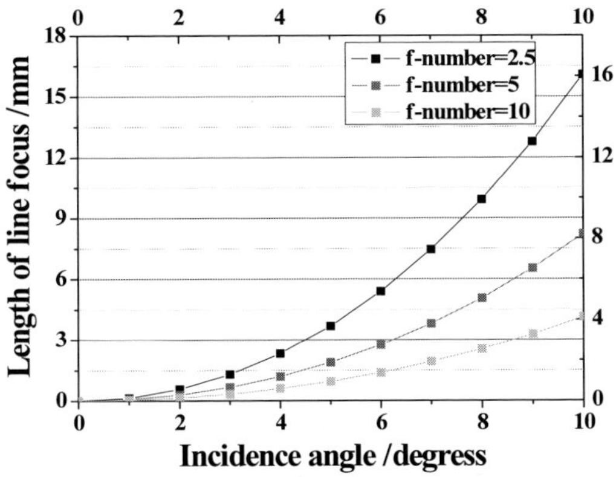

Fig. 2 Length of line focus as a function of mirror incidence angle for different f-number optics at a grazing incidence angle =20° produced by imaging a spot focus with a spherical mirror of radius of curvature 203.2 mm.

3 Parabolic mirrors

If the spot focus is moved far away from the spherical mirror, the diverging beam after the spot focus asymptotically approaches a parallel incident beam. A line focus can thus be produced by parallel incident rays with the line focus plane parallel to the incident beam (see Fig1 (b)). A beam incident onto a parabolic mirror placed at $\alpha/2$ from normal incidence causes focusing into a line at α grazing incidence angle. This new geometry is more straightforward. It is useful to investigate this as parabolic mirrors are already available for many experiments for the production of spot foci. To understand the ray tracing programme, we need to calculate the intersection points between rays and the parabolic surface. A general ray (x,y,z) incident onto the mirror can be represented in Cartesian co-ordinate as:

$$\begin{cases} x = x_0 + t\cos\beta \\ y = y_0 + t\sin\beta \\ z = z_0 \end{cases} \tag{2}$$

where x_0, y_0, z_0 are the initial points and $\beta=\pi/2-\alpha/2$. A parabolic surface with f as the effective focal length can be defined by:

$$y=(x^2+z^2)/4f \tag{3}$$

Substituting the linear function of ray equation (2) into the parabolic equation (3), we can find the parameter t in equation (2)

$$t=\frac{-(2x_0\cos\beta-4f\sin\beta)-\sqrt{(2x_0\cos\beta-4f\sin\beta)^2-4\cos^2\beta(x_0^2+y_0^2-4fy_0)}}{2\cos^2\beta} \tag{4}$$

We can substitute equation (4) into the linear function (2) to calculate the intersection points (x_1, y_1, z_1) of the incident rays with the mirror surface. Calculation of normal directions on parabolic surfaces depends on the coordinates used. If the parent parabola coordinate is chosen, the normal direction from point (x_1, y_1, z_1) on surface is

$$\begin{cases} N_x=\dfrac{dy}{dx}\Big/\sqrt{(\dfrac{dy}{dx})^2+(\dfrac{dy}{dz})^2+1} \\ N_y=-1 \\ N_z=\dfrac{dy}{dz}\Big/\sqrt{(\dfrac{dy}{dx})^2+(\dfrac{dy}{dz})^2+1}. \end{cases} \tag{5}$$

The incident ray $\mathbf{R_{inc}}$ is the vector which is reflected at the point on the mirror surface with the normal $\mathbf{N}$, the reflected ray can be written as [6]

$$\mathbf{R_{ref}}=2(\mathbf{N}\cdot\mathbf{R_{inc}})\mathbf{N}-\mathbf{R_{inc}} \tag{6}$$

Using the above equations, the reflected ray is defined by:

$$\begin{cases} x=x_1+L_x t' \\ y=y_1+L_y t' \\ z=z_1+L_z t' \end{cases} \tag{7}$$

where

$$\begin{cases} L_x=2\left(\dfrac{x_1/2f}{\sqrt{(x_1/2f)^2+(z_1/2f)^2+1}}\cos\beta-\sin\beta\right)\dfrac{x_1/2f}{\sqrt{(x_1/2f)^2+(z_1/2f)^2+1}}-\cos\beta \\ L_y=-2\left(\dfrac{x_1/2f}{\sqrt{(x_1/2f)^2+(z_1/2f)^2+1}}\cos\beta-\sin\beta\right)-\sin\beta \\ L_z=2\left(\dfrac{x_1/2f}{\sqrt{(x_1/2f)^2+(z_1/2f)^2+1}}\cos\beta-\sin\beta\right)\dfrac{z_1/2f}{\sqrt{(x_1/2f)^2+(z_1/2f)^2+1}}. \end{cases}$$

We assume the focal plane is defined by

$$\frac{x}{a}+\frac{y}{b}=0 \tag{8}$$

where a and b are the x and y intercepts on our co-ordinate plane. Substituting equation 7 into equation 8, we can calculate the coordinates of A'(x_2, y_2). By changing the initial point (x_0, y_0, z_0), we can trace all the rays and obtain the line focus profile. Fig.3 shows that a pumping beam with diameter of 60 mm is focused by a parabolic mirror of 762 mm focal length to achieve ~5.5 mm line focus at 20°grazing incidence angle. 700 rays have been traced to plot the line focus profile. The energy distribution along the line focus can be investigated from the statistics of the numbers of rays within set steps. The length of line focus is dependent on the pumping beam diameter d, the radius of curvature of mirror R and the grazing incidence angle α.

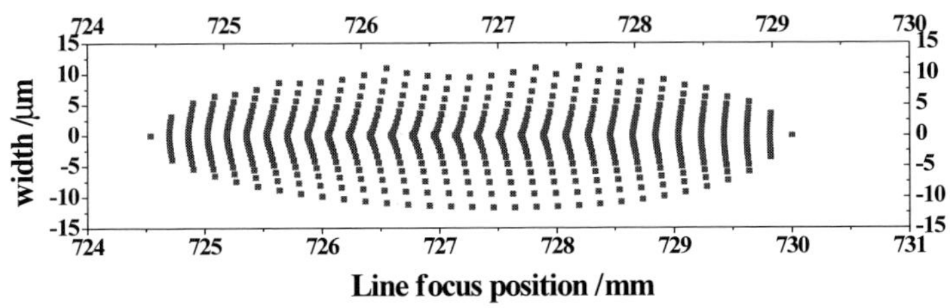

Fig. 3 Line focus profile from ray tracing. A parabolic mirror of 762 mm focal length is placed 10° from normal incidence which allows the beam to be incident onto the target at 20° grazing incidence.

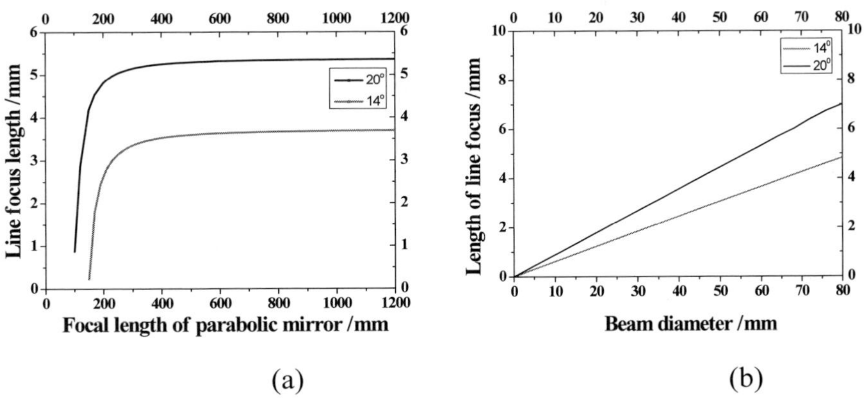

Fig. 4 (a) Length of line focus as a function of focal length with a beam diameter of 60 mm for grazing incidence angle of 14° and 20°. (b) Length of line focus as a function of the beam diameter for grazing incidence angle of 14° and 20°with the parabolic mirror focal length of 762 mm.

Fig.4 shows that various focal lengths of parabolic mirror do not affect the line length much, but the beam diameter is a key factor affecting the line focus length. To adjust the length of the line focus over a large range with a single parabolic or spherical mirror, it is clearly necessary to adjust the incoming beam diameter using a beam expansion technique.

In an ASTRA experiment [7], a parabolic mirror of 762 mm focal length is utilized to achieve a line focus of 3.4 mm at 14° grazing incidence angle and 5.5 mm at 20°. Fig.5 (a) shows that a grazing incidence line focus can be produced with calculated line focus uniformity close to the measured variation of soft X-ray emission. Some mismatch of the left edge may be introduced by the profile of the incoming beam. Interestingly, ray tracing shows that more uniform irradiation across the width can be achieved by de-focusing (see Fig.5(b)).

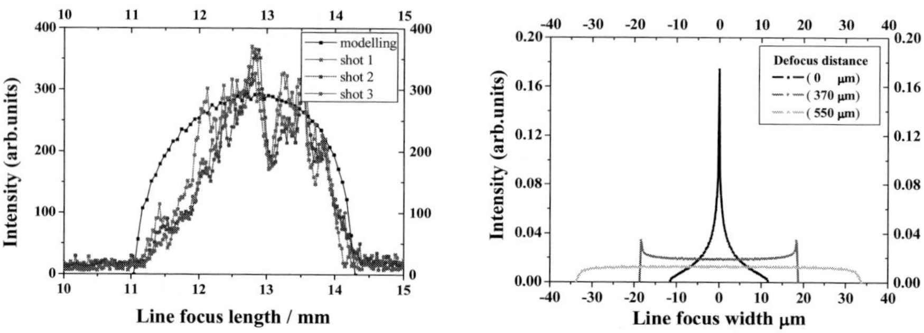

(a) (b)

Fig. 5 (a) Comparison of modelling and experimental irradiance distribution along the line focus. (b) Irradiance distributions across the line focus for different defocus distances as labelled. A 60 mm diameter pump beam with 20° grazing incidence angle is assumed.

4 Conclusion

Various optics can be used to produce the line foci for grazing incidence pumping of X-ray lasers. A combination of a spherical and a parabolic mirror is preferred when a longer line focus is needed. A single parabolic mirror is easy to setup and produces a sufficiently narrow line focus for experiments with shorter (< 6 mm) line focus lengths. A more uniform irradiation across the line focus width can be achieved by de-focusing when using a single parabolic mirror.

References

1. D. Matthews et al., Phys. Rev. Lett. **54**, 110 (1985).
2. J. Nilsen et al., Phys. Rev. Lett. **74**, 3376 (1995).
3. P. V. Nickles et al.,Phys. Rev. Lett. **78**, 2748 (1997).
4. Y. Wang et al., Phys. Rev. A **72**, 053807 (2005).
5. I.N.Ross et al., Applied Optics, **26** (1987) 1584
6. http://www-cs-students.stanford.edu/~adityagp/final/node3.html
7. Edwards M H et al. Proc. SPIE, Vol. **6702**, 670205 (2007);

Resolution and Feature Size Assessment in Soft X-Ray Microscopy Images

M.C. Marconi[1], P.W. Wachulak[1], C. Brewer[1], F. Brizuela[1], R. Bartels[1], C.S. Menoni[1], J.J. Rocca[1], E. Anderson[2], W. Chao[2]

[1] NSF Engineering Research Center for Extreme Ultraviolet Science and Technology and Electrical and Computer Engineering Department, Colorado State University, Fort Collins, Colorado
[2] Center for X Ray Optics. Lawrence Berkeley National Laboratory. Berkeley, California

Abstract. Soft X-ray microscopy images of nanostructures where analyzed with a method developed to simultaneously determine the object feature size and image resolution. This method is based on the correlation between the image and a set of templates of known resolution obtained from the original image. The analysis was applied to images obtained with a Fresnel zone plate microscope that uses 13.2 nm wavelength laser light for illumination. The object feature size and the resolution obtained with this method are shown to be in very good agreement with independent measurements of both magnitudes.

1 Introduction

A direct way to improve the resolution of imaging systems is by reducing the wavelength of the illumination. With this approach, soft x-ray (SXR) synchrotron light has been used to demonstrate full-field zone plate microscopy with a record spatial resolution of 13nm [1]. The development of compact SXR sources has made possible the realization of tabletop microscopes that can render images of nanoscale objects with exposures of few seconds and spatial resolution approaching that of synchrotron based microscopes [2-4].

The spatial resolution can be expressed in terms of the Rayleigh criterion and can be measured by analyzing images of specifically designed objects with well-established tests, such as the knife-edge or the grating test [5]. Under optimum focusing conditions, the image resolution coincides with the resolution of the imaging system and it is simple to obtain the size of an object from a microscope image when the object is significantly larger than the resolution limit of the instrument. However, this is not the case when the size of the object approaches the resolution limit or the image is not perfectly focused. When working close to the resolution limit, in full field SXR microscopes with objectives that have ~ 100 nm depth of focus it becomes imperative to use a robust image analysis algorithm that can identify specific object features and determine their dimension.

We present a method for the analysis of SXR microscopy images that can simultaneously extract from the image, the size of the object and the resolution. The method is based on the correlation between a raw EUV image and a series of templates with decreasing resolution. These templates are constructed applying Gaussian filters with different widths to a master binary template constructed from the original image. With this analysis it is possible to extract the object's dimensions even when they approach the spatial resolution of the microscope. The analysis of images obtained with a tabletop 13.2 nm wavelength laser-based microscope using this method gives object dimensions that are in excellent agreement with those obtained using scanning electron microscopy and resolution values that are consistent with those obtained by applying the Rayleigh test.

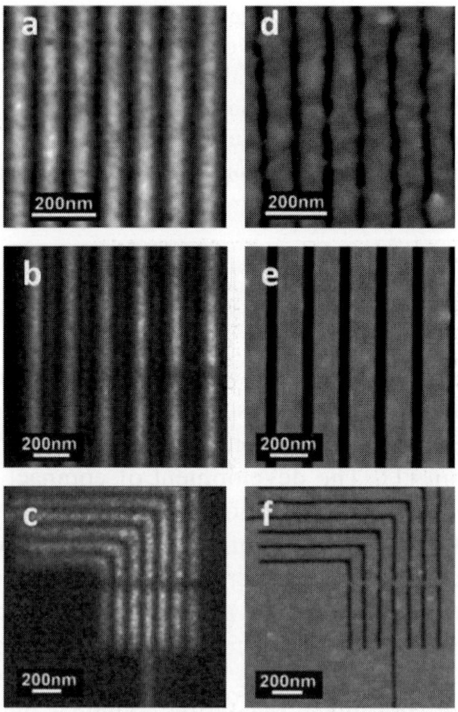

Fig.1. EUV microscope image of a) a 100 nm full period grating, b) 200 nm full period grating, and c) 100 nm full period elbow-shaped grating obtained with 13.2 nm wavelength laser illumination. d), e) and f) are the corresponding SEM images of the test objects

2 Nanoscale images at λ=13.2 nm

The SXR images used to test the correlation method described in the next section, were acquired with a full-field microscope operating in transmission mode at a wavelength of λ=13.2 nm [3]. The SXR microscope is based on Fresnel Zone plate (FZP) condenser and objective with numerical aperture NA= 0.07 and 0.132 respectively. The microscope is illuminated by the highly directional output from a λ=13.2 nm Ni-like Cd laser operating at a repetition rate of 5 Hz [6, 7]. The test objects were gratings and periodic elbow patterns fabricated by electron beam lithography on a 100 nm thick metalized silicon nitride membrane [8, 9]. The images were recorded using a 2048×2048 back illuminated CCD camera with 13.5 μm pixel size. Images of gratings with 100 and 200 nm period and an elbow-shaped grating with 100 nm period were obtained at a magnification of 1080×. At this magnification, the pixel size in the image plane corresponds to 12.5 nm at the object plane, which ensures that the image resolution is not limited by the detection system. The spatial resolution of the λ=13.2 nm microscope was obtained from the analysis of SXR images of gratings with periods down to 76 nm [3]. Figure 1 shows the SXR images used to test the correlation method described in Section 3, and compares them with scanning electron microscope (SEM) images from which the linewidths of the test patterns were independently assessed.

3 Description of the algorithm

We will illustrate the sequence of steps implemented in our image analysis algorithm with the SXR image of the 200 nm full period grating shown in Figure 1(b). The sequence is summarized in Fig. 2. First an intensity threshold filter is applied to the raw image data to homogenize the background. The threshold level is adjusted to maintain the features of the image unaltered

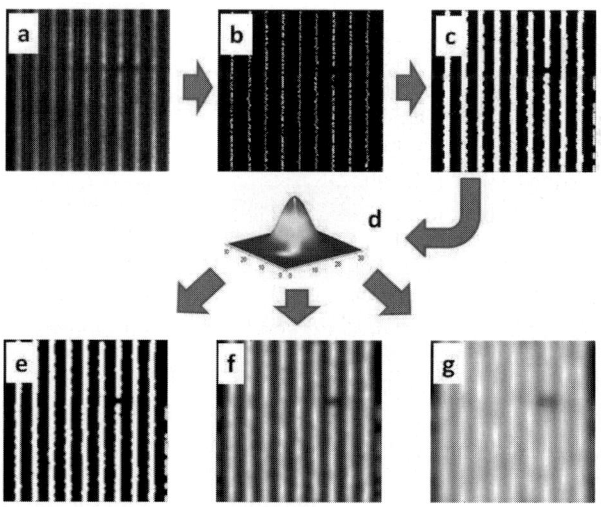

Fig. 2 Sequence of steps in the image processing algorithm. From the original image (a) the skeleton is obtained (b) that is convolved with different size circles to obtain a binary master template (c). After spatial filtering with Gaussian filters of different widths (d) a set of lower resolution templates is obtained (e-g)

while generating a new image with improved contrast. The thresholded image is shown in Fig 2(a). A skeletonizing algorithm is applied to the thresholded image [10] to produce a template with lines one pixel wide that follows the shape of the original image. The skeleton is shown in Fig 2(b). Third, the skeleton is convolved with *K* different circular templates of different diameters. The convolution between the skeleton and the *K* different circular templates generates a set of *K* binary templates, each one with a different linewidth, that resemble the original grating image. One of these *K* binary templates with a diameter of 50 nm is shown in Fig 2(c). The *K* templates are degraded in resolution by applying *L* different Gaussian filters with selected full width at half maximum (FWHM). Applying *L* different Gaussian filters to each one of the *K* binary templates produces a set of $K \times L$ templates. These $K \times L$ templates are individually correlated with the original image. The correlation coefficients plotted in $K \times L$ dimensional space generate a surface whose global maximum identifies the template that best resembles the original SXR image. We associate *K*- and *L*- indices of the template that maximize the correlation to the size of the feature and the resolution of the image respectively. The strict relationship between the resolution obtained by this method and the Rayleigh resolution can be derived by applying the Gaussian filtering to a set of synthesized images of two Airy disks. This analysis is presented in [11] and indicates that the image Rayleigh resolution can be obtained by dividing the resolution calculated by this algorithm by a factor $\beta = 0.786$.

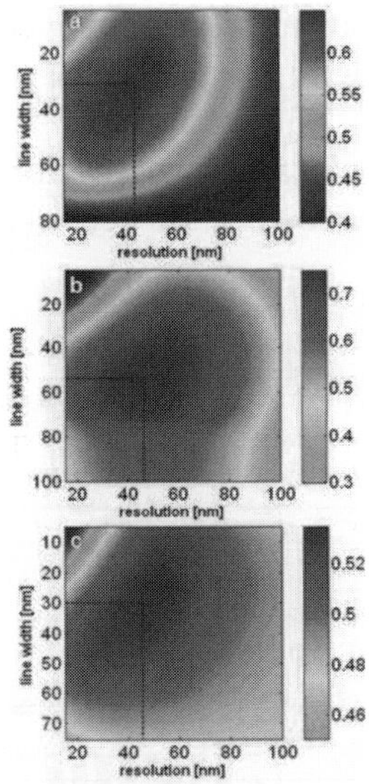

Fig. 3 Correlation coefficients plotted in the feature size-resolution space for each of the three different images of Fig. 1. The dashed lines indicate the coordinates of the global maxima for each of the data sets: a) 100 nm full period grating linewidth 31±5 nm, and resolution 54.8±5 nm. b) 200 nm full period grating, linewidth 53± 5 nm and 59.2 ± 5 nm resolution c) 100 nm full period elbow-shaped grating, linewidth 29 ± 5 nm and 58.3 ± 5 nm resolution.

The correlation plots generated by applying the algorithm to the three images of Fig. 1 are presented in Fig. 3. From these plots we estimated the feature size and the image resolution for each pattern. The error in the image resolution was conservatively assigned to one step size used to generate the plots shown in Fig. 3, in this case 5 nm. The grating linewidth obtained for all three patterns are in very good agreement with those obtained independently form the SEM images. The error bars in the linewidths, obtained from the SEM images, correspond to the standard deviation of the values measured at ten different locations in the image. Table 1 summarizes the results. Other images containing patterns with different shapes were also analyzed, resulting in similar degree of agreement. The Rayleigh resolution obtained from the image analysis is in agreement with the measured spatial resolution of the λ=13.2 nm microscope that yielded a resolution better than 76 nm (38 nm half period resolution) [3].

	Test Image		
	100nm period gratings	200nm period gratings	100nm period elbow-gratings
Linewidth from the SEM measurements	31 ± 2.6 nm	53.8 ±1.8 nm	31.3 ±2.4 nm
Linewidth from the algorithm	31 ± 5 nm	53 ± 5 nm	29 ± 5 nm
Image resolution from algorithm	54.8 ± 5nm	59.2 ± 5 nm	53.8 ± 5 nm
Instrument resolution	< 76 nm	< 76 nm	<76 nm

References

1. W.L. Chao, B.D. Harteneck, J.A. Liddle, E.H. Anderson, and D.T. Attwood, "Soft X-ray microscopy at a spatial resolution better than 15nm". *Nature*. 435 1210-1213 2005.
2. P.A.C. Takman, H. Stollberg, G.A. Johansson, A. Holmberg, M. Lindblom, and H.M. Hertz, "High-resolution compact X-ray microscopy". *Journal of Microscopy-Oxford*. 226 175-181 2007.
3. G. Vaschenko, C. Brewer, E. Brizuela, Y. Wang, M.A. Larotonda, B.M. Luther, M.C. Marconi, J.J. Rocca, and C.S. Menoni, "Sub-38 nm resolution tabletop microscopy with 13 nm wavelength laser light". *Optics Letters*. 31 1214-1216 2006.
4. H.M. Hertz, G.A. Johansson, H. Stollberg, J. de Groot, O. Hemberg, A. Holmberg, S. Rehbein, P. Jansson, F. Eriksson, and J. Birch, "Table-top X-ray microscopy: Sources, optics and applications". *Journal De Physique Iv*. 104 115-119 2003.
5. J. Heck, Attwood, D.T., Meyer-Ilse, W., Anderson, E.H., "Resolution determination in X-ray microscopy: an analysis of the effects of partial coherence and illumination spectrum". *Journal of X-Ray Science and Technology*. 8 95-104 1998.

6. Y. Wang, M.A. Larotonda, B.M. Luther, D. Alessi, M. Berrill, V.N. Shlyaptsev, and J.J. Rocca, "Demonstration of high-repetition-rate tabletop soft-x-ray lasers with saturated output at wavelengths down to 13.9 nm and gain down to 10.9 nm". *Physical Review A*. 72 2005.
7. J.J. Rocca, Y. Wang, M.A. Larotonda, B.M. Luther, M. Berrill, and D. Alessi, "Saturated 13.2 nm high-repetition-rate laser in nickellike cadmium". *Optics Letters*. 30 2581-2583 2005.
8. E.H. Anderson, "Specialized electron beam nanolithography for EUV and X-ray diffractive optics". *Ieee Journal Of Quantum Electronics*. 42 27-35 2006.
9. E.H. Anderson, D.L. Olynick, B. Harteneck, E. Veklerov, G. Denbeaux, W.L. Chao, A. Lucero, L. Johnson, and D. Attwood, "Nanofabrication and diffractive optics for high-resolution x-ray applications". *Journal Of Vacuum Science & Technology B*. 18 2970-2975 2000.
10. T. Yatagai, S. Nakadate, M. Idesawa, and H. Saito, "Automatic fringe analysis using digital image processing techniques". *Optical Engineering*. 21 432-435 1982.
11. P.W. Wachulak, Brewer, C.A., Brizuela, F., Chao, W., Anderson, E.H., Bartels, R.A., Menoni, C.S., Rocca, J.J., Marconi, M.C., "Simultaneous determination of feature size and resolution in soft x-ray microscopy images ". *Journal of the Optical Society of America B*. 25 B20-B26 2008.

An Approach to the Generation of Uniform Line Foci for Use in X-Ray Laser Experiments

T. W. J. Dzelzainis and C. L. S. Lewis

Centre for Plasma Physics, Queen's University Belfast, Belfast BT7 1NN.

Abstract. We report on the results of an investigation into the generation of uniform line foci suitable for use in the generation of X-ray lasers. Our approach involves the use of specially designed apertures, whose shape depends on the beam profile and the focussing optics, the apertures limit the intensity to a predetermined level at each point along the line focus, thus generating a uniform line focus at the expense of some of the overall beam energy. A ray tracing code was written specifically to find the required aperture shapes. The code currently supports only line foci generated by off-axis spherical mirrors but is able to accommodate arbitrary beam profiles. As high energy lasers become more accessible (and hence pump energy becomes less of an issue), then due to their low cost and ease of implementation, these apertures are an appealing method of controlling the intensity profile of a line focus when overall efficiency of the pump system is not a governing factor.

1 Introduction

It has long been recognised that there is a correlation between the uniformity of a line-focus and output quality of an X-Ray Laser (XRL)[1,2,3]. As the plasma temperature and expansion rate depend on the intensity of the irradiating laser, any modulations present in the intensity distribution along a line-focus will be imprinted onto the resultant plasma. The resulting variation in distance from target surface to optimum gain region along the plasma column means that spontaneous emission travelling in a relatively straight line from the end of the target will not experience optimum gain conditions. In extreme cases, the x-ray pulse may pass through areas of negative gain, thus reducing the effective gain-length. The quality of the line focus can also have an effect on other output characteristics such as divergence and beam profile. By the use of specially designed apertures that we plan to field in our initial XRL campaign, we hope to increase the uniformity of the line foci we are using and to study the effect on the output of the XRL. The required shapes for the apertures were found by use of a simple ray-tracing program.

2 Optical Systems for Achieving Line-Foci

The optical set-ups investigated were those that had been chosen for use in our upcoming XRL experiments. The first, shown in Fig. 1(a), is the well known set-up first proposed by I. N. Ross et al[4] using a lens and a spherical mirror to achieve a near normal incidence line-focus. In the initial paper, apertures are identified as one way in which the uniformity of the line-focus could be increased, but as of yet we know of no experiments where an aperture has been introduced. The second set-up, shown in Fig. 1(b), is one which has gained popularity recently with the demonstrated increased efficiency delivered via the GRazing Incidence Pumped (GRIP) XRL scheme[5]. This set-up allows the grazing incidence angle to be achieved and automatically gives a travelling wave geometry.

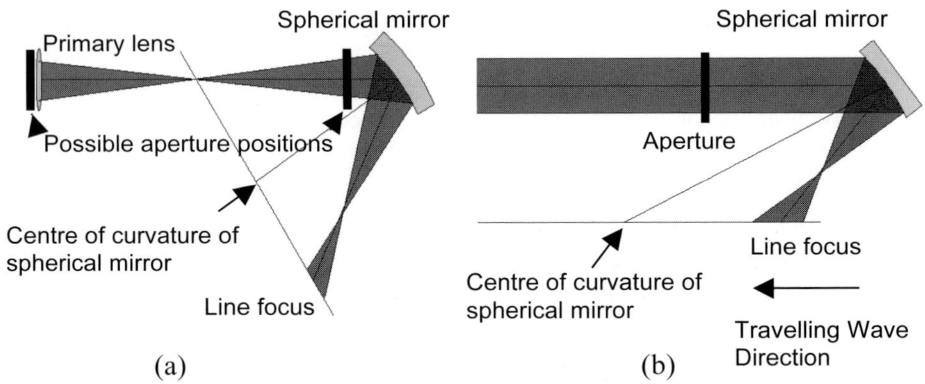

Fig. 1 Focussing geometry for producing a) near normal incidence line-foci and b) grazing incidence line-foci, with suitable positions identified for placement of apertures to modify the intensity distribution at the line focus.

3 Design of the Apertures

The program written for this task was a simple ray trace program, and therefore could not take into account diffraction effects which may indeed alter the intensity distribution at focus. The method for calculating the intensity profiles of the line-foci exploits the cylindrical symmetry inherent in both of the above systems which can both be treated as symmetric around the line-focus (also the XRL axis). The task is then to calculate the function which maps a point in the input beam profile onto the line focus. The mapping function for set-up shown in Fig. 1(a) is an ellipse and for the set-up in Fig. 1(b) is a circle; a visual representation of these mapping functions is given in Fig. 2. In order to calculate the intensity at the line-focus the program uses an assumed shape (i.e. rectangular or elliptical) and width for the line-focus itself, as geometrical calculations lead to a zero-width line. To find the aperture required to give a certain intensity level along the line the beam is split into cells centred on the points defined by the mapping function and then each cell is successively included in a summation until the required intensity at line focus has been met. The position of the last included cell defines the boundary position. A boundary defining a sharp cut-off at either end of the line-focus is defined by the first mapping functions where the required line-focus intensity was not reached.

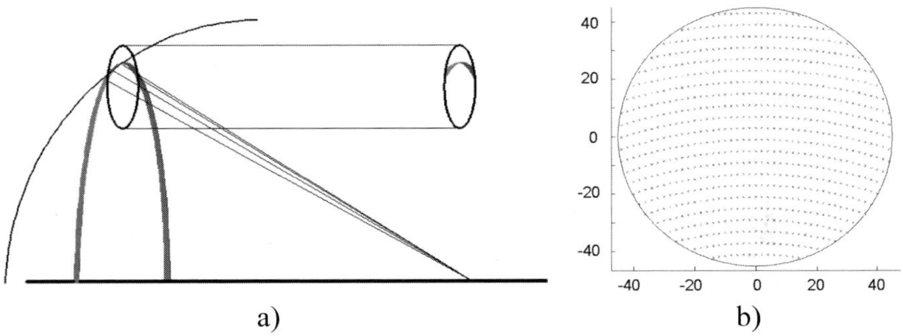

Fig. 2 a) A schematic showing the mapping function from the beam onto the line focus. b) Points along a set of mapping functions (each arc corresponding to a different point along the line focus) as calculated by the program.

4 Apertures Calculated for the QUB X-Ray Laser Set-up.

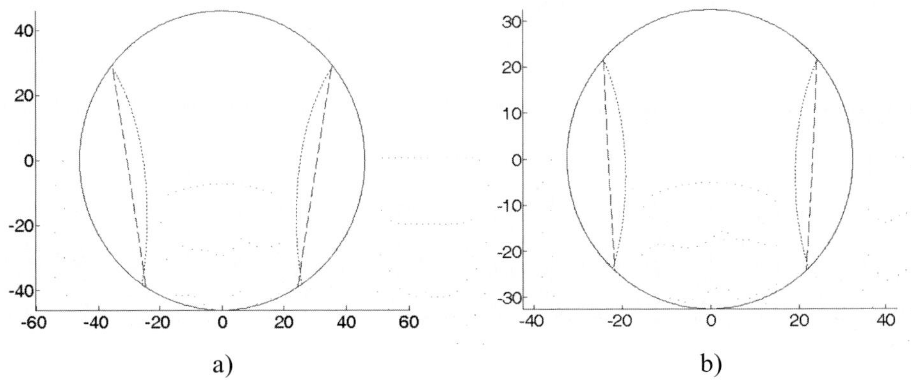

Fig. 3 The apertures designed for a) the heating pulse and b) the plasma forming pulse in the QUB GRIP XRL.

In an experimental campaign due to start in the near future, QUB's in-house laser system 'TARANIS' (Tera-watt Apparatus for Relativistic and Non-linear Interdisciplinary Science') will be used to pump a GRIP x-ray laser. In this section we present apertures required for uniform line-foci, using the TARANIS beams as input parameters for the program. The TARANIS laser system is described in detail elsewhere in these proceedings[6], so only a brief summary of the key parameters will be given here. TARANIS is a high power laser system consisting of two beams. Each beam can be run independently in either compressed (up to 20J in 800fs) or uncompressed (up to 30J in 800ps) mode. In Fig. 3(a) the dashed line shows an aperture calculated for the short heating pulse. The input parameters were a 5J, 1ps pulse in a beam of diameter 92mm. The aperture shown permits a 6mm×50μm line-focus with uniform irradiation of 1×10^{15}Wcm^{-2}. Fig. 3(b) the dashed line shows an aperture calculated for the long pre-pulse. The input parameters were 10J, 0.8ns pulse in a beam diameter of 65mm. The aperture permits a 5mm×50μm line-focus with uniform intensity of 2.5×10^{12}Wcm^{-2}The reason for the discrepancy between the line lengths lies in the inflexibility of the grazing incidence focus set-up. The line-focus length here depends only weakly on the focal length of the mirror but strongly on the beam diameter and grazing incidence angle, both of which are heavily constrained by either technical or experimental considerations. This means it is not possible to customise the line focus length for this set-up.

In order to illustrate the effect of changing beam profile, each diagram also shows the aperture required for a Gaussian beam with full width half max (FWHM) equal to the beam radius (dotted line). The energy beyond the FWHM is neglected in the simulation. It is noted here that for a Gaussian, more energy is needed within the FWHM than in the flat top profile to get the same length of line-focus. This is because Gaussian profiles produce more strongly peaked intensity profiles, making the aperturing process less efficient. TARANIS, like most lasers, has a near flat-top profile and hence will fall somewhere in between these two extremes. What is also recognised as a potential difficulty in the use of these apertures is the limitations of the geometrical methods used in their design. It is expected that the actual intensity profile of the un-apertured line focus will vary slightly from that for which the apertures are designed. It is likely that some iterations may be necessary in order to 'fine tune' the shapes after their initial installation, but the understanding of the mapping that takes place from the expanded beam to the line-focus, which we have gained from these studies, will be useful in this task.

5 Advantages and Disadvantages of Using Apertures

Aperturing the beam is by no means the only way to achieve a more uniform line focus, and each method has its merits. The main advantages of these apertures are the relative ease with which they can be installed into a pre-existing set-up and the small financial cost associated with them. The main drawback however is the energy lost by obscuring part of the beam, with losses of the order of 40% to be expected even for a relatively optimised aperture, as seen in Fig 4. If the pump energy requirements of the desired XRL are at the limit of that producible by the pump system, then a more suitable alternative to achieving a uniform line focus may lie in the costlier form of custom designed optics, or the use of phase plates to alter the intensity profile. But as the accessibility of high energy lasers increases due to reductions in size and cost, apertures may become more appealing as a way to improve the output of many XRLs. A secondary concern with these apertures is the possibility of propagating a diffraction effects down the beam path and potentially damaging subsequent optical components. This risk can be minimised by placing the aperture as close as possible to the final focussing optic. In the case of the set-up shown in Fig. 1(a), this would mean making a modification to the shape of the aperture to account for the primary lens. This however is not a major task and therefore does not present any real difficulties.

Conclusions

We have presented a study on the design of apertures for use in tailoring the profile of line-foci two popular focussing geometries. Practical examples of apertures designed for a GRIP XRL pumped by the TARANIS laser system were given. The losses associated with such apertures were investigated and they were found to be of an acceptable size in the case of the QUB XRL where the energy output of the pump laser is high and pump efficiency is not a limiting factor.

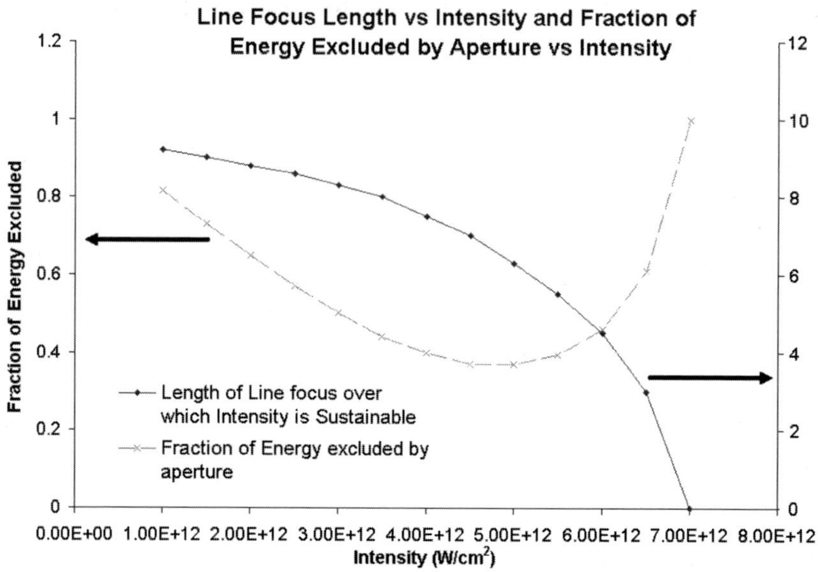

Fig. 4 Graph showing how the fraction of energy lost and the length of sustainable line-focus vary with required intensity

References

1. Elton, R. C.: *'X-Ray Lasers'*, Academic Press, 1990
2. Nantel, M.: 'Spectroscopy and gain dynamics issues in inhomogeneous x-ray laser plasmas', *J. Phys. B: At. Mol. Opt. Phys.*, 28, 2765-2780, 1995
3. La Fontaine, B.: 'Electron-temperature inhomogeneities along an x-ray laser plasma', *Phys. Rev. E.*, 47, 583-590, 1993
4. Ross, I. N.: 'Design and performance of a new line focus geometry for x-ray laser experiments', *App. Opt.*, 26, 1584-1588, 1987
5. Keenan, R.: 'High-repetition-rate grazing-incidence pumped x-ray laser operating at 18.9 nm', *Phys. Rev. Lett.*, 94, Art. No. 103901, 2005
6. Nersysian, G.: 'TARANIS: A Pump Source for X-Ray Lasers', *Submitted to 'Proceedings of ICXRL 2008'*, 2008.

Interferometric Lithography with a Desk-Top Size Soft X-Ray Laser

P.W. Wachulak[1,2], M.C. Marconi[1,2], W. Rockward[1,3], D. Hill[1,3], E.H. Anderson[1,4], C.S. Menoni[1,2], J.J. Rocca[1,2]

[1] NSF Engineering Research Center for Extreme Ultraviolet Science and Technology
[2] Electrical and Computer Engineering Department, Colorado State University, Fort Collins, Colorado
[3] Morehouse College, Physics Department, Atlanta, Georgia
[4] Center for X Ray Optics, Lawrence Berkeley National Laboratory, Berkeley, California

Abstract. We have demonstrated a compact interferometric lithography nanopatterning tool based on an amplitude division interferometer and a 46.9 nm wavelength desk-top size capillary discharge laser. The system is designed to print arrays of lines, holes, and dots with sizes below 100 nm on high resolution photoresists for the fabrication of arrays of nanostructures with physical and biological applications. The future combination of this amplitude division interferometer with high repetition rate table-top lasers operating at shorter wavelengths should allow the printing of arrays with features size only limited by the photoresist resolution in a table top set-up.

1 Introduction

Interferometric lithography (IL) is an attractive alternative to print in a relatively simple way arrays of nano-scale periodic features. This technique requires the use of a coherent light source and relies on the activation of a photoresist by the interference pattern generated by two or more mutually coherent light beams [1-5]. Two mutually coherent beams are combined in the surface of the photoresist and print periodic lines with a period p given by $p=\lambda/(2 \sin\theta)$, where λ is the wavelength of the illumination and θ is the incidence angle. Several schemes can be implemented based on the same idea but in all cases the ultimate resolution, $\lambda/2$, is limited by the wavelength of the illumination. Thus, reducing the wavelength is a direct path towards realizing interference patterns with dimensions of tens of nanometer and below. This characteristic has been the motivation for using extreme ultraviolet (EUV) and soft X-rays (SXR) synchrotron light for this application [6].

Feature sizes as small as 19 nm have been printed using a Lloyd's mirror interferometer with synchrotron illumination [5, 6]. However, the widespread use of IL in nanotechnology applications would benefit from the implementation of more compact and easily accessible set-ups. The increasing average power of compact plasma-based SXR lasers offers the possibility of implementing IL on a table-top. In previous work we demonstrated a compact nanopatterning systems based on a Lloyd's mirror interferometer configuration with a 46.9 nm wavelength table-top capillary discharge laser that allowed the printing of gratings and arrays of nanodots and holes [7-9].

In this paper we describe a compact interferometric nano-patterning tool based on the combination of an amplitude division interferometer (ADI) and a desk-top λ=46.9 nm capillary discharge laser. The ADI configuration described here has significant advantages as compared with the wave-front division Lloyd's mirror. First, with the ADI interferometer it is possible to print larger areas due to its relaxed spatial coherence requirements. Secondly, both beams have practically the same intensity and thus produce a higher contrast and more uniform interference pattern over the whole printed area.

2 ADI design

Figure 1 is a scheme of the ADI interferometric lithography tool. The -1st and 1st orders diffracted from the transmission diffraction grating, are used to form the two arms of the interferometer. The diffraction angle α depends on the wavelength λ and the period of the diffraction grating $\alpha = \arcsin(\lambda/d)$. The zero order is blocked from reaching the sample plane. By tilting the mirrors it is possible to change the incident angle of the beams at the sample plane. This is described by the angle β relative to the axis of the interferometer. Assuming that the period of the patterned structures is p, the two beams reflected from the mirrors will overlap and interfere in the sample with an incidence angle Θ:

$$\Theta = \arcsin\left(\frac{\lambda}{2p}\right) \tag{1}$$

The position z of the sample measured from the center of the mirrors depends on the assumed period as well as on the angle Θ:

$$z = \frac{h}{2\tan(\Theta)} \tag{2}$$

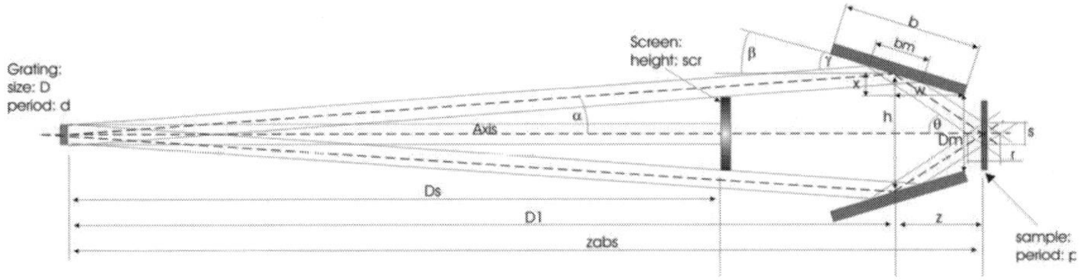

Fig. 1 Scheme of the ADI interferometric lithography tool with all design parameters indicated. The SXR laser impinges the interferometer from the left through the diffraction grating.

Where h is the distance between the two mirrors. At this location the overlap between two interfering beams should be maximal. However the interference will be visible at distances $\left(z-\frac{r}{2};z+\frac{r}{2}\right)$, where r defines the overlap between two beams:

$$r = \frac{D}{\sin(\Theta)} \tag{3}$$

Where D is the size of the grating. The parameter r may be considered as the sample position accuracy. All the parameters mentioned in this analysis are depicted in Figure 1.

In our experimental set up the separation between the centers of the mirrors was h=15 mm. The distance between the grating and the center of the mirrors was calculated to be $D1$=32 cm. For a printing period in the sample p=100nm, Θ is 13.6° . The beam size on the mirror was then calculated bm=4.6 mm. Placing the sample exactly at z=31 mm will produce the interference pattern with maximal overlap of the beams. The maximum displacement allowed from the optimum position will be r/2=1.3 mm.

3 Experimental details

For the demonstration of the ADI lithography tool we used a "desk top" size λ=46.9 nm Ne-like Ar capillary discharge laser configured to emit pulses with energy of approximately 10 μJ and about 1 ns FWHM duration. The EUV laser can operate at repetition rates up to 12 Hz producing pulses, corresponding to average powers up to 0.12 mW. Due to its short capillary plasma column length, 21 cm, its spatial coherence length is only a fraction of a millimeter at the sample location. The laser longitudinal coherence length is approximately 470 μm determined by its line width $\Delta\lambda/\lambda<1\times10^{-4}$. The laser unit is extremely compact, it occupies a footprint of 0.8 × 0.4 m^2 including its turbomolecular pump, and has a small power supply that can fit under the optical table [10].

The grating period d~2μm gives a diffraction angle α=1.34°. The transmission grating used to split the beam was fabricated in a 2 x 0.6 mm^2 Si membrane approximately 100 nm thick. A thin layer of Si was sputtered on top of a Si_3N_4 membrane deposited on a 550 μm thick Si wafer. Subsequently, a 350 nm thick photoresist layer was deposited by spin coating on top of the sputtered Si. The grating was patterned in the photoresist by electron beam lithography. Finally, the Si_3N_4 layer was removed by chemically assisted ion beam etching through the substrate opening leaving a self standing 100 nm thick Si membrane with the grating defined in the photoresist. The thin Si membrane provides 15% to 20% transparency to the λ=46.9 nm photons, while the photoresist that remains in the membrane acts as perfectly absorbent regions, constituting an amplitude transmission diffraction grating. The resulting open areas were measured to be 980 nm wide, and to have a period of 2.02 μm. The interferometer was positioned at 1.4 m from the exit of the capillary discharge plasma where the laser beam uniformly illuminates the whole grating.

The optical path difference between the two branches of the interferometer can be easily adjusted within a distance smaller than 470 μm, the longitudinal coherence length of the laser source. The required accuracy in the overlapping of the two beams at the sample's surface is defined by the spatial coherence of the beams which for the λ=46.9 nm laser used in this experiment is a fraction of a millimeter at the location of the sample. The system is very robust, requiring only minor adjustments after its initial alignment.

4 Results

Figure 2a shows a dense line grating pattern with a period of 145 nm (72.5 nm thick lines) printed on HSQ with the ADI set up. The printed area corresponds to the size of the grating beam splitter, in this experiment 2 × 0.6 mm^2. The penetration depth of the 46.9 nm light in the HSQ photoresist is approximately 120 nm. The images shown in Figure 2 are atomic force microscope (AFM) scans of the sample. To print gratings of smaller period we increased the angle between the two beams impinging at the sample by changing the angle of incidence at the two Si mirrors and by correcting the position of the sample. A dense line pattern with ~ 95 nm period (47.5 nm wide lines) printed on HSQ also over an area of 2 × 0.6 mm^2 was obtained in this case. An AFM scan of this dense line pattern is shown in Fig. 2b. Since the laser was operated at a repetition rate of only 3 Hz, exposures of more than 10 minutes were required. The exposures can be reduced by more than one order of magnitude by combining this ADI with the table-top version of the SXR laser that is capable of producing mW average powers [11, 12].

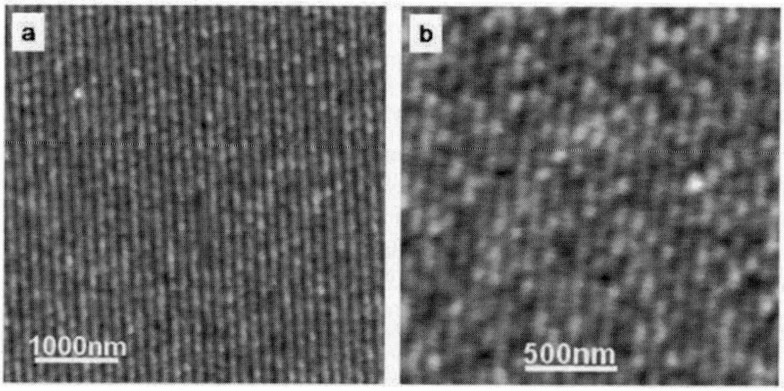

Fig.2 a) AFM image of a 145 nm period grating printed by interferometric lithography using an EUV laser. b) AFM image of a 95 nm period grating. Both images were obtained with the AFM working in "tapping" mode over a small section of the printed grating.

The 95 nm period lines show an increased noise as compared with the larger pitch sample. One possible explanation for this lower quality printing is the influence of vibrations during the exposure. In addition the print quality may possibly be affected by photo-resist scumming, that is particularly severe in 50% duty cycle lines with line widths approaching 50 nm, [13] or by the presence of scattering centers in the grating and mirrors that introduce a random noise background reducing the fringe visibility.

This proof-of-principle experiment can also be extended to demonstrate printing of more complex two dimensional motifs, such as arrays of dots or holes. This can be realized by performing a second exposure after rotating the sample. The shape of the printed features can be altered by selecting the rotation angle and the exposure. This requires the incorporation of suitable rotation mechanisms for the sample, similar to the experiment described in Ref. [9]. Larger areas are also possible by either utilizing a transmission grating beam splitter of larger dimensions, or by precise translating the sample and realizing multiple exposures.

5 Summary

In summary, we have demonstrated a compact EUV interferometric lithography tool that combines an amplitude division interferometer with a desk-top size SXR laser. In this proof of principle experiment we printed gratings over areas of 2×0.6 mm^2 with periods down to 95 nm. These capabilities would enable the demonstration of a practical SXR IL tool for the fabrication of large arrays of periodic features that so far were restricted to the use of large synchrotron facilities.

References

1. S.R.J. Brueck, "Optical and interferometric lithography - Nanotechnology enablers". *Proceedings Of The Ieee*. 93 1704-1721 2005.
2. A. Fernandez, J.Y. Decker, S.M. Herman, et al., "Methods for fabricating arrays of holes using interference lithography". *Journal Of Vacuum Science & Technology B*. 15 2439-2443 1997.
3. W. Hinsberg, F.A. Houle, J. Hoffnagle, et al., "Deep-ultraviolet interferometric lithography as a tool for assessment of chemically amplified photoresist performance". *Journal Of Vacuum Science & Technology B*. 16 3689-3694 1998.
4. J.A. Hoffnagle, W.D. Hinsberg, M. Sanchez, et al., "Liquid immersion deep-ultraviolet interferometric lithography". *Journal Of Vacuum Science & Technology B*. 17 3306-3309 1999.
5. H.H. Solak, D. He, W. Li, et al., "Nanolithography using extreme ultraviolet lithography interferometry: 19 nm lines and spaces". *Journal Of Vacuum Science & Technology B*. 17 3052-3057 1999.
6. H.H. Solak, D. He, W. Li, et al., "Exposure of 38 nm period grating patterns with extreme ultraviolet interferometric lithography". *Applied Physics Letters*. 75 2328-2330 1999.
7. P.W. Wachulak, M.G. Capeluto, D. Patel, et al., "Nanoscale patterning in high resolution HSQ photoresist by interferometric lithography with table top EUV lasers." *Journal of Vacuum Science & Technology B*. 25 2094 2007.
8. M.G. Capeluto, G. Vaschenko, M. Grisham, et al., "Nanopatterning with interferometric lithography using a compact lambda=46.9-nm laser". *Ieee Transactions On Nanotechnology*. 5 3-7 2006.
9. P.W. Wachulak, M.G. Capeluto, M.C. Marconi, et al., "Patterning of nano-scale arrays by table-top extreme ultraviolet laser interferometric lithography". *Optics Express*. 15 3465-3469 2007.
10. S. Heinbuch, M. Grisham, D. Martz, et al., "Demonstration of a desk-top size high repetition rate soft x-ray laser". *Optics Express*. 13 4050-4055 2005.
11. B.R. Benware, C.D. Macchietto, C.H. Moreno, et al., "Demonstration of a high average power tabletop soft X-ray laser". *Physical Review Letters*. 81 5804-5807 1998.
12. C.D. Macchietto, B.R. Benware, and J.J. Rocca, "Generation of millijoule-level soft-x-ray laser pulses at a 4-Hz repetition rate in a highly saturated tabletop capillary discharge amplifier". *Optics Letters*. 24 1115-1117 1999.
13. D.P. Mancini, K.A. Gehoski, E. Ainley, et al., "Hydrogen silsesquioxane for direct electron-beam patterning of step and flash imprint lithography templates". *Journal of Vacuum Science & Technology B*. 20 2896-2901 2002.

Time-Resolved Fluorescence Spectrum of Wide-Gap Semiconductors Excited by 13.9 nm X-Ray Laser

M. Tanaka[1], Y. Furukawa[2], T. Nakazato[3], T. Tatsumi[3], H. Murakami[3], T. Shimizu[3], N. Sarukura[3], M. Nishikino[1], T. Kawachi[1], Y. Kagamitani[4], D. Ehrentraut[4], T. Fukuda[4], H. Nishimura[3] and K. Mima[3]

[1] Advanced Photon Research Center, Japan Atomic Energy Agency
[2] RIKEN (The Institute of Physical and Chemical Research)
[3] Institute of Laser Engineering, Osaka University
[4] Institute of Multidisciplinary Research for Advanced Materials, Tohoku University

Abstract. The time-resolved fluorescence spectrum of wide-gap semiconductors, single crystals of zinc oxide and gallium nitride, excited by nickel-like silver x-ray laser with the wavelength of 13.9 nm were observed. The streak image of fluorescence at around 380 nm and 370 nm for each samples were obtained by one-shot excitation of the x-ray laser. The fluorescence lifetimes were compared with the 351 nm excitation cases and evaluated the samples for scintillation materials of EUV lithography.

1 Introduction

X-ray laser is a characteristic extreme ultraviolet (EUV) source with short pulse duration, narrow spectral width, and high coherence. At the Advanced Photon Research Center, the x-ray lasers generated with transient gain collisional excitation (TCE) method have been studied[1] and have achieved the spatially full coherent lasing of a Ni-like Ag x-ray laser at the wavelength of 13.9 nm.[2,3] The TCE scheme have several advantages, such as short pulse duration less than ten pico-seconds, and a high gain generation with low excitation energy of several tens of Joules. In particular, x-ray laser with the wavelength of around 13 nm is expected as a powerful tool of various research fields such as EUV lithography because the multilayer mirror with high reflectance is commercially prepared. However, spectroscopic studies of materials optically excited with x-ray laser are few while imaging and interference measurements are intensively studied. In this study, we have demonstrated time resolved emission spectroscopy of a solid-state material excited by Ni-like Ag x-ray laser.

As an application of the fluorescence measurement excited by x-ray laser, the scintillation properties of wide-gap semiconductors such as zinc oxide

(ZnO) and gallium nitride (GaN) are evaluated for EUV around 13 nm. These materials are prominent candidate of efficient and fast imaging scintillator devices for next-generation lithography applications, in aspects of high fluorescence yield and availability of large sized homogeneous crystal with a reasonable fabrication cost. For the evaluation of these materials, a Ni-like Ag x-ray laser operating at 13.9 nm is the ideal light source; having large pulse energy up to about micro-joules level and a sufficiently short pulse duration down to several picoseconds.

2 Experimental

The Ni-like Ag x-ray laser with the wavelength of 13.9 nm was generated by TCE method using a chirped pulse amplification Nd:glass laser system. [1] The beam divergence of the x-ray laser was improved to be 0.5×2.0 mrad by use of two gain media separated for 20 cm.[2-4] The typical pulse energy of the x-ray laser emission was 0.5 μJ and the pulse duration was 7 ps. The narrow divergence allowed us to transport the beam using a simple mirror system with a small diameter of 1 inch.

The single crystal samples of ZnO and GaN are grown by solvothermal method combined with a platinum inner container.[5-7] The x-ray laser was focused on the sample using a Mo/Si multilayer spherical mirror suitable for 13.9 nm. To eliminate continuous emission from the plasma, a 0.2 μm-thick zirconium foil was placed before the multilayer mirror. The fluorescence spectrum and the fluorescence lifetime of the samples were measured using a 25 cm-focal-length spectrograph coupled with a streak camera with the temporal resolution of 100 ps in the fastest scanning range. For comparison, the scintillation properties were also evaluated using the 351 nm third harmonics

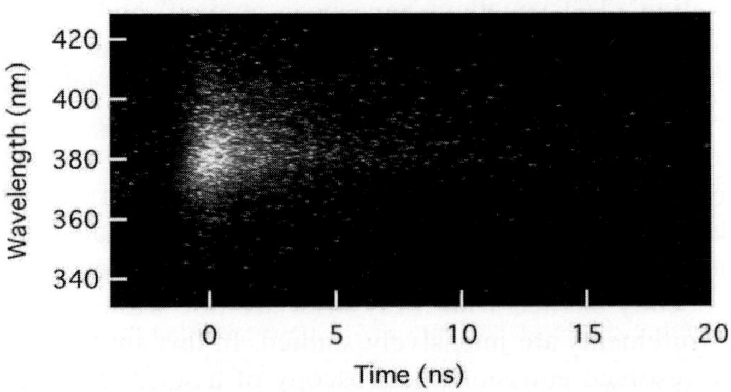

Fig. 1. The spectrum-resolved streak image of zinc oxide single crystal at room temperature excited by one shot of 13.9 nm x-ray laser.

from the 1053 nm chirped pumping source for the x-ray laser. In this case, the ZnO was excited at energy slightly above the bandgap. The pulse duration of the third harmonics is measured to be 110 ps.

3 Results

As shown in Fig. 1, one shot of x-ray laser was enough to obtain a streak image. For the ZnO single crystal, a prominent fluorescence peak of the exciton transition was observed at around 380 nm in both 13.9 nm excitation and 351 nm excitation.[8] Shown in Fig. 2 are temperature dependence of the fluorescence spectrum of ZnO crystal.[9] Falling the temperature of ZnO, the peak position of the fluorescence shifted blue and the spectral width became narrower. In generally, this emission originates from the emission of free exciton and banding exciton[10] and the blue shifts of the exciton emission with the decreasing temperature is attributed to the bandgap shift due to the thermal expansion and electron-phonon interaction.[11] These spectral properties of the x-ray laser excitation at measured temperatures were almost identical to that of UV excitation. The time profiles of fluorescence at room temperature are shown in Fig. 3(a). It could be expressed by a double exponential decay with time constants of 1 ns and 3 ns in the both cases. The two decay

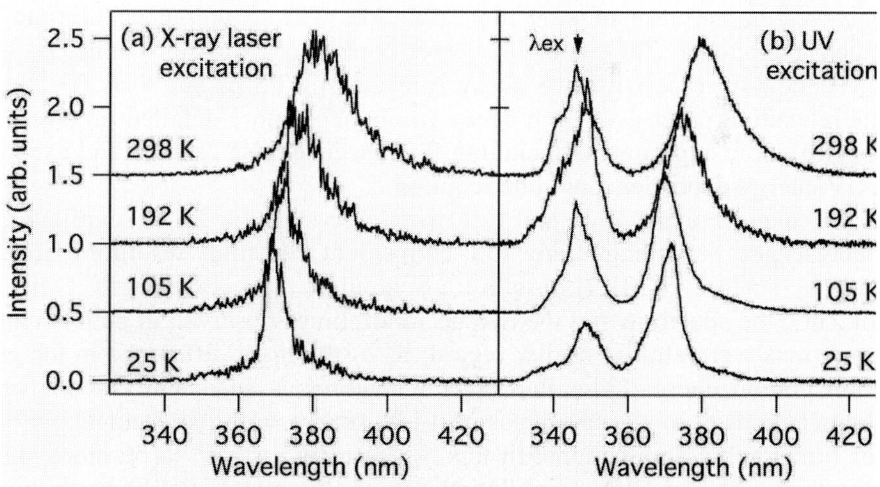

Fig. 2. The fluorescence spectrum of ZnO single crystal at the temperature of 298 K, 192 K, 105 K and 25 K. Excitation source were (a) 13.9 nm x-ray laser and (b) 351 nm UV. To reduce the noise level, each spectrum of the x-ray laser excitation was integrated for 3 shots. The peaks on left hand side of UV excitation data were the excitation light reflected on the surface of the sample.

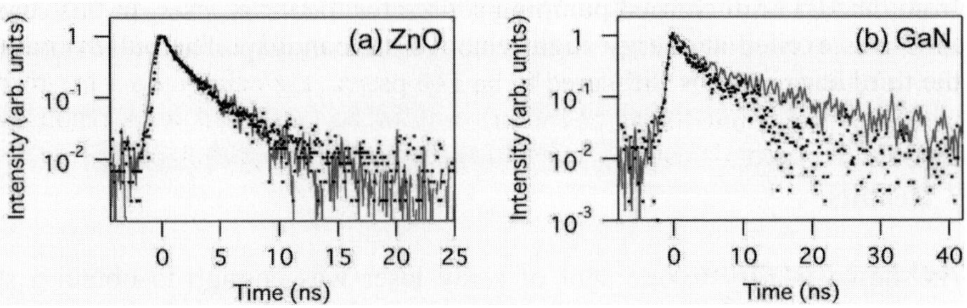

Fig. 3. The time profiles at the peak of the fluorescence spectra of (a) ZnO and (b) GaN single crystal. The grey lines indicate the results of 13.9 nm excitation and the dots are 351 nm excitation. To reduce the noise level, the time profiles of the x-ray laser excitation were integrated 3 shots for ZnO and 14 shots for GaN.

constants have been measured in several works for UV excited ZnO single crystals. The fast decay is the lifetime of free exciton and the slower decay is assigned to be trapped carriers.[12] The lifetimes and intensity ratio of the two decay components in the both cases were almost similar in spite of the huge difference in the excitation photon energy.

The same measurement was also held for GaN single crystal for room temperature. Shown in Fig. 3(b) are the time profiles at fluorescence peak at the wavelength of 370 nm. For GaN, two decay lifetimes of 2.5 ns and 15 ns were observed in the case of 13.9 nm excitation. The fluorescence lifetime was much longer than ZnO, and the slower component was not similar with the UV excitation case with two decay constants of 2.5 ns and 8 ns. To explain the relaxation process of each decay component, more detailed time-resolved spectroscopic experiments including fluorescence raise profiles and excitation EUV energy dependence will be required.

To consider using ZnO and GaN single crystals for EUV scintillator, the fluorescence wavelength are still convenient for high resolution imaging devices, since even BK7 glass is transparent at their wavelength. Especially for ZnO, the spectrum and the two decay lifetimes observed in both excitation conditions were almost similar regardless of the huge difference in the excitation photon energy. The fluorescence lifetime is sufficiently short for the characterization of the laser plasma EUV source with nanoseconds duration for lithographic applications. In this respect, ZnO is said to be more favourable than GaN for EUV scintillation device. In spite of the huge photon energy compared with the bandgap, no visible color center was observed on the ZnO sample after about 50 shots irradiation. Furthermore, a large-sized and homogeneous material is potentially attractive for EUV imaging applications including lithography.

As shown in this work, a Ni-like Ag x-ray laser is an excellent tool for time-resolved spectroscopy and characterization of materials intended for next-generation EUV lithography applications. Especially on time-resolved measurement, high-speed phenomenon can be observed with resolution of several picoseconds using a TCE x-ray laser as an excitation source.

Acknowledgements

This work was in part performed by auspice of MEXT (Japanese Ministry of Education, Culture, Sports, Science and Technology) project on "Mono-energetic quantum beam science with PW lasers" and "Development of Growth Method of Semiconductor Crystals for Next Generation Solid-State Lighting". The results were achieved under the Facilities Utilization system of Japan Atomic Energy Agency.

References

1. Kawachi T *et al.*, *Phys. Rev. A* **66**, 33815, 2002
2. Tanaka M, Nishikino M, Kawachi T, Hasegawa N, Kado M, Kishimoto M, Nagashima K, Kato Y, *Opt. Lett.* **28**, 1680, 2003
3. Nishikino M, Tanaka M, Nagashima K, Kishimoto M, Kado M, Kawachi T, Sukegawa K, Ochi Y, Hasegawa N, Kato Y, *Phys. Rev. A* **68**, 61802R, 2003
4. Nishikino M, Hasegawa N, Kawachi T, Yamatani H, Sukegawa K, Nagashima K, Appl. Opt. **47**, 1129, 2008
5. Ehrentraut D, Sato H, Kagamitani Y, Sato H, Yoshikawa A, Fukuda T, *Prog. Cryst. Growth Charact. Mat.* **52**, 280, 2006
6. Ohshima E, Ogino H, Niikura I, Maeda K, Sato M, Ito M, Fukuda T, *J. Cryst. Growth*, **260**, 166, 2004
7. Fukuda T, Ehrentraut D, *J. Crys. Growth*, **305**, 304, 2007
8. Tanaka M *et al.*, *Appl. Phys. Lett.* **91**, 231117, 2007
9. Furukawa Y *et al.*, *J.Opt. Soc. Am. B* **25**, 118, 2008
10. Chen Y, Bagnall D, Yao T, *Mater. Sci. Eng. B* **75**, 190, 2000
11. Hauschild R, Priller H, Decker M, Brucker J, Kalt H, Kilingshirn C, *Phys. Stat. Sol.* **3**, 976, 2006
12. Wilkinson J, Ucer K B and Williams R T, *Radiat. Meas.* **38**, 501, 2004

Part 8 – Alternative Approaches for Sources of Bright X-Rays

Application of Extremely Bright and Coherent Soft and Hard X-Ray Free-Electron Laser Radiation

Th. Tschentscher

European XFEL Project Team, c/o Deutsches Elektronen-Synchrotron DESY, 22607 Hamburg, Germany

Abstract. The first soft x-ray free-electron laser (FEL) is providing extremely bright and coherent radiation since 2005 for a wide range of scientific applications from atomic physics to life sciences. Successful experiments using the FLASH facility at DESY, Hamburg depend critically on radiation parameters such as pulse duration, photon number and coherence properties and, in most cases, request the combination of these parameters in a single experiment. At the same time the transport and preservation of x-ray FEL radiation parameters leads to extreme requirements for x-ray optical elements. The construction of soft and hard x-ray FEL facilities for user experiments with many instruments operating nearly simultaneously led to specialized facility layouts described here for FLASH and for the European XFEL, Hamburg.

1 Introduction

Scientific applications of free-electron laser (FEL) radiation in the soft and hard x-ray domain had been discussed for several years in preparation of the construction of these sources. Proposed applications originated from a wide spectrum of scientific disciplines and aimed at taking advantage of the beam properties offered by x-ray FEL radiation for the very first time. Investigations of ultrafast and non-linear phenomena on length scales down to the sub-nanometer range require in general ultrashort pulse duration, coherence or high pulse intensities. In order to penetrate matter, to investigate inner-shell atomic states and to enable atomic resolution photon energies in the soft to hard x-ray regime are needed. Tunability of the photon energy provides element and state selectivity and can also be used to tune the absorption coefficient. To furthermore enhance the request many proposed investigations require a combination of the above x-ray beam properties. FELs are a unique source of x-ray radiation providing these radiation properties simultaneously. Other light sources, like optical laser-based methods to produce x-ray radiation or storage ring based synchrotron radiation sources, provide x-ray radiation with high performance for many applications, but are facing strong limitations in delivering radiation combining the above properties. While for the VUV to XUV regime laser-based high harmonic generation methods can achieve performances close to that of a FEL, in the regime of hard x-ray radiation at one to ten keV FELs are currently without competition.

Since 2005 with FLASH the first user facility for soft x-ray FEL radiation is in operation at DESY, Hamburg[1-3]. Initial experiments using FLASH already came from a wide range of scientific disciplines. These experiments have explored a large number of state-of-the-art synchrotron radiation and laser experimental techniques and furthermore have introduced new techniques. During the next years the experimental program using FEL radiation and the user access will broaden due to the start of operation of new soft x-ray facilities as well as the first operation of FEL facilities for the hard x-ray regime. The experimental program at FLASH and the experiences drawn from it turn out to be very useful in the preparation of the initial experiments in the hard x-ray domain.

2 Requirements of scientific applications

Most present day x-ray experiments investigate equilibrium properties of matter. Synchrotron radiation provides experimental conditions well-suited for such studies and the number of facilities world-wide and the growing user community prove the success of this development. The improvement of x-ray beam properties such as pulse duration and emittance is limited for synchrotron radiation by equilibration of the electron bunches circulating in the storage ring. This process leads to a settling of the electron bunch lengths corresponding to pulse durations of 10–100 ps and to a limitation of the emittance in the horizontal plane affecting the spatial coherence. Operating with several MHz repetition rates the number of photons per pulse is much reduced compared to FEL sources and is typically far below 10^8. The investigation of dynamical processes and excited, non-equilibrium states of matter has been initiated through the development of femtosecond x-ray sources using laser sources[4] and storage rings[5]. While pulse durations of the order 100 fs have been obtained, scientific applications using these sources are limited due a relatively small number of photons per pulse.

Going beyond todays x-ray scientific applications has been proposed for many areas. In the following we discuss three major fields of application and introduce exploratory experiments carried out using vacuum-ultraviolett and soft x-ray FEL radiation. The x-ray investigation of the dynamics and of non-equilibrium states of matter with atomic resolution in space and time explores mostly the temporal domain and could involve many experimental techniques. Natural time scales of solid-state dynamics, like acoustic and optical phonons or chemical reactions request time resolutions of the order 100 fs. Investigation of solid-state electronic excitations in the time domain will require pulse durations typically of order femtoseconds. One challenging task for time-domain experiments is to measure the pulse duration of x-ray pulses using cross-correlation and auto-correlation techniques. Cross-correlation measurements involving ultrashort near-visible optical laser pulses and FEL radiation have been carried out by creating photo-electrons in the strong optical laser field[6,7] and by modifying the reflectance of an optical substrate[8,9]. These experiments have reached temporal resolutions limited by the pulse duration of the optical laser to ~150 fs[10]. Even more demanding are auto-correlation experiments using mirror optics to split the beam into two parts, to create a delay of one part with respect to the other, and to recombine the two pulses. Varying the delay and observing a non-linear response by the probe allowed first application of this technique for soft x-rays[11]. Among the first experiments to use this time-resolution the investigation of two-color above-threshold ionization of helium[12] and loss of long-range order in solids after interaction with an intense optical laser pulse[13] have been studied.

Scattering techniques using the interference of a coherent x-ray pulse will enable atomic resolution spatial resolved investigation of single nanoscale structures on solid substrates, e.g. quantum dots, or even of single particles, cluster or biomolecules injected into vacuum. In particular experiments on injected single particles are extremely demanding with respect to soft x-ray radiation properties[14]. High photon numbers are required in order to achieve significant scattering within a single pulse. These photons have to be focussed to a small spot maximizing the overlap with the particle. The reconstruction of the density distribution from the two-dimensional scattering pattern requires a known wavefront at the sample position. To fulfil this requirement the x-ray pulse should exhibit high spatial coherence which has to be preserved by the beam transport and focussing optics. In addition, the x-ray pulse needs to be as short as possible to overcome sample damage effects due to the interaction with the intense radiation. Although atomic resolution will require the use of hard x-ray radiation the principles of this technique could be studied for the first time using ion beam produced test structures[15] and have meanwhile shown to work for single particles[16].

Exciting sample systems using very intense and ultrashort x-ray pulses enables the investigation of fundamental multi-photon and non-linear processes and can be applied further to the formation of well-defined plasma states of matter. Using the ultrashort pulse duration, tight focusing and the high photon numbers per pulse it is possible to achieve intensities up to 10^{17} W/cm^2. Despite this value the interaction process is considerably different compared to near-visible radiation. Due to the high photon energy, respectively short wavelength, the ponderomotive energy $E_p \propto \lambda^2$ is smaller by a factor ~4×10^3 for 92 eV radiation and interaction occurs primarily by multiphoton processes leading to multiply ionized atoms[17]. Furthermore the high photon energies enable one-photon photoionization opening an interaction channel not accessible for near-visible radiation. Combining the high intensity with enhanced single-photon absorption due to resonances new absorption processes have been observed for Xe clusters[18] and atoms[19]. Interaction of FEL radiation at these intensities with liquids and solids leads to formation of near-solid density plasmas with temperatures of several electronvolt[20].

In many cases the application of high repetition rates is advantageous. Albeit being an average effect it occurs that several of the new scattering techniques lead to moderate count rates per event and sampling many events will improve the statistical accuracy. Most prominent example here are studies of extremely dilute systems. These experiments often require high peak fluxes in order to obtain signal-to-noise ratio in single pulses while high repetition rates are needed for statistical accuracy. Spectroscopic studies on trapped highly charged ions[21] and beams of molecular ions[22] furthermore take advantage of the narrow spectral bandwidth of FEL radiation. Experiments studying single particles injected into the vacuum in cross beam geometry require effectively many FEL pulses to obtain a good hit[16].

3 Description of x-ray FEL facilities

Free-electron lasers are based on relativistic electrons travelling through a periodic magnetic field. The interaction with the magnetic field leads to spontaneous emission of synchrotron radiation. This radiation co-propagates with the electron and in a very long undulator it can act back on the electron bunch by forming a density modulation with the period of the radiated wavelength. The bunched electrons emit in phase leading to a strong increase of the radiative field. The increased field enforces the bunching and this again increases the coherent radiation. This so-called FEL instability leads to an exponential increase of the radiation power (linear gain regime) until the majority of electrons are bunched and saturation occurs. Necessary requirement for this interaction are a normalized emittance of ~1 μm rad, a peak current of 1-5 kA and a bandwidth <0.1% of the electron beam. Electron bunches of such a performance can until now only be obtained using low emittance linear accelerators. Since in the x-ray regime mirrors have limited reflectivity it is needed to obtain high gain and saturation in a single pass. This scheme of self-amplified stimulated emission (SASE)[23] uses undulator lengths of 30-200 m for generation of soft to hard x-ray FEL radiation, respectively.

The FEL process leads to a very intense x-ray pulse produced by a highly collimated and small source. The length of the electron bunch determines in first order the duration of the x-ray pulse of typically tens of femtoseconds. The most appropriate number to describe such sources is their peak brightness

$$B_{peak} = \frac{N_{ph}}{\Delta_x \Delta_{x'} \Delta_y \Delta_{y'} \Delta_t \times BW} \quad \left[\frac{1}{mm^2 mrad^2 s} \right].$$

for the number of photons per pulse N_{ph}, the emittance of the FEL radiation radiation $\Delta_x \Delta_{x'} \Delta_y \Delta_{y'}$ for apparent source sizes $\Delta_{x,y}$ and source divergences $\Delta_{x',y'}$ in horizontal and vertical directions, respectively, the duration of the x-ray pulse Δ_t and its spectral bandwidth BW in units of 0.1% of the photon energy E_{ph}. Values are given as full width half maximum. For the emittance of FEL radiation the diffraction limit

$$\Delta_x \Delta_{x'} \Delta_y \Delta_{y'} = \left(\frac{\lambda}{2} \right)^2$$

with the wavelength $\lambda[nm] = 1.24 / E_{ph}[keV]$ is valid for soft x-rays and is nearly reached for hard x-ray radiation. In the spectral-temporal domain FEL radiation is in general not transform limited, but exhibits several longitudinal modes leading to a spiky temporal and spectral distribution. This difference can also be analysed in terms of the coherence of x-ray FEL radiation. While the angles of coherent radiation $\theta_{x,y}^{coh} = \lambda/(2\Delta_{x,y})$ are close to $\Delta_{x',y'}$ for all photon energies, the longitudinal coherence length $l^{coh} = \lambda^2/\Delta\lambda$ is usually shorter than the pulse length. For hard x-rays and 100 fs pulse duration a ratio $\Delta_z / l^{coh} \cong 300$ is calculated while for soft x-rays and 10 fs pulses[1] this ratio is ~2.5. In general, the coherence properties of FEL sources require a complex description[24] and have to be subject of measurements[25].

In the case of diffraction and bandwidth limited FEL radiation the peak brightness is determined by N_{ph} only. This number depends on the number of electrons per bunch and on the electron energy. The peak power of an x-ray FEL can reach several tens to hundreds of Gigawatt with typical source sizes of 100 μm and highly collimated beams. Table 1 lists typical values of peak power densities of soft and hard x-ray FEL radiation. Despite the large distance one obtains considerable peak power densities at x-ray optics and instruments. Stable operation of these elements leads to the requirement to use high reflectance optics to minimize the effects of energy deposition and optics damage. The repetition rate of FELs is determined by the accelerator repetition rate. Warm technology accelerators can operate up to 100 Hz and produce an average output power less than one Watt. No effects in terms of heat load are to be expected for these machines. Super-conducting technology allows increasing the frequency to several 10 kHz thus delivering few 10 W averaged output power. However, present-day technology allows operating super-conducting accelerators only in pulsed mode with a frequency of order 10 Hz. Within one radio-frequency pulse several thousand electron bunches can be accelerated using this technology. The averaged output power within one radio-frequency pulse of typically 1 ms duration is therefore of order few kW. Such a power combined with small FEL beam sizes at the optical elements lead to challenging heat load requirements for the optical elements.

Table 1 Peak power density values for soft and hard x-ray FEL radiation calculated for several photon energies. Distances for x-ray optics of 40/200/400 m and for instruments of 70/400/1000 m are taken for FLASH[1] and European XFEL[26], respectively. Focusing is assumed without losses.

Photon energy [eV]	Source [W/cm²]	X-ray optics [W/cm²]	Instrument [W/cm²]	1 μm focus [W/cm²]
92[1]	$3{,}5\times10^{13}$	$6{,}9\times10^{10}$	$2{,}2\times10^{10}$	$8{,}9\times10^{17}$
800[26]	$3{,}4\times10^{15}$	$3{,}2\times10^{12}$	$8{,}0\times10^{11}$	$1{,}7\times10^{19}$
12400[26]	$5{,}2\times10^{14}$	$1{,}5\times10^{13}$	$2{,}5\times10^{12}$	$2{,}5\times10^{18}$

The operation principle of the first generation of FELs is based on the SASE scheme amplifying from shot noise. This leads to statistical variation of the spectral and temporal distributions and to fluctuations of the output power. Improvements of these parameters are to be expected if the FEL process is seeded using coherent radiation. The difficulty lies in providing a coherent radiation seed with appropriate photon energy and intensity. Various schemes have been suggested and some are currently investigated[27]. Seeding is considered to be easier for soft x-rays and the first soft x-ray FEL facilities now plan to implement these schemes. However, for hard x-rays above $E_{ph} \cong 1\,keV$ yet no concept could be validated.

Most FEL facilities are built with the purpose to provide FEL radiation for scientific experiments carried out by internal or external users. This concept is analogue to the operation of synchrotron radiation user facilities. However, since the FELs are based on linear accelerators sending the beam into single undulators one has to develop new concepts to distribute the FEL radiation to a variety of instruments fulfilling user needs from a wide range of scientific communities. In the following we briefly describe the examples of the FLASH[28] and European XFEL[26] photon beam installations as possible realizations of x-ray user FEL facilities. Figure 1 shows the FLASH experimental hall with the photon beam distribution and the locations of the instruments. At

FLASH the electron beam with energies up to 1 GeV is steered through one fixed-gap undulator before being bent down to an absorber. Tuning the photon energy is achieved by tuning the electron energy in this scheme. The FEL radiation produced in the undulator is transported ~40 m to the experimental hall, where it is distributed by flat mirrors to the different beamlines.

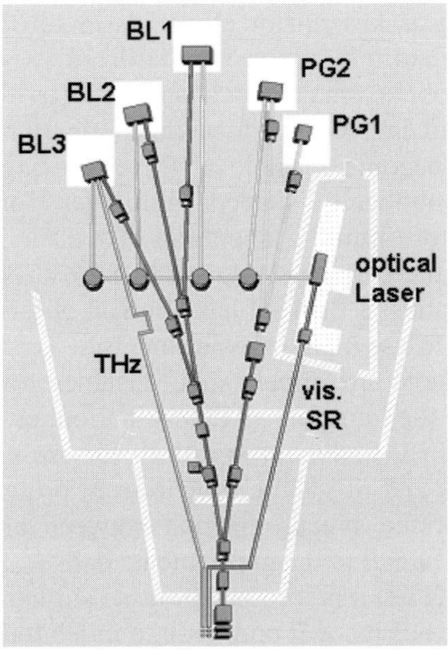

Fig. 1 Schematic view of the FLASH experiments hall with an area of 20×30 m^2. The FEL beam comes from below and is distributed by means of mirrors to five possible instruments. Distribution of optical laser radiation from the central laser room and the transport of THz and visible synchrotron radiation from the accelerator tunnel are indicated.

Entering into the hall the FEL beam first hits a double mirror system to select the two principal beamlines for high intensity and for monochromatic beam. The deflection angle is increased by a second mirror in both beamlines to total four degrees in the horizontal plane for both beams. In addition to separating the two beamlines the four mirror system has the function to stop highly energetic Bremsstrahlung radiation produced in the undulator section and transported together with the FEL beam. In the monochromator beamline a plane-grating monochromator is installed to provide a spectral resolution from few hundreds up to about ten-thousand. A port to use 0-th order radiation is provided[29]. Between monochromator grating and exit slit a focusing mirror is used to steer the FEL beam to two instrument locations. In the high intensity beamline the beam can be steered to three instrument locations using double mirror reflections to achieve a significant angular separation from the initial beam direction. At four of the five instrument locations no instrument is permanently mounted, instead user groups bring their own instruments and install them here. This concept provides large flexibility, but can lead to long setup times in case of complicated instruments. At the fifth location a high resolution spectrometer for Raman-type experiments is permanently installed. The focal spot properties at the five locations vary from 20 to 100 μm (FWHM) employing as last mirror ellipsoids or toroids for focusing. These mirrors inside the beamlines enable standardized vacuum interfaces towards the instruments. To obtain smaller focal spots shorter focal lengths are needed therefore requiring to include the focusing optics into the instrument itself. This approach has become feasible at 92 eV, using unfocused beam and spherically and parabolically shaped multilayer optics[19]. All mirrors are carbon coated therefore limiting operation of the facility in the present configuration to below the carbon K-edge at ~284 eV. Nevertheless the overall layout of the FEL optics at FLASH has allowed observing harmonics of FEL radiation[28] and even to use them in experiments.

FLASH allows due to a relatively small distance from the experiments hall to the electron beam the transport of THz radiation, produced by an additional radiator behind the FEL undulator and used for cross-correlation and pump-probe experiments[30], and of visible synchrotron, produced by a bend magnet and used to monitor the synchronization between optical laser and FEL[10], from the accelerator tunnel to the experiments hall.

At the European XFEL it is foreseen to use a similar concept for five beamlines. Each beamline has its own undulator, a much longer beam transport and an experiments area inside the experiments hall of approximately 15×40 m^2. Initially three of the five beamlines will be constructed with two FEL undulators for hard x-ray radiation and one for soft x-ray radiation (compare Tab. 2). It is foreseen to operate the accelerator at fixed electron energy of 17.5 GeV and to enable tuning the photon energy using gap adjustment of the undulators. The undulators for the hard x-ray regime have been optimized such that one device operates at fixed photon energy of 12.4 keV while the other can be tuned from 3.1-12.4 keV. The third undulator uses the spent beam of the fixed energy device to deliver FEL radiation in the soft x-ray regime tuneable in the range 0.775-3.1 keV. Operating the accelerator at reduced energy of 10 GeV the soft x-ray FEL undulator could provide FEL radiation in the range 0.25-1.0 keV. This low energy operation allows closing the gap with respect to photon energies provided by FLASH but a thorough design has to show the feasibility to transport this extremely intense soft x-ray FEL radiation.

Table 2 Design parameters of the FEL undulators at the European XFEL[26]. L_{sat} and D indicate saturation length and distance from undulator exit to experiments hall.

Source	Electron energy [GeV]	Photon energy [keV]	Period [cm]	L_{sat} [m]	D [m]	Prioritized instruments
SASE 1	17.5	12.4	35.6	133	1000	SPB, MID
SASE 2	17.5	3.1 – 12.4	48	174	1000	FDE, HED
SASE 3	17.5	0.77 – 3.1	80	81	400	SQS, SCS
	10.0	0.25 – 1.0				

The beam transport includes long flat mirrors to displace the FEL beam from the direct beam and to distribute the FEL beam to the various instruments. The mirror grazing angle for hard x-rays is with 1-5 mrad much smaller than at FLASH. Therefore locating the mirrors inside the long transport tunnels between undulators and experiments hall enables to efficiently reduce radiation background due to Bremsstrahlung and to maximize the separation of beam transport to the various instruments. Crystal and grating monochromators for hard and soft x-ray radiation, respectively, will also be placed inside the tunnels. Each of the experiments areas in the hall shall accommodate in the future up to three instruments. For initial user experiments six instruments have been prioritized (compare Fig. 2). Structure determination of single particles, like atomic clusters, biomolecules, virus particles or cells will be the prime target of an instrument dedicated to ultrafast coherent diffraction imaging from gaseous samples (SPB). The Materials Imaging and Dynamics (MID) instrument will focus on structure determination of nano-devices by coherent diffraction imaging techniques and on dynamical processes at the nanoscale using x-ray photon correlation spectroscopy techniques. Time-resolved investigations of ultrafast dynamics in solids, liquids and gases will be carried out using a large variety of methods at the Femtosecond Diffraction Experiments (FDE) instrument. Matter under extreme conditions, e.g. very high pressures and plasmas, will be the target of the High Energy Density (HED) physics instrument. A Small Quantum Systems (SQS) instrument will be specialized for the investigation of atoms, ions, molecules and clusters in intense fields and of non-linear phenomena. The Soft x-ray Coherent Scattering (SCS) instrument will bundle coherence applications using soft x-rays, like e.g. the investigation of structure and dynamics of nano-systems and of non-reproducible biological objects using soft X-rays using imaging and spectroscopy techniques.

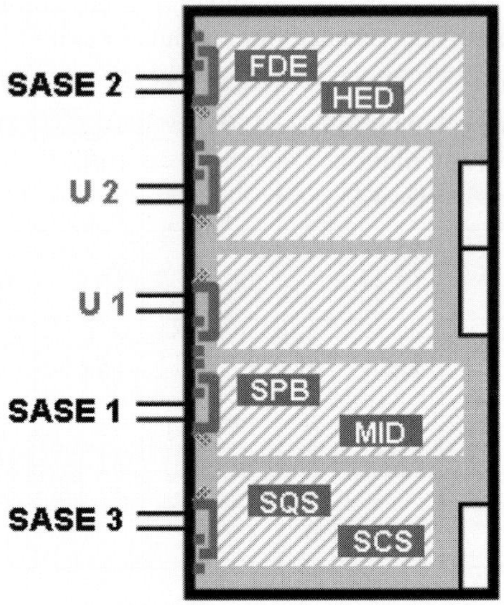

Fig. 2 Schematic layout of the European XFEL experiments hall with a size of $50\times100\ m^2$ and the five areas for each of the beamlines. Beamlines U1 and U2 will not be constructed initially. The FEL beams arrive in tunnels from the left. In addition to the six prioritized instruments (see text for abbreviations) one or two laser rooms will be placed on the experiments floor.

4 Conclusions

The investigation of dynamics of matter and of high-intensity or non-linear processes using x-ray radiation requires sources of very high brightness. First experiments using soft x-ray FEL radiation provided by FLASH at DESY, Hamburg, show the possibilities and challenges to use these new sources. Other soft x-ray FELs under construction will become accessible for user experiments in the coming years. Another big step will be the provision of hard x-ray FEL radiation by the LCLS[31] in Stanford, U.S.A., scheduled for 2010, the SCSS[32] in Harima, Japan, scheduled for 2011, and the European XFEL in Hamburg, Germany. Construction of the European XFEL as a large-scale user facility with international partners has started and first beam is scheduled for 2014.

References

1. Ackermann, W., et al., Nature Photonics, 1, 336-342, 2007
2. Ayvazyan, V., et al., Eur. Phys. J. D, 37, 297-307, 2006
3. Feldhaus, J., 'One year of user operation of FLASH, the free-electron laser in Hamburg', in Proc. Int. Conference on X-ray Lasers, Berlin 2006
4. Siders, C.W., et al., Science, 286, 1340-1342, 1999
5. Beaud, P., et al., Phys. Rev. Lett., 99, 174801, 2007
6. Meyer, M., et al., Phys. Rev. A, 74, 011401(R), 2006
7. Cunovic, S., et al., Appl. Phys. Lett., 90, 121112, 2007
8. Gahl, C., et al., Nature Photonics, 2, 165-169, 2008
9. Maltezopoulos, T, et al., N. Jour. Phys., 10, 033026, 2008
10. Azima, A., et al., submitted for publication, 2009
11. Mitzner, R., et al., Optics Express, 16, 19909-19919, 2008
12. Meyer, M., et al., Phys. Rev. Lett., 101, 193002, 2008
13. Barty A., et al., Nature Photonics, 2, 415-419, 2008
14. Neutze, R., et al., Nature, 406, 752-757, 2000
15. Chapman, H.N., et al., Nature Physics, 2, 839-843, 2006
16. Bogan, M.J., et al., Nano Lett., 8, 310-316, 2008
17. Moshammer, R., et al., Phys. Rev. Lett., 98, 203001, 2007
18. Wabnitz, H., et al., Nature, 420, 482-485, 2002
19. Sorokin, A.A., et al., Phys. Rev. Lett., 99, 213002, 2007
20. Zastrau, U., et al., Phys. Rev. E, 78, 066406, 2008
21. Epp, S.W., et al., Phys. Rev. Lett., 98, 183001, 2007
22. Pedersen, H.B., et al., Phys. Rev. Lett., 98, 223202, 2007
23. Saldin, E.L., et al.: *The Physics of Free Electron Lasers*, Springer-Verlag, Berlin, 1999
24. Saldin, E.L., et al., Opt. Commun., 281, 1179-1188, 2008
25. Singer, A., et al., Phys. Rev. Lett., 101, 254801, 2008
26. Altarelli, M., et al. (Edts.), *The European X-ray Free-Electron Laser – Technical Design Report*, DESY 2006-097, DESY, Hamburg, 2006; see also http://www.xfel.eu
27. Lambert, G., et al., Nature Phys., 4, 296-300, 2008
28. Tiedtke, K., et al., N. Jour. Phys., accepted for publication, 2009
29. Martins, M., et al., Rev. Sci. Instr., 77, 115108, 2006
30. Gensch, M., et al., Infrared Phys. Technol., 51, 423-425, 2008
31. see http://lcls.slac.stanford.edu
32. see http://www-xfel.spring8.or.jp

Design Study of Compact Thomson X-Ray Sources for Material and Life Sciences Applications

E. G. Bessonov, M. V. Gorbunkov, P. V. Kostryukov, Yu. Ya. Maslova, V. G. Tunkin, A. A. Postnov, A. A. Mikhailichenko, V. I. Shvedunov, B. S. Ishkhanov, A. V. Vinogradov

P.N. Lebedev Physical Institute of the Russian Academy of Sciences, 53, Leninskiy Prospekt, Moscow, 119991, Russia

Abstract. X-ray generators utilizing Thomson scattering fill in the gap existing between conventional and synchrotron based X-ray sources. They are expected to be more intensive than X-ray tubes and more compact, accessible and cheap than synchrotrons. In this work two operation modes of Thomson X-ray source: quasi CW (QCW) and pulsed are considered for material sciences and medical applications being implemented now at synchrotron radiation (SR) facilities. The system contains a ~50 MeV linac and a few picoseconds, few hundred W average power laser. Thomson X-ray source can provide up to $5 \cdot 10^{11}$ photons in a millisecond flash and average flux 10^{12}–10^{13} phot/sec. To achieve these parameters with existing optical and accelerator technology the system must also contain a ring for storage of e-bunches for 10^3–10^5 revolutions and an optical circulator for storage of laser pulses for 10^2 passes. As possible applications of the considered X-ray source XAFS spectroscopy, small animal angiography and human noninvasive coronary angiography are discussed.

1 Introduction

An infrared photon scattered by relativistic, ~50 MeV, electron in a head-on collision changes its energy from $\hbar\omega_L$~1 eV to $\hbar\omega$~20–30 keV depending on the scattering angle. This process of relativistic Thomson scattering can be used for efficient generation of X-rays if the reliable synchronization of dense electron bunches with high power laser pulses in space-time domain is provided. The radiation sources based on this principle can occupy the position between commercial tube based X-ray sources and conventional synchrotrons. The related technology as well as application field for such laser-electron X-ray generators (LEXG) was extensively studied during last 10–15 years [1-12]. In the next section we will give a brief overview of existing X-ray generators to find out whether there is a niche for LEXG among present and developing X-ray facilities. In section 3 the temporal structure, X-ray flux, and main technical parameters of QCW LEXG are determined. In section 4 the application of QCW LEXG in XAFS spectroscopy and 3D imaging are discussed. It is shown that 3D imaging technology can benefit considerably if LEXG is used as an X-ray source for microtomography. New equipment will combine high density resolution of medical tomographs with high spatial resolution of microtomographs based on microfocus sources. Section 5 deals with pulse repetition mode of LEXG which is necessary for study of live objects. Possible applications of pulsed LEXG are cardiovascular diagnostics and the study of very small vessels and capillary. These problems pose the highest requirements on the generated X-ray flux.

2 Overview and motivation

X-ray generation is among the key and frontier technologies in such important fields as medical imaging, material studies, drug development and public security. Conventional laboratory sources and user X-ray facilities utilize radiation of electrons in metals or magnetic fields. Few tens keV X-ray photons are produced in ~100 keV X-ray tubes or in bending magnets and insertion devices installed at one – few GeV electron accelerators. Further we will briefly give and compare some characteristics of tube and accelerator based X-ray generators.

(a) X-ray tubes are inexpensive, compact and robust devices which convert electric to X-ray power with the efficiency 0.1–1%. Their radiation is nearly isotropic but the main disadvantage is that it is not monochromatic and gives poor possibilities to manipulate with the spectrum. However the advanced apparatus for absorption spectroscopy has been developed. An example of commercially available XAFS-spectrometer [13] capable to analyze the compounds of 60 elements is presented in Table 1.

Table 1

# elements	Spectral resolution	Source type	Power	Sizes	Time of 400 points scan
60	$3 \cdot 10^{-4}$	Rotating anode	18 kW	2x3 m^2	12 hours

The main problem is that achieving spectral resolution and elemental sensitivity needed for a wide class of applications often require too long acquisition times due to low X-ray photon flux.

(b) Conventional accelerator based sources – dedicated synchrotrons offer at least 4 orders of magnitude higher average X-ray flux in the beam and allow spectroscopic analyses of diluted samples and investigation of fast processes in live objects. The spectra requiring hours of acquisition time at the X-ray tube based apparatus can be taken in minutes at the synchrotrons. Table 2 demonstrates statistics of 2 largest and 2 middle size synchrotrons.

Table 2

Facility	ESRF	SPring-8	Canadian Light Source	Australian Synchrotron Storage Ring
Energy GeV	6.4	8	2.9	3
Circumference, m	844	1436	147.4	216
Current, mA	200	99.8	250	200
Cost, $\$10^6$	> 1000	> 1000	194	170

There are 70 synchrotrons in operation or construction stage in about 30 countries. However their cost, sizes, power supply exceed corresponding values of advanced X-ray tubes by a factor of 200–500.

(c) Higher brightness and femtosecond time structure are provided by free electron lasers (FEL). The experiments at the first X-ray FEL FLASH (DESY, see Table 3) designed for the wavelength range of 50–6.5 nm started in 2005 [14]. Construction of the first user FEL facility, XFEL begins also in Hamburg in 2008 [15]. It will combine extreme peak and very high average brilliance in hard X-rays. Six more European countries have projects for building FELs in the nearest future. Compared with conventional synchrotrons these facilities present the next step in the technology of accelerators and also their sizes and cost (see Table 3).

(d) Meanwhile the European Union makes efforts to further develop FEL-based research infrastructure [15]. The three year project has been supported as a Design Study – the European FEL (EUROFEL) and completed in 2007. EUROFEL is focused on the basic technologies for the next generation FELs: electron injectors, synchronization, high duty-cycle and CW superconducting accelerators, beam dynamics, seeding and harmonic generation [16, 17]. The goal is high-repetition-rate FELs in soft and hard X-rays (see also [18]). The development of FELs is motivated by one of the fundamental and eternal scientific challenges: to understand and control the structure and properties of materials and living matter.

Table 3

Facility	FLASH	XFEL
X-ray range	50–6.5 nm	0.1 nm
Electron energy	730 MeV	17.5 GeV
Overall length	0.33 km	3 km
Bunch charge	1 nC	1nC
Bunch duration	400 fs	88fs
# of bunches per beam pulse	7200	3000
Repetition rate	10 Hz	10 Hz
Emittance	2 mm mrad	1.4 mm mrad
Beam pulse current	5 mA or 2.5 kA in a bunch	5 mA or 10 kA in bunch
Cost		1081.6 m€

At the same time much of fundamental and applied research is routinely made now not only at large facilities but also in universities, academic and industrial labs, medical schools etc. However a middle size university presently cannot afford even conventional synchrotron. The problem is recognized by the European Commission that approved the COST Action MP0601 "Short Wavelength Laboratory Sources" [19], claiming "…Although expansion in application areas is due largely to modern synchrotron sources, many applications will not become widespread, and therefore routinely available as analytical tools, if they are confined to synchrotrons. This is because synchrotrons require enormous capital and infrastructure costs and are often, of necessity, national or international facilities. This seriously limits their scope for applications in research and analysis, in both academia and industry. How many universities, research institutes or even industrial laboratories would have electron microscopes if electron sources cost 100 M€ or more? Hence there is a need to develop bright but small and (relatively) cheap x-ray sources, not to replace synchrotrons but to complement them."

LEXG based on relativistic Thomson scattering have the potential to become such a new X-ray source and fill in the gap between X-ray tubes and synchrotrons for a wide variety of fundamental research and practical applications. In the following sections we will consider the LEXG schemes meeting requirements of material composition analysis, high density resolution microtomography, small animal angiography and human noninvasive coronary angiography.

3 Quasi CW mode

Consider a simplified generator scheme shown in Figure 1. A laser and a source of relativistic electrons (linac) synchronically direct laser pulses and e-bunches of approximately equal duration 2–10 ps into an interaction chamber IC. To enhance the number of collisions the laser pulses are trapped for $n_C \approx 100$ revolutions in an optical circulator and electron bunches are stored in a ring of the same perimeter ≈ 3 m for $n_S = 10^3$–10^5 revolutions. The values n_C and n_S are limited by radiation damage of optical materials and the lifetime of the e-bunch emittance in a storage ring. Thus linac injects a bunch into a ring with the rate ν_e and laser with the same rate $\nu_L = \nu_e$ emits a train of $n_L = n_S/n_C$ pulses into a circulator. Each head-on collision of a laser pulse (photon energy $\hbar\omega_L = 1.16$ eV) and e-bunch (electron energy $E_e = 43$ MeV) results in generation of an X-ray burst with a maximum photon energy $\hbar\omega \approx 33$keV. A number of X– ray photons generated in one collision equals:

$$N = N_L N_e \frac{\sigma_T}{s_e + s_L}, \tag{1}$$

where N_e, N_L are the total numbers of electrons and photons, $\sigma_T = 6.6 \cdot 10^{-25}$ cm^2, $s_e = 2\pi\sigma_e^2$, $s_L = 2\pi\sigma_L^2$; σ_e, σ_L are transverse sizes of electron and laser beams, $\sigma_e = \sqrt{\varepsilon\beta}$, where ε is the emittance and β is beta function of electron bunch in the interaction point. The most part of photons (1) is contained in a narrow angle ~γ^{-1}, where $\gamma = E_e/mc^2 \approx 0.01$.

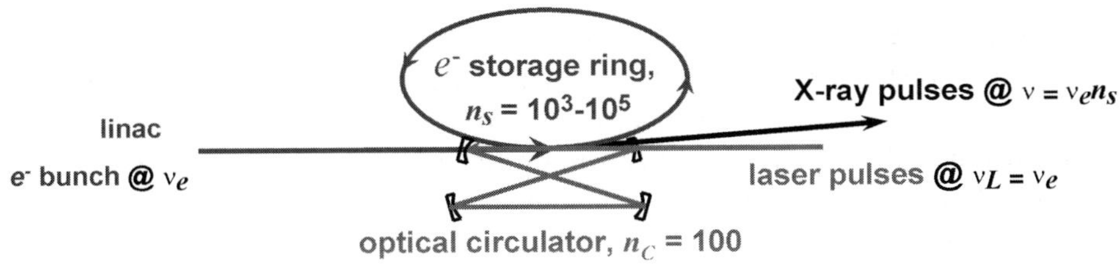

Fig. 1 A principal scheme providing efficient arrangement for e-bunch – laser pulse collisions.

For further estimations we will take the following parameters of existing and being projected low emittance linacs: $N_e = 6 \cdot 10^9$ (1 nC bunch charge) ε_n = 1 mm·mrad – normalized emittance, $\sigma_e = \sigma_L = 20$ μm. Then by the reasons of damage of circulator mirrors and reliability and commercial availability of the laser we restrict the design by laser micropulse energy 10 mJ and laser average power $P_L \leq 0.5$ kW. Anyway the X-ray beam consists of picosecond bursts radiated in the collisions of 1 nC bunches with 10 mJ (or less) laser pulses. The total amount of X-ray photons emitted in one “building brick” collision for IR ($\hbar\omega_L = 1.16$ eV) laser is:

$$N = 1.3\times10^{-20} N_L N_e = 7.8\times10^{-11} N_L = 5\times10^6. \qquad (2)$$

The goal of the design is to arrange these standard collisions in a compact, cost effective system so as to meet the requirements of specific application.

The X-ray flux **Φ** provided by the generator in Figure 1 equals to:

$$\Phi = N\nu_e n_s = N\nu_e n_L n_C, \quad P_L = E_L \nu_e n_L = E_L \nu_e \frac{n_S}{n_C}. \qquad (3)$$

The 3 m circulator (and ring) perimeter allows the product $\nu_e n_S$ up to 100 MHz. This is the value of linac frequency considered in the design of the next generation CW light sources involving warm and cryogenic accelerator structures [16, 18]. To achieve it the problems connected with radiation and electrical damage of cathodes and availability of photocathode UV lasers (see [11]) are to be solved. The storage ring in LEXG reduces the linac frequency down to 0.5–5 kHz (see further this section) and no problems with cathode and photocathode laser arise.

To find the design features resulting from limitations on the laser power $P_L = 0.5$ kW and laser micropulse energy $E_L \leq 10$mJ consider two relations following from (2) and (3):

$$\Phi = n_C N_e \frac{\sigma_T}{s_e + s_L} \frac{P_L}{\hbar\omega_L} = 2.5\times10^{13} \frac{phot}{\sec}, \quad \nu_e n_S = \frac{P_L}{E_L} n_C \geq 5 \text{ MHz}. \qquad (4)$$

Firstly, eq. (4) gives maximum X-ray photon flux in a QCW mode which is proportional to average laser power. Secondly it determines the lower limit of the product $\nu_e n_S$. On the other hand ν_e and n_S by technical reasons cannot be too large to avoid overloading of the photocathode and to achieve the system compactness and reasonable cost. For $E_L = 10$ mJ the realistic compromise is:

$$\nu_e = 500 \text{ Hz}, \quad n_S = 10^4. \qquad (5)$$

Note that for fixed amount of laser pulses per second $\nu_L n_L = \nu_e n_S / n_C$ some freedom remains for the time structure of laser beam, i.e. for the choice of ν_L and n_L. Various approaches to laser design to produce trained laser beams for circulators of laser electron X-ray generators are considered in [6, 11].

So formulas (4) and (5) and other relations of this section determine the output flux and main parameters of QCW LEXG. In the next section we will briefly discuss possible applications.

4 Applications of QCW LEXG

As was shown in the previous section the LEXG is expected to provide a photon flux ~$2\cdot10^{13}\text{sec}^{-1}$ that is sufficient for applications in material sciences. For example a standard method to determine material chemical composition and structure is X-ray Absorption Fine Structure Spectroscopy (XAFS) [20]. One XAFS spectrum typically contains ~600 measurements in narrow spectral intervals $\delta\lambda/\lambda \approx 10^{-3}$–$3\cdot10^{-4}$ and for practical reasons must take not more than 1–2 hours. From this the photon flux before monochromator can be estimated as 10^{12} sec^{-1} [20, 21] that is well below the value given by (4).

Another possible field for application of LEXG is X-ray computer tomography (CT) which is widely used now for 3D imaging in medicine, industry, applied research etc. Normally for 3D image reconstruction 100–200 projections are required. Medical CTs provide very high accuracy ~few tenths of 1 percent of the contrast measurement that is of principal importance for diagnostics of many human diseases. The spatial resolution provided by medical CT reaches ~100 μm whereas standard value is four–five times larger. In practice spatial resolution is a compromise of several factors: source and pixel sizes, source power, exposure time, absorbed radiation dose etc. However some kinds of applications require higher spatial resolution than that provided by medical CTs (see the next section). The problem is overcome by *microtomographs* (micro CTs), that are based on combination of table-top microfocus tubes and CCD detectors [22]. Micro CT is a compact device designed for imaging of ~1 mm–5 cm objects. Due to the reduced source size its spatial resolution reaches ~ few microns. In fact it is reached at the expense of accuracy of contrast measurements that is limited by relatively low average photon flux produced by microfocus sources – air-cooled < 50W power X-ray tubes with a 1–10 μm focal spot. So the density resolution of micro CT is now by far lower than that of medical CTs. Therefore it was always a challenge for micro CT to visualize most essential for biomedical research low contrast tissues (vessels, membranes, blood, nerves, muscles, brains etc.). This can be achieved by illuminating every voxel with a sufficient enough amount of X-ray photons to obtain high signal-to-noise ration and improve density resolution. Hence the X-ray tube has to be replaced by another also microfocus but more powerful X-ray source.

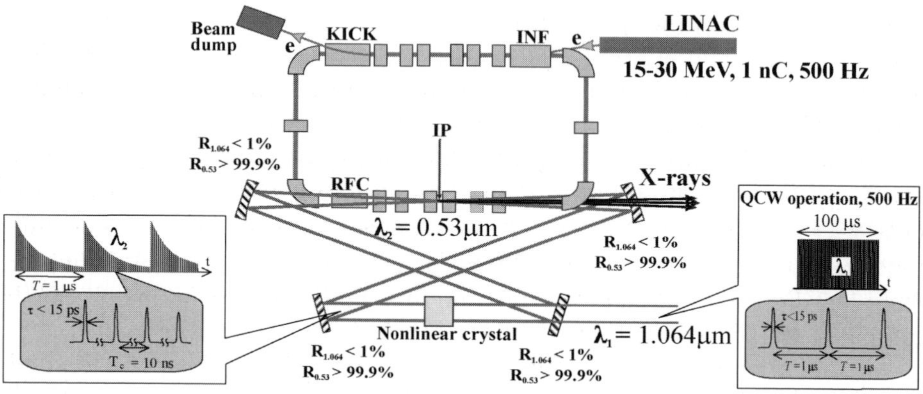

Fig. 2 QCW Laser-electron X-ray Generator for micro-CT or XAFS applications: circulator based on second harmonic generation in a high-Q cavity.

Taking this into account, the requirements for the new source are the following. To keep the acquisition time of one projection in the range of 0.1–10 sec (as in standard micro CT) and increase the accuracy of the absorption measurements the source must provide 100–1000 times more photons delivered and registered in every single pixel of the detector. Then the number of photons incident on the object and needed for 0.1% accuracy of contrast measurement can be estimated as 10^{12} per one projection (10^5 pixels). The flux needed to collect one tomographic dataset (200 frames) in one hour is $\sim 2\cdot 10^{11}\,\text{sec}^{-1}$ which is also in reach of LEXG as estimated in the previous section. Figure 2 shows the principle scheme of such X-ray generator with an optical circulator based on frequency doubling of a pumping IR laser radiation (see [11]).

5 Pulsed operation mode for life sciences. Coronary angiography

For some medical and industrial research and applications the pulsed X-ray beam is more advantageous. Average X-ray power is determined by the same equation (4), but the linac rate ν_e depends on the specific application. For example to observe breath, heart beat and other physiological motion, the exposure 0.1–1 ms and the rate of frames $\nu = 30$ Hz are needed. In LEXG this is achieved by collision of pulse trains. The resulting 30 Hz X-ray signal consists of trains of n_S micropulses separated by 10 ns. If consider life sciences than $\nu_e = 30$ Hz that limits laser power (see (3)) and hence the X-ray output:

$$P_L = E_L \frac{\nu_e n_S}{n_C} \approx 300W, \quad \Phi = \nu_e N_e \frac{\sigma_T}{s_e + s_L} \frac{P_L}{\hbar\omega_L} = 1.5\times 10^{13} \frac{phot}{\sec}. \quad (6)$$

As optimistic values $E_L = 10$ mJ and $n_S = 10^5$ have been used in estimate (6) it gives the maximum X-ray flux that can be reached in the pulsed repetition mode. Further we will discuss two possible applications of such a source in angiography that are currently made at synchrotron beamlines.

A number of new medical technologies for imaging and therapy for which the X-ray flux of conventional sources is not sufficient are being developed during 1–2 decades on synchrotron radiation sources [23, 24]. Partly with the hope that if successful they may be implemented on a new intense, compact and available source appropriate for application in clinics. This could be an important and promising field for expansion of LEXG.

One of these new technologies is aimed at selective coronary angiography which is the most widespread tool for diagnostics of cardiac diseases. In this invasive method X-ray images of arteries expected for affection or pathology are produced at arteries catheterization for delivering a contrast agent directly to the investigated area. Millions of patients a year undergo this investigation. The procedure is not absolutely safe for patients due to (small) probability of arteries damage and sometimes negative effects caused by contrast agent, and for medical personnel acquiring radiation dose. So the screening is not always recommended though the diagnostics is very desirable at the earliest stages of ischemic disease. Therefore several noninvasive methods of coronary arteries imaging are being developed [25]. However presently selective coronary angiography remains the "gold standard" for diagnostics of most heart diseases as other methods do not provide comparable image quality.

A new noninvasive method utilizing synchrotron radiation was suggested to replace selective coronary angiography. It was suggested that the high power of a synchrotron beam allows improving image quality by digital subtracting of images obtained on two sides of absorption K-edge of iodine contrast agent [26]. This substantially reduces the concentration of the contrast agent and gives possibility to inject it intravenously (noninvasively). After ~400 tests at synchrotron DORIS that showed very good image quality and acceptance of the new noninvasive procedure by patients [27]. Then a project of 1.6 GeV storage ring was published as a step towards facilities for noninvasive coronary angiography to be installed in medical centers and clinics [28]. However further progress in this field came slower. Probably the reason is that everyday medical exploitation of a facility with GeV-scale storage ring and 42.2x37.6 m sizes is not economically justified.

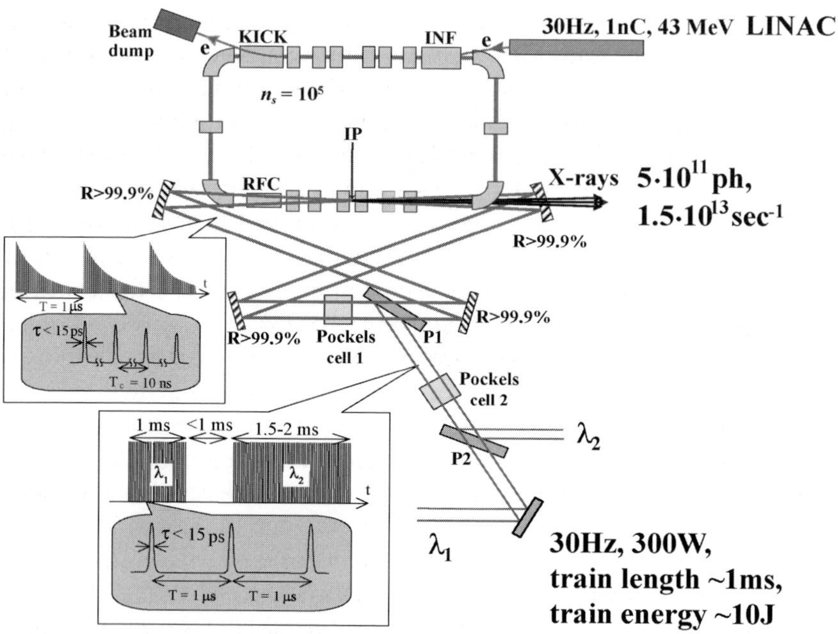

Fig. 3 A scheme of LEXG designed for human K-edge subtraction imaging angiography.

A conceptual design of LEXG for noninvasive coronary angiography based on subtraction imaging was considered in [29]. The design contained a ~50 MeV linac, 1 m radius storage ring and a picosecond laser. The storage ring increases average current and makes easier the requirements to photo-injector. The usage of trained (~10^2–10^4 micropulses) laser beams for Thomson scattering allows to decrease X-ray photon flux employed for video screening in human angiography (~10^{13}) and also to reduce the radiation dose on a patient. Though the obtained value is in agreement with estimations of other authors [3] it is difficult to say now without the experiment whether this flux will be sufficient for practical angiography. An updated LEXG scheme for noninvasive coronary angiography is given in Figure 3. Two lasers are needed for taking images on both sides of iodine K-edge.

Synchrotron radiation due to high X-ray photon flux is being applied to animal studies for new treatments in medical practice. It enables an ***ex vivo*** and ***in vivo*** research with high spatial and temporal resolution. Special synchrotron radiation microangiography (SRM) systems have been developed for visualization of vessels and cardiac pathologies in small animals: mice, rats, dogs and rabbits [30-32]. The latter, at SPring-8, has the following parameters: field of view 7x7 mm or 4.5x4.5 mm; pixel size 7x7 μm or 4.5x4.5 μm; spatial resolution 6 μm; minimum detectable vessel diameter 20 μm; shortest exposure time 2 ms; rate 30 frames/sec. Note that conventional angiography does not allow to visualize vessels of less than 200 μm diameter. This SRM imaging system was used to evaluate morphometric and physiological evaluation of coronary collateral microvessels development at stem cells transplantation in rats. The 6 μm resolution images were firstly obtained on beating hearts that was not previously available for conventional angiography as well as for micro CT. For example coronary collateral growth was earlier studied by micro CT but with spatial resolution three times larger – 18 μm, and as was discussed in section 4 this could be performed only at postmortem examination [33] with long enough exposures. Now the efforts are being done to improve the system in two respects: (a) to reduce the exposure time down to 0.1 ms in order to obtain blur-free images and completely use the resolving potential of the detector (b) to switch from ex-vivo to in-vivo experiments with individual animals. For this purpose the photon flux will be increased with X-ray optical system.

Let us discuss the design of LEXG system for microangiography meeting the above mentioned requirements. As SRM utilize radiation above the iodine K-edge and does not use the subtraction of images only one laser of two shown in Figure 3 is needed. Then the estimation of the number of photons to obtain one frame gives $\sim 10^{10}$ which leads to average photon flux $\sim 3\cdot 10^{11}$ sec^{-1}. Again we see that this type of research is within the reach of LEXG. Special measures can be taken to provide 0.1 ms exposure and reduce the source size below 10 microns.

6 Discussion and summary

Main characteristics of two Thomson X-ray sources (LEXG) designed for material science applications (quasi CW) and imaging of live objects (pulsed repetitive) are presented in Table 4. The first four rows characterize the output beam of 33 keV photons. The last three concern the IR pumping laser.

Table 4

Operation mode	quasi CW	Pulsed repetitive
Average flux, sec^{-1}	$2.5\cdot 10^{13}$	$1.5\cdot 10^{13}$
Rate of flashes = linac operation rate, Hz	500	30
X-ray pulses in a flash	10^4	10^5
# of X-ray photons in a flash	$5\cdot 10^{10}$	$5\cdot 10^{11}$
Laser pulse energy, mJ	10	10
Laser average power, W	500	300
Laser pulses in a train	10^2	10^3

The schemes of X-ray sources considered in this paper contain an optical circulator and a compact storage ring. Both are necessary to make LEXG competitive with conventional tube based sources. The optical circulator with $n_C = 10^2$ revolutions was described in the previous paper [11]. Figure 4 displays the scheme of a compact storage ring.

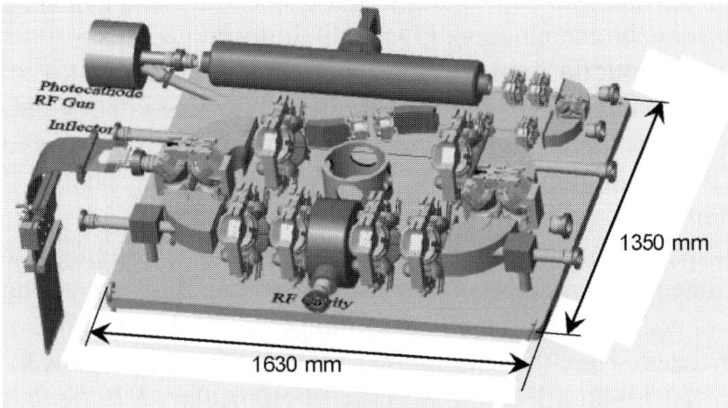

Fig. 4 Compact storage ring for Thomson X-ray generator.

The main e-bunch parameters correspond to those used in sections 3–5: charge is 1 nC, diameter σ_e = 14 μm, emittance ε = 1 mm·mrad, β = 2 cm, emittance lifetime > 1 ms. Calculations have been made with the code BEAMOPTICS created by H. Wiedeman.

In conclusion the designs of two types of laser-electron X-ray generators based on Thomson scattering are presented for control of materials and imaging of live objects. To be realistic the pulse energy of IR laser was restricted by ≤ 10 mJ and average power ≤ 0.5 kW. Other requirements of this study to lasers, accelerators and their synchronization are also close to current technology and well correspond to the level stated in [17]. The realization of compact Thomson scattering X-ray source will make many modern methods developed on synchrotron radiation facilities available to a wide community of material and medical scientists. This perspective and technical feasibility attract more and more researches to the field [34].

The work was supported by the Program "Laser systems" of Russian Academy of Sciences and RFBR Grant 08-08-00108-a.

References

1. Crane, J. K., LeSage, G. P., Ditmire, T. et al.: '600 eV Falcon-Linac Thomson X-ray Source', UCRL-ID-141613, December 15, 2000
2. Carroll, F. E.: 'Tunable Monochromatic X Rays: A New Paradigm in Medicine', *AJR* 179, 583-590, 2002
3. Loewen, R. J.: 'A Compact Light Source: Design and Technical Feasibility Study of a Laser-Electron Storage Ring X-Ray Source', SLAC-R-632, June 2003 (Ph.D.thesis, Stanford University, Stanford CA).
4. Hartemann, F. V., Brown, W. J., Anderson, S. G. et al.: 'Compton Scattering and its Applications: The PLEIADES Femtosecond X-Ray Source at LLNL', Quantum Aspects of Beam Physics 2003, edited by Pisin Chen & Kevin Reil (Stanford Linear Accelerator Center, USA), Proceedings of the Joint 28th ICFA Advanced Beam Dynamics and Advanced & Novel Accelerators Workshop, Hiroshima, Japan, 7 - 11 January 2003, 49-62
5. Agafonov, A., Androsov, V., Botman, J. I. M. et al.: 'Status of Kharkov x-ray generator NESTOR', *Proc. SPIE*, 5917, 97-104, 2005
6. Bessonov, E.G., Vinogradov, A. V., Gorbunkov, M. V. et al.: 'Laser electron-beam X-ray source for medical applications', *Physics – Uspekhi*, 46, 872-876, 2003; Gorbunkov, M.V., Maslova, Yu.Ya., Shabalin, Yu.V., Vinogradov, A.V.: 'Laser Physics Research Relevant to Laser-Electron X-Ray Generator', Proc. of the 10th International Conference on X-Ray Lasers, Aug. 20-25, 2006, Springer Proc. in Physics, P.V. Nickles, K.A. Janulewicz (Eds.), 115, 619-629, 2006
7. Dobashi, K., Fukasawa, A., Uesaka, M. et al.: 'Design of Compact Monochromatic Tunable Hard X-Ray Source Based on X-band Linac', Japanese Journal of Applied Physics, 44, No.4A , 1999-2005, 2005
8. Yanagida, T., Nakajyo, T., Ito, S., Sakai, F.: 'Development of high-brightness hard x-ray source by Laser-Compton scattering', Proc. SPIE, 5918, 231-238, 2005
9. Sakaue, K., Washio, M., Araki, S., et al.: 'Development of pulse laser super-cavity for compact high flux x-ray sources', Proc. of EPAC 2006, Edinburgh, Scotland, 3155-3157, 2006.
10. Kaertner, F. X., Graves, W. S., Moncton, D. E., Ilday, F.O.: 'Compact, High-Flux, Short-Pulse X-ray Source', US Patent 2006/0251217 A1, Pub. Date: Nov.9, 2006.
11. Bessonov, E. G., Gorbunkov, M. V., Maslova, Yu. Ya. et al.: 'Relativistic Thomson scattering in compact linacs and storage rings: a route to quasi-monochromatic tunable laboratory-scale X-ray sources', *Proc. SPIE*, 6702, OE-1 – OE9, 2007
12. Sakamoto, F., Uesaka, M., Yamamoto, T. et al.: 'Beam Generation and Acceleration Experiments of X-Band LINAC and Monochromatic keV X-Ray Source of the University of Tokyo', Proceedings of PAC07, 2784-2786, Albuquerque, New Mexico, USA, 2007
13. Taguchi, T., Harada, J., Tohji, K., Shinoda, K.: 'An Innovated Laboratory XAFS Apparatus', Advances in X-ray Analysis, 45, 397-401, 2002.
14. http://flash.desy.de/

15. Tschentscher, Th., Altarelli, M., Brinkmann, R. et al.: 'The European X-ray Free-Electron Laser Facility: A New Infrastructure for Research Using Ultrashort, Coherent X-ray Pulses of Extreme Brightness', Synchrotron Radiation News, 19, Issue 6, 13-19, 2006
16. Feldhaus, J., Ferianis, M., Ferrario, M. et al.: 'EUROFEL: Europe Tackles Key Issues of Free Electron Lasers', *Synchrotron Radiation News*, 21, Issue 2, 28-36, 2008
17. http://www.eurofel.org
18. Belkacem, A., Byrd, J. M., Corlett, J. et al.: 'Design Studies for a High-Repetition-Rate FEL Facility at LBNL', *Synchrotron Radiation News*, 20, Issue 6, 20 – 27, 2007
19. http://short-wavelength-sources.net/html/cost_mp0601.html
20. *X-Ray Absorption: Principles, Applications, Techniques of EXAFS, SEXAFS, and XANES*, edited by D.C. Konningsberger and R. Prins, Wiley, New York, 1988
21. Fetisov, G.V.: *Synchrotron Radiation. Methods of Structural Studies* (in Russian), Fizmatlit, Moscow, 2007
22. http://www.skyscan.be/home.htm
23. Suortti, P., Thomlinson, W.: 'Medical applications of synchrotron radiation', *Physics in Medicine and Biology*, 48, No 13, R1-R35(1), 2003
24. Baruchel, J., Bleuet, P., Bohic, S. et al.: 'State of the Art and Perspectives of Biomedical Imaging at the ESRF', *Synchrotron Radiation News*, 21, Issue 4, 30-41, 2008
25. Greenland, P.: 'Who Is a Candidate for Noninvasive Coronary Angiography?', *Annals of Internal Medicine*, 145, Issue 6, 466-467, 2006
26. Rubinshtein, E., Hofstadter, R., Zeman, H.D. et al.: 'Transvenous coronary angiography in humans using synchrotron radiation', *PNAS*, 83 no. 24, 9724-9728, 1986
27. Dill, T., Dix, W.-R., Hamm, C.W. et al.: 'Intravenous coronary angiography with synchrotron radiation', *Eur. J. Phys.*, 19, 499-511, 1998
28. Brinker, F., Febel, A., Hemmie, G. et al.: 'ARI - A STORAGE RING FOR NON-INVASIVE CORONARY ANGIOGRAPHY', *Proc. EPAC* 2000, 610-612, Vienna, Austria, 2000
29. Gorbunkov, M.V., Tunkin, V.G., Bessonov, E.G. et al.: 'Proposal of a Compact Repetitive Dichromatic X-ray Generator with Millisecond Duty Cycle for Medical Applications', *Proc. SPIE*, 5919, OU1-OU6, 2005; Artyukov, I.A., Bessonov, E.G., Vinogradov, A.V. et al.: 'Laser Electron Generator of X-ray Radiation', Proc. of the 10th International Conference on X-Ray Lasers, Aug. 20-25, 2006, Springer Proc. in Physics, P.V. Nickles, K.A. Janulewicz (Eds.), 115, 631-642, 2006
30. Mori, H., Hyodo, K., Tanaka, E. et al.: 'Small vessel radiography in situ with moniochromatic synchrotron radiation', *Radiology*, 201, 173-177, 1996
31. Iwasaki, H., Fukushima, K., Kawamoto, A. et al.: 'Synchrotron radiation coronary microangiography for Morphometric and Physiological Evaluation of Myocardial Neovuscularization Induced by Endothelial Progenitor Cell Transplantation', *Arteriosclerosis, Thrombosis, and Vascular Biology*, 27, 1326-1333, 2007

32. Kidoguchi, K., Tamaki, M., Mizobe, T. et al.: ' In vivo X-ray angiography in the mouse brain using synchrotron radiation', *Stroke*, 37, 1856-1861, 2006
33. Toyota, E., Waritier, D.C., Brock, T. et al.: 'Vascular endothelial growth factor is required for coronary collateral growth in the rat', *Circulation,* 112, 2108-2113, 2005
34. http://www.ca.infn.it/alghero2008/, Workshop 'Compton Sources for X/gamma Rays: Physics and Applications', Alghero, 7-12 September, 2008

An Attempt to Generate an Inner-Shell Photo-Ionisation Pumped X-Ray Laser Using the ASTRA Laser at RAL

T. W. J. Dzelzainis[1], M. Streeter[2], F. Y. Khattak[3], R. Ferrari[1], C. L. S. Lewis[1], D. Riley[1], R. Tommasini[4], and G. Gregori[2,5]

[1]Centre for Plasma Physics, Queen's University of Belfast, Belfast, UK
[2]Central Laser Facility, Rutherford Appleton Laboratory, Didcot, UK
[3]Kohat University of Science and Technology, Kohat, Pakistan
[4]Lawrence Livermore National Laboratory, USA
[5]Clarendon Laboratory, University of Oxford, Oxford, UK

Abstract. An attempt was made to use the ASTRA laser system at the Rutherford Appleton Laboratory to generate an X-ray Laser (XRL) via an inner-shell photo-ionisation pumping scheme, in which a laser produced plasma acts as an x-ray flash-lamp to excite electrons from the inner-shell of the lasant material and hence generate a population inversion. The rise time of the x-ray source must be extremely fast in order that lasing can dominate over other relaxation routes such as Auger decay. The lasing line chosen was the K-alpha line of carbon at 44.6 A. The targets consisted of a volume of low density carbon foam (ranging between 3-10 mg/cc) over-coated with 500 nm of Al_2O_3 as a high pass x-ray filter, and an Au flash-coating between 70-100 nm thick on the surface as the x-ray pump. Evidence of lasing was seen, but the results were not conclusive. Detailed simulations exploring the feasibility of such an XRL scheme were carried out to optimise the pump design requirements. During the experiment the ASTRA laser delivered 0.6 J in ~60 fs pulses which was only marginal in terms of achieving significant gain according to our simulations. With the advent of new generation laser systems with shorter pulse duration and higher energies, we should then be able to access significant gain within this scheme.

1 Introduction

The inner-shell photo-ionisation (ISPI) scheme, first proposed by Duguay and Rentzepis in 1967[1] is, schematically, a very simple method of achieving a population inversion across an energy gap of X-ray wavelength. An x-ray flash-lamp providing a pulse with photon energy just above an absorption edge of a material causes the ejection of inner-shell electrons, automatically generating a population inversion. In practical application however, many problems exist preventing laser action from taking place. One of the biggest obstacles is overcoming the competing relaxation mechanism of Auger decay, which happens on a much shorter time-scale than radiative decay. Other problems include the quenching of the lower lasing state by ionisation of upper levels via electron impact or photo-ionisation by lower energy photons. The problems can be somewhat solved by the design of the x-ray flash lamp and the target. An x-ray flash-lamp with an extremely fast rise time is needed to beat Auger decay and low density targets can reduce the effect of electron impact ionisation. The problem of photo-ionisation by lower energy photons can be mitigated by placing a high pass filter between the generator layer and the target material. A study by Moon et al in 1998[2] provided a set of pump laser requirements for achieving lasing on the K_α line of carbon which were 1 J of energy in less then 50 fs in a line focus 1 cm × 10 μm. These laser parameters have become available recent years by the implementation of high energy Ti:Sapphire lasers such as ASTRA[3]. In the experiment that we have performed, the ASTRA system was capable of delivering 500 mJ to the target chamber in pulses of duration down to 60 fs.

2 Experimental Layout

A plan of the experiment is shown in Figure 1. The line focus was generated through the combination of an off-axis parabola (a) and a tilted spherical mirror (b), which gave a line focus of absolute length 12 mm with a travelling wave speed of 1±0.1 c. The principle diagnostic observing the X-ray laser output was a flat-field spectrometer (c). Diagnostics used to monitor the line focus uniformity and x-ray flash lamp spectrum was a differentially filtered 4-channel pinhole camera (d).

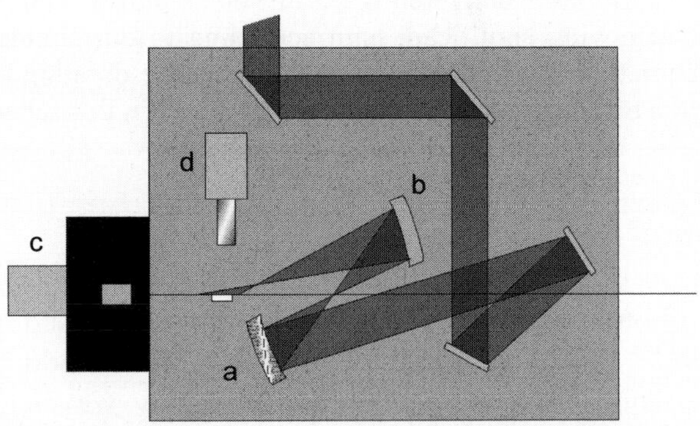

Fig. 1 Chamber layout for the experiment.

In order to minimise hot electron production via resonance absorption, the laser was incident on the target in an s-polarised state. The targets were fabricated by filling grooved substrates with low density foam (3, 5, 8 or 10 mg/cc) and then over-coating this with 500 nm of Al_2O_3 as a high pass filter. For the x-ray generator layer, a flash coating of Au, of thickness 50, 70 or 100 nm was deposited. In the majority of cases the targets also had a thin layer of smoked (foam-like) gold. Smoked Au forms a structured target, which it is believed to improve absorption of a laser pulse by modifying the surface properties. This is produced by depositing gold onto the target surface in the presence of an inert gas, causing the gold to form into nanometre scale clumps with large cavities between them thus giving a large surface area for laser absorption to take place and causing the laser to undergo multiple reflections on impact, rather than a single reflection as is the case for smooth targets. They have been shown to increase the x-ray output of laser plasma sources by up to 20 times[4].

3 Results and Discussion

3.1 X-ray Flash Lamp

A typical image from the differentially filtered pinhole is shown in Figure 2.

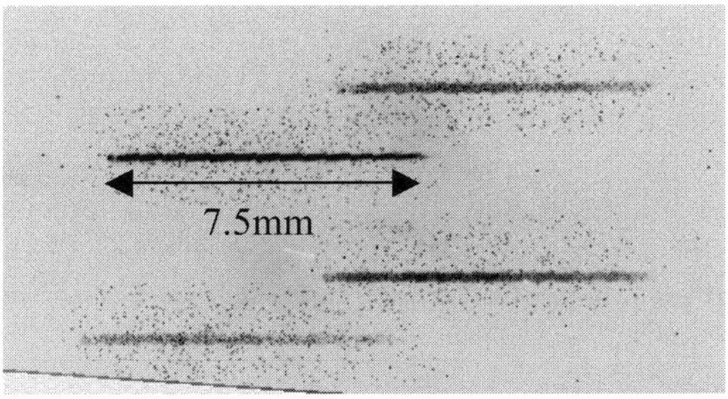

Fig. 2 Pin-hole-array image of linefocus. The filters on the pin-holes, from top to bottom are 1 μm Zn, 1 μm CH + 0.2 μm Al, 1 μm Ag and 6 μm Al.

The different filters were chosen to transmit various regions of the spectrum in the energy range we were expecting. The transmission coefficients for the filters are shown in Figure 3. The count numbers on each channel can be used to estimate a temperature for a blackbody emission spectrum. Equation 1[5] was used for this purpose.

$$\chi^2 = \sum_{i=1}^{4}\left(\frac{\left[F_i - A\int_0^{\infty}\frac{E^3}{e^{E/k_bT_e}-1}T_i(E)R(E)dE\right]}{\sigma_i}\right) \tag{1}$$

where i = 1 to 4 and represents the filter, F_i is the experimental data through filter i, E is photon energy, $T_i(E)$ is the transmission coefficient of filter i, R(E) is the detector response and σ_i, is the uncertainty in the experimental data. χ^2 was minimised by performing a scan over a large parameter space for A and T_e. A temperature of 400 eV is estimated for the x-ray flash lamp.

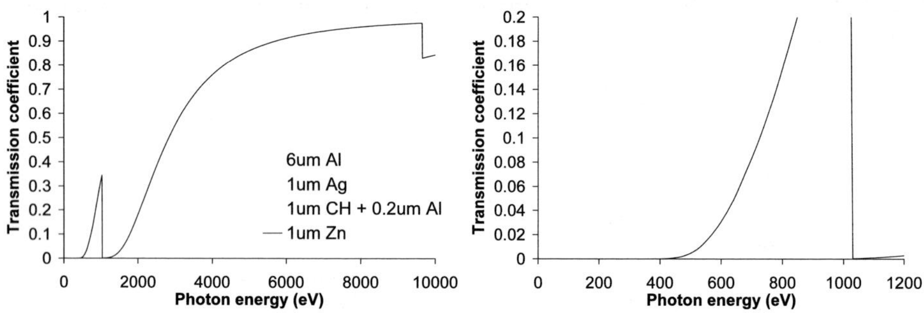

a) b)

Fig. 3 Graphs showing the transmission coefficient for the filters used on the pinhole camera (a) for the entire response range of the CCD and (b) in the region where the majority of the emission from the plasma was expected.

Figure 4 shows the estimated blackbody spectrum and the spectrum after passing through the Al_2O_3 filter. It can be seen that the flashlamp-filter combination is optimised spectrally, emitting very little below the C K-edge, but heavily above.

The intensity profile along the line focus was seen, in many shots, to exhibit strong spatial modulations which varied from shot to shot. These were thought to arise from modification to the foam during the coating process for the x-ray filter and generator. Previous work with the foam had shown that the coating process caused collapse of the foam close to the surface. This caused a slight increase in the density of the foam in the gain region and also would be able to produce spatial modulations as seen in the pinhole images. Both of these effects would have a negative impact on the gain conditions.

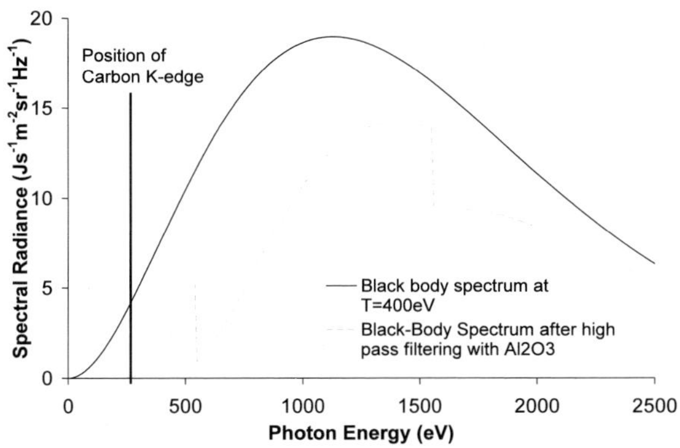

Fig. 4 The estimated blackbody spectrum and the spectrum after passing through the filter layer.

3.2 X-ray Laser Output.

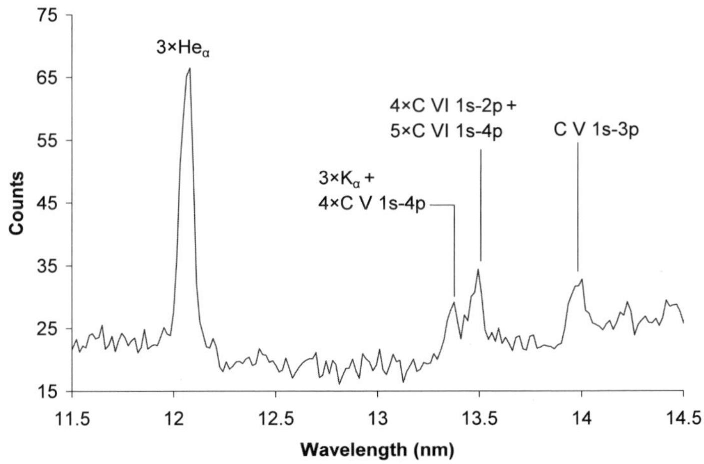

Fig. 5 Spectral lineout of an image taken from the flat-field spectrometer observing the axial output integrated over 5 mrads The position of the potential lasing line is indicated.

It was necessary to observe the 3rd order of the K_{α} line due to conflicts with other lines and reduced grating efficiency of the lower and higher orders. A typical spectrum of the axial output from the plasma column is shown in Figure 5. Just on the longer wavelength side of the potential lasing line, the 4th orders of the C VI 2p-1s and C V 3p-1s lines can be seen. The 3rd order K_{α} is almost perfectly overlapped with 4th order C V 1s-4p line, and so special care must be taken when measuring the contribution to the signal from K_α emission. In order to do this the ratio between the 3rd orders of the C V 1s-4p and 1s-3p line is taken. The intensity of the 4th order C V 1s-3p line can then be used to calculate a contribution to the lasing line from the C V 1s-4p transition. The C VI 1s-2p line, which is closer to the potential lasing line can not be used for this purpose as it is itself overlapped with 5th order C VI 1s-4p emission. Using this method, the K_α signal strength integrated over 5mrad was estimated at $\sim 10^5$ photons.

4 Conclusions

We have presented the results of an investigation of the viability of using the ASTRA laser to generate an X-ray laser on the K_α transition of carbon. The data show that the x-ray flash-lamp produced was generating pump radiation in the desired wavelength region, while the results regarding the demonstration of lasing action are still inconclusive due to the low gain. The rise time of the x-ray pulse, due to unfavourable laser conditions, combined with unavoidable inhomogeneities in the target surface which are present by virtue of the target material and structure are identified as problems which need to be overcome. It is thus concluded that further investigation into this scheme is warranted.

References

1. Duguay, M. A.: 'Some Approaches to Vacuum UV and X-Ray Lasers', *App. Opt. Lett.*, 10, 350-352, 1967
2. Moon, S. J.: 'Theoretical investigation of an ultrashort-pulse coherent x-ray source at 45Å', *Phys. Rev. A*, 57, 1391-1394, 1998
3 http://www.clf.rl.ac.uk/Facilities/AstraWeb/AstraTour1.htm
4. Khattak, F. Y.: 'Enhanced He-α emission from "smoked" Ti targets irradiated with 400nm, 45fs laser pulses', *Europhys. Lett.,* 72, 242-248, 2005
5. Jung, D.: 'Experimental characterisation of picosecond laser interaction with solid targets', *Phys. Rev. E*, 77, Art. No. 056403, 2008

Electron Self-Injection and Radiation in the Laser Plasma Accelerator

M. R. Islam[1], S. Cipiccia[1], B. Ersfeld[1], A. Reitsma[1], J. L. Martin[2], L. Silva[2], D. A. Jaroszynski[1]

[1] Department of Physics, University of Strathclyde, Glasgow, G4 0NG, UK
[2] Centro de Fisica dos Plasmas, Instituto Superior Tecnico, 1049-001 Lisboa, Portugal

Abstract. The origin of electron trapping at the back of a 'bubble', which is formed when the electrons are expelled from an under-dense plasma by an intense laser pulse, is investigated using a reduced model wakefield. The subsequent acceleration and transverse oscillation produces betatron radiation. Such electromagnetic radiation emitted by trapped electron in an ionic bubble is estimated using the Lienard-Wiechert potential.

1 Introduction

Laser-wakefield accelerator is a promising candidate for a compact electron accelerator. Very high acceleration gradients are produced by an ion bubble-like structure following laser pulse over a few centimeters in an under-dense plasma [1]. The intense laser pulse drives a large amplitude plasma density wake that produces huge electric field of the order of 100 GV/m which enable the acceleration of particles to relativistic energies (GV), which is three orders of magnitude higher than what can be achieved with the same length conventional RF accelerator. The intense laser pulse expels all the plasma electrons from its axis of propagation leaving a completely evacuate tens of micron sized ion cavity behind. This highly nonlinear regime is sometimes called the 'bubble' regime [2]. The expelled electrons then flow backwards, with respect to the bubble motion, along the surface of the bubble and may be trapped at the rear of the bubble. These trapped electrons are then accelerated by the electric field of the bubble while undergoing transverse motion called betatron oscillation. This transverse motion gives rise to copious radiation in the forward direction. A compact next-generation source of radiation based on betatron emission could find use in many areas of science and technology [3]. One of the central issues to be resolved is to pinpoint the mechanism of electron trapping and subsequent acceleration in a plasma channel [4-6]. In this paper we show that the mechanism of electron trapping can be described using a simple modelled static wakefield potential. The model also naturally incorporates the subsequent betatron oscillation and radiation. We compare our reduced model with the 2-D particle-in-cell (PIC) simulation undertaken using OSIRIS [7]. Section 2 describes the result of electron trapping, and the structure of static wakefield using the PIC simulation. Section 3 shows how our model can be used to simulate on electron trapping and betatron motion. Section 4 shows the calculation of X-ray emission from the trapped electrons using our model potential.

2 2D PIC simulation

We focus on laser plasma interaction in the 'bubble' regime [5] and have undertaken 2D particle-in-Cell (PIC) simulations of the interaction of the laser with the plasma using the OSIRIS code [7-8]. The incident laser pulse is chosen to be circularly polarized, and have a Gaussian envelope and a peak intensity $a_0 = eA_0 / mc^2 = 4$, where A_0 is the laser vector potential. The plasma density is given by $n = n_c / \gamma_g^{\,2}$, where $n_c = 1.7 \times 10^{21}$ cm^{-3} is the critical plasma density for a laser wavelength of 0.8 μm, and γ_g is the Lorentz factor for the laser group velocity. In our simulation we have considered $n \cong 1.0 \times 10^{18}$ cm^{-3} for $\gamma_g = 40$. Fig. 1 shows a snapshot of the spatial distribution of the electron density. The void shape is determined by the ponderomotive potential of the laser pulse. From Fig. 1 we can trace a moving void which can be divided into: (i) A void core region, i.e., the cavity with totally depleted electron density and a large positive space charge, (ii) the transverse boundary region where the electrons are compressed and form the radial sheath around the cavity, and (iii) the front (nose of the laser pulse) and rear (end of the void) cavity boundaries. As time evolves we observe that electrons are trapped by the strong wakefield, and acquire a relativistic energy and accelerate forward as a bunch. A model for particle trapping in spherically symmetric space charge has already been discussed by Kostyukov *et al* [5]. However, in their model, the application of the laser perturbation in the bubble, in addition to the onset of coupling with the plasma, was not considered.

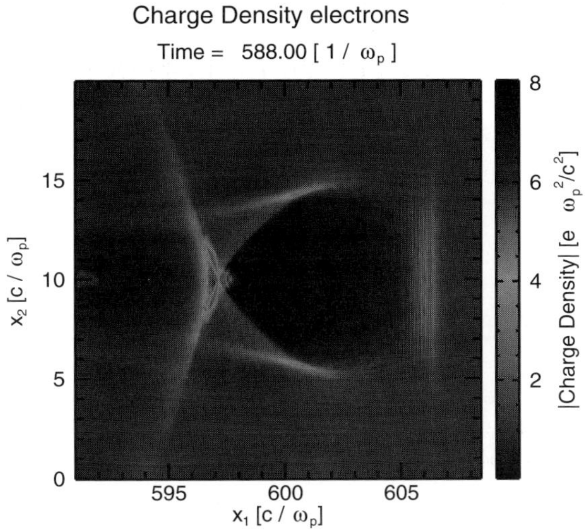

Fig. 1 PIC simulation: Snapshot (at time 588 $1/\omega_p$) of the electron density and formation of the bubble.

3 Model

Here, we discuss a simple model of a predefine shape of the space charge void and use this to investigate the mechanism of electron trapping. We considered that the laser pulse is linearly polarized, and travels along the x-axis. Single electron trajectories are described in the co-moving coordinate $\xi = x - v_g t$ where v_g is velocity of the wakefield, which is equal to the moving coordinate is

$$\mathrm{H} = \gamma - v_g p_x - \phi, \tag{1}$$

where, $\gamma = \sqrt{1 + (\vec{P} + \vec{A})^2 + a^2}$ is the canonical momentum of electron, a is the vector potential of the laser field, and $\vec{A}$ and φ are the slowly varying vector and scalar potentials, respectively [5,9]. Here, p_x is the longitudinal momentum, $\Psi = \phi - v_g A_x$ is the static normalized wakefield potential for our calculation. We approximate the void region by a sphere to obtain an analytical result [10].

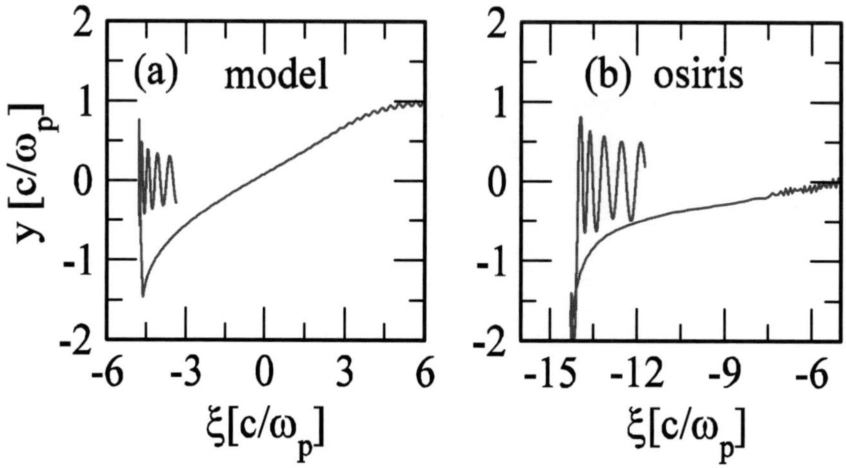

Fig. 2 Trajectories of trapped electrons: Using our model (a), one of the selected tracks from OSIRIS simulation (b).

The electron sheath around the void screens the ion in the surrounding plasma. We model the wakefield potential as

$$\Psi(\Delta x) = f_1(\Delta x, R, r) + \mathrm{erf}[\Delta X, R, r] f_2(\Delta X, R, r), \tag{2}$$

where, ΔX is the width of electron sheath, R is the radius of the void, $r = \sqrt{(x - v_g t)^2 + y^2 + z^2}$, and f_i is a function of r.We solve the coupled equations of motion in the co-moving frame for a spherical bubble of radius $R = 6c/\omega_p$, and a sheath width $\Delta X = 0.6c/\omega_p$. Initially the electrons are placed at the front of the bubble, where the Gaussian linearly polarized laser profile is situated. Our model calculation self trapping is shown in Fig. 2(a). One can estimate the turning points where the electrons get trapped by the bubble as $\partial \mathrm{H}/\partial p_x = \mathrm{d}\xi/\mathrm{d}t = 0$. To compare our model calculation we show one of the selected electron tracks using OSIRIS (see Fig. 2(b)).We observe that the electrons undergo betatron motion soon after they get trapped with a betatron frequency $\omega_\beta = \omega_p/\sqrt{2\gamma}$, where ω_p is the plasma frequency, and γ is the Lorentz factor associated with the electrons. As we know that an accelerating charge radiates electromagnetic radiation. Very recent experimental observation of synchrotron X-ray radiation from relativistic electrons produced from a gas target illuminated by a petawatt laser pulse has been discussed in [11]. Below we discuss on the theoretical estimate of X-ray radiation by electron undergoing betatron using our model.

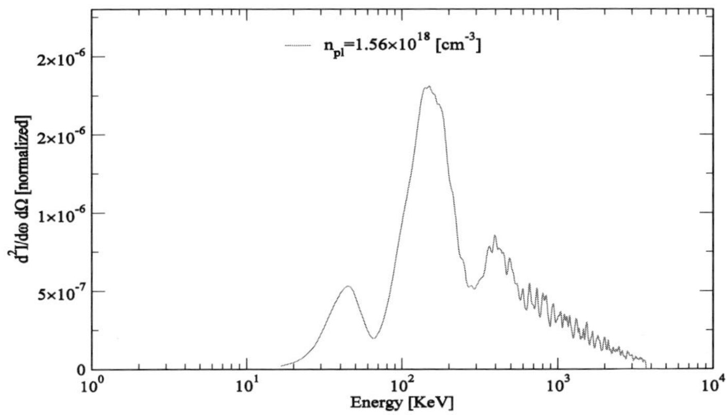

Fig. 3 X-ray emission spectrum from betatron oscillation of self-trapped electrons inside the bubble.

4 Betatron radiation

The electromagnetic radiation from electrons, observed in the far field, can be calculated from the betatron oscillation of the electron using the Lienard-Wiechert potentials [12], where the velocity field is neglected and only the accelerated field is considered, and is given by

$$P(\omega,\Omega)=\frac{e^2}{4\pi^2 c}\left|\int_0^{t_b}\frac{\vec{n}\times\{(\vec{n}-\vec{\beta})\times\dot{\vec{\beta}}\}}{(1-\vec{n}.\vec{\beta})^2}\mathrm{e}^{i\omega(t-\vec{n}.\vec{r}/c)}\mathrm{d}t\right|^2, \tag{3}$$

where $P(\omega,\Omega)=\dfrac{\mathrm{d}I}{\mathrm{d}\omega\mathrm{d}\Omega}$ is the energy radiated into unit solid angle Ω at frequency ω; $\vec{n}$ is a unit vector along the direction of observation point, $\vec{r}$ is the position of electron, $\vec{\beta}$ is the normalized velocity of electron normalized to velocity of light, $\dot{\vec{\beta}}$ is the acceleration, and t_b is the total time over which the electron oscillates after they are trapped. The characteristics of the radiation depend strongly on the betatron amplitude, i.e., for near-axis oscillation ($<< 1\mu m$), only the fundamental wavelength of the undulator like radiation is emitted, which is given by $\lambda=\lambda_\beta/2\gamma$, where λ_β is the betatron wavelength. When the amplitude of betatron oscillation is a few microns, then high harmonics are radiated with a broadband wiggler-like spectrum rising to a peak and then dropping exponentially [13]. Fig. 3 shows a typical X-ray spectrum with peak at roughly 150 KeV.

We gratefully acknowledge the OSIRIS Consortium, consisting of UCLA, IST (Lisbon, Portugal), and USC, for the use of OSIRIS, and IST for providing access to the OSIRIS 2.0 framework. We also acknowledge financial support from the EPSRC and the EU Euroleap Nest contract 028514.

References

1. W P Leemans *et al.*, *Nature Phys.* **2**, 696 (2006)
2. S P Mangles *et al.*, *Nature* **431**, 535 (2004)
3. D A Jaroszynski *et al.*, *Phil. Trans. R. Soc. A* **364**, 689 (2006)
4. T Tajima and J M Dawson, *Phys. Rev. Lett.* **43**, 267 (1979)
5. I. Kostyukov *et al.*, *Phys. of Plasmas.* **11**, 5256 (2004)
6. J M Dawson, *Rev. Mod. Phys* **55**, 403 (1989)
7. L O Silva *et al.*, *Phys. Rev. Lett.* **92**, 015002 (2004)
8. R A Fonseca *et al.*, *Lecture Notes in Computer Science* **2331**, 342 (2002).
9. I Y Dodin and N J Fisch, *Phys. Rev. E* **68**, 056402 (2003)
10. M R Islam *et al.*, *EPS conference on plasma Phys.* **32**, P-4.151 (2008)
11. S Kneip *et al.*, *Phys. Rev. Lett.* **100**, 105006 (2008)
12. J D Jackson, *Classical Electrodynamics 2nd edn (new York: Wiley),* chapter 14 (1975)
13. A Rousse *et al.*, *Phys. Rev. Lett.* **93**, 135005 (2004)

Emission Spectroscopy from an XUV Laser Irradiated Solid Target

T. W. J. Dzelzainis[1], F.Y. Khattak[1,12], B. Nagler[2], S. Vinko[2], T. Whitcher[2], A. J. Nelson[3], R.W. Lee[3], S Bajt[4], S. Toleikis[4], R. Fäustlin[4], T. Tschentscher[4], L. Juha[5], M. Kozlova[5], J Chalupsky[5], V. Hajkova[5], J. Krzywinski[6], R. Soberierski[7], M. Jurek[7], M. Fajardo[8], F.B. Rosmej[9,10], P. Heinmann[11], J. S. Wark[2], and D. Riley[1]

[1]School of Mathematics and Physics, Queen's University Belfast, BT7 1NN
[2]Clarendon Laboratory, University of Oxford, South Parks Road OX1 3PU
[3]Lawrence Livermore National Laboratory, Livermore, CA USA
[4]HASYLAB DESY, Hamburg
[5]Institute of Physics, AS CR, Prague 8
[6]SLAC, *Menlo Park CA, USA*
[7]Inst. of Physics, Polish Academy of Sciences
[8]GoLP - Instituto de Plasmas e Fusão Nuclear, Instituto Superior Técnico
[9]Université Pierre et Marie Curie, LULI, UMR 7605, case 128, 4 Place Jussieu, 75252 Paris Cedex 05
[10]Ecole Polytechnique, LULI-PAPD, 91128 Palaiseau, France
[11]Lawrence Berkeley National Laboratory, Berkeley, CA USA
[12]Kohat University of Science and Technology, Kohat, Pakistan

Abstract. We have used the XUV FLASH laser at DESY to irradiate solid targets with intense XUV pulse at 13.5nm and $\sim 10^{16}$ Wcm^{-2}. XUV emission spectroscopy using a grating spectrometer has been used to observe both continuum radiation and line emission from Al IV. We present some preliminary results that indicate time integrated temperatures of below 20eV while simulation indicates higher initial temperatures.

1 Introduction

The use of intense XUV and X-ray pulses to create samples of warm dense matter (WDM) has been proposed by several different authors[1-3], being motivated by the development of free electron laser technology that has pushed operation into the XUV and, soon, the X-ray regime. One of the advantages of such heating is that not only are the pulse lengths of the proposed sources short (generally ~100fs or less) but they do not suffer from the critical density issue that occurs with optical short pulse lasers. In the latter, the energy is absorbed at critical density, which is about two orders of magnitude lower than solid density. The transport of energy to high density by either thermal conduction or fast electron transport is then needed to create warm dense matter samples. With XUV or X-ray heating it is expected that we may be able to create foil samples of greater than a micron thickness at solid density and temperatures in excess of 1eV, the hydrodynamic timescale for expansion of such a sample is such that investigation is possible with time resolved absorption spectroscopy and expansion measurement techniques such as Fourier domain interferometry (FDI)[4]. Such states of matter are of intrinsic interest because of the combination of strong coupling and partial degeneracy that means neither classical plasma physics nor solid state physics can be used to describe them. They are also of interest because of their occurrence in planetary science and in fusion capsules.

In this paper, we report some initial experiments in which we have irradiated solid targets with an XUV-FEL beam in order to create WDM samples as record subsequent emission in the XUV wavelength regime.

2 Experimental Layout

A schematic plan of the experiment is shown in Figure 1. The FEL beam at 13.5nm was focused using a MoSi multi-layer optic with a focal length of 250mm. After an adjustable aperture, the beam size was ~5mm resulting in f/50 focusing. Initially, the reflectivity was ~60% at 13.5nm with a full-width at half-maximum of ~0.5nm. Contamination of the surface was noted after the experiments, due to the chamber environment. On further analysis, the reflectivity was found to have dropped to ~45% at the centre of the optic. During the experiment, typically ~5μJ on target per shot was achieved in a pulse width of ~15fs. The target mount allowed multiple flat foil samples to be held, including PMMA as well as metallic foils and SiN. The target drive was fitted with externally controlled motorized actuators capable of micron level control of motion.

A Questar QM1 long distance microscope coupled to a high quality mirror was used to image the front of the target surface during the experiment. The damage spots created by the incident FEL beam could be readily observed to check that during a shot series a fresh surface was exposed for each shot. This telescope was also used to establish, prior to the experimental runs, that the target surface remained within a few microns of the pre-set focal position was the target was rastered a distance of ~20mm across the focus- thus focussing conditions were consistently well within the Rayleigh depth for a particular run.

The sample foil could be moved in and out of the focal plane to vary the focal spot. By changing the incident energy and varying the focus onto the PMMA we were able to make a focal spot analysis from the ablation craters. Figure 2 shows the results from using a Normarski differential interference contrast (DIC) microscope to analyse 5 craters from each of the focal spot positions and an atomic force microscope (AFM) to image the spots.

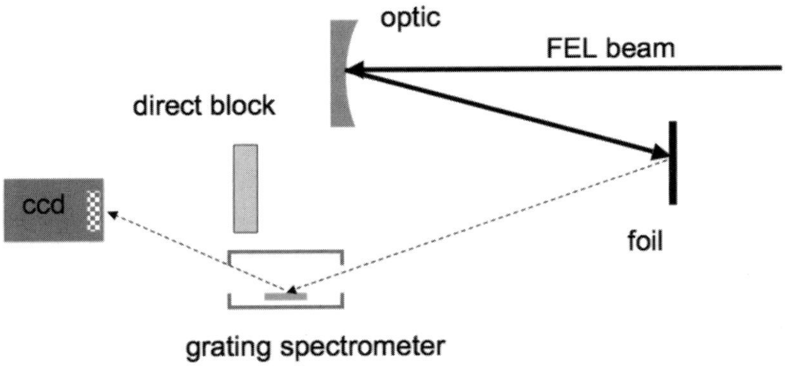

Fig. 1 Schematic layout for the experiment.

At best focus, we get a focal spot full-width at half maximum (FWHM) of ~2 microns. This gives a fluence of ~160 Jcm^{-2} and an intensity on target of ~10^{16} Wcm^{-2}. Gaussian beam optics would predict that for this spot size and focus the Rayleigh range is ~ 330 microns. The lower figure seems to indicate that the crater area doubles when the focus is moved by ~160 microns- Gaussian optics would therefore yield a best focal spot of ~1.5 microns FWHM, roughly consistent with the best estimate from craters at best focus.

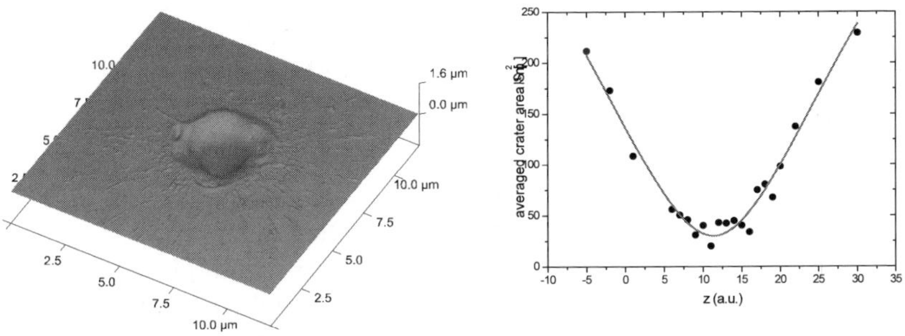

Fig. 2 *Left*: An AFM image of a typical crater from near best focus exposure. *Right*: Average crater area for ablation of PMMA. The distance scale on the right is in units of 1/10 rotation of the actuator (=31.75 microns).

The emission from the irradiated target was recorded with a grating spectrometer situated with the grating ~50cm from the target. The grating was a 1200 l/mm variable spacing grating that produced a flat spectral focus onto a back-thinned CCD (Andor DX420-BN) cooled to -10° C. The spectral range was from ~12-35 nm with resolution of $\lambda/\Delta\lambda$ ~150. The cooling of the CCD allowed us to record hundreds of data shots accumulated into one image whilst moving the target foil across the focal plane.

3 Results and Discussion

In Figure 3, we can see some typical collected data from an Al target. A total of 1800 shots are integrated. In the top figure we see the resultant raw CCD image. In the lower figure, we can see a line-out showing scattered FEL radiation in both first and second order. We also see an emission feature that we identify[5,6] as the Al IV $2p^6$-$2p^5\,3s\ ^3P$ and ^1P doublet at 16.01nm and 16.17nm respectively. We can also see this in second order. We do not see strong evidence of Al V lines and indeed, these are not expected to be significant compared to Al IV below ~20eV. This puts an upper limit on the temperature.

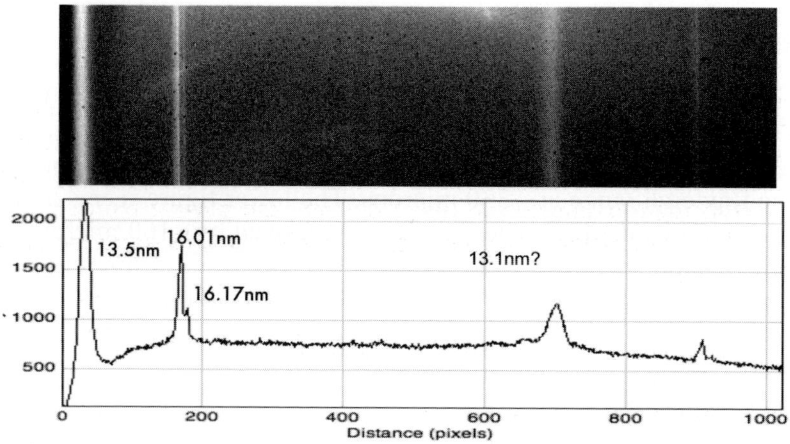

Fig. 3 Spectrum recorded by integrating 1800 shots on an Al target at ~10^{16} Wcm^{-2}. *Top* raw image from CCD; *bottom:* The emission line-out showing the FEL in 1st and 2nd order as well as Al IV emission at ~16nm and at ~13nm in 2nd order.

We can see this latter point illustrated in Figure 4, which shows Saha-LTE spectra derived from data on the NIST online database[5,6]. The spectra are derived for N_e~$3x10^{21}$ cm^{-3} which is in the estimated density range, based on the line widths. Clearly, the Al V lines would start to be noticed in the spectrum somewhere between 10-20eV.

We can compare this to the temperature estimated by considering energy absorbed by the target. For 13.5nm radiation the linear absorption depth[7] is approximately 30nm for both Al and the few nm oxide layer expected to be on the surface. This means that, for example, within an absorption depth of the surface we expect an average of >10^{10} JKg^{-1} absorbed. If we assume the density of the warm dense matter created to be solid density for Al, and allow the Saha-Boltzmann equation to govern ionisation with only ground states accounted for and the Stewart-Pyatt[8] model to account for continuum lowering then we estimate that the plasma reaches T_e~150eV on initial creation with Z*~9. At this temperature we should certainly expect to see Al V emission as well as higher ionisation stage emission.

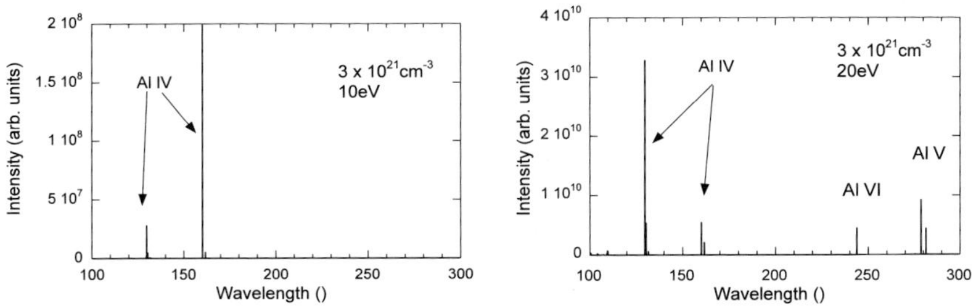

Fig. 4 Saha-LTE spectra with Doppler broadening for Al showing how Al IV dominates below ~20eV

One explanation for this apparent discrepancy could be that at densities of around solid, the plasma continuum lowering and strong Stark broadening of emission features does not allow clear emission lines of higher ion stages to be seen. After rapid expansion to lower density, the plasma temperature may have cooled to a significant degree. For example, let us assume an adiabatic expansion where;

$$TV^{\gamma-1} = \text{constant}$$

where T is temperature, V is specific volume and γ is the adiabatic index, equal to 5/3 for a monatomic gas. For solid density with Z*=9 we expect an initial N_e~$5\text{x}10^{23}$ cm^{-3}. From the width of the lines seen in the spectrum, we can estimate that the effective electron density of the emission is broadly between ~10^{21}-10^{22} cm^{-3}. With the equation above, we can then estimate a range of temperatures of 2-11 eV, which is certainly in agreement with the observation of no Al V lines.

Since there is some uncertainty about processes such as inverse bremsstrahlung in the XUV regime for initially cold metals such as Al, we have run a 1-D HYADES[9] code simulation of FLASH XUV-FEL heated Al by assuming a dump of energy into electrons and allowing evolution to occur. Some results are presented in Figure 5. In this simulation we have simulated the energy absorption as decreasing with Beer's law into the target. The average energy absorbed per atom in the top 30nm is ~2.7keV. We can see from the figure on the right, that the electron temperature is predicted to drop below 20eV after approximately 50ps. At this time much of the layer is between N_e~10^{21} -10^{22} cm^{-3}. Thus we might say that, for a time integrated diagnostic, it is no surprise if the Al IV emission dominates.

Clearly a more detailed analysis with time dependent ionisation dynamics would be needed to correctly estimate the averaged emission density and temperature that would be observed when the emission is integrated over the entire emission history. Nevertheless, the preliminary analysis indicates that Al IV may dominate because of rapid adiabatic cooling of a thin initially heated layer and the effect of high density Stark broadening and continuum lowering on the higher ionisation stage emission at earlier times.

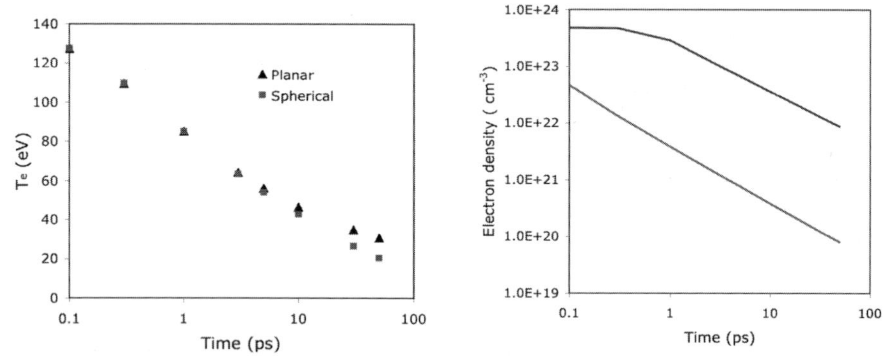

Fig. 5 Hyades simulation of the top 30nm of a solid Al target. *Left*: average electron temperature in layer for planar and spherical expansions. *Right*: Upper and lower bound to electron density in planar case.

4 Conclusions

We can conclude that we can generate warm dense matter samples using an intense XUV laser pulse such as the FLASH facility in DESY. It is possible also that longer established alternatives may also be used, such as plasma based X-ray lasers. However, the short pulse and tunability of the XUV-FEL facility, combined with 5Hz operation, make it a valuable, flexible tool for WDM research. The modest temperature of below ~20 eV seen from the existence of Al IV but absence of Al V emission is consistent with strong adiabatic cooling combined with integration of the spectrum over the entire emission duration. It is expected that in future experiments, it will be possible to integrate many shots using a streak camera synchronised with the FEL beam to provide time resolved data averaged over many shots with time resolution of a few picoseconds or better. This will help resolve the issue of the maximum initial temperature reached. At 150eV, we might expect to look for K-shell emission from low-Z elements up to perhaps Al, using crystal spectrometers.

Acknowledgements

The authors gratefully acknowledge support for access to FLASH by DESY and the European Community under contract RII3-CT-2004-506008 (IA-SFS).

References

1. Richard W. Lee, et al , JOSA B, 20(4), pp770-778, (2003)
2. M Fajardo et al The European Physical Journal D 29 (1), 2004, pp69-7
3. RW Lee *et al* , Laser and Particle Beams **20** (3) pp527-536 (2002)
4. P Audebert *et al* Phys. Rev. E **64** 056412, 2001
5. Ralchenko, Yu., Kramida, A.E., Reader, J., and NIST ASD Team (2008). NIST Atomic Spectra Database:http://physics.nist.gov/asd3, National Institute of Standards and Technology, Gaithersburg, MD.
6. V Kaufman, M-C Artru, and WL Brillet, J. Opt. Soc. Am. 64, 197-201, 1974
7. LBNL website at www.cxro.lbl.org
8. J. C. Stewart and K. D. Pyatt, Astrophys. J. **1203**, 144 1966
9. JT Larsen and SM Lane, J.Quant. Spectrosc. Radiat. Transfer **51** p179, 1994

Innershell X-Ray Laser in Sodium Vapor: Final Steps Towards Experimental Verification

J. Nejdl[1], T. Mocek[1], B. Rus[1], S. Sebban[2], B. Wellegehausen[3]

[1]Institute of Physics/ PALS Centre, Na Slovance 2, 18221 Prague, Czech Republic
[2]Laboratoire d´Optique Appliqué, ENSTA, 91761 Palaiseau Cedex, France
[3]Institut für Quantenoptik, Universität Hannover, Welfengarten1, D-30167 Hannover, Germany

Abstract. Another attempt to verify the innershell x-ray laser scheme in sodium is about to run at the end of 2008 at PALS Centre in Prague. In this contribution we will present new experimental setup that provides more pumping power at the target. Also a numerical code solving the rate equations in 1D has been developed. This code enables to find optimum density of the sodium (controlled by its temperature) for different pumping conditions to fit with the parameters of the x-ray laser driver available at PALS.

1 Introduction

Innershell photoionization scheme is one of the promising ways to reach x-ray lasers (XRL) at wavelengths that are not achievable with the commonly used electron collisional excitation or recombination. However, creating the population inversion by photoionization of an inner electron was so far demonstrated only for wavelengths in visible and VUV spectral range with the shortest wavelength of 90.7 nm for Auger-type laser in Kr III [2]. One of the very first proposals for an x-ray laser scheme is the Duguays' innershell scheme in Na vapor at 37.2 nm [1]. There have been few attempts to realize this scheme, however all without success. The Ne-like Zn XRL at 21.2 nm at PALS facility seems to be the ideal pumping source for Duguays' scheme. The new experimental arrangement that ensures more pumping power and optimum gain medium properties found by new 1D numeric code are presented below.

2 Physical principles

The Duguays' innershell laser scheme in sodium is based on creation of inversion population by ionization of 2p electron in atomic sodium by soft x-ray radiation ($h\nu$>38.5 eV) without significant influence on population of the 3s level. For radiation near 20 nm (i.e. $h\nu \approx$ 60 eV) the photoionization cross-sections ratio of 2p and 3s electron is $\sigma_p/\sigma_s \approx 200$. The $2p^5 3s$ state is four-times degenerated and only radiative decay $2p^5 3s\ (^1P_1) \rightarrow 2p^6\ (^1S_0)$ at 37.2 nm is allowed. There is no other possible depletion of the upper laser level (like Auger type transitions).

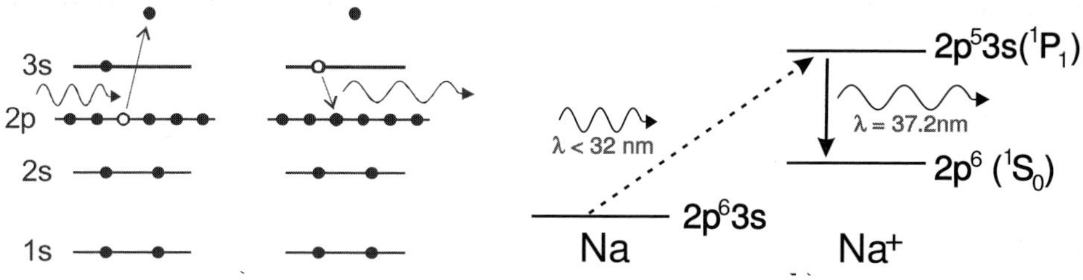

Fig. 1 a) Schematic diagram of sodium electronic structure: photoionization of an innershell electron and radiative decay of the electron from the upper level.
b) Simplified energy level diagram of the transitions.

3 Experimental setup

In the planned experiment at PALS facility, the Ne-like Zn XRL (λ= 21.2 nm; $h\nu$ = 58.5 eV) will be used for pumping. It delivers up to 10 mJ in 200 ps pulses in double pass mode with beam divergence 3×5 mrad (horizontal × vertical) [4]. Its beam will be focussed with a toroidal mirror into the sodium vapor generated in the oven.

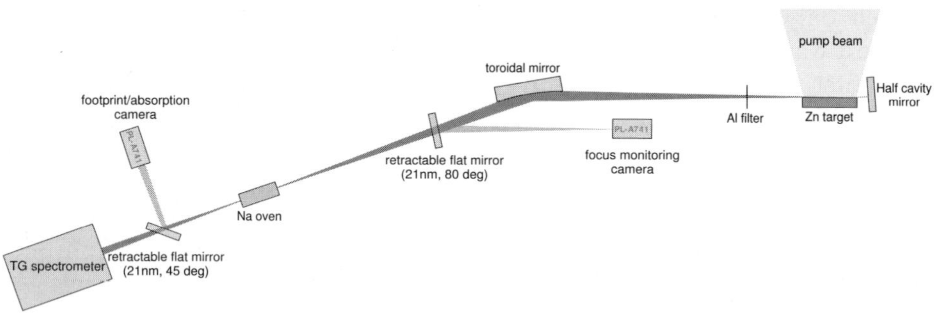

Fig. 2 Draft of the experimental arrangement of the incoming PALS experiment.

For focusing the Zerodur toroidal mirror at incidence angle of 10 deg (f = 1500 mm) imaging the source with demagnification factor of 1.5 will be used. The diameter of focus is expected to be $\approx$ 80 μm. The XRL focus will be monitored on the PixeLINK camera PL-A741 covered by P43 phosphor (using 1 retractable flat mirror at 21 nm for 80 deg. incidence).

The vapor streams out of a nozzle head, which has 800 μm holes on both sides and a length of 10 mm. The oven is located in a vacuum chamber, which will be attached to the target chamber of the x-ray laser.

Behind the oven there will be a chamber containing diagnostics (flat field spectrometer or transmission grating spectrometer). In front of spectrometer there will be another chamber to check the footprint and/or absorption in sodium of 21 nm XRL.

4 Modelling of population inversion

4.1 1D code for solving rate equations

We have developed a new one–dimensional code solving the system of rate equations (described in [3] in detail) along the axis of pumping (z-axis) that takes into account all physical parameters of the future experiment at PALS facility, i.e. neutral sodium atoms density distribution, pumping pulse shapes, and pumping beam intensity distribution that is affected by focusing and absorption in gain medium.

Typical values of physical conditions considered are shown in Fig. 3 and 4.

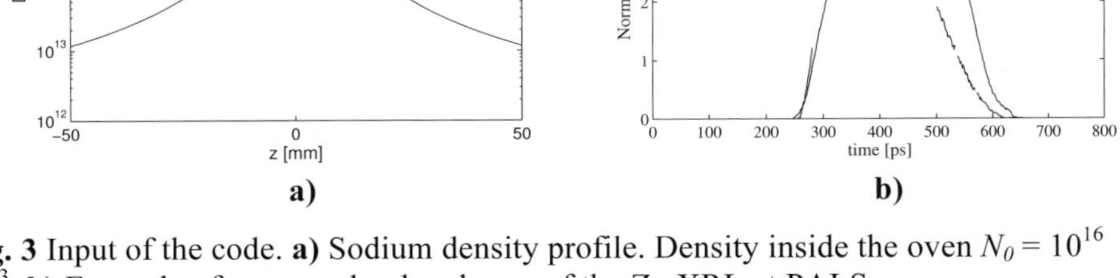

Fig. 3 Input of the code. **a)** Sodium density profile. Density inside the oven $N_0 = 10^{16}$ cm^3. **b)** Example of measured pulse shapes of the Zn XRL at PALS.

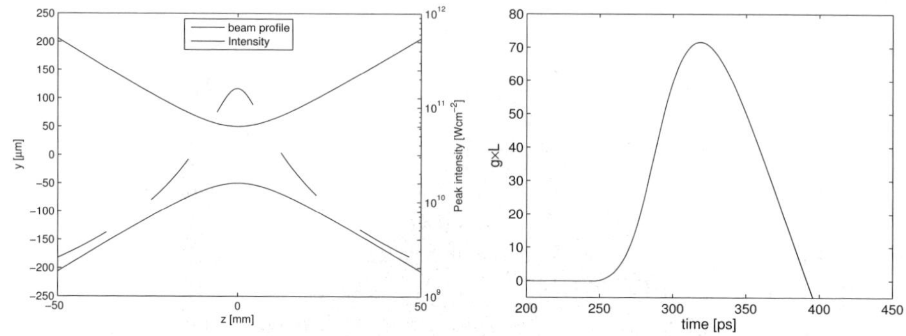

a) b)

Fig. 4 Input and output of the code. **a)** Solid line: Beam profile along the z-axis (beam waist $w_0 = 50$ µm, divergence $\theta = 8$ mrad); dashed line: pumping intensity distribution. Gaussian beam is considered. **b)** Time dependent gain-length product for double pass pump pulse with energy 3 mJ in focal spot.

It was revealed that that maximum gain is strongly dependent on the rising edge of the pumping pulse, because for the considered pumping intensities the population of the ground state is depleted very quickly (all sodium atoms are ionized) in times that are much shorter then x-ray pumping pulse duration. To generate high gain a steep rising edge in conjunction with sufficiently high intensity is necessary, therefore the gain in the single pass pumping regime is comparable with the gain reached by the double pass pumping pulse, although the energy of the double pass pulse is higher approximately by one order of magnitude.

4.2 Search for optimum pumping conditions

Sodium density and energy of the pumping pulse were scanned to obtain the optimum conditions (high gL) of the Na x-ray laser. The gain-length product was observed for various focus diameters (pumping beam waists) and real pulse shapes.

The results of this modelling for pumping by single pass pulse are shown in Fig. 5. The optimum sodium density in dependence on energy of the pumping pulse can be easily found. Sodium vapors density is determined by temperature with empiric formula [5]

$$N_0 = \exp\left[11.946 - \frac{12634}{T} - 0.4672 \ln T\right] \frac{1}{k_B T} \tag{1}$$

Optimum conditions for different diameter of the foci are plotted in Fig. 6

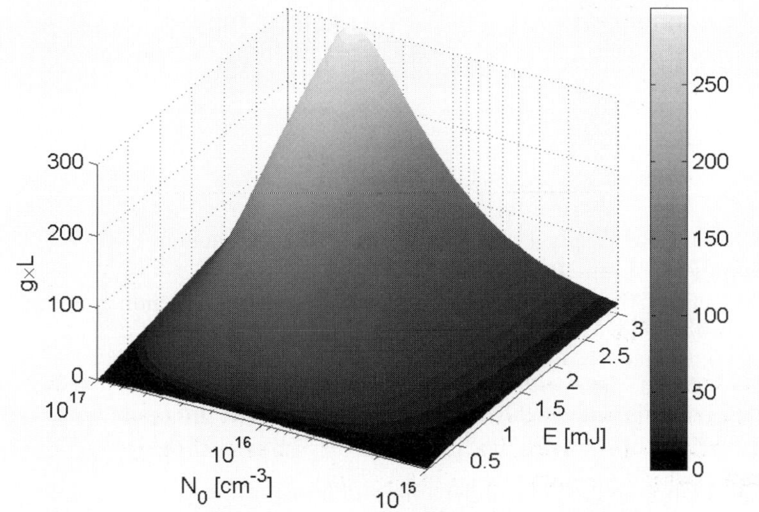

Fig. 5 Dependence of the peak gain-length product on the energy of pumping pulse in single pass mode ($w_0 = 50$ µm, θ = 8 mrad) and on the sodium density inside the oven.

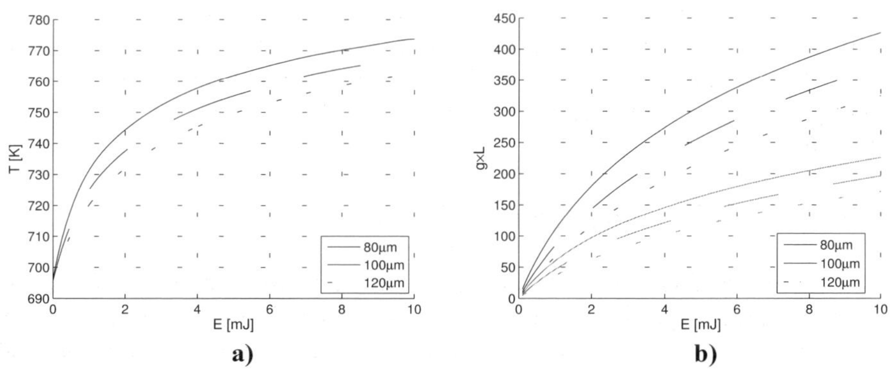

Fig. 6 a) Optimum temperature of sodium (determined by the optimum density of sodium vapors) in dependence on pulse energy and focus diameter for double pass pulses. **b)** Peak gain-length product (black lines) and time averaged gain-length product (grey lines) at optimum sodium density.

5 Conclusion

According to the results of the detailed numerical simulations with regard to relatively high pumping source shot-to-shot stability and mechanical stability improvements of the sodium apparatus, there is a very good chance to verify the original Duguays' innershell scheme within the next experimental campaign at PALS that will take place in the first half of 2009. The sodium XRL is expected to produce up to 0.5 μJ of radiation at 37.2 nm in 100 ps pulse.

References

1. Duguay, M. A. and Rentzepis, P. M. 'Some approaches to vacuum UV X-ray lasers', Appl. Phys. Lett. 10, 350, 1967
2. Kapteyn, H.C. et al. 'Auger-pumped short-wavelength lasers in xenon and krypton', Phys. Rev. A 37, 2033, 1988
3. Wellegehausen, B. et al. 'Novel attempts to realize an innershell x-ray laser in Na', X-Ray Lasers 2006, Springer Proc. in Phys. Ser. 115, p. 649
4. Rus B. et al. 'Multimillijoule, highly coherent x-ray laser at 21 nm operating in deep saturation through double-pass amplification', Physical Review A 66, 063806 (2002)
5. Fink J. K. and Leibowitz L. 'Thermodynamic and transport properties of sodium liquid and vapor', ANL/RE-95/2 (1995)